CELL MEMBRANES
Methods and Reviews

Volume 3

CELL MEMBRANES

Methods and Reviews

Volume 3

Edited by

Elliot Elson
William Frazier

Washington University School of Medicine
St. Louis, Missouri

and
Luis Glaser

University of Miami
Coral Gables, Florida

PLENUM PRESS • NEW YORK AND LONDON

The Library of Congress cataloged the first volume of this title as follows:

Cell membranes, methods and reviews.—Vol. 1.　　—New York: Plenum Press, c1983.
　　v.: ill.; 24 cm.
　　Annual.
　　Includes bibliographies.
　　Editors: Elliot Elson, William Frazier, and Luis Glaser.
　　Continues: Methods in membrane biology.

　　ISSN 0740-784X = Cell membranes, methods and reviews.
　　1. Cell membranes—Periodicals. 2. Cytology—Methodology—Periodicals. I. Elson, Elliot, 1937–　　. II. Frazier, William (William A.) III. Glaser, Luis, 1932–　　.
　　[DNLM: 1. Cytological Technics—periodicals. 2. Cell Membrane—periodicals. W1CE128L]
QH601.C37　　　　　　　　574.87'5'05—dc19

83-646233
AACR 2 MARC-S

Library of Congress　　　　　　　[8611r84]rev

ISBN-13: 978-1-4612-9065-0　　　e-ISBN-13: 978-1-4613-1915-3
DOI: 10.1007/ 978-1-4613-1915-3

© 1987 Plenum Press, New York
Softcover reprint of the hardcover 1st edition 1987
A Division of Plenum Publishing Corporation
233 Spring Street, New York, N.Y. 10013

CONTRIBUTORS

Robert M. Bell   Department of Biochemistry, Duke University Medical Center, Durham, North Carolina 27710

Chitra Biswas   Department of Anatomy and Cellular Biology, Tufts University Schools of Medicine, Dental Medicine and Veterinary Medicine, Boston, Massachusetts 02111

Richard W. Bond   Department of Anatomy and Neurobiology, Washington University School of Medicine, St. Louis, Missouri 63110

Paul Bornstein   Department of Biochemistry, University of Washington, Seattle, Washington 98195

Jean-Claude Boucaut   Laboratory of Experimental Biology, University of Paris 6, 75005 Paris, France

Habib Boulekbache   Developmental Biology Group, University of Paris 7, 75221 Paris Cedex, France

Gregory J. Cole   Department of Anatomy and Cell Biology, Medical University of South Carolina, Charleston, South Carolina 29425

Thierry Darribère   Laboratory of Experimental Biology, University of Paris 6, 75005 Paris, France

Jean-Loup Duband   Laboratory of Developmental Physiopathology, Ecole Normale Supérieure and National Center for Scientific Research, 75230 Paris Cedex 05, France

John H. Exton   Howard Hughes Medical Institute and Department of Molecular Physiology and Biophysics, Vanderbilt University School of Medicine, Nashville, Tennessee 37232

Barry R. Ganong   Department of Biochemistry, Duke University Medical Center, Durham, North Carolina 27710

Kathleen L. Gould   Molecular Biology and Virology Laboratory, The Salk Institute, San Diego, California 92138

Yusuf A. Hannun   Departments of Biochemistry and Medicine, Duke University Medical Center, Durham, North Carolina 27710

Tony Hunter   Molecular Biology and Virology Laboratory, The Salk Institute, San Diego, California 92138

Carson R. Loomis   Department of Biochemistry, Duke University Medical Center, Durham, North Carolina 27710

Richard A. Majack   Department of Biochemistry, University of Washington, Seattle, Washington 98195. *Present address:* Atherosclerosis and Thrombosis Research, The Upjohn Company, Kalamazoo, Michigan 49001

Paul L. McNeil   Department of Biological Sciences and Center for Fluorescence Research in Biomedical Sciences, Carnegie–Mellon University, Pittsburgh, Pennsylvania 15213. *Present address*: Department of Anatomy and Cellular Biology, Harvard Medical School, Boston, Massachusetts 02115

D. Lansing Taylor   Department of Biological Sciences and Center for Fluorescence Research in Biomedical Sciences, Carnegie–Mellon University, Pittsburgh, Pennsylvania 15213

Jean Paul Thiery   Laboratory of Developmental Physiopathology, Ecole Normale Supérieure and National Center for Scientific Research, 75230 Paris Cedex 05, France

Bryan P. Toole   Department of Anatomy and Cellular Biology, Tufts University Schools of Medicine, Dental Medicine and Veterinary Medicine, Boston, Massachusetts 02111

James R. Woodgett   Molecular Biology and Virology Laboratory, The Salk Institute, San Diego, California 92138

PREFACE

This volume assembles reviews on topics in two major related areas. One of these concerns the interactions of cells with substrata and with other cells, which are mediated by the extracellular matrix and soluble molecules. As described in this volume, these interactions are responsible for controlling cell functions ranging from embryogenesis and neural development to blood clotting. Moreover, important properties of the extracellular matrix can be modulated by the interdependent actions of tumor cells and fibroblasts. The other major area of interest concerns the response of cells to extracellular signals. Recent work has begun to reveal how a remarkable diversity of cellular functions, including neuronal, proliferative, membrane–cytoskeletal, and many other kinds of responses, are elicited through the mediation of a relatively small and interdependent set of second messenger systems. These include both changes in cytoplasmic ionic balances and activation of various kinds of protein kinases. Both subjects are covered in this volume. The two areas are linked by the common theme of cellular response to an external environment that is sensed through cellular interactions with informational molecules, which are soluble agents, as well as those that are components of insoluble matrices. It is only recently that we have come to appreciate the complex interplay between the matrix surrounding a cell and the cell's response to hormones and growth factors. Thus, we have tried to select examples in which this type of extracellular integration may play a role.

We thank all of the contributors for providing these excellent reviews and for doing so in a timely fashion.

Elliot L. Elson
William A. Frazier
Luis Glaser

St. Louis, Missouri

CONTENTS

Chapter 1

Regulation of Development by the Extracellular Matrix

Jean-Loup Duband, Thierry Darribère, Jean-Claude Boucaut,
Habib Boulekbache, and Jean Paul Thiery

Chapter 2

Thrombospondin: A Multifunctional Platelet and Extracellular Matrix
Glycoprotein

Richard A. Majack and Paul Bornstein

Chapter 3

Neuronal Antigens Involved in Cell Adhesion and Cell Recognition

Gregory J. Cole and Richard W. Bond

Chapter 6

Protein Kinase C and Its Role in Cell Growth

James R. Woodgett, Tony Hunter,
and Kathleen L. Gould

Chapter 7

Modulation of the Extracellular Matrix by Tumor Cell–Fibroblast Interactions

Chitra Biswas and Bryan P. Toole

Chapter 8

Early Cytoplasmic Signals and Cytoskeletal Responses Initiated by Growth Factors in Cultured Cells

Paul L. McNeil and D. Lansing Taylor

REGULATION OF DEVELOPMENT BY THE EXTRACELLULAR MATRIX

Jean-Loup Duband, Thierry Darribère, Jean-Claude Boucaut, Habib Boulekbache, and Jean Paul Thiery

1. INTRODUCTION

In the vertebrates, cells are considerably displaced during the processes leading to the shaping of the embryo. Extensive morphogenetic transformations, including changes of cell shape, cell migrations, distortions, remodelings, and dissociations of sheets of cells, are essential in establishing the basic structures of the embryo. These movements allow cells of different areas of the embryo to interact transiently, a necessary step for the transduction of inductive signals. Such events are responsible for the segregation of embryonic cells into the endoderm, mesoderm, and ectoderm, and then each layer becomes, in turn, regionalized.

At any time, cells can be engaged in one or more of the five primary processes of division, migration, adhesion, differentiation, and death. Division, differentiation, and death of cells are necessary for the construction of the body plan but are far from sufficient, since they cannot account for the transient interactions of cells and their final localization. Therefore, these processes can be considered as secondary compared with cell migration and adhesion in the shaping of the embryo.

Cell adhesion and migration are influenced in part by the relationships of the cells to their immediate environment and particularly by the interactions with the adhesive molecules of the extracellular matrix (ECM). The early appearance

Jean-Loup Duband and Jean Paul Thiery ● Laboratory of Developmental Physiopathology, Ecole Normale Supérieure and National Center for Scientific Research, 75230 Paris Cedex 05, France. Thierry Darribère and Jean-Claude Boucaut ● Laboratory of Experimental Biology, University of Paris 6, 75005 Paris, France. Habib Boulekbache ● Developmental Biology Group, University of Paris 7, 75221 Paris Cedex, France.

1

of fibronectin and laminin during embryogenesis and their further distribution as ubiquitous components of the ECM are essential for many different morpho-genetic mechanisms. In this chapter, we will outline some of the biochemical properties and potential functions of fibronectin and laminin. We will then concentrate on two major morphogenetic events that occur during embryogenesis: cell migration and epithelium–mesenchyme interconversion.

2. MOLECULAR ANALYSIS, LOCALIZATION, AND FUNCTIONS OF CELL-TO-SUBSTRATE ADHESION MOLECULES

A common feature of fibronectin and laminin is that they are composed of separable functional domains, each specialized for specific binding activities. These domains are largely responsible for the different functions of the molecules.

2.1. Laminin

2.1.1. Structure of Laminin

Laminin (LN) is a high-molecular-weight glycoprotein (950 Kd) with a cross-shaped structure composed of a long arm and three short arms, as seen on electron microscopic images (Timpl *et al.*, 1979; Engel *et al.*, 1981) (Figure 1).The functional domains of LN consist of binding domains for cells, heparin, and type IV collagen (Terranova *et al.*, 1980; Rao *et al.*, 1982). The major heparin-binding site is located in the globular terminal domain in the long arm. In contrast, the cell-binding domain and the collagen-binding domain have not been precisely mapped; the cell-binding domain is probably present in the center of the molecule and extends along the short arms, whereas the collagen-binding domains seem located at the ends of the short arms (Rao *et al.*, 1983; Timpl *et al.*, 1983).

LN contains three distinct polypeptide chains of 200 Kd (B2 chain), 230 Kd (B1 chain), and 420 Kd (A chain). The precise arrangement of these chains is not known. However, preliminary results obtained from the cloning of the genes coding for the different chains suggest that the three chains are associated in the central part of the long arm into a coiled-coil alpha helix (Barlow *et al.*, 1984). In addition, regions with repeated cysteine residues are observed in the short arms (Ott *et al.*, 1982). LN does not form fibrils but can self-assemble into large complexes through its globular end domains (Yurchenko *et al.*, 1985).

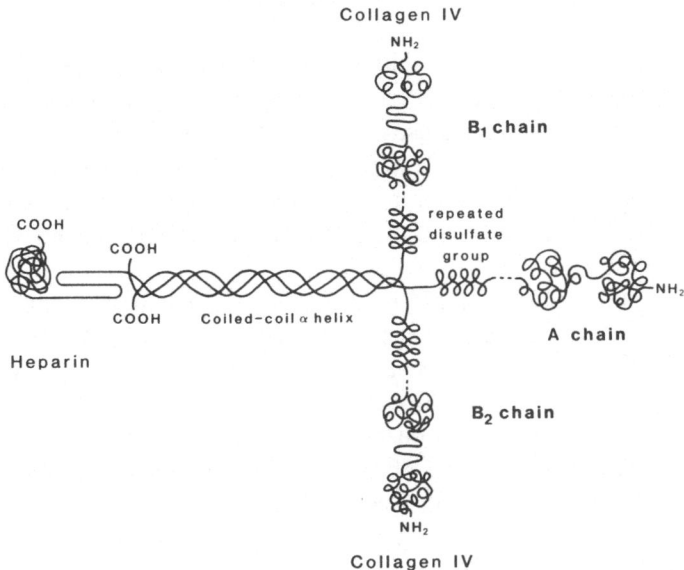

FIGURE 1. Schematic representation of LN: LN is a cross-shaped molecule composed of three chains (A, B_1, and B_2 chains). The chains are associated in the center of the cross into a coiled-coil alpha helix. Two globular domains on each B chain and three on the A chain are supposed to contain the binding sites for heparin and type IV collagen. The cell-binding site is suspected to distribute in the center of the cross, but it has not been precisely mapped. Finally, particular regions very rich in a repeated disulfate group have been represented in the center of the cross; they probably participate in the typical shape of the molecule.

2.1.2. Receptors for Laminin

A high-affinity receptor for LN ($K_D = 10^{-9}$ M) has been isolated from carcinoma and sarcoma cells, as well as from myoblasts (Rao *et al.*, 1983; Malinoff and Wicha, 1983; Lesot *et al.*, 1983; S. S. Brown *et al.*, 1983). It consists of a disulfide-linked glycoprotein complex that contains subunits with apparent molecular weights of 68 Kd. However, it remains to be determined whether all cell types interact with LN through the same receptor. A low-affinity receptor ($K_D = 10^{-6}$ M) has also been isolated that is, in fact, the receptor for FN (Horwitz *et al.*, 1986). Although the high-affinity receptor could participate in the permanent anchorage of epithelial cells, the low-affinity receptor might participate in dynamic processes during embryogenesis such as cell migration and tissue remodeling (see Section 3.4 and 4). Finally, LN can interact indirectly with the cell membrane through its heparin-binding domain (Edgar *et al.*, 1984).

2.1.3. Distribution and Functions of Laminin

LN appears very early during development, one polypeptide chain being synthesized during oogenesis whereas the second and third polypeptides appear in the 2- to 8-cell-stage mouse embryo. LN is finally expressed on the surface of blastomeres at the 16-cell stage (Leivo *et al.*, 1980; Cooper and McQueen, 1983). Later on in development, LN is present in the basal lamina of epithelia, but its precise distribution in the different compartments of the basal lamina is still controversial.

The primary role of LN is to mediate the adhesion of epithelial and endothelial cells to type IV collagen (Terranova *et al.*, 1980; Carlson *et al.*, 1981; Palotie *et al.*, 1983; Donaldson and Mahan, 1984). However, fibroblastic cells can also interact with LN (Couchman *et al.*, 1983), probably through the FN receptor. LN is also found on the surface of Schwann cells along the nerves (Cornbrooks *et al.*, 1983; Duband and Thiery, 1987), and transiently in the central nervous system of the mouse (Liesi, 1985). In this respect, LN seems to play an important role during the ontogeny of the nervous system (see Section 3.4).

2.2. Fibronectin

2.2.1. Structure of Fibronectin

Fibronectin (FN) constitutes a class of high-molecular-weight glycoproteins that consist of two similar, but not identical, polypeptide subunits of 220–250 Kd linked by disulfide bonds (for reviews, see Ruoslahti *et al.*, 1981; Hynes and Yamada, 1982; Yamada, 1983; Yamada *et al.*, 1985) (Figure 2). FN is present in large amounts in the plasma (0.3 mg/ml) as a soluble dimer (plasma FN), around mesenchymal cells in 10-nm fibrils, and on the basal surface of epithelia (cellular FN). Plasma FN is synthesized by hepatocytes and endothelial cells, whereas cellular FN is produced by a large variety of cell types (reviewed by Hynes and Yamada, 1982).

The primary structure of FN has now been completely established (Petersen *et al.*, 1983; Kornblihtt *et al.*, 1983, 1984a,b, 1985). Each FN chain contains approximately 2350 amino acids and contains three different types of internal repeating units called type I, II, and III sequence homologies.

Different forms of FN have been identified, and, in fact, each FN dimer consists of a characteristic set of distinguishable subunits differing slightly in apparent molecular weight. These differences are due in part to the variability

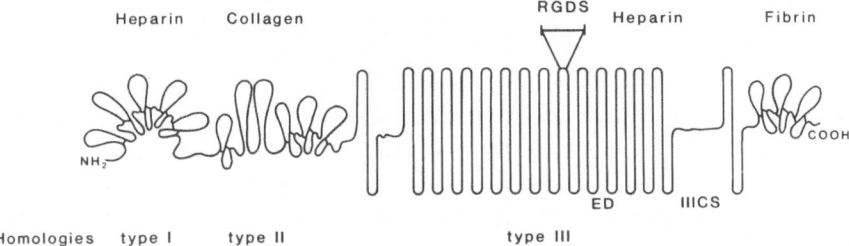

FIGURE 2. Schematic representation of one of the polypeptide chains of FN. The secondary structure of FN can be obtained from the arrangements of the different types of homologies (type I, II, and III). The structure of the homologies derives mostly from the sequence of the gene obtained in mammals. There are 12 type I homologies and 2 type II homologies, each being encoded by one exon; in contrast, each of the 16 type III homologies is probably generated by 2 exons. The binding domains mainly determined by partial proteolytic cleavage of the molecule correspond roughly to the association of several homologies. The cell-binding site (RGDS) has been precisely mapped; it is located on the top of the tenth type III homology. In fact, this domain corresponds to the insert of the DNA sequence coding for the RGDS peptide into a type III homology. Finally, the regions concerned by splicing, termed ED (extra domain) and IIICS, are indicated on the molecule.

of internal amino acid sequences. The various FN molecules derive from a single large gene covering 48 kb; a common mRNA precursor may undergo alternative splicing (Hirano *et al.*, 1983; Schwarzbauer *et al.*, 1983; Vibe-Pedersen *et al.*, 1984). The splicing events concern two DNA sequences coding for the extra domain (ED) and the IIICS domain. Complete and incomplete splicing in the IIICS region generates five different mRNA sequences; two mRNA populations are generated by the excision of the entire ED exon in a fraction of the primary transcripts (Kornblihtt *et al.*, 1983, 1985, Schwarzbauer *et al.*, 1983). The combination of the two splicing events could provide as many as ten different mature messages from a single gene. Interestingly, the ED sequence is totally deleted in hepatocyte mRNAs that code for plasma FN, whereas it is maintained in a fraction of mRNAs from other cell types (Kornblihtt *et al.*, 1983, 1984a).

The functional domains of FN comprise binding sites for a variety of molecules such as collagen, heparin, fibrin, actin, and possibly DNA (for reviews see Furcht, 1983; Yamada, 1983). The cell-binding sequence has been precisely mapped. It consists of a peptide in which the very hydrophilic sequence Arg-Gly-Asp-Ser (RGDS) is absolutely required for the adhesion of cells to FN (Pierschbacher and Ruoslahti, 1984a; Yamada and Kennedy, 1984). Variants of this sequence have been tested for their ability to inhibit binding of cells to FN or binding of the FN cell-binding region to the cell surface. Most substitutions, except at the carboxyl terminus, inactivate the peptide; in fact, RGDA, RGDV, and RGDT sequences also retain binding properties (Piersbacher and Ruoslahti, 1984b; Yamada and Kennedy, 1985). Specific spacing between the two charged amino acids (Arg and Asp) is required; these charges must be located at a specific distance from the peptide backbone (Yamada and Kennedy, 1985). The RGDS sequence is found only once on the molecule, and is probably externally exposed on the top of a type III domain loop (Pierschbacher *et al.*, 1985). The RGDS sequence is found in several other molecules, such as fibrinogen, the receptor for phage on *Escherichia coli*, and the coat protein of Sindbis virus. The RGDA sequence is present in discoidin I, thrombin, collagen alpha 1(I), and alpha 2(I) chains, whereas the RGDV sequence is found in vitronectin (Pierschbacher and Ruoslahti, 1984b; Pierschbacher *et al.*, 1985; Dufour *et al.*, 1987). Interestingly, most of these molecules are known to have cell-binding properties. The wide distribution of the RGD sequence suggests that it constitutes an ancient recognition signal common to many proteins. However, it should be noted that the specificity of binding to the receptor may be ensured by the sequences surrounding the tetrapeptide, since adjacent sequences can also modify its binding specificity (Yamada and Kennedy, 1985).

In contrast to plasma FN, cellular FN self-associates to form polymers and frequently codistributes with collagens in fibers (Furcht *et al.*, 1980). So far, the sequences responsible for fibrillogenesis have not been yet completely identified. A possible candidate is the 90-amino-acid peptide coded by the ED sequence (Schwarzbauer *et al.*, 1983; Kornblihtt *et al.*, 1983, 1984a) but no "polymerization sequence" has been detected in the ED region. The frequent codistribution of FN and collagens in fibers suggests that collagen could constitute a framework for FN polymerization (Furcht *et al.*, 1980). However, this statement is contradicted by *in vitro* experiments which suggest that it is FN that is responsible for the organization of collagen fibers and not the opposite (McDonald *et al.*, 1982). Recently, a 70-Kd domain of FN located in the amino-terminus part of the molecule has been shown to participate to FN fibrillogenesis; this domain does not appear to contain neither the ED nor the cell-binding domain (McKeown-Longo and Mosher, 1985).

2.2.2. Receptors for Fibronectin

The modest affinity of FN for the cell membrane ($K_D = 10^{-6}$ M; Akiyama and Yamada, 1985) has greatly hampered the isolation of FN receptor(s). Recently, two different approaches—one using monoclonal antibodies that interfere with cell attachment (Greve and Gottlieb, 1982; P. J. Brown and Juliano, 1985; Chen *et al.*, 1985a, Damsky *et al.*, 1983, 1985; Knudsen *et al.*, 1985) and the other based on affinity between the cell-binding sequence of FN and solubilized membrane proteins (Akiyama *et al.*, 1985)—have lead to the identification, in the chick, of a 140-Kd complex composed of three glycoproteins of approximately 120, 140, and 160 Kd, involved in the interaction of cells with FN. Similarly, a glycoprotein of 140 Kd has also been isolated from mammalian cells (Pytela *et al.*, 1985). This 140-Kd complex inhibits the attachment and spreading of fibroblasts into FN-coated substrates (Akiyama *et al.*, 1985). This complex is located on fibroblasts, myoblasts, and other cell types at the cell-to-substratum contact sites and codistributes with FN fibers and stress fibers (Chen *et al.*, 1985a,b; Damsky *et al.*, 1985). The proteins of the 140-Kd complex are integral membrane proteins that are linked to talin in the internal side of the membrane (Horwitz *et al.*, 1986). They thus must serve as a bridge between FN and the cytoskeleton. Interestingly, the 140-Kd complex is phosphorylated in transformed cells resulting in the disruption of the coupling between the complex and talin molecules (Hirst *et al.*, 1986). The arrangement of the three proteins within the cell membrane and the mode of interaction with FN remains to be determined. However, the three proteins are required to provide full binding with FN molecules (Buck *et al.*, 1986). The number of receptor molecules on fibroblasts has been estimated to approximately 3×10^5 molecules/cell (Akiyama and Yamada, 1985; Akiyama *et al.*, 1985). Finally, as already mentioned, this complex has also been shown to act as a low-affinity receptor for LN (Horwitz *et al.*, 1986). Even though the 140-Kd complex is now considered as the leading candidate for a functional FN receptor, one cannot neglect other possible receptors, such as a 47-Kd glycoprotein (Aplin *et al.*, 1981), cell-surface heparan sulfate (Laterra *et al.*, 1983), and gangliosides (Kleinman *et al.*, 1979; Spiegel *et al.*, 1985). These various receptors could each be of importance at different stages of development as a cell exhibits different behaviors. Interestingly, platelets interact with FN through a specific receptor, which is a different member of a class of 140-Kd-like complexes. This receptor is in fact the IIb–IIIa glycoprotein complex on the platelet surface that also binds fibrinogen and von Willebrand factor (Gardner and Hynes, 1985).

2.2.3. Functions of Fibronectin

The multiple cellular responses to FN have somewhat obscured its primary function. In addition, the behavior of a cell depends on the synthesis of FN, on the relative concentration of FN in the milieu, and on the mode of interaction of FN with the cell membrane (Couchman *et al.*, 1982; see Sections 3.2.2 and 3.2.3). It seems that the primary role of cellular FN is to promote cell-to-substrate adhesion; nevertheless, plasma FN can also mediate *in vitro* cell adhesion. A wide variety of cells, including fibroblastic and epithelial cells, can attach and spread on FN. A consequence of this function is an effect of FN on cell morphology. For example, FN and FN-containing ECM can partially restore the normal morphology of transformed cells (Yamada *et al.*, 1976; Fairbairn *et al.*, 1985). In addition, the synthesis and accumulation of FN on the cell surface induce the arrest of cell motility, the paralysis of cells, and changes in the organization of their cytoskeleton (Couchman and Rees, 1979; Couchman *et al.*, 1982).

FN also influences cell division. Indeed, several cell types were successfully grown on FN (Orly and Sato, 1979; Baron-Van Evercooren *et al.*, 1982) or in the presence of ECM previously deposited by other cells (Gospodarowicz *et al.*, 1982; Rojkind *et al.*, 1980). On the other hand, the subcutaneous injection of FN in rats results in a higher rate of proliferation of precartilagenous cells (Weiss and Reddi, 1981). *In vitro*, 3T3 cells are able to reach a higher degree of confluency when grown on FN (Yamada *et al.*, 1982), and embryonic heart fibroblasts proliferate when they start to synthesize FN (Couchman *et al.*, 1982).

FN has been implicated in the mechanism of cell migration. In particular, it induces migration of previously immobile cells (Ali and Hynes, 1978). The role of FN during migration of embryonic cells will be discussed extensively in the following sections.

In several instances, levels of FN were shown to decrease during cell differentiation (Dessau *et al.*, 1980), and the addition of FN can temporarily block (Pennypacker *et al.*, 1979; Podleski *et al.*, 1979) or even reverse differentiation (West *et al.*, 1979). Conversely, FN can favor the differentiation of trunk crest cells into adrenergic neurons (Sieber-Blum *et al.*, 1981) or maintain the differentiated state of endothelial cells (Maciag *et al.*, 1982).

FN has a crucial role in tissue organization. *In vitro*, destabilization of an epithelial structure can be induced after degradation of its basement membrane; conversely, the integrity of an epithelium can be restored after addition of different ECM components, including FN (Greenburg and Hay, 1982; Sugrue and Hay, 1981, 1982). However, it is not always possible to evaluate the relative influence of FN compared to that of other components of the ECM, since FN is associated with them and they all serve to modify the behavior of cells. Furthermore, cells themselves contribute macromolecules to the ECM to which they later respond. In this respect, it is interesting to note that collagen can modulate the interaction of FN with the cell surface (Nagata *et al.*, 1985).

Plasma FN has functions very different from those of cellular FN. However, these functions are also based on the binding properties of FN. For example, plasma FN is involved in wound healing and hemostasis through its binding to fibrin (Ruoslahti and Vaheri, 1975) and in opsonization (Van de Water *et al.*, 1981).

3. INVOLVEMENT OF CELL–SUBSTRATUM ADHESION MOLECULES IN EMBRYONIC CELL MIGRATION

3.1. Gastrulation

In the vertebrate embryo, gastrulation takes place very soon after initial cleavage and division of the egg into a blastula. It consists of complete reorganization of the distribution of cells into a three-layered embryo without any noticeable increase in the size of the embryo. After gastrulation, neurulation and organogenesis can start; thus, gastrulation involves crucial morphogenetic movements during embryonic development. The precise movements that comprise gastrulation depend on the shape of the embryo, but it almost always proceeds by the ingression of cells beneath a superficial layer.

3.1.1. Fish Embryo

The process of gastrulation varies considerably within the class of fishes (Ballard, 1981). We will examine the case of the *Salmo* embryo here (this genus includes the salmon and most species of freshwater trout). After fertilization, the cytoplasm of the egg accumulates at the animal pole, where it surrounds the nucleus. Then, the blastodisc undergoes cleavage into a blastoderm consisting of a single outer enveloping layer of cells covering a number of deeper blastomeres (deep cells or hypoblast). Cleavage also gives rise to a cell layer located beneath the enveloping layer and deep cells, the yolk syncytial layer (YSL) or periblast. It is a syncytium that covers the yolk and is separated from the other two layers by a cavity, the blastocoele (see Figure 3). The blastoderm then flattens and spreads over the yolk, eventually surrounding it completely; this process is called epiboly. During epiboly, the margin of the blastoderm thickens and forms the germ ring. Deep cells in the germ ring converge toward a point along the ring and form a cellular aggregate, the embryonic shield (see Figure 3). Apparently, in contrast to other vertebrate embryos, there is no blastopore in the *Salmo* embryo and no involution of cells at the margin of the blastoderm. The embryonic shield increases in mass and elongates as epiboly progresses. Concomitant with epiboly, cells in the embryonic shield undergo migration under the blastocoele roof; they will give rise to the notochord and the mesodermal cells, whereas the rest of the hypoblast in the embryonic shield will constitute the neurectoderm. The origin of the endoderm is not yet precisely determined, but it probably derives from a complex movement of cells located in the germ ring. Thus, the embryo derives exclusively from the hypoblast, which engages

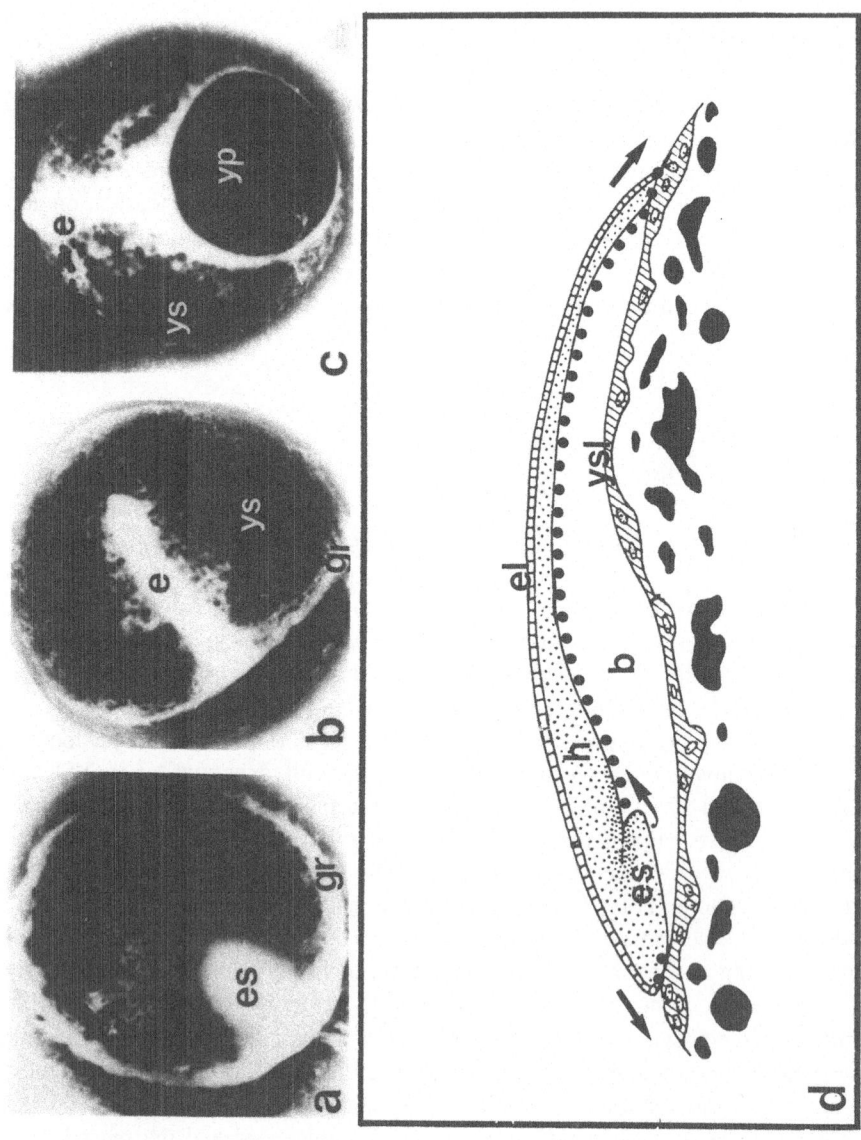

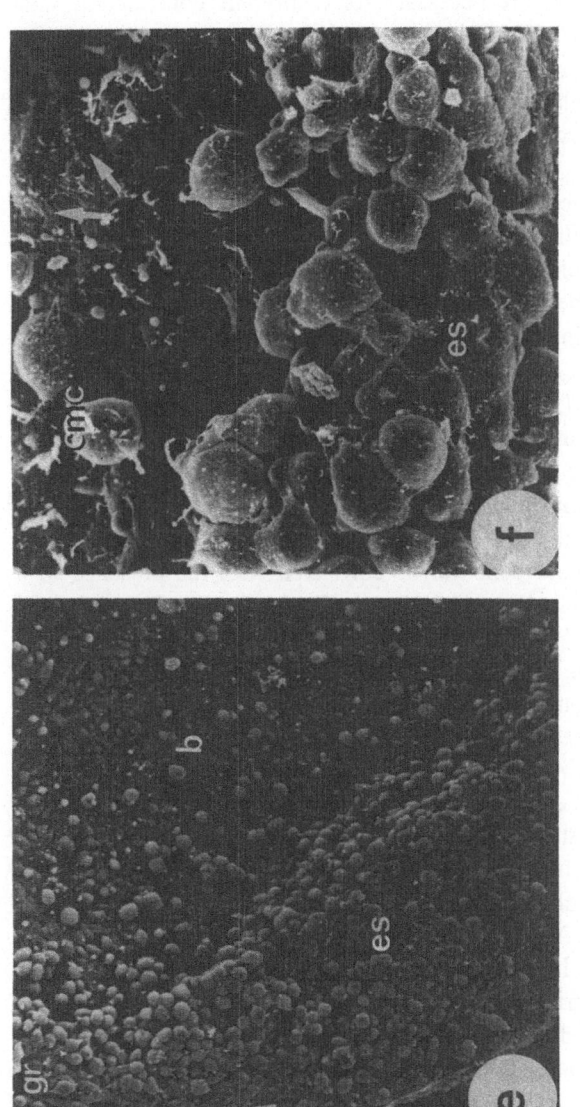

FIGURE 3. Mechanism of gastrulation in the trout embryo. (a–c) The external aspect of the egg at three typical steps of gastrulation and epiboly. (a) Gastrula stage showing the embryonic shield (es) and the germ ring (gr). (b) Epiboly stage; the embryonic axis (e) is now clearly visible, whereas the germ ring is extending over the yolk. (c) Late epiboly stage; the yolk is almost entirely covered by the yolk sac (ys), the yolk plug (yp) being in the closing process. (d) Schematic representation of a sagittal section of an early gastrula. Arrows indicate the various cell movements, i.e., epiboly over the yolk and movement of cells from the embryonic shield under the hypoblast (h). Large dots indicate the localization of FN both under the blastocoele (b) roof and at the edge of the expending germ ring. (e,f) General view and detail of the embryonic shield as seen by scanning electron micrography. Note the presence of numerous FN fibrils (arrows) in front of migrating cordomesodermal cells (cmc). (el) Enveloping layer; (ysl) yolk syncytial layer.

in extensive morphogenetic movements, including formation of the germ ring and the embryonic shield (Pasteels, 1936; Ballard, 1966).

Two sorts of movements are necessary for embryonic construction: convergence and epiboly. Convergence is the process leading to the formation of the germ ring and the embryonic shield; it only concerns deep cells. The precise mechanism of this cell displacement is not known. The mechanism of epiboly has been studied extensively by Trinkaus in the *Fundulus* embryo (Trinkaus, 1984a,b). He suggests that the leading force of epiboly resides in the YSL. Indeed, prior to epiboly, the YSL spreads well beyond the border of the blastoderm, and it exhibits contractile properties at the margin of the blastoderm. The contractions could result from the presence of a microfilament meshwork located at the boundary of the blastoderm. Trinkaus thus proposes a model based exclusively on the contraction of the YSL. Cells of the enveloping layer would be attached to the YSL and would be pulled by the contraction of the YSL to the vegetal pole of the egg (Trinkaus, 1984a,b).

It appears, however, that the extension of the germ ring to the vegetal pole is not only passive but is also governed by active autonomic movements of germ ring cells themselves (Boulekbache *et al.*, 1984; H. Boulekbache, unpublished results). Indeed, cells from the hypoblast and the enveloping layer can spread over the YSL. Such a displacement can be correlated with the presence of a dense meshwork of FN fibers under germ ring cells (Figure 4). This FN meshwork could act as a "purse string," thus mediating the extension of the germ ring and the closing of the yolk plug. It remains to be determined which cell type is responsible for the deposition of this FN meshwork.

The mechanism of migration of mesodermal cells under the blastocoele roof also appears to be mediated by FN (Boulekbache *et al.*, 1984; H. Boulekbache, unpublished results). At the early blastula stage, FN cannot be detected in the blastoderm. At the late blastula–early gastrula stage, a starry network of FN fibrils settles down all over the blastocoele roof surface (Figures 3 and 4); these fibrils seem to be bound to cellular processes such as lamellipodia and filopodia, which appear at this stage. As the embryonic shield appears, the FN network becomes denser and thicker in front of the migrating mesodermal cells (Figure 4). The spatiotemporal distribution of FN in early fish development is thus indicative of the role of FN during the various morphogenetic movements (epiboly, migration of mesodermal cells) that occur during gastrulation.

3.1.2. Amphibian Embryo

In amphibians, after the phase of segmentation, the blastula appears as a hollow sphere containing a blastocoelic cavity. The polarity of the embryo (i.e., the animal and vegetal poles) is already established at that stage. The first sign

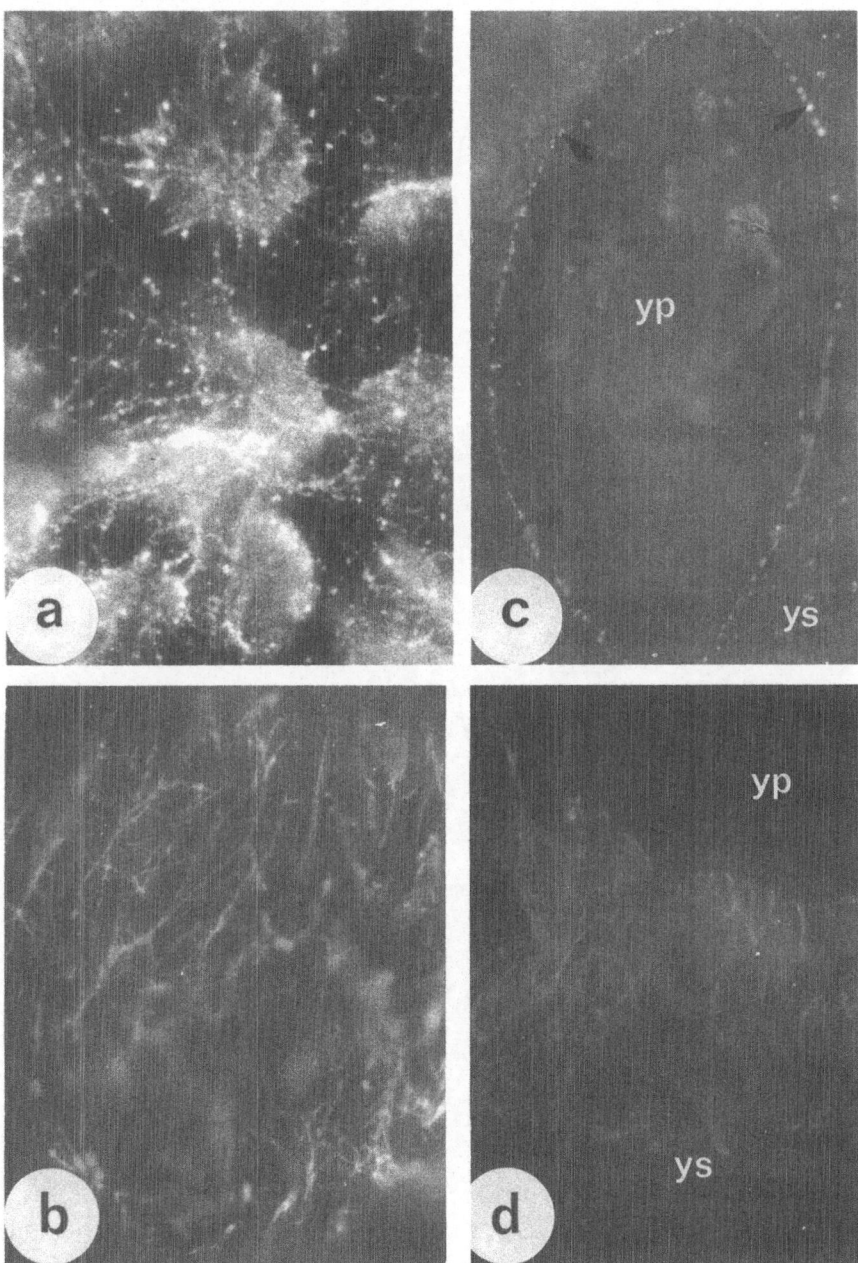

FIGURE 4. Distribution of FN during gastrulation and epiboly of the trout embryo. (a) Blastocoele roof at the early gastrula stage. A starry network of FN fibrils can be detected on the surface of the blastocoele roof. (b) YSL at the early epiboly stage. Note that FN forms a very dense meshwork. (c,d) General view and detail of the yolk plug edge. A springlike FN meshwork is located all around the yolk plug (yp) on the border of the yolk sac (ys). (c) Arrows indicate the edge of the yolk sac.

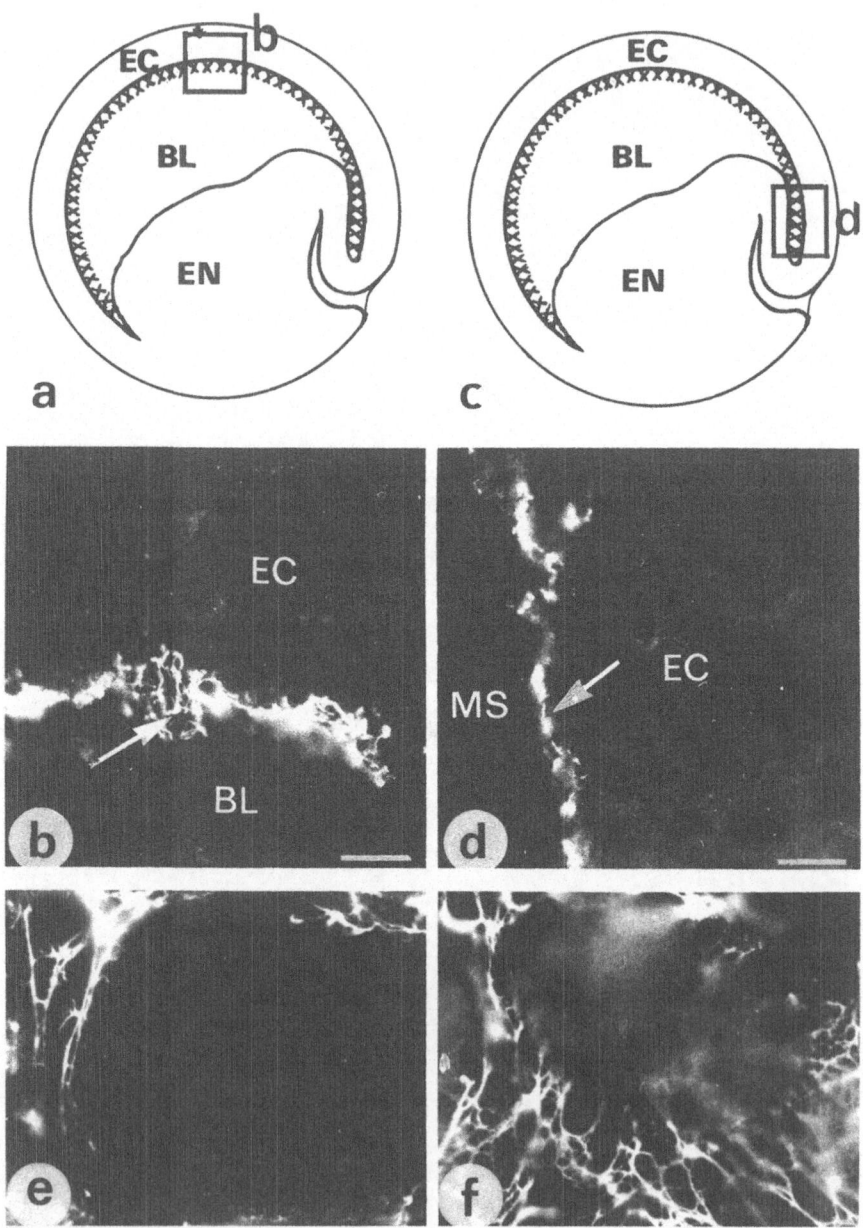

FIGURE 5. Immunofluorescent distribution of FN during gastrulation in amphibian embryo. (a–d) Distribution of FN in two different areas of the blastocoele roof. (b,d) Arrows indicate the FN meshwork. Note that the matrix is already assembled in regions not yet occupied by mesodermal cells. (e,f) Distribution of FN in whole mount embryos. The ectoderm was surgically removed and immunostained for FN in order to visualize the blastocoele roof. (e) Late blastula: Radially ordered FN fibrils cover part of the surface of ectodermal cells. (f) Early gastrula: A rapid increase in the density of FN fibrils occurs just prior to the invagination of mesodermal cells. (BL) Blastocoelic cavity; (EC) ectoderm; (EN) endoderm.

of gastrulation is a slit, the blastopore, which appears in the marginal zone between the vegetal and the animal poles, and defines the future dorsal side of the embryo (see Figure 5).

Scanning electron microscopy, time-lapse microcinematography, and cultures of tissue explants have shown that gastrulation involves several coordinated cellular movements (Nakasutji, 1975; Keller, 1975, 1978, 1980). In *Xenopus*, cells of the animal pole undergo epiboly and cover the entire surface of the sphere; they give rise to the neurectoderm and the skin. On the edge of the blastopore, presumptive mesodermal cells invaginate into the blastopore and actively migrate under the basal surface of the blastocoelic roof. Finally, most of the presumptive endoderm (i.e., cells from the vegetal pole) are carried passively inside. The triggering of the invagination of cells into the sphere remains unknown, but it may involve changes in the shape of cells close to the future blastopore (Cooke, 1975).

Migration of cells from the blastoporal lip along the basal surface of the blastocoelic roof has been studied by histological methods, electron microscopy, and time-lapse microcinematography (Nakasutji, 1975; Keller and Schoenwolf, 1977; Keller, 1975, 1978, 1980). Cells released from the superficial layer emit fine filopodia, allowing anchorage and locomotion on a network of fibrils (Nakasutji *et al.*, 1982). Biochemical studies and immunocytochemistry with fluorescent or gold-coupled antibodies have revealed the presence of both FN and LN in this ECM (Boucaut and Darribère, 1983; Darribère *et al.*, 1985; Nakasutji *et al.*, 1985a,b,; see also Figure 5). FN is synthesized at a low rate from maternally derived mRNA during oogenesis; translation then increases rapidly at the late blastula and early gastrula stages (Darribère *et al.*, 1984; Lee *et al.*, 1984). LN becomes clearly detectable in the early gastrula (Nakasutji *et al.*, 1985b; Darribère *et al.*, 1986).

Although the precise role of LN in gastrulation has not been established, the role of FN in the adhesion and migration of mesodermal cells has been assessed by three types of perturbation experiments, which have led to the following conclusions (Figure 6): (1) When part of the blastocoelic roof is inverted, mesodermal cells avoid the area now lacking an ECM (Boucaut *et al.*, 1984a); (2) microinjection of monovalent anti-FN antibodies into the blastocoelic cavity of late blastulae or early gastrulae blocks gastrulation (Boucaut *et al.*, 1984a); (3) similarly, when Arg-Gly-Asp-Ser-containing peptides are injected, gastrulation is also arrested (Boucaut *et al.*, 1984b). Thus, competitive inhibition of the receptors and an immunological steric hindrance effect, both of which prevent the interaction between the cell surface and FN, interfere with the movement of mesodermal cells during gastrulation.

In contrast to the requirement for FN in the formation of the mesoderm, FN is unlikely to be involved in the mechanism of neural induction. *In vitro*, mesodermal cells associated with the apical surface of the ectoderm devoid of ECM can induce the appearance of neural elements (Duprat and Gualandris,

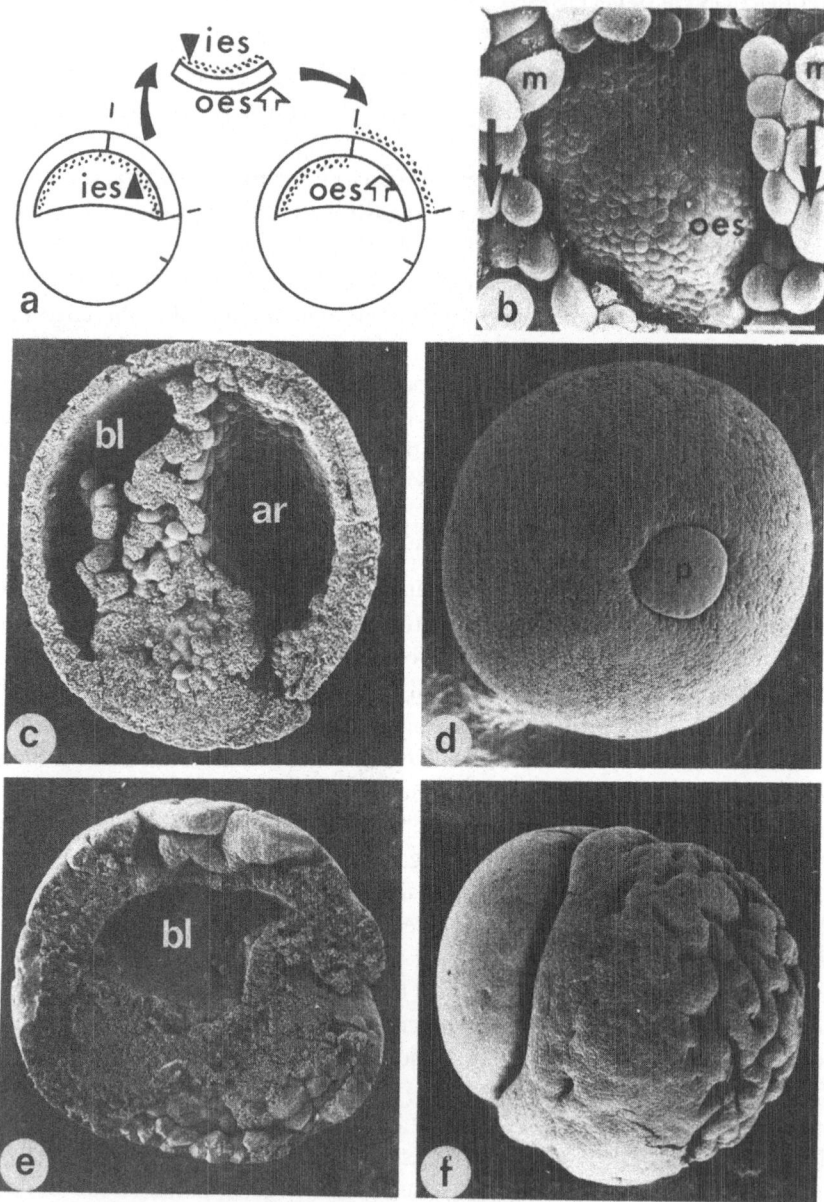

FIGURE 6. Perturbation of amphibian gastrulation by inhibition of cell interaction to FN. Cell attachment and migration can be altered by two different procedures. (a,b) When part of the ectoderm is inverted upside down, the migrating mesodermal cells (m) avoid the outer ectodermal surface (oes) but remain in contact with the inner ectodermal surface (ies), which contains FN. (b) Arrows indicate the direction of cell displacement. (c–f) When antibodies directed against the cell-binding domain of FN or RGDS-containing decapeptides are injected into the blastocoelic cavity, most mesodermal cells remain on the outside and cannot gastrulate; in addition, the ectoderm does not undergo proper epiboly and, instead, becomes highly convoluted (e,f). In contrast, control embryos gastrulate normally when injected with nonimmune antibodies or several peptides, including a peptide contained in the collagen-binding domain (c,d). (ar) archenteron; (bl) blastocoele.

1984). *In vivo*, when either the antibodies or the peptides are introduced during or at the end of gastrulation, a partial or a complete neural plate forms (Boucaut ét äl., 1984a,b).

3.1.3. Avian Embryo

In avian embryo, gastrulation is somewhat different from that in the amphibian, since the shape of the embryo is not spherical but discoidal. However, it also proceeds by ingression of cells, the blastopore being replaced by a long growing slit in the axis of symmetry of the embryo. This slit, called the primitive streak, appears in the posterior area of the blastoderm and then grows anteriorly. The primitive streak is the site at which cells from the superficial layer invaginate (see Figure 7). Ingressing cells then actively move under the basal surface of the upper layer. In contrast to the amphibian, no epiboly is observed, and the definitive endoderm does not derive from the cells in the deep layer but from gastrulating cells. Indeed, three deep layers form sequentially, each replacing the previous one, which is pushed away to the periphery of the blastodisc (for a review, see Vakaet, 1970, 1984).

After segmentation, cells from the blastoderm do not exhibit a typical epithelial organization. Gastrulation is just preceded by the appearance of an epitheliallike organization among cells from the upper layer, which is then underlain by a basal lamina that contains LN (Mitrani, 1982), FN (Duband and Thiery, 1982a; Mitrani and Farberov, 1982; Sanders, 1982; Harrisson *et al.*, 1984), type I collagen (Duband and Thiery, 1987), and glycosaminoglycans (Vanroelen *et al.*, 1980). Whole mount embryos reveal the overall distribution of FN during gastrulation; abundant quantities of FN are visualized at the periphery of the expanding blastoderm and as a crescent at the anterior portion of the area pellucida. Lower amounts are found all over the upper layer, sometimes with fibers running perpendicularly from the primitive streak (Critchley *et al.*, 1979). Under the appearing primitive streak, the basal lamina disrupts and disappears progressively when cells from the upper layer sink toward the deep layer (Sanders, 1979, 1982; Duband and Thiery, 1982a; see also Figures 7 and 8). The destruction of the basal lamina is observed under the rapidly expanding primitive streak. In contrast, the basal lamina remains intact in regions not subject to the ingression of cells (Duband and Thiery, 1982a; see also Figures 7 and 8).

The process leading to the formation of the primitive streak is not understood at the molecular level. In particular, one cannot yet explain how cells from the upper layer move toward the primitive streak. Very interestingly, radioactive labeling of FN in portions of the upper layer showed that the cells move to the primitive streak together with their basal lamina (Sanders, 1984). This raises the intriguing problem of the concomitant movement of a cell along with its basal lamina.

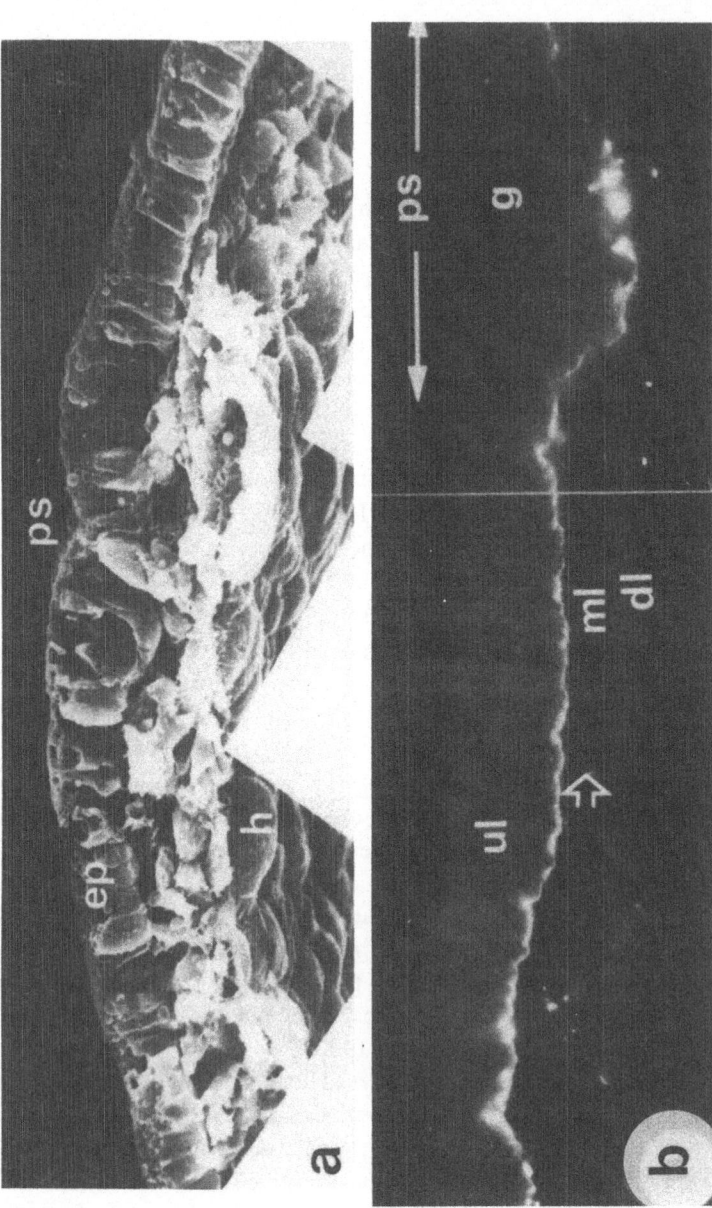

FIGURE 7. Gastrulation in the chick embryo. (a) Scanning electron micrograph of a late gastrula at the level of the primitive streak. Cells in the primitive streak (ps) are disorganized and send processes in the cavity underneath. Under the upper layer or epiblast (ep), the mesodermal cells occupy the full space and form a multilayered cell population. (b,c) Distribution of FN at the midgastrula stage. (b) immunofluorescence; (c) phase contrast. The middle layer cells (ml), i.e., the mesodermal and endodermal cells, migrate as a single layer along the FN-containing basement

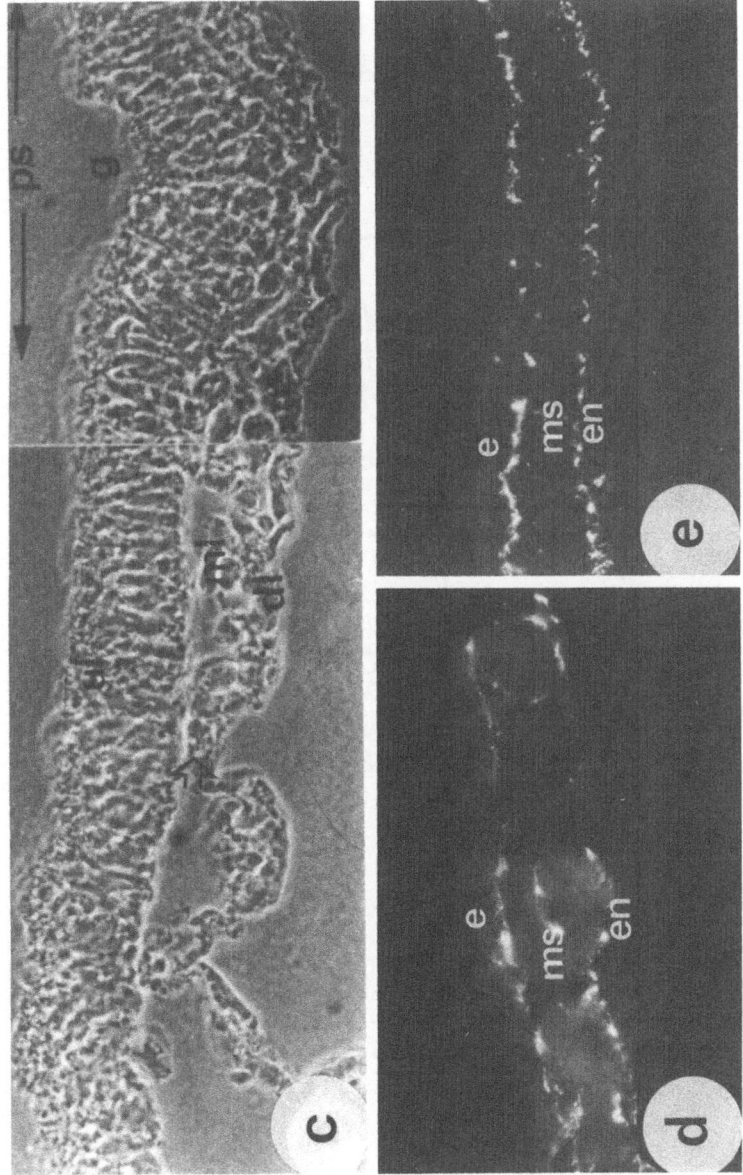

FIGURE 7. (*Continued*) membrane of the upper layer (ul). The open arrow indicates the front of migration of ml cells. (d,e) Distribution of FN at the end of gastrulation. When mesodermal cells (ms) have stopped migrating, FN appears among them. When the definitive endoderm (en) is formed, it is underlain by a basal lamina. (dl) Deep layer; (e) ectodermal layer; (g) groove; (h) hypoblast.

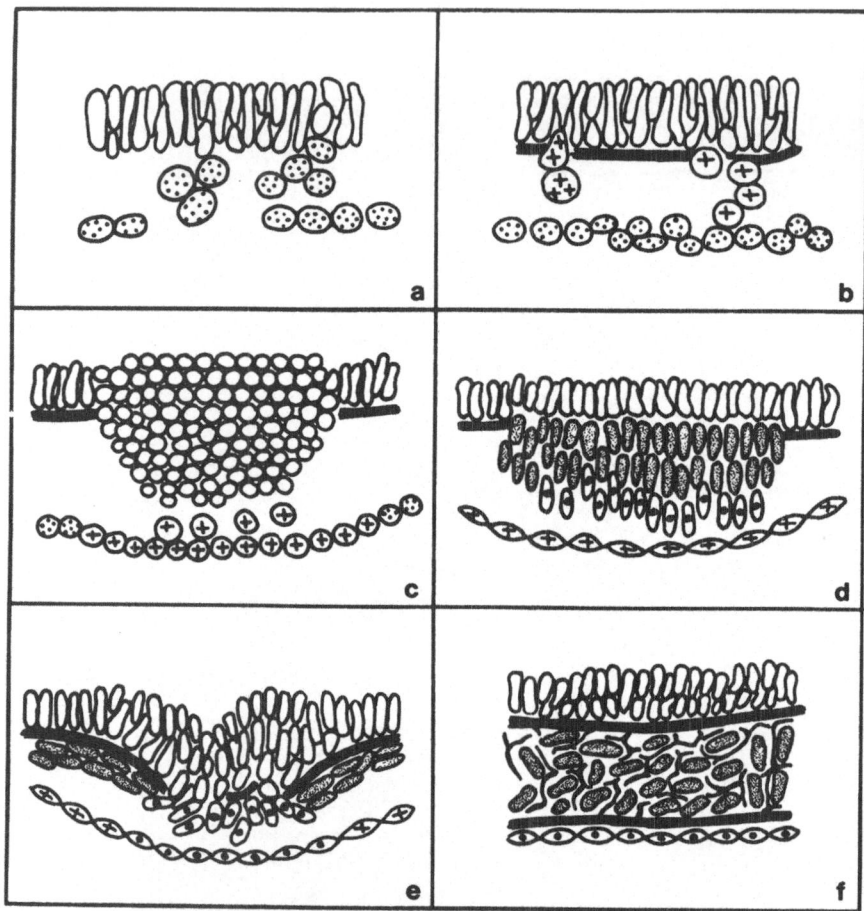

FIGURE 8. Diagram showing the possible role of FN during the main steps of gastrulation in the chick embryo. (a) Appearance of the cells of the endophyll. (b) Completion of the endophyll; appearance of FN under the upper layer; segregation of the hypoblast from the upper layer. (c) Accumulation of cells in the primitive streak; disappearance of FN under the primitive streak. (d) Invagination of mesodermal cells between the upper layer and deep layer. (e) Lateral migration of the mesoblast; formation of the definitive endoderm. (f) Stabilization of ml cells within an FN network.

The invasion by ingressing cells of the space under the upper layer is understood more clearly and involves active cell movement. During this process, gastrulating cells closely adhere to the FN-rich basement membrane of the superficial layer (Duband and Thiery, 1982a; Sanders, 1982), but FN is not found around the moving cells until the end of their migration (Figures 7 and 8). A parallel can be drawn with heart mesenchymal cells, which stop migrating when they produce large amounts of FN (Couchman *et al.*, 1982). The mechanism of cell migration during avian gastrulation is thus very similar to that during amphibian gastrulation; however, the exact role of FN during the migratory process of gastrulating cells has not yet been assessed by perturbation experiments.

The formation of the successive deep layers remains obscure. However, they all derive from the superficial layer either by passive release of cells or by active migration during gastrulation. During segmentation, the formation of the first deep layer (endophyll) is likely to be facilitated by the absence of a basement membrane under the upper layer (Figure 8). In contrast, the release of the cells that will form the second (hypoblast) and third (endoblast) deep layers could be governed by the local disruption of the basal lamina under the upper layer. The replacement of a deep layer by the successive one cannot yet be explained. Interestingly, the deep layers are not lined by a basal lamina until the definitive endoderm is formed (Duband and Thiery, 1982a), suggesting that the absence of a basal lamina can allow cells to intercalate within the cell sheet (Figures 7 and 8).

3.2. Migration of Avian Neural Crest Cells

A huge number of cell types derive from a single structure lying along the entire dorsal border of the neural axis, the neural crest (for a review, see Le Douarin, 1982). For example, in the head and neck, most of nonnervous tissues, such as muscles, bones, cartilages, and connective tissues, are of crest origin; in the trunk, the entire peripheral nervous system originates from the neural crest. Such a situation can only be achieved by the conjunction of the migration of crest cells and intense morphogenetic movements in the surrounding tissues. In this respect, the neural crest offers a remarkable model system of morphogenesis.

After their separation from the neural epithelium, crest cells encounter an extracellular milieu that greatly influences their progression and final distribution. At the end of their migration, neural crest cells often aggregate into dense structures exhibiting a transient epithelial organization. Thus, a very dynamic pattern of cell adhesion is observed throughout the development of crest cells.

3.2.1. Structure and Organization of the ECM in Neural Crest Pathways of Migration

The routes of migration are greatly influenced by the morphology of the embryo; however, it appears that their structure presents similarities throughout the embryo. Some ECM components are always associated with crest migration pathways and, at least during the initial phases of migration, pathways of migration are cell-free spaces limited by one or two basal laminae of epithelia (Figure 9).

FN is a major component of the pathways of crest migration. The presence of migrating crest cells is always correlated with the presence of FN (Figure 9), and, in some cases, cessation of movement is accompanied by the disappearance of FN among the cells (Duband and Thiery, 1982b; Thiery et al., 1982a; Duband et al., 1985). However, FN is not the only molecule found in crest pathways; collagen type I (Duband and Thiery, 1987), hyaluronic acid (Derby, 1978; Pratt et al., 1975), and cytotactin (Crossin et al., 1986) are also abundant. LN is present in the basal laminae bordering the pathways, but it cannot be detected among migrating crest cells (Duband and Thiery, 1987). On the other hand, there are areas in the embryo that exhibit high quantities of FN but are not invaded by crest cells. The presence of FN in acellular spaces is thus necessary but not sufficient for crest migration. For example, crest cells do not invade the cephalic mesenchyme, the notochordal area, and the space between the ectoderm and the dermatome, areas particularly rich in chondroitin sulfate (Derby, 1978; Thiery et al., 1982a; Duband and Thiery, 1982b; Brauer et al., 1985); this component has been shown not to favor cell displacement (Newgreen et al., 1982; Tucker and Erickson, 1984).

It is very likely that spatial and temporal variations in the composition of the ECM may influence neural crest cell migration (Derby, 1978; Löfberg et al., 1985). However, it has been impossible so far to detect any gradient in the concentration of the ECM components along the pathways (Duband et al., 1985). It thus seems improbable that either a haptotactic or a chemotactic mechanism, as suggested by Greenberg et al. (1981), is responsible for the orientation or migration of crest cells. The three-dimensional structure of the ECM does not appear to influence the direction of crest migration by a mechanism of contact guidance (Weston and Butler, 1966; Erickson et al., 1980; Erickson, 1985). Indeed, with the possible exception of the amphibian embryo (Löfberg et al., 1980), the meshwork does not seem to exhibit any particular orientation (Tosney, 1978). However, reorganizations of the ECM close to the neural tube occur prior to crest cell emigration and thus may favor the detachment of crest cells from the neural tube (Löfberg et al., 1985). Finally, during their migration, crest cells tend to reorganize their matrix (Rovasio et al., 1983), a process which may require some degradation of matrix components (Valinsky and Le Douarin, 1985).

The morphology of the embryo and the organization of the tissues are responsible for the structure of the pathways of migration. Various tissues may constitute obstacles or guides to crest migration. In the head, the presence of local thickenings of the neural tube (optic vesicle) or of the ectoderm (ectodermal placodes) prevents ventral migration of crest cells and forces them to move either anteriorly or caudally (see Figure 9; Noden, 1975; Duband and Thiery, 1982b; Cochard and Coltey, 1983). In the trunk, the migration of crest cells is guided mainly by meta-merized structures, the somites (Thiery *et al.*, 1982a; Duband *et al.*, 1985; see also Figure 9). Depending on their location with respect to the somite, crest cells move between two consecutive somites or between the somite and neural tube. Those using the first pathway rapidly reach the aorta where they will provide the sympathetic ganglia. Crest cells facing the somites accumulate only in the anterior part of each segment or migrate more ventrally along the basement membrane of the my-otome; they will give rise to the spinal ganglia and to the Schwann cells along the motor nerves (Weston, 1963; Thiery *et al.*, 1985).

3.2.2. Role of ECM Components in Neural Crest Adhesion and Migration

The effect of the major ECM components on neural crest cell adhesion and migration has been examined in an *in vitro* system (Figure 10; Greenberg *et al.*, 1981; Newgreen *et al.*, 1982; Rovasio *et al.*, 1983; Tucker and Erickson, 1984). FN alone, in association with collagen in two- and three-dimensional lattices, or deposited by fibroblasts, greatly promotes the spreading and movement of crest cells (Greenberg *et al.*, 1981; Newgreen *et al.*, 1982; Rovasio *et al.*, 1983; Tucker and Erickson, 1984). *In vitro*, crest cells do not attach to pure collagen deposited on two-dimensional substrata (Newgreen *et al.*, 1982; Rovasio *et al.*, 1983), but the cells are able to move in three-dimensional collagen gels, with the restriction that the collagen must be native and at a relatively low concen-tration. However, the speed of movement is low as compared to that in the presence of FN (Tucker and Erickson, 1984). Likewise, hyaluronate in two- and three-dimensional gels is a very poor substrate for migration (Newgreen *et al.*, 1982; Tucker and Erickson, 1984), but due to its hydration properties, it expands spaces (Pratt *et al.*, 1975; Pintar, 1978) and indirectly enhances the speed of migration (Tucker and Erickson, 1984). From this point of view, it is interesting to note that crest cells synthesize hyaluronate at the onset of their migration (Greenberg and Pratt, 1977; Pintar, 1978). *In vitro*, chondroitin sulfate is not a good substrate for movement (Newgreen *et al.*, 1982; Tucker and Erickson, 1984); it is present in low quantities in crest pathways (Derby, 1978; Brauer *et al.*, 1985), and its increase in amount has been correlated with the arrest of crest cells (Derby, 1978). Finally, LN is also a poor substrate for crest migration but rather induces crest cell aggregation (Rovasio *et al.*, 1983). Interestingly, when crest cells are confronted with alternative stripes of FN and coated serum proteins,

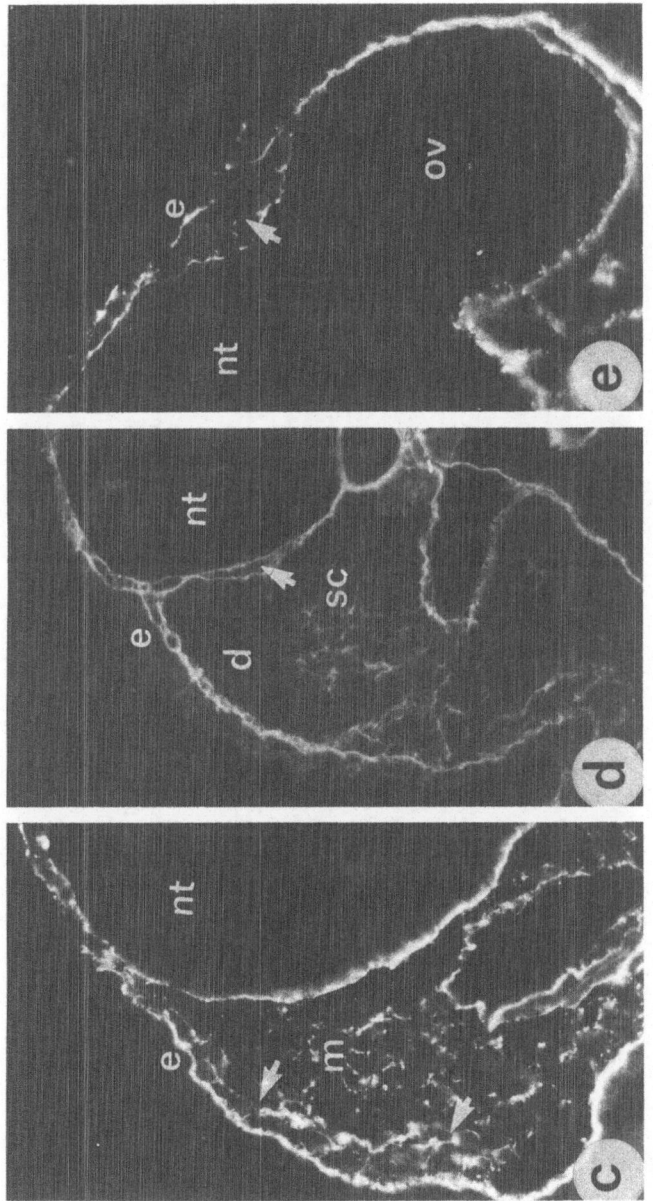

FIGURE 9. Substrate and pathways of migration of neural crest cells in the chick. (a) External view of a the head of a 12-somite-old embryo showing cephalic neural crest cells migrating laterally under the ectoderm. (b) Scanning electron micrograph showing in cephalic neural crest cells (nc) migrating in a fibrillar meshwork under the ectoderm (e). Arrowheads indicate the basement membrane of the ectoderm. (c–e) Distribution of FN in neural crest pathways at different axial levels and role of the environment on neural crest distribution. (c) At the mesencephalic level, neural crest population (arrows) expends in an FN-rich ECM between the ectoderm (e) and the cephalic mesenchyme (m). (d) In the trunk, crest cell (arrow) migration is influenced by the metamerized lateral structure, the dermomyotome (d), and the sclerotome (sc). (e) At the level of the optic vesicle (ov), the lateral migration of crest cells (arrow) is prevented by the close attachment of the vesicle to the ectoderm (e). (nt) Neural tube; (s) somite.

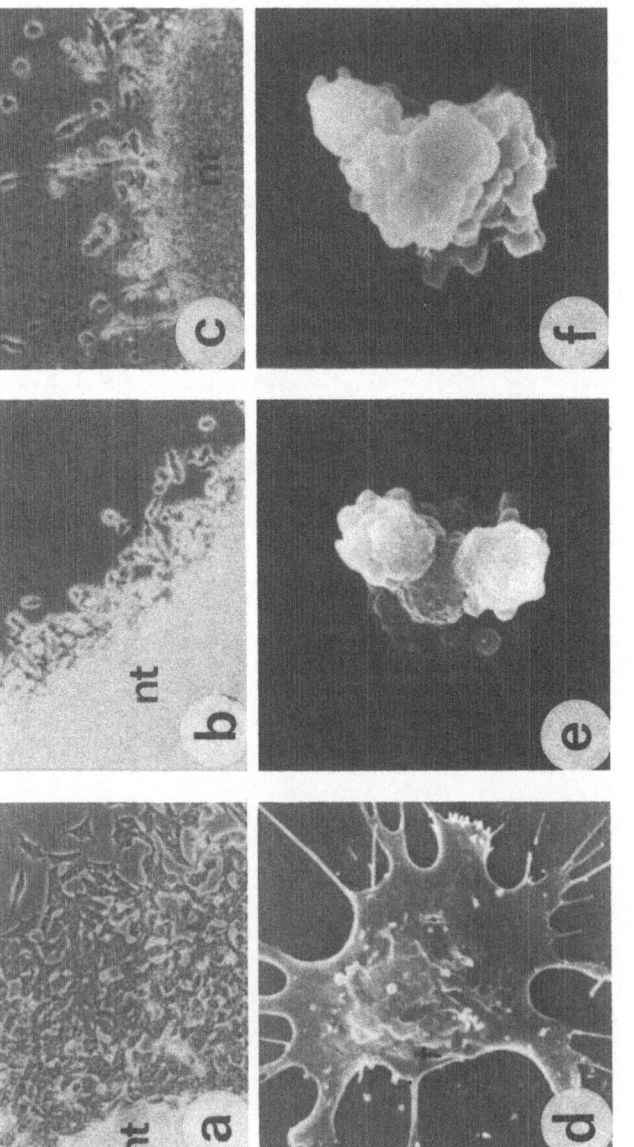

FIGURE 10. *In vitro* migration (a–c) and spreading (d–f) of neural crest cells on different substrata. In the migration assay, neural tubes (nt) were explanted onto FN (a), LN (b), and type I collagen (c), and neural crest cells were allowed to emigrate from the neural tubes for 18 hr. In the spreading assay, 24-hr cultured crest cells were dissociated and incubated for 1 hr in bacterial plastic dishes coated with FN (d), LN (e), and type I collagen (f). As opposed to LN and collagen, FN greatly promotes both the spreading and the migration of neural crest cells.

they migrate exclusively on the FN ones. When stripes of LN were used instead, few crest cells could move on LN and tended to form aggregates (Rovasio *et al.*, 1983).

The crucial role of FN in crest cell displacement has been approached in perturbation experiments (Figures 11 and 12). Monovalent antibodies directed against the cell-binding site of FN reversibly block the migration of crest cells both *in vitro* and *in vivo* (Rovasio *et al.*, 1983; Poole and Thiery, 1987). When a decapeptide containing the RGDS sequence is used to compete with the cell-binding site of FN, crest cells do not move *in vitro*; *in vivo*, they are not seen on the sides of the neural tube, but rather form a bulk in the neural tube lumen (Boucaut *et al.*, 1984b). Finally, antibodies to the 140-kd FN-receptor complex inhibit the adhesion and migration of motile neural crest cells *in vitro* (Duband *et al.*, 1986). Together, these studies strongly suggest that a direct interaction with FN is necessary for active crest cell migration.

3.2.3. Behavior of Neural Crest Cells

For several reasons, the final distribution of crest cells at the end of their migration does not result solely from the structure of the pathways and from the properties of FN: (1) Crest cells, or pigment cells, when grafted into normal crest migratory pathways, distribute in the normal sites of arrest, in contrast to fibroblasts from somite, limb bud, and lateral plate, which remain intact at the site of the graft. Interestingly, other motile cells, such as tumor cells, exhibit the same behavior as crest cells (Erickson *et al.*, 1980; Erickson, 1985). (2) Nonmigratory embryonic cells exhibit the same requirement for FN as crest cells do for their attachment and spreading, and nevertheless are unable to move. These data strongly suggest that crest cells exhibit a specific migratory behavior (Figures 13 and 14).

In contrast to stationary somitic and notochordal fibroblasts, migratory crest cells do not show any polarized shape, but rather have many cell processes. Crest cells display very few organized microfilament fibers (Tucker *et al.*, 1985; Duband *et al.*, 1986), low numbers of focal contacts, and limited amounts of localized vinculin and alpha-actinin (Duband *et al.*, 1986). In addition, crest cells do not synthesize FN (Newgreen and Thiery, 1980; Sieber-Blum *et al.*, 1981) and exert a weak tractional force on their substratum (Tucker *et al.*, 1985). This behavior has been suggested to be necessary for efficient displacement on a FN substrate (Couchman and Rees, 1979; Couchman *et al.*, 1982; Koliga *et al.*, 1982). Finally, the 140-Kd FN-receptor complex is present on crest cells both *in vivo* and *in vitro*; however, in contrast to mesenchymal cells, where it is enriched in the cell-to-substratum contact sites, the receptor complex is diffusely organized on motile crest cells. This allows a labile adhesion of the cell

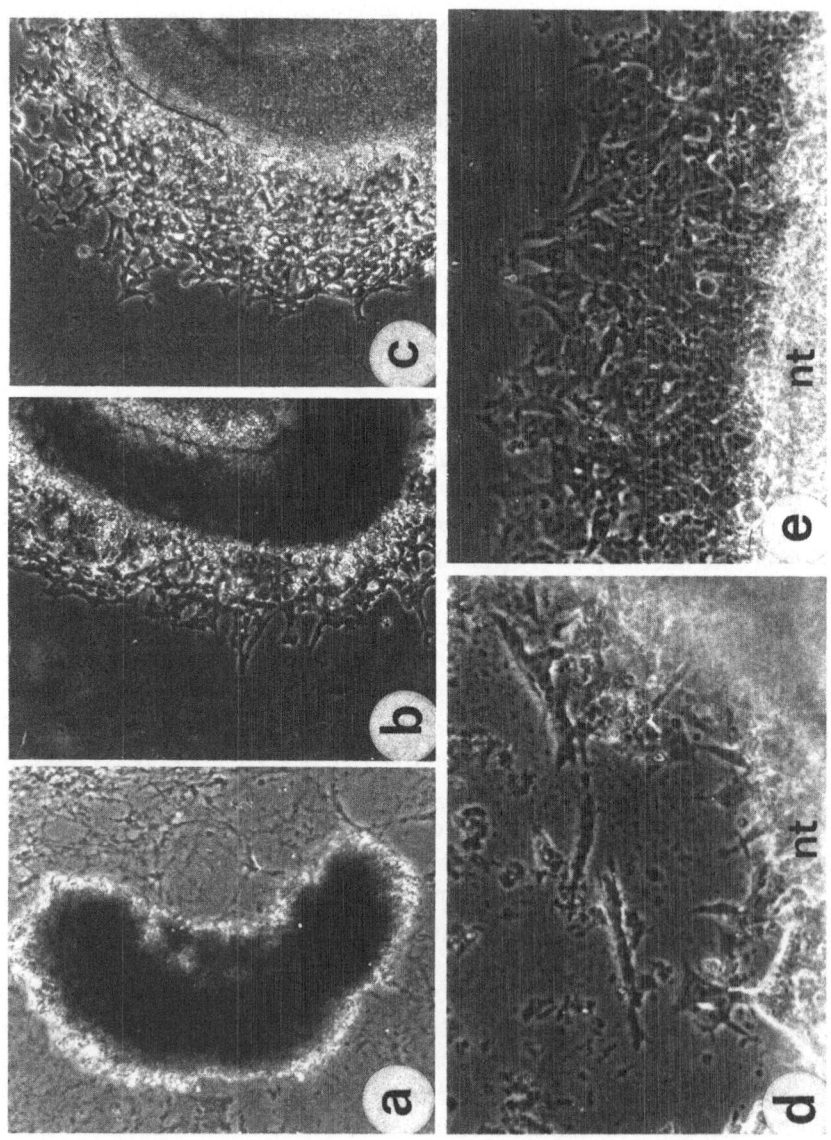

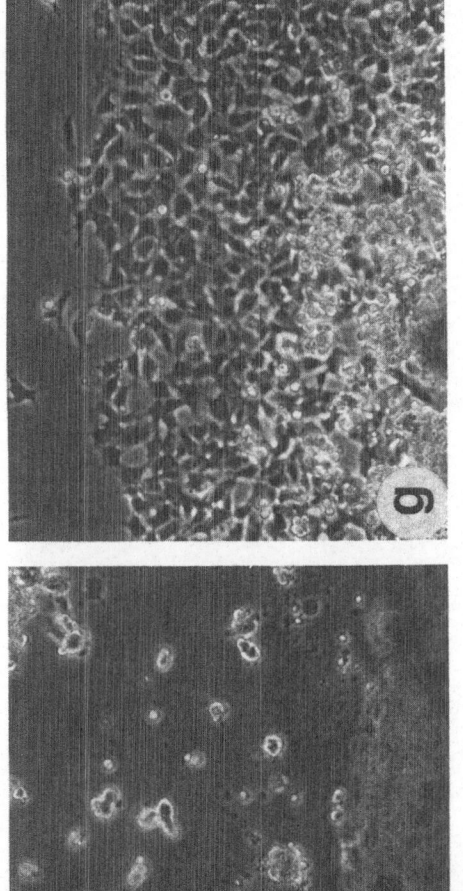

FIGURE 11. *In vitro* effect of the inhibition of cell-to-FN interaction on neural crest cell migration. The binding of the cell membrane to FN can be abolished by antibodies to the cell-binding site of FN (a–c), by RGDS-containing peptides (d,e), and by antibodies to the receptor for FN (f,g). In all cases, neural crest cell migration is significantly inhibited in contrast to control antibodies and peptides that have no effect (c,e,g). Note that the inhibitory effect can be reversed by removing the antibodies and peptides (b). (nt) Neural tube.

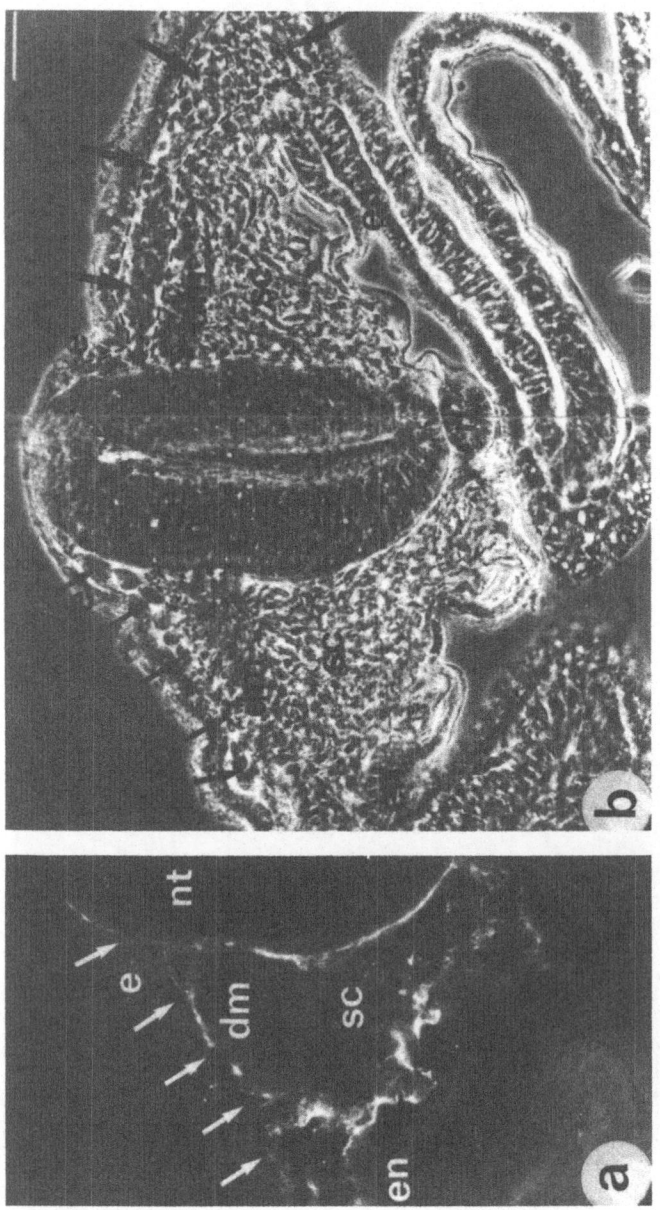

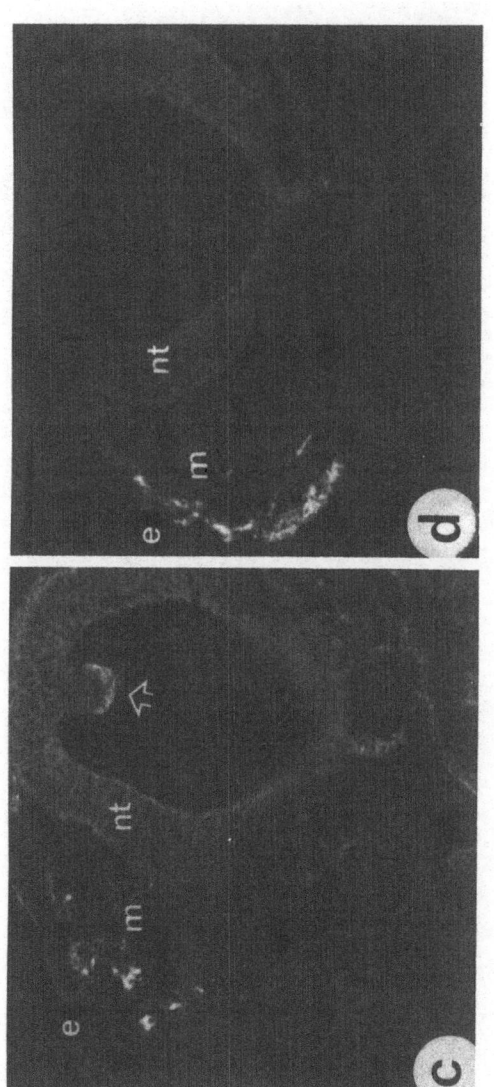

FIGURE 12. *In vivo* inhibition of neural crest migration by antibodies to the cell-binding site of FN and RGDS-containing peptides. (a,b) Monovalent anti-FN antibodies were microinjected between the first two somites in 10-somite-stage embryos prior to neural crest cell emigration. In the contralateral side, neural crest cells migrate between the ectoderm (e) and the dermomyotome (dm) as a very dense mass of confluent cells (limited by dotted lines). In the injected side, evidenced by a fluorescent antibody directed against the anti-FN antibodies (a), very few crest cells were found, and the front of migration was not as advanced as that in the control side. Arrows indicate crest cells. (c,d) RGDS-containing decapeptides injected in the mesencephalon of 7-somite embryos result in drastic inhibition of neural crest cell migration which, instead, remain as a bulk in the neural tube (nt) lumen. In contrast, control peptides have no effect on crest cell migration. (en) Endoderm; (m) mesenchyme; (n) notochord; (sc) sclerotome.

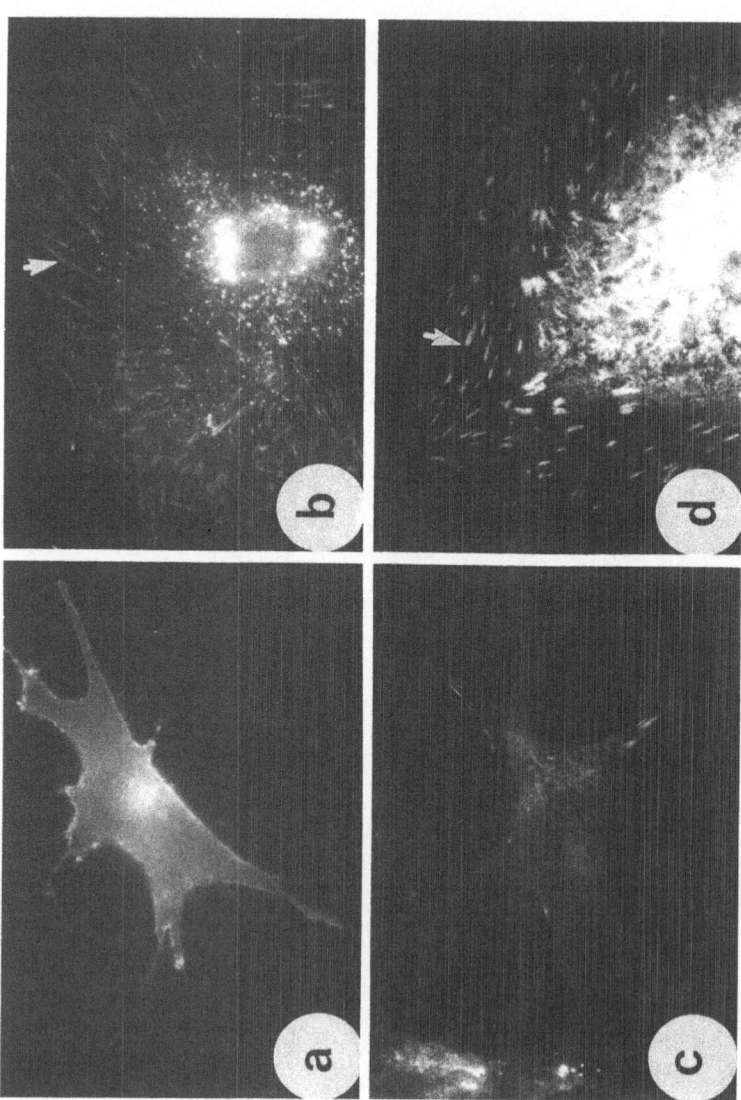

FIGURE 13. Comparison of some properties of migrating neural crest cells and stationary somitic cells. The 140-Kd FN receptor complex is diffuse on neural crest surface (a), and, in somitic fibroblasts (b), it is concentrated in regions close to focal contact sites (arrows in b and d) and stress fibers. As seen with vinculin staining, neural crest cells display very few focal contact sites (c), in contrast to somitic fibroblasts, which are firmly anchored to the substratum (d). Although stationary

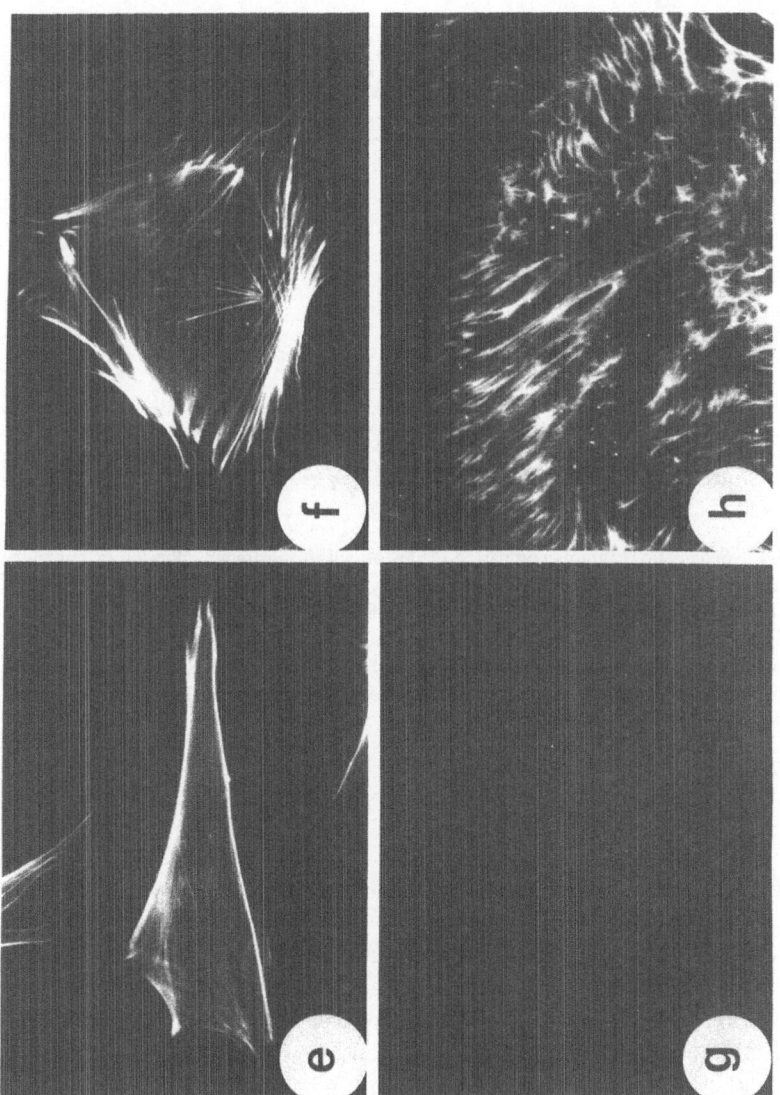

FIGURE 13. (*Continued*) cells exhibit a very organized actin microfilament meshwork (f), neural crest cells show only actin fibers in the cell cortex (e). Finally, in contrast to somitic cells (h), neural crest cells fail to deposit a FN meshwork on their substratum (g).

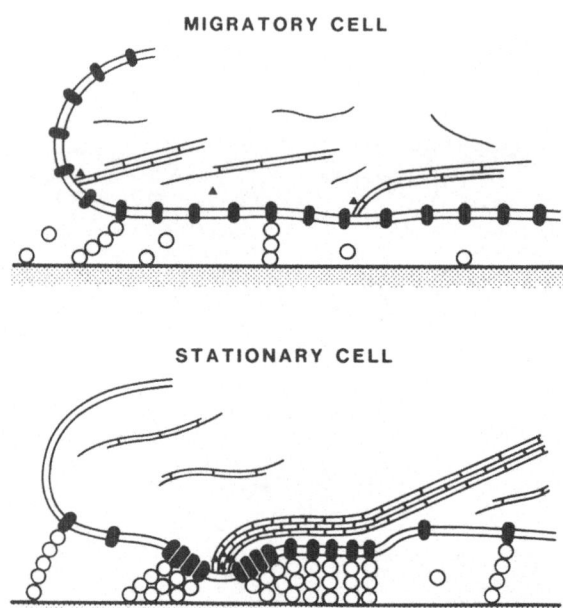

FIGURE 14. Schematic diagrams comparing modes of cell-to-substratum adhesion of migratory and stationary cells. Migratory cells are characterized by nonorganized microfilaments of actin and numerous FN receptors organized diffusely in the membrane. The receptor may be mobile, and a number of them not bound to FN. In contrast, in stationary cells, FN receptors are concentrated in clusters close to focal contact sites and microfilament bundles. They are linked both to the cytoskeleton and FN fibers. These receptors have poor membrane mobility, and most of them are bound to FN deposited by the cell.

to the substratum and confers upon the cell the ability to establish rapidly new contacts with the substratum (Duband *et al.*, 1986).

Isolated crest cells move very actively on FN, but their effective distance is very small in contrast to crest cells among a cell population that move in a precise direction (Newgreen *et al.*, 1979; Rovasio *et al.*, 1983; Erickson and Oliver, 1983). The unidirectional translocation of crest cells could thus result from a population pressure in limited pathways of migration and from contact inhibition of movement (Newgreen *et al.*, 1982; Rovasio *et al.*, 1983; Erickson, 1985). In some cases, however, crest cells behave quite differently. Migration within the gut is an example of a process in which several distinct mechanisms may be involved in the directed movement of enteric precursors. At first, crest cells remain in close contact with a delaminating epithelium; subsequently, they intermingle within the mesenchyme before regrouping into plexuses. The analysis of their movement is complicated by the intrinsic displacement of the gut during its closure and by the formation of a progressively more differentiated and complex environment (Tucker *et al.*, 1986).

3.2.4. Arrest of Crest Cell Migration

The final loss of motility of crest cells may result from a sudden modification of the cytoplasmic motile machinery, such as increased formation of microfilament bundles. Also, cells may interact more tightly with the ECM, particularly with FN, if FN receptors become organized in clusters or in strands. The expression of new adhesive properties involved in the aggregation of crest cells could be triggered by a chemical modification in the ECM. As crest cells migrate, their local environment is progressively transformed: Epithelia dissociate into mesenchymes, which expand and, in consequence, obstruct the pathways. In some cases, the ECM itself is modified. This latter alteration is particularly true in the area where the spinal ganglia form; chondroitin sulfate increases in amount; FN, hyaluronate, and type I collagens disappear; and, finally, LN appears among crest cells (Derby, 1978; Thiery *et al.*, 1982a; Duband and Thiery, 1987). *In vitro*, LN can induce crest aggregation. When crest cells are cultured for long periods, they develop a greater capacity to bind to LN than to FN (Rovasio *et al.*, 1983). *In vivo*, LN appears within the dorsal root ganglion rudiments. However, modifications of the environment are not solely responsible for the arrest of crest cell migration; modifications of the cell–cell adhesion properties of crest cells are also strongly implicated in the arrest of neural crest cells (Thiery *et al.*, 1982b; Duband *et al.*, 1985; Aoyama *et al.*, 1985).

3.3. Migration of Primordial Germ Cells

In the vertebrates, the primordial germ cells (PGCs) originate extragonadally and, in some species (e.g., in the avian), extraembryonally. Thus, PGCs have to reach the gonads by migration. PGCs of early amphibian embryos provide a useful system since they can be easily recognized *in vivo* and can be transplanted *in vitro* (see Wylie and Heasman, 1982). This migration occurs in two steps: first, through the endodermal mass, and then, through the mesenchyme of the dorsal mesentery in the area of the developing gonadal ridge. Whereas the first step of migration is achieved by a combination of active movements of the PGCs themselves and passive transport by the morphogenetic movements of the gut, the second phase of the migration has been clearly shown to be an active migration controlled by the ECM (Heasman and Wylie, 1981; Heasman et al., 1981; Wylie and Heasman, 1982; Heasman et al., 1985). We will thus focus only on this second step.

In vitro, isolated Xenopus PGCs, seeded onto cellular substrates of epithelial cells from mesenteries, can adhere to these and spread upon them. Under such conditions, PGCs become oriented along the substratum, send filopodia, and move easily (Heasman and Wylie, 1981; Wylie and Heasman, 1982). Like the neural crest cells, PGCs do not provide their own substrate-adhesive molecules, but rather use those deposited in the ECM by mesenteric cells (Heasman et al., 1981). Indeed, epithelial mesenteric cells synthesize large amounts of FN and lay it down in fibrils that align with intracellular microfilament bundles. PGCs that lack the ability to deposit FN are closely associated with FN. In addition, monovalent antibodies to FN strongly inhibit the adhesion and spreading of PGCs on the mesentery (Heasman et al., 1981). In vivo, the mesenteric cells are surrounded by FN and show a preferential orientation toward the gonadal ridge (Wylie and Heasman, 1982). These results suggest that, at least in amphibians, PGCs migrate on a substrate and use a contact guidance mechanism. Finally, it should be noted, that PGCs are incapable of penetrating the basement membrane of the intestine, which thus prevents their return into it (Heasman et al., 1985). It remains to be determined whether avian PGCs exhibit such behavior, since they are first transported in the blood and then have to cross the endothelial barrier to reach the gonadal ridge.

3.4. Elongation of Neurites

During normal development, neurons form processes that sometimes have to grow long distances to connect with their targets. Depending on the type of neurons, growing axons can be guided by a variety of cues. Diffusible factors, such as nerve growth factor, or the factors produced by glial and muscle cells have a chemotactic effect on growth cones in vitro (Gundersen and Barret, 1980; Ebendal, 1981; Gundersen and Park, 1984; Lumdsen and Davis, 1983; Barde et al., 1982). Other neurons or other cell types can be specifically recognized by growth cones and thus serve as "guidepost" cells. A striking example of such a mechanism is the insect nervous system; early in development, a set of pioneer neurons arise and project axons to the central nervous system. The trajectories of these axons become the routes of the major afferent nerves (Bate, 1976; Keshishian and Bentley, 1983; Taghert et al., 1982). In addition, the axons of the pioneer neurons are guided toward the central nervous system by a gradient of adhesion (haptotaxis) to the surface epithelium (Berlot and Goodman, 1984). In the vertebrate central nervous system, glial cells are thought to guide extending neurites. For example, in the cerebellar cortex, granule cells use Bergmann glial cells as a substratum (for a review, see Rakic, 1985), and this migration is perturbed when the interaction between neurites and glial cells is inhibited by antibodies to the cell-adhesion molecule L-1 (Lindner *et al.*, 1983).

ECM molecules have also been shown to serve as a substratum to neurites. Both FN and LN promote the *in vitro* attachment and elongation of growth cones of peripheral neurons (Rogers *et al.*, 1983, 1985). In contrast, FN does not support the migration of neurites or central neurons, but LN has been found to be an excellent substratum for various central neurons, including motoneurons and retinal neurons (Manthorpe *et al.*, 1983; Rogers *et al.*, 1983; Smalheiser *et al.*, 1984; Adler *et al.*, 1985). LN has been detected transiently in the central nervous system at sites where axon elongations are observed. For instance, LN is present along the inner limiting membrane of the retina, i.e., in the area where retinal axons grow (Adler *et al.*, 1985). In the cerebellum, LN is detected around external granule cells, Purkinje cells, and in punctate deposits along the radial Bergmann glial fibers. When the migration of external granule cells is achieved, LN disappears from these areas (Liesi, 1985). These data strongly suggest that LN plays a major role in neurite development, at least in the central nervous system, whereas FN could also serve as a substratum for peripheral neurons. Interestingly, LN has been found to be associated with the "neurite outgrowth-promoting factors" present in a variety of conditioned media from bovine corneal endothelium, chicken muscle, PC 12, and rat Schwannoma cells (Lander *et al.*, 1985; Davis *et al.*, 1985).

Like many other migrating embryonic cells, neurites may not provide their own substratum. This seems to be the role of the cells or glial cells in the central nervous system. As already mentioned, LN is present along Bergmann glial cells (Liesi, 1985). *In vitro*, Schwann cells and astrocytes synthesize LN (Cornbrooks *et al.*, 1983; Liesi *et al.*, 1983) and, *in vivo*, LN is intimately associated with these cells (Cornbrooks *et al.*, 1983; Liesi *et al.*, 1984).

The mechanisms by which elongating neurites use LN or FN as a substrate are not known. It seems that neuron attachment and growth cone extension are mediated by the heparin-binding domain of LN and FN (Edgar *et al.*, 1984; Rogers *et al.*, 1985). The role of the FN and LN receptors in these processes remains to be determined, but recent data suggest that the 140-Kd complex might be involved in neurite extension (Buck *et al.*, 1986).

The comparison of the migration of gastrulating cells, neural crest cells, PGCs, and neurite growth cones shows that active cell migration in early vertebrate embryogenesis depends on the presence of the cell–substrate adhesion molecules FN and LN. Migratory cells all exhibit a common and specific behavior; they do not provide their own substrate adhesive molecules and interact transiently with adhesion molecules, in contrast to stationary cells, which are firmly anchored to the matrix they deposit (see Figure 14).

4. INVOLVEMENT OF CELL–SUBSTRATUM ADHESION MOLECULES IN TISSUE REMODELING

During the first steps or organogenesis, tissues undergo successive transformations into epithelia and mesenchymes. Such modifications are associated with complete reorganizations of the ECM components and their interactions with cells. The prevalence and localization of cell–substrate adhesion molecules are consistent with a direct role in the conversion between the epithelial and mesenchymal states.

4.1. Dissociation of Epithelia

During gastrulation, the epiblast delaminates in the primitive streak region; after neurulation, neural crest cells detach from the neural tube, and the somites partially dissociate into sclerotome. The study of these various examples shows that they share the same mechanism of loss of the epithelial structure (an example is given in Figure 15). The disappearance of the basal lamina under disintegrating epithelia is one of the earliest events in the dislocation of epithelia; the disruption of the basal lamina occurs in defined areas of the epithelium (Duband and Thiery, 1982a,b; Thiery *et al.*, 1982a; Sanders, 1982; Mitrani, 1982). Concomitant with the local destruction of the basal lamina, intercellular junctions disappear among cells. For example, gap junctions are lost among the premigratory crest cells, as shown by the absence of electrical coupling (Revel and Brown, 1975), and the cell-adhesion molecule N-cadherin is no longer expressed (Hatta *et al.*, 1987). In addition, transmission and scanning electron microscopy reveal that cells are irregular in shape, do not show tight junctions, and are frequently separated by acellular spaces (Wakely and England, 1977; Solursh and Revel, 1978; Tosney, 1978, 1982; Newgreen and Gibbins, 1982). Progressively, the disruption of the basal lamina extends to the neighboring cells, and it then completely disappears (Newgreen and Gibbins, 1982; Tosney, 1978, 1982). Thereafter, cells send projections out of the epithelium and are progressively surrounded by FN (Thiery *et al.*, 1982a). The cells are then released from the epithelium and acquire a typically mesenchymal morphology; debris of the basal lamina is found among these cells (Duband and Thiery, 1987). At the end of the dissociation, LN has completely disappeared from the environment of the mesenchyme and has been replaced by FN. Depending on their environment, dissociating cells may either undergo migration (e.g., cells in the primitive streak and neural crest cells at the onset of their individualization) or remain as a mesenchyme close to their site of release (e.g., sclerotomal cells deriving from the somite). In an *in vitro* system, it has been possible to show that epithelial cells tend to lose their

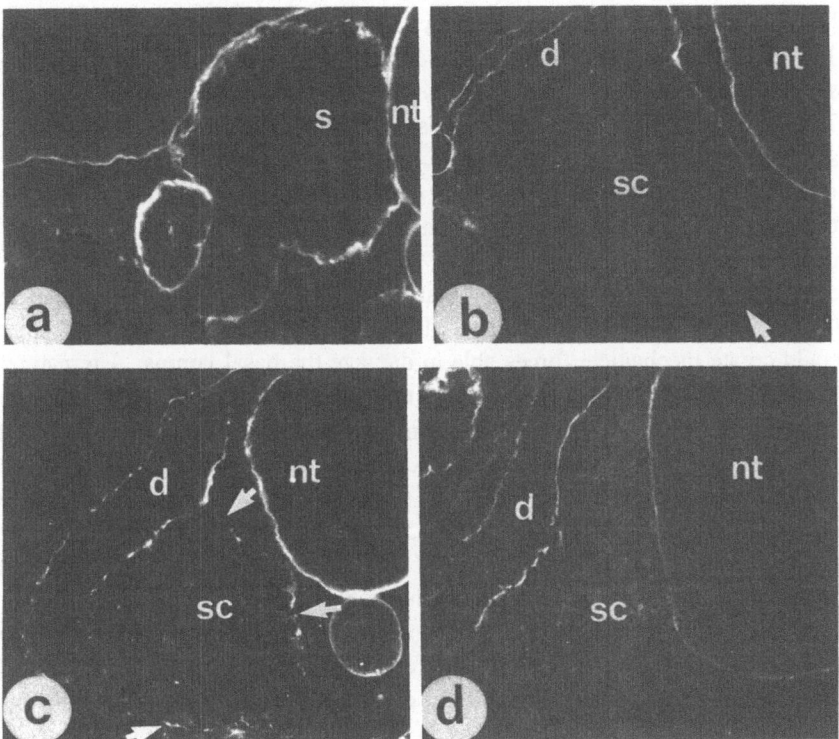

FIGURE 15. Dissociation of epithelia. Distribution of LN during the disruption of the somites (s) into dermomyotome (d) and sclerotome (sc). (a) Compacted somite prior to dissociation. LN is distributed in a thick basal lamina around the somite. (b) Early dissociation of the somite. The laterodorsal part of the somite retains a typical epithelial structure, and the medioventral part has lost its basal lamina (arrow). (c) Further dissociation of the somite. The ventral portion of the somite is completely dissociated but remains surrounded by LN strands, remainings of the disrupted basal lamina (arrows). (d) Postdissociation stage. The dermomyotome, which retains an epithelial organization, is completely surrounded by a basal lamina in contrast to the mesenchymal sclerotome, which is devoid of LN. (nt) Neural tube.

intercellular junctions and to emigrate when their basal surface is damaged (Banerjee *et al.*, 1977; Greenburg and Hay, 1982; Sugrue and Hay, 1981, 1982). These studies strongly suggest that the disappearance of the basal lamina is a prerequisite for the disruption of an epithelium.

However, the underlying mechanisms that trigger the disruption of the basal lamina are not known. A wide variety of factors may participate in this phenomenon. Local modifications of the ECM under an epithelium could induce its disruption, as suggested by experiments on amphibian neural crest cells (Löfberg et al., 1985). However, no direct evidence of such changes in the ECM has been found so far in the avian embryo (Tosney, 1978, 1982; Thiery et al., 1982a). Local pressure could result either from intense cell proliferation, as suggested for the epiblast and crest cells (Spratt and Haas, 1965; Duband and Thiery, 1982a,b), or from the expansion of the acellular space following the synthesis of hyaluronate by cells, as seems to be the case for somitic cells (Solursh et al., 1979). The release of cells from epithelia often occurs in regions of intense morphogenetic movements. This is particularly true for presumptive crest cells in the head; they are located in the neural fold at the boundary of both the ectoderm and neural tube, a region that is subject to folding during neurogenesis (Di Virgilio et al., 1967; Karfunkel, 1974; Jacobson, 1981). These movements could create mechanical forces able to damage the basal lamina. Alternatively, local proteolytic activity (plasmin and collagenases) could digest LN and type IV collagen of the basal lamina. Plasminogen is stored in large quantities in the yolk and can diffuse into the embryo before the blood circulation is established (Valinsky and Reich, 1981). Crest cells synthesize plasminogen activator at least in vitro (Valinsky and Le Douarin, 1985), and metastatic cells secrete collagenases to destroy basement membranes of epithelia (Liotta et al., 1982). Thus, the local destruction of the basal lamina of the ectoderm during trunk crest emigration could result from the action of proteases released by crest cells. Finally, Newgreen and Gibbins (1982) have shown that crest cells leave the neural tube only after they reach a proper stage. They proposed that the precise schedule of the onset of crest cell migration correlates with a decrease in the intercellular adhesion of crest cells. In addition, it has recently been found that both the emigration of crest cells from the neural tube and the dissociation of the somite are under the control of a calcium-dependent mechanism (Newgreen and Gooday, 1985; Gillespie et al., 1986). This suggests that the detachment of cells from an epithelium is under genetic control of cell-to-cell adhesion mediated by molecules such as L-CAM and N-cadherin (for reviews, see Edelman, 1985; Hatta et al., 1985, 1987). Another observation supporting this statement is that the disruption of the basal lamina always occurs in very precise areas of the epithelium and frequently remains confined to them. This is particularly true for the primitive streak in the epiblast and for the somites, part of which remain epithelial.

4.2. Formation of Epithelia

The observation of some examples of formation of epithelia, e.g., somites (Duband et al., 1987), kidney tubules (Ekblom et al., 1980; Ekblom, 1981), and peripheral ganglia (Thiery et al., 1982a; Duband et al., 1985), reveals a sequence of events common to all of these embryonic epithelia (an example is

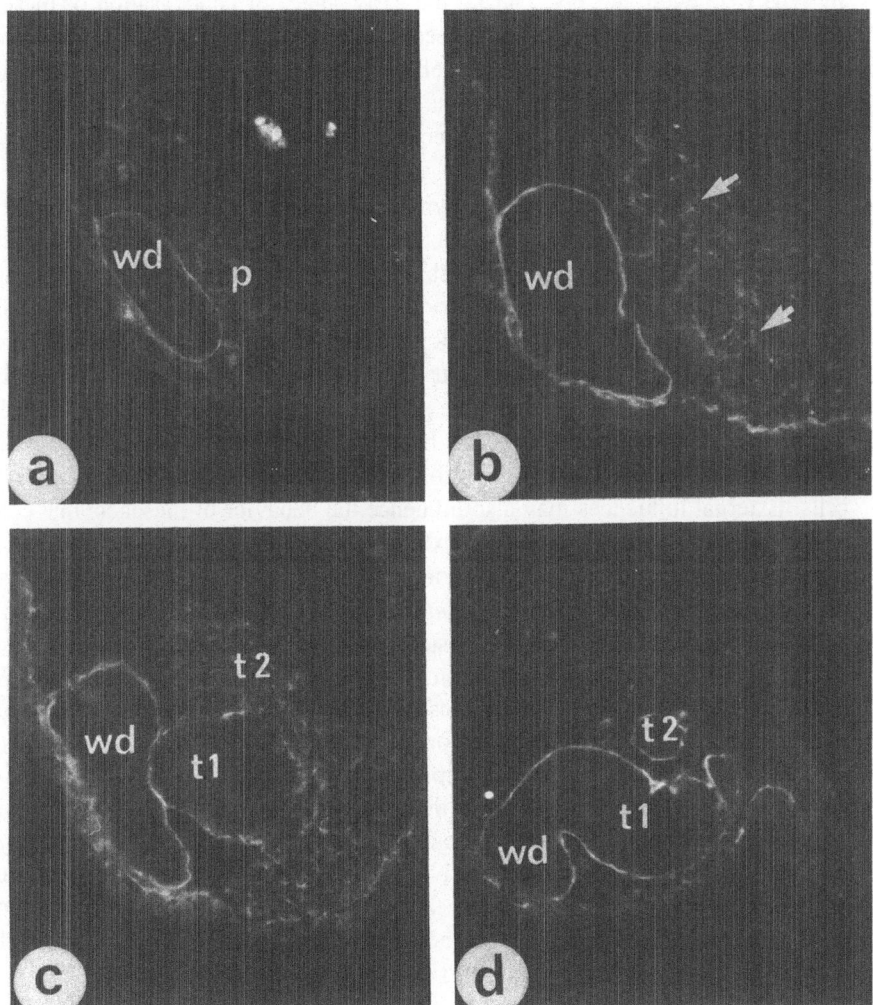

FIGURE 16. Formation of epithelia. Distribution of LN during the formation of the mesonephric tubules from the nephrogenous blastema. (a) Preaggregation stage showing the location of the mesonephric primordium (p). The Wolffian duct (wd), forming a solid cord, is limited by a basal lamina that stains brightly for LN. (b) Radial arrangement of the nephrogenous cells proximal to the Wolffian duct. LN appears as strands among the cells (arrows), but cells are not yet organized into an epithelium. (c) Primary mesonephric tubule formation. A basal lamina is appearing around the nephrogenic cells located close to the Wolffian duct. (d) Curving of the primary tubules (t1) into the typical S-shape tubules and formation of the secondary tubules (t2). At this stage, the primary and some of the secondary tubules are completely surrounded by LN.

given in Figure 16). Prior to their aggregation, mesenchymal cells are surrounded by type I collagen and FN. During the early stages of condensation of mesenchymal cells, the intercellular spaces become depleted of FN, and LN is deposited randomly within cell clusters. As cell aggregation is achieved, LN becomes codistributed with FN in a basal lamina underlying the newly formed epithelium. Thus, reorganizations of the ECM precisely follow the reassociation of the cells into an epithelium.

The factors that induce mesenchymal cells to aggregate are not known. The formation of epithelia may result from the loss of cell-free space among a rapidly dividing cell population. Such a mechanism is particularly suitable for the aggregation of neural crest cells into sensory ganglia (Thiery *et al.*, 1982a; Rovasio *et al.*, 1983). Indeed, spinal ganglia form in a pouch created by the neural tube, myotome, and sclerotome, which constitutes a barrier for migrating crest cells. The disappearance of both a substrate suitable for movement and available cell-free space, i.e., the loss of FN and hyaluronate (Derby, 1978; Thiery *et al.*, 1982a; Duband *et al.*, 1985), is possibly sufficient to provoke the arrest of crest cells. External influences may also influence the behavior of mesenchymal cells and force them to aggregate. For example, the formation of kidney tubules undoubtedly seems to require the action of an inductive tissue, i.e., the Wolffian duct (Grobstein, 1955; Wartiovaara *et al.*, 1974; Saxen and Lehtonen, 1978). A similar mechanism has been suggested to govern the formation of the somites (Nicolet, 1970; Packard and Jacobson, 1979) but is still controversial (Bellairs and Vieni, 1984). Finally, as suggested earlier for the somitic mesenchyme (Cheney and Lash, 1984), modulations of cell-to-cell adhesion are good candidates to explain the formation of epithelia. Segmentation of somites, for example, is related to the expression modulation of the calcium-dependent and calcium-independent adhesion molecules N-cadherin and N-CAM (Duband *et al.*, 1987). The general occurence of N-CAM and N-cadherin in the early phases of aggregation prompt us to suggest that epithelium formation mainly depends on the expression of these two cell-adhesion molecules (Thiery *et al.*, 1982b; Hatta *et al.*, 1987; Duband *et al.*, 1987; J.-L. Duband, S. Dufour, M. Takeichi, and J. P. Thiery, in preparation).

4.3. Remodeling of Epithelia

Once definitive epithelia are formed, they frequently undergo remodelings, such as lobulations and foldings. In this section, we will concentrate on the mechanism of lobulation of the mouse salivary epithelium, but the model described here can be applied to other lobulated epithelia, such as lung, mammary, and kidney epithelia. The salivary gland arises from a sheet of epithelial cells forming a bud that protrudes into the surrounding mesenchyme. The epithelium undergoes branching, resulting from the repetitive formation of clefts and intervening lobules. This morphogenesis is primarily due to changes in cell shape and localized differences in mitotic rates and is under the control of the surrounding mesenchyme (Grobstein, 1953).

In vitro cultures of isolated epithelia revealed that the basal lamina is required for maintaining the morphology of the epithelium in contrast to fibrous collagen located in the periphery (Bernfield and Banerjee, 1982). The surrounding mesenchyme does not appear to be involved in the synthesis and deposition of the basal lamina, but rather in its turnover (Banerjee *et al.*, 1977). In fact, glycosaminoglycans (GAGs) in the basal lamina undergo rapid dynamic changes. Newly synthesized GAGs are mainly deposited at the distal ends of the lobules, a region characterized by rapid changes in cell shape and intense cell proliferation (Bernfield and Banerjee, 1982). In contrast, the turnover of GAGs is slow in the clefts. This differential turnover rate is due to the action of the mesenchyme that is responsible for the degradation of the GAGs (Smith and Bernfield, 1982). This local degradation of GAGs could allow changes in cell shape, and, by virtue of the intercellular junctions, the population could invaginate. The different GAGs could also exert direct inhibitory or stimulatory effect on cell proliferation.

5. CONCLUDING REMARKS

Morphogenesis involves a limited repertoire of cellular behaviors; this review indicates that cell–substratum adhesion molecules and their receptors are intimately involved in these processes. It appears that different cell behaviors can be governed by the same ECM molecule, depending on the mode of interaction of the cell with this molecule. In addition, the comparison of the behavior of cell types of different embryonic origin and of different species reveals that these cells use very similar mechanisms of migration.

The major phases of embryonic development can be schematically reduced to a series of epithelium–mesenchyme interconversions—cells exhibiting either a mesenchymal or an epithelial morphology at each stage. Thus, a cell's behavior is governed by a continuous modulation of its adhesion to the substratum and to its neighboring cells. In this review, we have only considered the modulations of cell–substrate adhesion. It should be stressed that these modulations are intimately linked to modulations of cell–cell adhesion. A detailed study of the fate of cell–substrate adhesion molecules and cell–cell adhesion molecules during tissue remodeling shows a defined pattern and reveals periodicities in the prevalence of each of these components (for reviews, see Edelman and Thiery, 1985).

ACKNOWLEDGMENTS. The authors are particularly grateful to their colleages involved in the original work described in this paper. Research by the authors is supported by grants from INSERM (CRL 83-4017), CNRS (ATP 950 906), MRT (84-C1312), the Ligue Nationale Francaise contre le Cancer, the Foundation pour la Recherche Médicale, the Association pour la Recherche contre le Cancer (ARC 6455), and the Ministère de l'Education Nationale.

REFERENCES

Adler, R., Jerdan, J., and Hewitt, T. A., 1985, Responses of cultured neural retina cells to substratum-bound laminin and other extracellular matrix molecules, *Dev. Biol.* **112**:100–114.

Akiyama, S. K., and Yamada, K. M., 1985, The interaction of plasma fibronectin with fibroblastic cells in suspension, *J. Biol. Chem.* **260**:4492–4500.

Akiyama, S. K., Yamada, S. S., and Yamada, K. M., 1986, Characterization of a 140-kd avian cell surface antigen as a fibronectin-binding molecule, *J. Cell Biol.* **102**:442–448.

Ali, I. U., and Hynes, R. O., 1978, Effect of LETS glycoprotein on cell motility, *Cell* **14**:439–446.

Aoyama, H., Delouvée, A., and Thiery, J. P., 1985, Cell adhesion mechanisms in gangliogenesis studied in avian embryo and in a model system, *Cell Differ.* **17**:247–260.

Aplin, J. D., Hughes, R. C., Jaffe, C. L., and Sharon, N., 1981, Reversible cross-linking of cellular components of adherent fibroblasts to fibronectin and lectin-coated substrata, *Exp. Cell Res.* **134**:488–494.

Ballard, W. W., 1966, History of the hypoblast in *salmo, J. Exp. Zool.* **168**:211–220.

Ballard, W. W., 1981, Morphogenetic movements and fate maps of vertebrates, *Am. Zool.* **21**:391–399.

Banerjee, S. D., Cohn, R. H., and Bernfield, M. R., 1977, Basal lamina of embryonic salivary epithelia: Production by the epithelium and role in maintaining lobular morphology, *J. Cell Biol.* **73**:445–463.

Barde, Y. A., Edgar, D., and Thoenen, H., 1982, Purification of a new neurotrophic factor from mammalian brain, *EMBO J.* **1**:549–553.

Barlow, D. P., Green, N. M., Kurkinen, M., and Hogan, B. L. M., 1984, Sequencing of laminin B chain cDNAs reveals C-terminal regions of coiled-coil alpha helix, *EMBO J.* **3**:2355–2362.

Baron-Van Evercooren, A., Kleinman, H. K., Seppä, H., Rentier, B., and Dubois-Dalcq, M., 1982, Fibronectin promotes rat Schwann cell growth and motility, *J. Cell Biol.* **93**:211–216.

Bate, C. M., 1976, Pioneer neurons in an insect embryo, *Nature* **260**:54–56.

Bellairs, R., and Vieni, M., 1984, Experimental analysis of control mechanisms in somite segmentation in avian embryos II. Reduction of material in the gastrula stages of the chick, *J. Embryol. Exp. Morphol.* **79**:183–200.

Berlot, J., and Goodman, C., 1984, Guidance of peripheral pioneer neurons in the grasshopper: Adhesive hierarchy of epithelial and neuronal surfaces, *Science* **223**:493–496.

Bernfield, M., and Banerjee, S. D., 1982, The turnover of basal lamina glycosaminoglycan correlates with epithelial morphogenesis, *Dev. Biol.* **90**:291–305.

Boucaut, J. C., and Darribère, T., 1983, Fibronectin in early amphibian embryo, *Cell Tissue Res.* **234**:135–145.

Boucaut, J. C., Darribère, T., Boulekbache, H., and Thiery, J. P., 1984a, Antibodies to fibronectin prevent gastrulation but do not perturb neurulation in gastrulated amphibian embryos, *Nature* **307**:364–367.

Boucaut, J. C., Darribère, T., Poole, T. J., Aoyama, H., Yamada, K. M., and Thiery, J. P., 1984b,

Biological active synthetic peptides as probes of embryonic development: A competitive peptide inhibitor of fibronectin function inhibits gastrulation in amphibian embryos and neural crest cell migration in avian embryos, *J. Cell Biol.* **99:**1822–1830.

Boulekbache, H., Darribère, T., Joly, C., Boucaut, J. C., and Thiery, J. P., 1984, Immunolocalization of FN in fish embryo: Involvement in gastrulation and epiboly, *J. Embryol. Exp. Morphol.* **82:**25.

Bozyczko, D., and Horwitz, A. F., 1986, The participation of putative cell-surface receptor for laminin and fibronectin in peripheral neurite extension, *J. Neuroscience* **6:**1241–1251.

Brauer, P. R., Bolender, D. L., and Markwald, R. R., 1985, The distribution and spatial organization of the extracellular matrix encountered by mesencephalic neural crest cells, *Anat. Rec.* **211:**57–68.

Brown, P. J., and Juliano, R. L., 1985, Selective inhibition of fibronectin-mediated cell adhesion by monoclonal antibodies to a cell-surface glycoprotein, *Science* **228:**1448–1450.

Brown, S. S., Malinoff, H. L., and Wicha, M. S., 1983, Connectin: Cell surface protein that binds both laminin and actin, *Proc. Natl. Acad. Sci. USA* **80:**5927–5930.

Buck, C. A., Shea, E., Duggan, K., and Horwitz, A. F., 1986, Integrin (the CSAT antigen): Functionality requires oligomeric integrity, *J. Cell Biol.* **103:**2421–2428.

Carlson, R., Engvall, E., Freeman, A., and Rouslahti, E., 1981, Laminin and fibronectin in cell adhesion: Enhanced adhesion of cells from regenerating liver to laminin, *Proc. Natl. Acad. Sci. USA* **78:**2403–2406.

Chen, W. T., Hasegawa, E., Hasegawa, T., Weinstock, C., and Yamada, K. M., 1985a, Development of cell surface linkage complexes in cultured fibroblasts, *J. Cell Biol.* **100:**1103–1114.

Chen, W. T., Greve, J. M., Gottlieb, D. I., and Singer, S. J., 1985b, Immunocytological localization of 140 kd cell adhesion molecules in cultured chicken fibroblasts and in chicken smooth muscle and intestinal epithelial tissues, *J. Histochem. Cytochem.* **33:**576–586.

Cheney, C. M., and Lash, J. W., 1984, An increase in cell–cell adhesion in the chick segmental plate results in a meristic pattern, *J. Embryol. Exp. Morphol.* **79:**1–10.

Cochard, P., and Coltey, P., 1983, Cholinergic traits in the neural crest: Acetylcholinesterase in crest cells of the chick embryo, *Dev. Biol.* **98:**221–238.

Cooke, J., 1975, Local autonomy of gastrulation movements after dorsal lip removal in two anuran amphibians, *J. Embryol. Exp. Morphol.* **33:**147–157.

Cooper, A. R., and McQueen, A., 1983, Subunits of laminin are differentially synthesized in mouse eggs and early embryos, *Dev. Biol.* **96:**467–471.

Cornbrooks, C. J., Carey, D. J., McDonald, J. A., Timpl, R., and Bunge, R .P., 1983, *In vivo* and *in vitro* observations on laminin production by Schwann cells, *Proc. Natl. Acad. Sci. USA* **80:**3850–3854.

Couchman, J. R., and Rees, D. A., 1979, The behaviour of fibroblasts migrating from chick heart explants: Changes in adhesion, locomotion, and growth, and in the distribution of actomyosin and fibronectin, *J. Cell Sci.* **39:**149–165.

Couchman, J. R., Rees, D. A., Green, M. R., and Smith, C. G., 1982, Fibronectin has a dual role in locomotion and anchorage of primary chick fibroblasts and can promote entry into the division cycle, *J. Cell Biol.* **93:**402–410.

Couchman, J. R., Höök, M., Rees, D. A., and Timpl, R., 1983, Adhesion, growth, and matrix production by fibroblasts on laminin substrates, *J. Cell Biol.* **96:**177–183.

Critchley, D. R., England, M. A., Wakely, J., and Hynes, R. O., 1979, Distribution of fibronectin in the ectoderm of gastrulating chick embryo, *Nature* **280:**498–500.

Crossin, K. L., Hoffman, S., Grumet, M., Thiery, J. P., and Edelman, G. M., 1986, Site-restricted expression of cytotactin during development of the chicken embryo, *J. Cell Biol.* **102:**1917–1930.

Damsky, C. H., Richa, J., Solter, D., Knudsen, K., and Buck, C. A., 1983, Identification and purification of a cell surface glycoprotein mediating intercellular adhesion in embryonic and adult tissue, *Cell* **34:**455–466.

Damsky, C. H., Knudsen, K. A., Bradley, D., Buck, C. A., and Horwitz, A. F., 1985, Distribution of the cell substratum attachment (CSAT) antigen on myogenic and fibroblastic cells in culture, *J. Cell Biol.* **100**:1528–1539.

Darribère, T., Riou, J.-F., Shi, Di L., Delarue, M., and Boucat, J.-C., 1986, Synthesis and distribution of laminin-related polypeptides in early amphibian embryos, *Cell Tissue Res.* **246**:45–51.

Darribère, T., Boucher, D., Lacroix, J. C., and Boucaut, J.-C., 1984, Fibronectin synthesis during oogenesis and early development of the amphibian *Pleurodeles waltlii*, *Cell Differ.* **14**:171–177.

Darribère, T., Boulekbache, H., Shi, Di L., and Boucaut, J.-C., 1985, Immuno-electron-microscopic study of fibronectin in gastrulating amphibian embryos, *Cell Tissue Res.* **239**:75–80.

Davis, G. E., Manthorpe, M., Engvall, E., and Varon, S., 1985, Isolation and characterization of rat schwannoma neurite-promoting factor: Evidence that the factor contains laminin, *J. Neurosci.* **5**:2662–2671.

Derby, M. A., 1978, Analysis of glycosaminoglycans within the extracellular environments encountered by migrating neural crest cells, *Dev. Biol.* **66**:321–336.

Dessau, W., von der Mark, H., von der Mark, K., and Fisher, S., 1980, Changes in patterns of collagen and fibronectin during limb bud chondrogenesis, *J. Embryol. Exp. Morphol.* **57**:51–60.

Di Virgilio, G., Lavenda, N., and Worden, J. L., 1967, Sequence of events in neural tube closure and the formation of the neural crest in the chick embryo, *Acta Anat.* **68**:127–146.

Donaldson, D. J., and Mahan, J. T., 1984, Epidermal cell migration on laminin-coated substrates: Comparison with other extracellular matrix and non-matrx proteins, *Cell Tissue Res.* **235**:221–224.

Duband, J. L., and Thiery, J. P., 1982a, Appearance and distribution of fibronectin during chick embryo gastrulation and neurulation, *Dev. Biol.* **94**:337–350.

Duband, J. L., and Thiery, J. P., 1982b, Distribution of fibronectin in the early phase of avian cephalic neural crest cell migration, *Dev. Biol.* **93**:308–323.

Duband, J.-L., and Thiery, J. P., 1987, Distribution of caminin and collagens during avian neural crest development, *Development* (in press).

Duband, J. L., Tucker, G. C., Poole, T. J., Vincent, M., Aoyama, H., and Thiery, J. P., 1985, How do the migratory and adhesive properties of the neural crest govern ganglia formation in the avian peripheral nervous system? *J. Cell Biol.* **27**:189–203.

Duband, J. L., Rocher, S., Chen, W. T., Yamada, K. M., and Thiery, J. P., 1986, Cell adhesion and migration in the early vertebrate embryo: Location and possible role of the putative fibronectin receptor complex, *J. Cell Biol.* **102**:160–178.

Duband, J.-L., Dufour, S., Hatta, K., Takeichi, M., Edelman, G. M., and Thiery, J. P., 1987, Adhesion molecules during somitogenesis in the avian embryo, *J. Cell Biol.* **104**:1361–1374.

Dufour, S., Duband, J.-L., and Thiery, J. P., 1987, Role of a major cell-substratum adhesion in cell behavior and morphogenesis, *Biol. Cell* **58**:1–14.

Duprat, A. M., and Gualandris, L., 1984, Extracellular matrix and neural determination during amphibian gastrulation, *Cell Differ.* **14**:105–112.

Ebendal, T., 1981, Control of neurite extension by embryonic heart explants, *J. Embryol. Exp. Morphol.* **61**:289–301.

Edelman, G. M., 1985, Cell adhesion and the molecular processes of morphogenesis, *Annu. Rev. Biochem.* **54**:135–169.

Edelman, G. M., and Thiery, J. P. (eds.), 1985, *The Cell in Contact: Adhesions and Junctions as Morphogenetic Determinants*, John Wiley and Sons, New York.

Edgar, D., Timpl, R., and Thoenen, H., 1984, The heparin-binding domain of laminin is responsible for its effect on neurite outgrowth and neuronal survival, *EMBO J.* **3**:1463–1468.

Ekblom, P., 1981, Formation of basement membranes in the embryonic kidney: An immunohistological study, *J. Cell Biol.* **91**:1–10.

Ekblom, P., ALitalo, K., Vaheri, A., Timpl, R., and Saxen, L., 1980, Induction of a basement membrane glycoprotein in embryotic kidney: Possible role of laminin in morphogenesis, *Proc. Natl. Acad. Sci. USA* **77:**485–489.

Engel, J., Odermatt, E., Engel, A., Madri, J. A., Furthmayer, H., Rohde, H., and Timpl, R., 1981, Shapes, domain organizations, and flexibility of laminin and fibronectin, two multi-functional proteins of the extracellular matrix, *J. Mol. Biol.* **150:**97–120.

Erickson, C. A., 1985, Control of neural crest cell dispersion in the trunk of the avian embryo, *Dev. Biol.* **111:**138–157.

Erickson, C. A., and Olivier, K. R., 1983, Negative chemotaxis does not control quail neural crest cell dispersion, *Dev. Biol.* **96:**542–551.

Erickson, C.A., Tosney, K. W., and Weston, J A., 1980, Analysis of migratory behavior of neural crest and fibroblastic cells in embryonic tissues, *Dev. Biol.* **77:**142–156.

Fairbairn, S., Gilbert, R., Ojakian, G., Schwimmer, R., and Quigley, J. P., 1985, The extracellular matrix of normal chick embryo fibroblasts: Its effect on transformed chick fibroblasts and its proteolytic degradation by the transformants, *J. Cell Biol.* **101:**1790–1798.

Furcht, L. T., 1983, Structure and function of the adhesive glycoprotein fibronectin, *Mod. Cell Biol.* **1:**53–117.

Furcht, L. T., Smith, D., Wendelschafer-Crabb, G., Woodbridge, P. A., and Foidart, J. P., 1980, Fibronectin presence in native collagen fibrils of human fibroblasts: Immunoperoxidase and immunoferritin localization, *J. Histochem. Cytochem.* **28:**1319–1333.

Gardner, J. M., and Hynes, R. O., 1985, Interaction of fibronectin with its receptor on platelets, *Cell* **42:**439–448.

Gillespie, L. L., Armstrong, J. B., and Steinberg, M. S., 1985, Experimental evidence for a proteinacceous presomitic wave required for morphogenesis of axolotl mesoderm, *Dev. Biol.* **107:**220–226.

Gospodarowicz, D., Cohen, D. C., Fujii, D. K., 1982, Regulation of cell growth by the basal lamina and plasma factors: Relevance to embryonic control of cell proliferation and differentiation, *Cold Spring Harbor Conf. Cell Prolif.* **9:**95–124.

Greenberg, J. H., and Pratt, R. M., 1977, Glycosaminoglycan and glycoprotein synthesis by cranial neural crest cells in vitro, *Cell Differ.* **6:**119–132.

Greenberg, J. H., Seppä, A., Seppä, H., and Hewitt, T. A., 1981, Role of collagen and fibronectin in neural crest cell adhesion and migration, *Dev. Biol.* **87:**259–266.

Greenburg, G., and Hay, E. D., 1982, Epithelia suspended in collagen gels can lose polarity and express characteristics of migrating mesenchymal cells, *J. Cell Biol.* **95:**333–339.

Greve, J. M., and Gottlieb, D. I., 1982, Monoclonal antibodies which alter the morphology of cultured chick myogenic cells, *J. Cell Biol.* **18:**221–229.

Grobstein, C., 1953, Epithelial–mesenchyme specificity in the morphogenesis of the mouse sub-mandibular rudiments *in vitro*, *J. Exp. Zool.* **124:**383–404.

Grobstein, C., 1955, Inductive interaction in the development of the mouse metanephros, *J. Exp. Zool.* **130:**319–340.

Gundersen, R. W., and Barrett, J. N., 1980, Characterization of the turning response of dorsal root neurites towards nerve growth factor, *J. Cell Biol.* **87:**546–665.

Gundersen, R. W., and Park, K. H. C., 1984, The effect of conditioned media on spinal neurites: Substrate-associated changes in neurite direction and adherence, *Dev. Biol.* **104:**18–27.

Harrisson, F., Vanroelen, C., Foidart, J. M., and Vakaet, L., 1984, Expression of different regional patterns of fibronectin immunoreactivity during mesoblast formation in the chick blastoderm, *Dev. Biol.* **101:**373–381.

Hatta, K., Okada, T. S., and Takeichi, M., 1985, A monoclonal antibody disrupting calcium-dependent cell–cell adhesion of brain tissues: Possible role of its target antigen in animal pattern formation, *Proc. Natl. Acad. Sci. USA* **82:**2789–2793.

Hatta, K., Tadagi, S., Fijisawa, H., and Tadeichi, M., 1987, Spatial and temporal expression pattern of N-cadherin cell adhesion molecules correlated with morphogenetic processes of chicken embryos, *Dev. Biol.* **120:**215–227.

Heasman, J., and Wylie, C. C., 1981, Contact relations and guidance of primordial germ cells on their migratory route in embryos of *Xenopus laevis in vitro, Proc. R. Soc. London Ser. B* **213:**41–58.

Heasman, J., Hynes, R. O., Swan, A. P., Thomas, V. A., and Wylie, C. C., 1981, Primordial germ cells of *Xenopus* embryos: The role of fibronectin in their adhesion during migration, *Cell* **27:**437–447.

Heasman, J., Wylie, C. C., and Holwill, S., 1985, The importance of basement membranes in the migration of primordial germ cells in *Xenopus laevis, Dev. Biol.* **112:**18–29.

Hirano, H., Yamada, Y., Sullivan, M., De Crombrugghe, B., Pastan, I., and Yamada, K. M., 1983, Isolation of genomic DNA clones spanning the entire fibronectin gene, *Proc. Natl. Acad. Sci. USA* **80:**46–50.

Hirst, R., Horwitz, A. F., Buck, C. A., and Rohrschneider, L., 1986, Phosphorylation of the fibronectin receptor complex in cells transformed by oncogenes that encode tyrosine kinases, *Proc. Natl. Acad. Sci. USA* **83:**6470–6474.

Horwitz, A., Duggan, K., Greggs, R., Decker, C., and Buck, C., 1985, The CSAT antigen has properties of a receptor for laminin and fibronectin, *J. Cell Biol.* **101:**2134–2144.

Horwitz, A. F., Duggan, K., Buck, C. A., Beckerle, M., and Burridge, K., 1986, Interaction of a plasma membrane fibronectin receptor with talin: a transmembrane linkage, *Nature* **320:**531–533.

Hynes, R. O., and Yamada, K. M., 1982, Fibronectins: Multifunctional modular glycoproteins, *J. Cell Biol.* **95:**369–377.

Jacobson, A. G., 1981, Morphogenesis of the neural plate and tube, in: *Morphogenesis and Pattern Formation* (T. G. Connelly, ed.), Raven Press, New York, pp. 233–263.

Karfunkel, P., 1974, The mechanism of neural tube formation, *Int. Rev. Cytol.* **38:**245–271.

Keller, R. E., 1975, Vital dye maping of the gastrula and neurula of *Xenopus laevis.* I. Prospective areas and morphogenetic movements of the superficial layer, *Dev. Biol.* **42:**222–241.

Keller, R. E., 1978, Time-lapse cinematographic analysis of superficial cell behaviour during and prior to gastrulation in *Xenopus laevis, J. Morphol.* **157:**223–248.

Keller, R. E., 1980, The cellular basis of epiboly: An SEM study of deep-cell rearrangement during gastrulation in *Xenopus laevis, J. Embryol. Exp. Morphol.* **60:**201–234.

Keller, R. E., and Schoenwolf, G. C., 1977, An SEM study of deep-cell rearrangement during gastrulation in *Xenopus laevis, Roux's Arch. Dev. Biol.* **182:**165–186.

Keshishian, H., and Bentley, D., 1983, Embryogenesis of peripheral nerve pathways in grasshopper leg, *Dev. Biol.* **96:**89–124.

Kleinman, H. K., Martin, G. R., and Fishman, P. H., 1979, Ganglioside inhibition of fibronectin mediated cell adhesion to collagen, *Proc. Natl. Acad. Sci. USA* **76:**3367–3371.

Knudsen, K. A., Horwitz, A. F., and Buck, C. A., 1985, A monoclonal antibody identifies a glycoprotein complex involved in cell–substratum adhesion, *Exp. Cell Res.* **157:**218–226.

Koliga, J., Shure, F. M., Chen, W. T., and Young, N. D., 1982, Rapid cellular translocation is related to close contact formed between various cultured cells and their substratum, *J. Cell Sci.* **44:**23–34.

Kornblihtt, A. R., Vibe-Pedersen, K., and Baralle, F. E., 1983, Isolation and characterization of cDNA clones for human and bovine fibronectins, *Proc. Natl. Acad. Sci. USA* **80:**3218–3223.

Kornblihtt, A. R., Vibe-Pedersen, K., and Baralle, F. E., 1984a, Human fibronectin: Molecular cloning evidence for two MRNA species differing by an internal segment coding for a structural domain, *EMBO J.* **3:**221–226.

Kornblihtt, A. R., Vibe-Pedersen, K., and Baralle, F. E., 1984b, Human fibronectin: Cell specific alternative mRNA splicing generates polypeptide chains differing in the number of internal repeats, *Nucl. Acid Res.* **12**:5853–5868.

Kornblihtt, A. P., Umezawa, K., Vibe-Pedersen, K., and Baralle, F. E., 1985, Primary structure of human fibronectin: Differential splicing may generate at least 10 polypeptides from a single gene, *EMBO J.* **4**:1755–1759.

Lander, A. D., Fujii, D. K., and Reichardt, L. F., 1985, Laminin is associated with the "neurite outgrowth-promoting factors" found in conditioned media, *Proc. Natl. Acad. Sci. USA* **82**:2183–2187.

Laterra, J., Siebert, J. E., and Culp, L. A., 1983, Cell surface heparan sulfate mediate some adhesive responses to glycosaminoglycan-binding matrices, including fibronectin, *J. Cell Biol.* **95**:340–344.

Le Douarin, N. M., 1982, *The Neural Crest*, Cambridge University Press, Cambridge.

Lee, G., Hynes, R. O., and Kirschner, M., 1984, Temporal and spatial regulation of fibronectin in early *Xenopus* development, *Cell* **36**:729–740.

Leivo, I., Vaheri, A., Timpl, R., and Wartiovaara, J., 1980, Appearance and distribution of collagens and laminin in the early mouse embryo, *Dev. Biol.* **76**:100–114.

Lesot, H., Kühl, U., and Von der Mark, K., 1983, Isolation of a laminin binding protein from muscle cell membranes, *EMBO J.* **2**:861–865.

Liesi, P., 1985, Do neurons in the vertebrate CNS migrate on laminin? *EMBO J.* **5**:1163–1170.

Liesi, P., Dahl, D., and Vaheri, A., 1983, Laminin is produced by early rat astrocytes in primary culture, *J. Cell Biol.* **96**:920–924.

Liesi, P., Kaakkola, S., Dahl, D., and Vaheri, A., 1984, Laminin is induced in astrocytes of adult brain by injury, *EMBO J.* **3**:683–686.

Lindner, J., Rathjen, F. G., and Schachner, M., 1983, L1 mono- and polyclonal antibodies modify cell migration in early postnatal mouse cerebellum, *Nature* **305**:427–430.

Liotta, L. A., Thorgeirsson, U. P., and Garbisa, S., 1982, Role of collagenases in tumor cell invasion, *Cancer Metastasis Rev.* **1**:277–288.

Löfberg, J., Ahlfors, K., and Fällstrom, C., 1980, Neural crest cell migration in relation to extracellular matrix organization in the embryonic axolotl trunk, *Dev. Biol.* **75**:148–167.

Löfberg, J., McCoy, A. N., Olsson, C., Jönsson, L., and Perris, R., 1985, Stimulation of initial neural crest cell migration in the axolotl embryo by tissue grafts and extracellular matrix transplanted on microcarriers, *Dev. Biol.* **107**:442–459.

Lumdsen, A. G. S., and Davis, A. M., 1983, Earliest sensory nerve fibers are guided to peripheral targets by attractants other than nerve growth factor, *Nature* **306**:786–788.

Maciag, T., Kadish, J., Wilkins, L., Stemerman, M. B., and Weinstein, R., 1982, Organizational behavior of human umbilical vein endothelial cells, *J. Cell Biol.* **94**:511–520.

Malinoff, H. L., and Wicha, M. S., 1983, Isolation of a cell surface receptor protein for laminin from murine fibrosarcoma cells, *J. Cell Biol.* **96**:1475–1479.

Manthorpe, M., Engvall, E., Ruoslahti, E., Longo, F. R., Davis, G. E., and Varon, S., 1983, Laminin promotes neuritic regeneration from cultured peripheral and central neurons, *J. Cell Biol.* **97**:1882–1890.

McDonald, J. A., Kelley, D. G., and Broekelmann, T. J., 1982, Role of fibronectin in collagen deposition: Fab' to the gelatin-binding domain of fibronectin inhibits both fibronectin and collagen organization in fibroblast extracellular matrix, *J. Cell Biol.* **92**:485–492.

McKeown-Longo, P. J., and Mosher, D. F., 1985, Interactions of the 70,000-mol-wt amino-terminal fragment of fibronectin with the matrix-assembly receptor of fibroblasts, *J. Cell Biol.* **100**:364–374.

Mitrani, E., 1982, Primitive streak-forming cells of the chick invaginate through a basement membrane, *Roux's Arch. Dev. Biol.* **191**:320–324.

Mitrani, E., and Farberov, A., 1982, Fibronectin expression during the process leading to axis formation in the chick embryo, *Dev. Biol.* **91**:197–201.

Nagata, K., Humphries, M. J., Olden, K., and Yamada, K. M., 1985, Collagen can modulate cell interactions with fibronectin, *J. Cell Biol.* **101**:386–394.

Nakatsutji, N., 1975, Studies on the gastrulation of amphibian embryos: Light and electron microscopic observation of an urodele *Cynops pyrrhogaster*, *J. Embryol. Exp. Morphol.* **34**:669–685.

Nakasutji, N., Gould, A. C., and Johnson, K. E., 1982, Movement and guidance of migrating mesodermal cells in *Ambystoma maculatum* gastrulae, *J. Cell Sci.* **56**:207–222.

Nakasutji, N., Smolira, M. A., and Wylie, C. C., 1985a, Fibronectin visualized by scanning electron microscopy immunocytochemistry on the substratum for cell migration in *Xenopus laevis* gastrulae, *Dev. Biol.* **107**:264–268.

Nakasutji, N., Hashimoto, K., and Hayashi, M., 1985b, Laminin fibrils in newt gastrulae visualized by the immunofluorescent staining, *Dev. Growth Differ.* **27**:639–643.

Newgreen, D. F., and Gibbins, I. L., 1982, Factors controlling the time of onset of the migration of neural cells in the fowl embryo, *Cell Tissue Res.* **224**:145–160.

Newgreen, D. F., and Gooday, D., 1985, Control of the onset of migration of neural crest cells in avian embryos: Role of Ca^{++}-dependent cell adhesions, *Cell Tissue Res.* **239**:329–336.

Newgreen, D., and Thiery, J.P., 1980, Fibronectin in early avian embryos: Synthesis and distribution along the migration pathways of neural crest cells, *Cell Tissue Res.* **211**:269–291.

Newgreen, D. F., Ritterman, M., and Peters, E. A., 1979, Morphology and behaviour of neural crest cells of chick embryo in vitro, *Cell Tissue Res.* **203**:115–140.

Newgreen, D. F., Gibbins, I. L., Sauter, J., Wallenfels, B., and Wütz, R., 1982, Ultrastructural and tissue-culture studies on the role of fibronectin, collagen, and glycosaminoglycans in the migration of neural crest cells in the fowl embryo, *Cell Tissue Res.* **221**:521–549.

Nicolet, G., 1970, Is the presumptive notochord responsible for somite genesis in the chick? *J. Embryol. Exp. Morphol.* **24**:467–478.

Noden, D. M., 1975, An analysis of the migratory behavior of avian cephalic neural crest cells, *Dev. Biol.* **42**:106–130.

Orly, J., and Sato, G., 1979, Fibronectin mediates cytokinesis and growth of rat follicular cells in serum-free medium, *Cell* **17**:295–305.

Ott, U., Odermatt, E., Engel, J., Furthmayer, H., and Timpl, R., 1982, Protease resistance and conformation of laminin, *Eur. J. Biochem.* **123**:63–72.

Packard, D. S., and Jacobson, A. G., 1979, Analysis of the physical forces that influence the shape of chick somites, *J. Exp. Zool.* **207**:81–92.

Palotie, A., Peltonen, L., Ristelli, L., and Risteli, J., 1983, Effect of the structural components of basement membranes on the attachment of teratocarcinoma-derived cells, *Exp. Cell Res.* **144**:31–37.

Pasteels, J., 1936, Etudes sur la gastrulation des vertébrés méroblastiques. I. Téléostéens, *Arch Biol.* **47**:205–308.

Pennypacker, J. P., Hassell, J. R., Yamada, K. M., and Pratt, R. M., 1979, The influence of an adhesive cell surface protein on chondrogenic expression *in vitro*, *Exp. Cell Res.* **121**:411–415.

Petersen, T. E., Thogersen, H. C., Skortengaard, K., Vibe-Pedersen, K., Sahl, P., Sottrup-Jensen, L., and Magnusson, S., 1983, Partial primary structure of bovine plasma fibronectin: Three types of internal homology, *Proc. Natl. Acad. Sci. USA* **80**:137–141.

Pierschbacher, M. D., and Ruoslahti, E., 1984a, Cell attachment activity of fibronectin can be duplicated by small synthetic fragments of the molecule, *Nature* **309**:30–33.

Pierschbacher, M. D., and Ruoslahti, E., 1984b, Variants of the cell recognition site of fibronectin that retain attachment-promoting activity, *Proc. Natl. Acad. Sci. USA* **81**:5985–5988.

Pierschbacher, M. D., Hayman, E. G., and Ruoslahti, E., 1985, The cell attachment determinant in fibronectin, *J. Cell Biochem.* **28**:115–126.

Pintar, J. E., 1978, Distribution and analysis of glycosaminoglycans during quail neural crest morphogenesis, *Dev. Biol.* **67**:444–464.

Podleski, T. R., Greenberg, I., Schlessinger, J., and Yamada, K. M., 1979, Fibronectin delays the fusion of L6 myoblasts, *Exp. Cell Res.* **123**:104–126.

Poole, T. J., and Thiery, J. P., 1987, Antibodies and synthetic peptides that block cell-fibronectin adhesion arrest neural crest cell migration in vivo, in: *New Discoveries and Technologies in Developmental Biology* (H. C. Slavkin, ed.), Alan R. Liss, New York (in press).

Pratt, R. M., Larsen, M. A., and Johnston, M. C., 1975, Migration of cranial neural crest cells in a cell-free hyaluronate-rich matrix, *Dev. Biol.* **44**:298–305.

Pytela, R., Pierschbacher, M. D., and Ruoslahti, E., 1985, Identification and isolation of a 140 kd cell surface glycoprotein with properties expected for a fibronectin receptor, *Cell* **40**:191–198.

Rakic, P., 1985, Mechanisms of neuronal migration in developing cerebellar cortex, in: *Molecular Basis of Neural Development* (G. M. Edelman, W. E. Gall, and W. M. Cowan, eds.), John Wiley and Sons, New York, pp. 139–160.

Rao, N. C., Margulies, I. M. L., Goldfarb, R. H., Madri, J. A., Woodley, D. R., and Liotta, L. A., 1982, Differential proteolytic susceptibility of laminin alpha and beta subunits, *Arch. Biochem. Biophys.* **219**:65–70.

Rao, N. C., Barsky, S. H., Terranova, V. P., and Liotta, L. A., 1983, Isolation of a tumor cell laminin receptor, *Biochem. Biophys. Res. Comm.* **111**:804–808.

Revel, J. P., and Brown, S. S., 1975, Cell junctions in development with particular reference to the neural tube, *Cold Spring Harbor Symp. Quant. Biol.* **40**:433–455.

Rogers, S. L., Letourneau, P. C., Palm, S. L., McCarthy, J., and Furcht, L. T., 1983, Neurite extension by peripheral and central nervous system neurons in response to substratum-bound fibronectin and laminin. *Dev. Biol.* **98**:212–220.

Rogers, S. L., McCarthy, J. B., Palm, S. L., Furcht, L. T., and Letourneau, P. C., 1985, Neuron-specific interactions with two neurite-promoting fragments of fibronectin, *J. Neurosci.* **5**:369–378.

Rojkind, M., Gatmaitan, Z., Mackensen, S., Giambrione, M. A., Ponce, P., and Reid, L. M., 1980, Connective tissue biomatrix: Its isolation and utilization for long-term cultures of normal rate hepatocytes, *J. Cell Biol.* **87**:255–263.

Rovasio, R. A., Delouvée, A., Yamada, K. M., Timpl, R., and Thiery, J. P., 1983, Neural crest cell migration: Requirement for exogenous fibronectin and high cell density, *J. Cell Biol.* **96**:462–473.

Ruoslahti, E., and Vaheri, A., 1975, Interaction of soluble fibroblast antigen with fibrinogen and fibrin. Identity with cold insoluble globulin of human plasma, *J. Exp. Med.* **141**:497–501.

Ruoslahti, E., Engwall, E., and Hayman, E. G., 1981, Fibronectin: Current concepts of its structure and functions, *Coll. Rel. Res.* **1**:95–128.

Sanders, E. J., 1979, Development of the basal lamina and extracellular material in the chick embryo, *Cell Tissue Res.* **198**:527–537.

Sanders, E. J., 1982, Ultrastructural immunocytochemical localization of fibronectin in the early chick embryo, *J. Embryol. Exp. Morphol.* **71**:155–170.

Sanders, E. J., 1984, Labelling of basement membrane constituents in the living chick embryo during gastrulation, *J. Embryol. Exp. Morphol.* **79**:113–123.

Saxèn, L., and Lehtonen, E., 1978, Transfilter induction of kidney tubules as a function of the extent and duration of intercellular contacts, *J. Embryol. Exp. Morphol.* **47**:97–109.

Schwarzbauer, J. E., Tamkun, J. W., Lemischka, I. R., and Hynes, R. O., 1983, Three different fibronectin mRNAs arise by alternative splicing within the coding region, *Cell* **35**:421–431.

Sieber-Blum, M., Sieber, F., and Yamada, K. M., 1981, Cellular fibronectin promotes adrenergic differentiation of quail neural crest cells *in vitro*, *Exp. Cell Res.* **133**:285–295.

Smalheiser, N. R., Crain, S. M., and Reid, L. M., 1984, Laminin as a substrate for retinal axons *in vitro*, *Dev. Brain Res.* **12**:136–140.

Smith, R. L., and Bernfield, M., 1982, Mesenchyme cells degrade epithelial basal lamina glycos-aminoglycan, *Dev. Biol.* **94:**378–390.

Solursh, M., and Revel, J. P., 1978, A scanning electron microscope study of the cell shape and cell appendages in the primitive streak region of the rate and chick embryo, *Differentiation* **11:**185–190.

Solursh, M., Fischer, M., Meier, S., and Singley, C. T., 1979, The role of extracellular matrix in the formation of the sclerotome, *J. Embryol. Exp. Morphol.* **54:**75–98.

Spiegel, S., Yamada, K. M., Hom, B. E., Moss, J., and Fishman, P. H., 1985, Fluorescent gangliosides as probes for the retention and organization of fibronectin by ganglioside-deficient mouse cells, *J. Cell Biol.* **100:**721–726.

Spratt, N. T., and Haas, M., 1965, Germ layer formation and the role of the primitive streak in the chick, *J. Exp. Zool.* **158:**9–38.

Sugrue, S. P., and Hay, E. D., 1981, Response of basal epithelial cell surface and cytoskeleton to solubilized extracellular matrix molecules, *J. Cell Biol.* **91:**45–54.

Sugrue, S. P., and Hay, E. D., 1982, Interaction of embryonic corneal epithelium with exogenous collagen, laminin, and fibronectin: Role of endogenous protein synthesis, *Dev. Biol.* **92:**97–106.

Taghert, P. H., Bastiani, M. J., Ho, R. K., and Goodman, C. S., 1982, Guidance of pioneer growth cones: Filopodial contacts and coupling revealed with an antibody to lucifer yellow, *Dev. Biol.* **94:**391–399.

Terranova, V. P., Rohrbach, D. H., and Martin, G. R., 1980, Role of laminin in the attachment of PAM 212 (epithelial) cells to basement membrane collagen, *Cell* **22:**719–726.

Thiery, J.P., Duband, J. L., and Delouvée, A., 1982a, Pathways and mechanism of avian trunk neural crest cell migration and localization, *Dev. Biol.* **93:**324–343.

Thiery, J. P., Duband, J. L., Rutishauser, U., and Edelman, G. M., 1982b, Cell adhesion molecules in early chicken embryogenesis, *Proc. Natl. Acad. Sci. USA* **79:**6737–6741.

Thiery, J. P., Delouvée, A., Grumet, M., and Edelman, G. M., 1985, Initial appearance and regional distribution of the neuron glia cell adhesion molecule (Ng CAM) in the chick embryo, *J. Cell Biol.* **100:**442–456.

Timpl, R., Rohde, H., Robey, P. G., Rennard, S. I., Foidart, J. M., and Martin, G. R., 1979, Laminin. A glycoprotein from basement membranes, *J. Biol. Chem.* **259:**9933–9937.

Timpl, R., Johansson, S., Van Delden, V., Oberbaumer, I., and Höök, M., 1983, Characterization of protease-resistant fragments of laminin mediating attachment and spreading of rat hepatocytes, *J. Biol. Chem.* **258:**8922–8927.

Tosney, K. W., 1978, The early migration of neural crest cells in the trunk region of the avian embryo. An electron microscopic study, *Dev. Biol.* **62:**317–333.

Tosney, K. W., 1982, The segregation and early migration of cranial neural crest cells in the avian embryo, *Dev. Biol.* **89:**13–24.

Trinkaus, J. P., 1984a, Mechanism of *Fundulus* epiboly: A current view, *Am. Zool.* **24:**673–688.

Trinkaus, J. P., 1984b, *Cells into Organs: The Forces that Shape the Embryo*, 2nd ed., Prentice-Hall, Englewood Cliffs, New Jersey.

Tucker, R. P., and Erickson, C. A., 1984, Morphology and behavior of quail neural crest cells in artificial three dimensional extracellular matrices, *Dev. Biol.* **104:**390–405.

Tucker, R. P., Edwards, B. F., and Erickson, C. A., 1985, Tension in the culture dish: Microfilament organization and migratory behavior of quail neural crest cells, *Cell Motility* **5:**225–237.

Tucker, G. C., Ciment, G., and Thiery, G. P., 1986, Pathways of avian neural crest cell migration in the developing gut, *Dev. Biol.* **116:**439–450.

Vakaet, L., 1970, Cinematographic investigations of gastrulation in the chick blastoderm, *Arch. Biol.* **81:**387–426.

Vakaet, L., 1984, Early development of birds, in: *Chimeras in Developmental Biology*, (N. M. Le Douarin and N. McLaren, eds.), Academic Press, London, pp. 71–88.

Valinsky, J., and Le Douarin, N. M., 1985, Production of plasminogen activator by migrating cephalic neural crest cells, *EMBO J.* **4:**1403–1406.

Valinsky, J., and Reich, E., 1981, Plasminogen in the chick embryo: Transport and biosynthesis, *J. Biol. Chem.* **256:**12470–12475.

Van de Water, L., Schroeder, S., Creshaw, E.B., and Hynes, R. O., 1981, Phagocytosis of gelatin-latex particles by a murine macrophage line is dependent on fibronectin and heparin, *J. Cell Biol.* **90:**32–39.

Vanroelen, C., Vakaet, L., and Andries, L., 1980, Localization and characterization of acid mucopolysaccharides in the early chick blastoderm, *J. Embryol. Exp. Morphol.* **56:**169–178.

Vibe-Pedersen, K., Kornblihtt, A. R., and Baralle, F. E., 1984, Expression of a human alpha-globin/fibronectin gene hybrid generates two mRNAs by alternative splicing, *EMBO J.* **3:**2511–2516.

Wakely, J., and England, M. A., 1977, Scanning electron microscopy of the chick embryo primitive streak, *Differentiation* **7:**181–186.

Wartiovaara, J., Nordlying, S., Lehtonen, E., and Saxèn, L., 1974, Transfilter induction of kidney tubules: Correlation with cytoplasmic penetration into Nucleopore filters, *J. Embryol. Exp. Morphol.* **31:**667–686.

Weiss, R. E., and Reddi, A. H., 1981, Role of fibronectin in collagenous matrix-induced mesenchymal cell proliferation and differentiation *in vivo*, *Exp. Cell Res.* **133:**247–254.

West, C. M., Lanza, R., Rosenbloom, J., Lowe, M., Holtzer, H., and Avdalovic, N., 1979, Fibronectin alters the phenotypic properties of cultured chick embryo chondroblasts, *Cell* **17:**491–501.

Weston, J. A., 1963, A radioautographic analysis of the migration and localization of trunk crest cells in the chick, *Dev. Biol.* **6:**279–310.

Weston, J. A., and Butler, S. L., 1966, Temporal factors affecting localization of neural crest cells in the chicken embryo, *Dev. Biol.* **14:**246–266.

Wylie, C. C., and Heasman, J., 1982, Effects of the substratum on the migration of primordial germ cells, *Philos. Trans. R. Soc. London Ser. B* **299:**177–183.

Yamada, K. M., 1983, Cell surface interactions with extracellular materials, *Annu. Rev. Biochem.* **52:**761–799.

Yamada, K. M., and Kennedy, D. W., 1984, Dualistic nature of adhesive protein function: Fibronectin and its biologically active peptide fragments can auto-inhibit fibronectin function, *J. Cell Biol.* **99:**29–36.

Yamada, K. M., and Kennedy, D. W., 1985, Amino acid sequences specificities of an adhesive recognition signal, *J. Cell Biochem.* **28:**99–104.

Yamada, K. M., Yamada, S. S., and Pastan, I., 1976, Cell surface protein partially restores morphology, adhesiveness, and contact inhibition of movement in transformed fibroblasts, *Proc. Natl. Acad. Sci. USA* **73:**1217–1221.

Yamada, K. M., Kennedy, D. W., and Hayashi, M., 1982, Fibronectin in cell adhesion, differentiation, and growth, *Cold Spring Harbor Conf. Cell Prolif.* **9:**131–143.

Yamada, K. M., Akiyama, S. K., Hasegawa, T., Hasegawa, E., Humphries, M. J., Kennedy, D. W., Nagata, K., Urrushihara, H., Olden, K., and Chen, W. T., 1985, Recent advances in research on fibronectin and other cell attachment proteins, *J. Cell Biochem.* **28:**79–97.

Yurchenko, P. D., Tsilibary, E. C., Charonis, A. S., Furthmayer, H., 1985, Laminin polymerization *in vitro*: Evidence for a two step assembly with domain specificity, *J. Biol. Chem.* **260:**7636–7644.

2

THROMBOSPONDIN

A Multifunctional Platelet and Extracellular Matrix Glycoprotein

Richard A. Majack and Paul Bornstein

1. INTRODUCTION

Thrombospondin (TS) is a major glycoprotein secretory product of activated platelets and a component of a number of cell-derived extracellular matrices. Like fibronectin, TS appears to be a "modular" glycoprotein (Odermatt *et al.*, 1985), capable of a variety of molecular interactions and involved in a diverse spectrum of platelet and cellular functions (Hynes and Yamada, 1982). TS (thrombin-sensitive protein, glycoprotein G) (Baenziger *et al.*, 1971; George *et al.*, 1980; Lawler *et al.*, 1978) was first described as a surface component of thrombin-activated platelets (Baenziger *et al.*, 1971, 1972) and is now known to be synthesized by megakaryocytes (McLaren, 1983) and stored in platelet alpha granules (Gartner *et al.*, 1981a; Wencel-Drake *et al.*, 1984). Following platelet activation, TS can be detected in serum in concentrations ranging from 10 to 30 μg/ml (Dawes *et al.*, 1983; Saglio and Slayter, 1982). Platelet-derived TS appears to play functional roles in platelet aggregation, agglutination and adhesion, in fibrin clots, and in fibrinolysis. In addition, a variety of cell types in culture secrete TS and incorporate it into their extracellular matrix. The function(s) of cell-derived TS in the extracellular matrix is at present largely unknown but appears to be related to the facilitation of cell proliferation and/or motility. In this chapter we review the structure, molecular interactions, and known functions of TS, with emphasis on its role as a component of the extracellular matrix.

Richard A. Majack and Paul Bornstein ● Department of Biochemistry, University of Washington, Seattle, Washington 98195. *Present address for R.A.M.:* Atherosclerosis and Thrombosis Research, The Upjohn Company, Kalamazoo, Michigan 49001.

2. STRUCTURE AND MOLECULAR INTERACTIONS OF THROMBOSPONDIN

2.1. Structure of Thrombospondin

TS is a large (M_r approximately 450,000) glycoprotein composed of three apparently identical (Coligan and Slayter, 1984; Dixit *et al.*, 1984a,b; Raugi *et al.*, 1984) disulfide-bonded peptide chains. Estimates of the molecular weight of the constituent chains have ranged between 140,000 (Margossian *et al.*, 1981) and 190,000 (McPherson *et al.*, 1981). These differences can be largely attributed to the different mobilities of TS chains in the Weber–Osborn and Laemmli sodium dodecyl sulfate–polyacrylamide gel electrophoresis (SDS–PAGE) systems (Lawler *et al.*, 1982). Pertinent known characteristics of the TS molecule are presented in Figure 1.

Electron microscopic studies of rotary-shadowed TS molecules (Coligan and Slayter, 1984; Galvin *et al.*, 1985; Lawler *et al.*, 1982, Lawler *et al.*, 1985) have revealed that TS exists as a tripartite structure consisting of globular amino (NH_2) and carboxyl (COOH) termini connected by a long (M_r 65,000) linear domain (Coligan and Slayter, 1984). Extensive interchain disulfide bonding

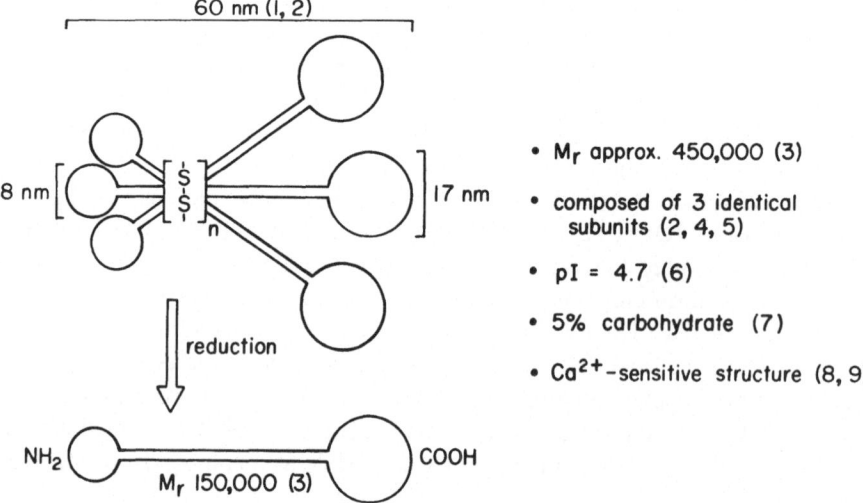

- M_r approx. 450,000 (3)
- composed of 3 identical subunits (2, 4, 5)
- pI = 4.7 (6)
- 5% carbohydrate (7)
- Ca^{2+}-sensitive structure (8, 9)

FIGURE 1. Molecular features of human platelet TS, as elucidated by physical, chemical, and electron microscopic studies. Estimates of molecular dimensions differ in different studies. Numbers in parentheses refer to references: (1) Lawler *et al.*, 1982; (2) Coligan and Slayter, 1984; (3) Lawler *et al.*, 1977; (4) Dixit *et al.*, 1984a,b; (5) Raugi *et al.*, 1984; (6) Lawler *et al.*, 1978; (7) Baenziger *et al.*, 1972; (8) Lawler and Simons, 1983; (9) Lawler *et al.*, 1985.

appears to occur at a point along the linear domain close to the NH_2-terminal globular end. The overall configuration of the TS molecule is somewhat suggestive of a bola, a South American weapon used (by hurling) to entangle the legs of animals (Coligan and Slayter, 1984).

2.2. Interactions with Ca^{2+}, Heparin, and Other Proteins

TS is known to interact with a variety of molecular ligands present in blood, on cell surfaces, and in the extracellular matrix. The wide range of interactions documented to date is listed in Table I and reflects the possible involvement of TS in a number of diverse cellular functions. Binding of TS to fibronectin, platelet surfaces, and fibrinogen appears to mediate the activity of TS in platelet aggregation (Leung, 1984). Binding of TS to histidine-rich glycoprotein (HRGP) (see Lijnen et al., 1983) and plasminogen may reflect a modulatory role for TS in thrombolysis and/or extracellular proteolysis (Leung et al., 1984; Silverstein et al., 1984, 185a,b). TS in the extracellular matrix may interact with cell surfaces, fibronectin, collagen, and possibly laminin, suggesting a role for TS in the adhesion of platelets to the subendothelium and in cell–cell or cell–substratum interactions.

Interactions between TS and Ca^{2+} and between TS and heparin may be especially important with regard to the functional properties of TS. The TS molecule is known to have at least 12 Ca^{2+}-binding sites (Lawler and Simons, 1983) and depends on Ca^{2+} for conformational stability (Lawler et al., 1982). In the absence of Ca^{2+}, eg., when TS is prepared in the presence of EDTA, the pattern of peptides generated in response to trypsin (Lawler and Simons, 1983) or thrombin (Lawler et al., 1982, 1985) is altered. Lawler et al. (1982) have compared the physical characteristics of TS prepared in the presence or absence of 2 mM Ca^{2+}. In the presence of EDTA, the sedimentation coefficient was 8.6 S compared to 9.7 S in the presence of Ca^{2+}. Similarly, the intrinsic viscosity increased from 21 ml/g for Ca^{2+}-replete TS to 40 ml/g for TS prepared in the presence of EDTA, indicative of a less asymmetric structure in the latter state. By electron microscopy, Ca^{2+}-depleted TS also appeared less asymmetric (Lawler et al., 1982, 1985). Significantly, Lawler and Simons (1983) showed that treatment of TS with EDTA altered the circular dichroism spectrum and caused a small change in secondary structure. These data indicate that the concentration of Ca^{2+} in local environments may be an important determinant of TS structure and, possibly, TS function. Although the monovalent binding of TS to cell surfaces does not require divalent cations (Gartner and Dockter, 1983), the agglutinating activity of TS (Gartner and Dokter, 1983; Phillips et al 1980; Haverstick et al., 1984, 1985) does show certain dependency. The interaction of TS with fibrinogen has been reported as being cation-dependent (Dixit et al.,

TABLE I. Molecular Interactions of TS

Ligand	Possible functional significance	Selected references
Ca^{2+}	Conformational stability	Lawler et al. (1982), Lawler and Simons (1983)
Cell surfaces	Cell–cell or cell–matrix interactions; growth facilitation of mesenchymal cells	Roberts et al. (1985a,b), Booth et al. (1985), Majack et al. (1985, 1986)
Collagen types I and V	Platelet adherence to the subendothelium; ECM[a]	Lahav et al. (1982), Mumby et al. (1984b)
Fibrin	Modulation of clot structure	Bale et al. (1985), Murphy-Ullrich and Mosher (1985)
Fibrinogen	Secretion-dependent irreversible platelet aggregation	Gartner et al. (1981b), Jaffe et al. (1982), Leung and Nachman (1982), Lahav et al. (1984), Tuszynski et al. (1985)
Fibronectin	Platelet aggregation and adherence to the subendothelium; ECM	Lahav et al. (1982), Lahav et al. (1984), Leung (1984), Dixit et al. (1985b)
Heparin	Binding of TS to cell surfaces or other molecules; modulation of TS-binding properties	Gartner et al. (1984b), McKeown-Longo et al. (1984), Majack et al. (1985), Dixit et al. (1985a)
Histidine-rich glycoprotein (HRGP)	Augmentation of thrombosis; inhibition of fibrinolysis; may form ternary complex with TS and plasminogen	Leung et al. (1984), Silverstein et al. (1985a)
Plasmin	Facilitation of extracellular proteolysis	Silverstein et al. (1984)
Plasminogen	Modulation of thrombolysis and/or extracellular proteolysis; may form ternary complex with TS and HRGP or PA	Silverstein et al. (1984, 1985a,b)
Sulfated glycolipids	Binding of TS to cell membranes	Roberts et al. (1985a)
Thrombin	Not known; may represent enzyme: substrate complex	Danishefsky et al. (1984), Hagen et al. (1983)
TS (self aggregation)	Hemagglutin activity; platelet agglutination; cell–cell interactions	Lahav et al. (1984), Booth et al. (1985), Gartner et al. (1984a)

[a] TS participates in this interaction as a structural component of the extracellular matrix (ECM), but the functional significance of this interaction is not known.

1984a) or -independent (Tuszynski *et al.*, 1985). TS binds to heparin in the presence or absence of Ca^{2+}.

Baenziger *et al.* (1971) first showed that TS, present on platelet surfaces following activation, may be proteolyzed upon further incubation with thrombin. Subsequent studies have established that the released peptide contains the heparin-binding domain of TS. The M_r 25,000–35,000 NH_2-terminal heparin-binding region is a common proteolytic product of TS and can be easily generated by a variety of proteases (see Figure 2), implying a physiologic role for the removal of the heparin-binding fragment by proteolysis. Recent data have confirmed the importance of heparin:TS interactions with regard to the binding of TS to other molecules. Gartner *et al.* (1984b) demonstrated that antibodies

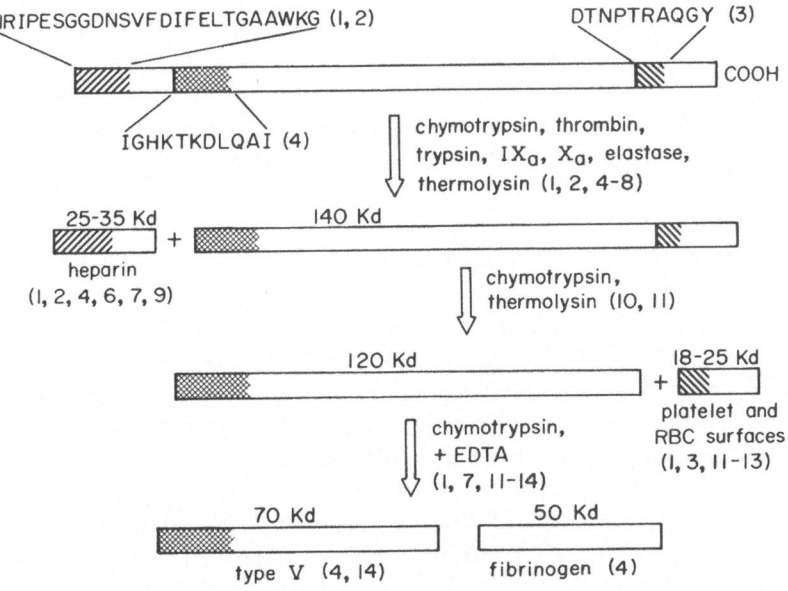

FIGURE 2. Preliminary linear mapping of several binding domains on TS. Positions of binding sites for heparin, fibrinogen, type V collagen, and platelet surfaces were localized to protease cleavage products by protein sequencing, antibody epitope mapping, and functional assays (see text for details). Hatching indicates sequence identity between regions of the TS chain and protease-produced peptides; sequences, using the single-letter amino acid code, are presented at top. Other binding domains exist for a variety of other ligands (see Table I) but have not yet been mapped. Molecular weights of fragments are based on those cited by Dixit *et al.* (1984b) and Galvin *et al.* (1985). Numbers in parentheses refer to references: (1) Dixit *et al.*, 1984b; (2) Coligan and Slayter, 1984; (3) Dixit *et al.*, 1985b; (4) Galvin *et al.*, 1985; (5) Baenziger *et al.*, 1971; (6) Lawler and Slayter, 1981; (7) Raugi *et al.*, 1984; (8) Takahashi *et al.*, 1984; (9) Lawler *et al.*, 1985; (10) Dixit *et al.*, 1984a; (11) Dixit *et al.*, 1985a; (12) Haverstick *et al.*, 1984; (13) Haverstick *et al.*, 1985; (14) Mumby *et al.*, 1984b. (Figure modified from Galvin *et al.*, 1985.)

prepared against an M_r 23,000 heparin-binding fragment of TS could inhibit platelet aggregation. Other workers (Dixit *et al.*, 1985a,b; Galvin *et al.*, 1985) have subsequently shown that the COOH terminus of the TS molecule can bind to platelets and red blood cells (see Figure 2), thus suggesting a heterotropic cross-linking of different receptors by the two globular regions of TS. Binding of a monoclonal antibody to the heparin-binding domain of TS nonetheless inhibited the agglutination of red blood cells and platelets (Dixit *et al.*, 1985a), apparently by inducing a conformational change in the molecule, as demonstrated by electron microscopy (Galvin *et al.*, 1985). Thus, binding of heparin to TS, or removal by proteolysis of the heparin-binding region, may significantly influence the functional properties of TS by altering its tertiary structure. Further studies are required to characterize and verify these phenomena.

In addition to the interactions of TS with Ca^{2+} and heparin, TS may interact with a variety of other ligands (Table I). We predict that future molecular analysis of TS and its gene will reveal that TS belongs to a growing class of molecules such as fibronectin (Odermatt *et al.*, 1985), proteases of the blood coagulation and fibrinolysis system (Patthy, 1985), and cell membrane receptor molecules (Sudhof *et al.*, 1985), which have arisen evolutionarily through the assembly of a number of gene (exon) modules, each encoding a specific functional domain.

2.3. Organization of Functional Domains

Much remains to be learned about the ordered arrangement of functional domains within the TS molecule and about the relationship of these domains to TS gene structure and to homologous domains in other proteins. However, recent studies using protease cleavages and functional assays have allowed a tentative linear mapping of several functional domains along the TS chain, as presented in Figure 2.

Treatment of intact TS monomers with a wide variety of proteases (for references see Figure 2) results in the removal of an M_r 25,000–30,000 peptide. Limited amino acid sequencing has shown that the NH_2 terminal sequence of this fragment is identical with the NH_2 terminus of intact TS (Dixit *et al.*, 1984b; Galvin *et al.*, 1985), thus establishing its origin from the NH_2 terminus of the TS chain. This M_r 25,000–30,000 peptide contains the heparin-binding domain of TS (Coligan and Slayter, 1984; Dixit *et al.*, 1984b; Galvin *et al.*, 1985; Lawler and Slayter, 1981; Lawler *et al.*, 1985; Raugi *et al.*, 1984). Epitope mapping using a monoclonal antibody against the heparin-binding fragment has verified the location of this heparin-binding domain in the small NH_2 terminal globular portion of the intact TS molecule (Figure 1) (Galvin *et al.*, 1985). Haverstick *et al.* (1984) have reported the possible presence of an additional heparin-binding site on TS. The M_r 140,000 fragment remaining after removal of the heparin-binding region retains hemagglutinating (Haverstick *et al.*, 1984) and platelet-agglutinating activity (Haverstick *et al.*, 1985). Additional proteolysis of the M_r 140,000 fragment removes an M_r 18,000–25,000 domain shown by functional antibody assays to participate in platelet aggregation (Dixit *et al.*, 1985b), platelet agglutination (Haverstick *et al.*, 1984), and hemagglutination (Dixit *et al.*, 1985a; Haverstick *et al.*, 1985). Further sequence analysis has shown that this fragment is derived from the COOH terminus of the TS subunit (Galvin *et al.*, 1985). The M_r 120,000 fragment, which also results from this proteolysis, appears to be devoid of these activities (Haverstick *et al.*, 1984, 1985). Raugi *et al.* (1984) have identified a binding site for the type V collagen in an M_r 70,000 chymotryptic fragment whose linear positions along the TS chain has recently been established since the NH_2-terminal sequences of the M_r 120,000 and the M_r 70,000 fragments are identical (Figure 2) (Galvin *et al.*, 1985). The binding domain for fibrinogen has been tentatively identified in the COOH end of the M_r 120,000 fragment (Galvin *et al.*, 1985). TS contains at least five additional known binding sites (for fibronectin, glycolipids, plasminogen, type I collagen, and HRGP) whose locations in the TS molecule have not yet been determined.

3. ROLES FOR THROMBOSPONDIN IN PLATELET FUNCTION

TS is a major glycoprotein secretion product of stimulated platelets and comprises 20–30% of the protein released from alpha granules (Baenziger *et al.*, 1972). TS appears to play several roles in platelet physiology, as diagramed schematically in Figure 3. TS has been identified as the endogenous platelet lectin (Jaffe *et al.*, 1982), by virtue of its ability to bind to fibrinogen, and is responsible for the stabilization of platelet:fibrinogen interactions (Leung, 1984). TS thus functions during the secondary, secretion-dependent phase to facilitate the irreversible macroaggregation of platelets (Agam *et al.*, 1984; Leung, 1984). Both polyclonal (Gartner *et al.*, 1984b; Leung, 1984) and monoclonal (Dixit *et al.*, 1985b) antibodies against TS have inhibitory effects on platelet aggregation.

TS, derived from either platelets or the vascular endothelium, may also mediate binding of platelets to the exposed subendothelium (Lahav *et al.*, 1982), although evidence for a role of TS in platelet adhesion was not observed in studies using a perfusion chamber (Houdijk *et al.*, 1986). In addition, TS is transiently incorporated into fibrin clots and can influence their structure (Bale *et al.*, 1985; Murphy-Ullrich and Mosher, 1985). Finally, by virtue of its interactions with HRGP, plasmin, and plasminogen (Silverstein *et al.*, 1984, 1985a,b), TS may be involved in the modulation of thrombolysis. The reader is referred to Silverstein *et al.* (1986) for a more comprehensive review of the role of TS in platelet functions.

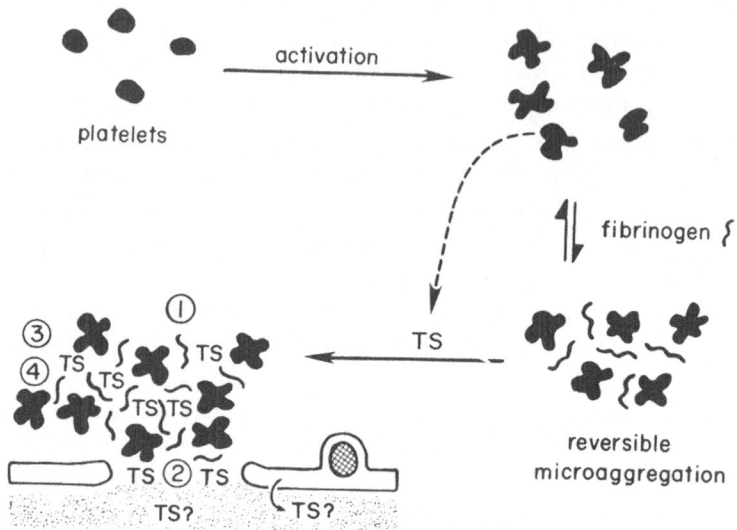

FIGURE 3. Roles for TS in platelet functions. Platelets release TS from alpha granules upon activation; the secreted TS may play several important roles in thrombosis. Platelets form reversible microaggregates using fibrinogen bound to glycoprotein IIb/IIIa on the platelet surface. (1) TS stabilizes the interaction between fibrinogen and the platelet surface and promotes platelet macroaggregation by virtue of its lectin activity (Leung, 1984). (2) TS may also function to bind the platelet plug to the exposed subendothelial matrix. (3) TS incorporates into fibrin clots and may modify clot structure (Bale *et al.*, 1985). (4) TS may also play a role in modulating fibrinolysis (Silverstein *et al.*, 1984, 1985a,b).

4. PRESENCE OF THROMBOSPONDIN IN EXTRACELLULAR MATRICES

4.1. Synthesis of Thrombospondin by Cells in Culture

In 1981, McPherson *et al.* demonstrated that cultured aortic endothelial cells synthesized a protein structurally and immunologically similar or identical to platelet TS. Subsequent studies reported the synthesis and secretion of TS by other endothelial cells, normal and transformed fibroblasts, smooth muscle cells, granular pneumocytes, macrophages, and monocytes (detailed, with references, in Table II). Surprisingly, synthesis of TS by smooth muscle cells and fibroblasts was up to sevenfold greater (on a per-cell basis) than that of endothelial cells (Raugi *et al.*, 1982). This finding, together with concurrent observations that TS produced by cells in culture was incorporated into the extracellular matrix

TABLE II. Biosynthesis of TS by Cells in Culture

Cell type	Species	References
Vascular endothelium		
Aortic	Bovine	McPherson et al. (1981), Sage et al. (1981), Raugi et al. (1982),* Mumby et al. (1984a)
Umbilical vein	Human	Mosher et al. (1982), Sage and Bornstein (1982), Hunter et al. (1984), Reinders et al. (1985)
Saphenous vein	Human	Hunter et al. (1984)
Skin microvascular	Human	Kramer et al. (1985)*
Fibroblasts		
Foreskin	Human	Raugi et al. (1982), Jaffe et al. (1983), Mumby et al. (1984a)
Embryonic lung	Human	Jaffe et al. (1983),* Mumby et al. (1984a)
SV-40-transformed WI-38	Human	Mumby et al. (1984a)
Smooth muscle		
Aortic	Human	Raugi et al. (1982)*
Aortic	Rat	Majack et al. (1985)*
Others		
Granular pneumocytes	Rat	Sage et al. (1983)
Macrophages, monocytes	Human, mouse	Jaffe et al. (1985)

*Studies also demonstrating incorporation of TS into the extracellular matrix.

(Jaffe et al., 1983; Kramer et al., 1985; Majack et al., 1985; Raugi et al., 1982), strongly indicated that TS may serve some extracellular function not necessarily related to platelet aggregation.

4.2. Immunolocation of Thrombospondin in Tissues

Wight et al. (1985) have used immunofluorescence techniques to examine the distribution of TS in human tissues. TS was detected in the peritubular connective tissue of the renal cortex, at the dermal–epidermal junction in skin, beneath glandular epithelia, in the interstitium between skeletal muscle bundles, and in small blood vessels throughout many tissues. Staining was also detected in the intima of atherosclerotic aortas. These data suggest that TS may be a normal constituent of some basement membranes and extracellular matrices.

5. POSSIBLE FUNCTIONAL ROLES FOR THROMBOSPONDIN IN THE EXTRACELLULAR MATRIX

5.1. Organizational and Other Roles

A definitive role for TS in the extracellular matrix of cells has not been elucidated to date. By analogy to its diverse roles in platelet aggregation, hemagglutination, and thrombolysis, TS may play several functional roles in the extracellular milieu.

TS undoubtedly serves a structural role in the matrix, by virtue of its interactions with collagen types I and V, heparin, fibronectin, and possibly other matrix components. The specificity of localization of TS to particular basement membranes and matrices (Wight *et al.*, 1985) suggests that the presence of TS may determine certain functional or informational properties of the matrix as a whole. For example, TS has been localized underlying glandular epithelia in developing, but not adult, lung and in pertibular, but not glomerular, basement membranes in the kidney (Wight *et al.*, 1985). In addition, as discussed in the following section, TS may serve a role in the extracellular matrix of proliferative cells, which is not required during quiescence. TS may act as a focus for the assembly of new matrix components or may function during matrix degradation; supportive data for either of these possibilities are lacking.

By analogy to the ability of TS to agglutinate platelets and red blood cells, which depends on the ability of TS to self-aggregate (Booth *et al.*, 1985), TS may play a role in cell–cell or cell–matrix interactions. Lahav *et al.* (1982), using affinity cross-linking techniques, demonstrated that TS became linked to fibronectin or collagenous matrices when platelets were activated on them. McKeown-Long and co-workers (1984) have demonstrated that exogenously added platelet TS can incorporate into a fibroblast-derived extracellular matrix. Thus, TS in the subendothelium (derived from either platelets or endothelial cells) may serve to anchor platelet plugs at sites of vascular injury (see Figure 3). However, experiments in this laboratory have failed to establish a role for TS in the cell–substratum adhesion of cultured endothelial cells (Mumby, 1984). Roberts *et al.* (1985b) have suggested that TS may mediate the adherence of erythrocytes infected with *Plasmodium falciparum*, a malarial parasite, to the venular endothelium. The binding of infected erythrocytes to immobilized TS or to a TS-secreting melanoma cell line was inhibited by excess TS or in the presence of specific antisera against TS.

Recent work by Silverstein *et al.* (1984, 1985a,b) has elucidated an extra-vascular role for TS that could be of some importance if operative in the extracellular matrix. The conversion of plasminogen to plasmin by plasminogen activator (PA) is an important reaction leading to the eventual breakdown of fibrin clots. The interaction between soluble plasminogen and PA is not kinetically favorable. However, in the presence of fibrin, a ternary (fibrin–plasminogen–PA) complex forms that results in a manyfold increase in plasmin generation (Hoylaerts *et al.*, 1982). Thus, the binding of plasminogen to fibrin serves to localize enzyme activity to the fibrin clot and, in addition, increases the kinetics of conversion of plasminogen to plasmin. Plasminogen-dependent proteolysis has been shown in a wide variety of normal and transformed cells and has been postulated as a mechanism whereby tumorigenic cells may degrade and invade connective tissues (Reich, 1978). Silverstein *et al.* (1985a,b) have shown that TS may serve as a focus for the activation of plasminogen, by virtue of formation of a trimolecular complex of TS, plasminogen, and PA. When plasminogen is bound to TS, the kinetics of conversion to plasmin are forty-fold higher than that of soluble plasminogen (see Figure 4). The rate of plasminogen activation was further enhanced in the presence of HRGP (Silverstein *et al.*, 1986). In addition, up to two thirds of the plasmin generated by these complexes was found to remain bound to TS and was protected from inhibition by α_2-plasmin inhibitor (Silverstein *et al.*, 1986). Thus, one role of TS in the extracellular matrix may be to serve as a focus for plasminogen activation. Other aspects of this phenomenon, such as the possible role of these complexes in the modulation of fibrinolysis, have been reviewed recently (Silverstein *et al.*, 1986). The physiologic significance of these observations, especially as they relate to cell migration and proliferation, is not clear at present.

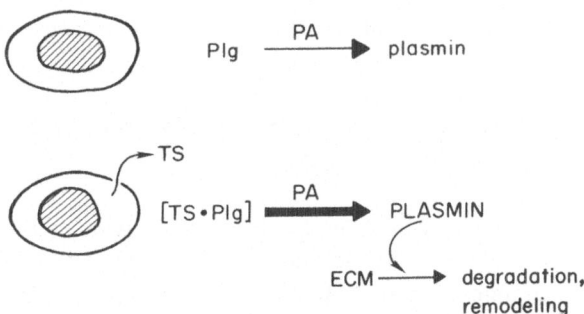

FIGURE 4. Possible role for TS in the extravascular activation of plasminogen (Plg). In the absence of complex formation (top), extracellular Plg is converted to plasmin with low efficiency, since soluble Plg is a poor substrate for PA. In the presence of TS (or fibrin in intravascular sites), a multimolecular complex of Plg, TS, and PA forms, which may result in elevated generation of plasmin (for details, see Silverstein *et al., 1986*). Thus, production of TS by a cell (as in growth-factor-stimulated SMC) may increase plasmin generation by facilitation of PA activity, leading to increased breakdown of nonfibrous extracellular matrix (ECM) components.

5.2. Evidence for an Involvement of Thrombospondin in Cell Proliferation

Several lines of evidence support the hypothesis (Majack *et al.*, 1985, 1986) that TS production may be related to cell growth. Hunter *et al.* (1984), in a quantitative *in vitro* study of TS production by adult and fetal saphenous vein endothelial cells, reported that fetal cells synthesized and secreted fivefold more TS per cell than did their adult counterparts. Similarly, Kramer *et al.* (1985) found that cultured newborn skin capillary endothelium produced an extracellular matrix rich in TS; capillary endothelium from adult skin (albeit from a different anatomical location) did not. These studies indicated that TS production may be related, in some fashion, to the proliferative potential of cells in culture.

A more direct demonstration of the involvement of TS in cell growth was presented by Mumby *et al.* (1984a), who described the density-dependent regulation of TS synthesis by cultured fibroblasts, aortic endothelium, and vascular smooth muscle cells. Human embryonic lung fibroblasts at sparse density, for example, were found to produce fivefold more TS than did cells at confluent densities. Synthesis of collagen and fibronectin were not similarly regulated by cell density, suggesting that TS synthesis may be controlled by mechanisms independent of those for other matrix components. TS synthesis therefore appears to be preferentially enhanced in actively growing cells.

Consistent with this concept, Wight *et al.* (1985) have shown that TS may be localized *in situ* in human tissues in areas of expected cell proliferation, such as along the basal aspect of stratified or glandular epithelia and in the budding alveolae of embryonic lungs. Staining of TS was also observed in the intima of atherosclerotic lesions but not in the underlying vascular media.

5.3. A Role for Thrombospondin in the Regulation of Proliferation of Vascular Smooth Muscle Cells

We have been investigating the mechanisms by which components of the extracellular matrix regulate the biosynthetic phenotype, growth, and migration of vascular smooth muscle cells (SMC) in culture (Majack and Bornstein, 1984, 1985, 1986; Majack and Clowes, 1984; Majack *et al.*, 1985, 1986). The available evidence suggests that TS may play an important regulatory role in the control of SMC migration and growth. We base this statement on the following observations:

1. TS is present in the matrix of SMC grown in culture but cannot be detected in the vascular media in situ. As shown in Figure 5b, we have been

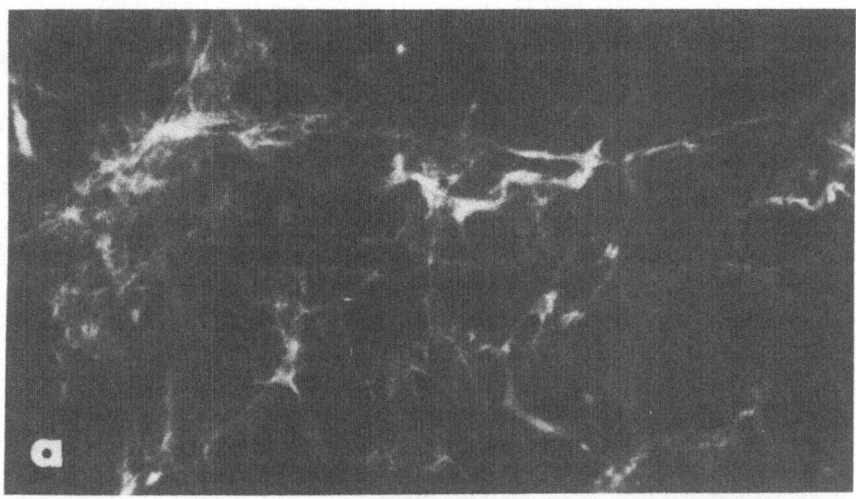

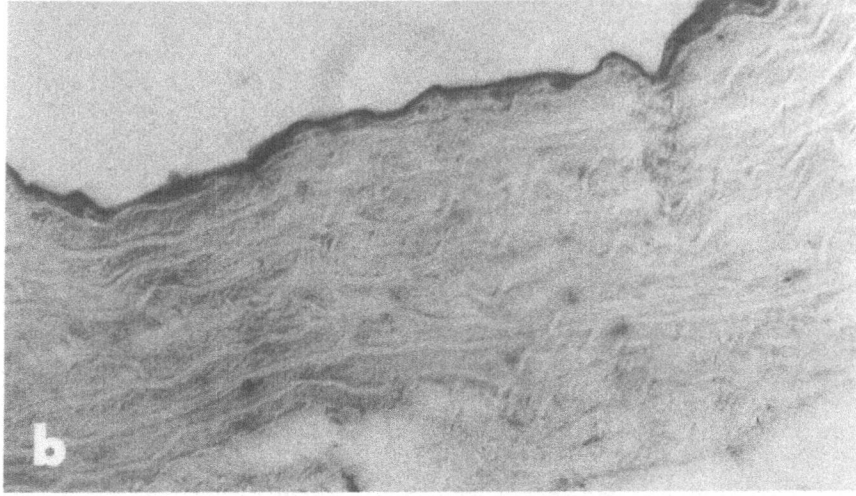

FIGURE 5. Immunolocation of TS in vascular SMC in culture (a) or *in situ* (b). (a) Immunofluorescence staining of cultured rat aortic SMC. Intense staining of TS can be seen throughout the extracellular matrix. (b) Immunoperoxidase staining of a rat aorta following experimental balloon catheterization to allow platelet deposition. We were unable to detect TS immunostaining in SMC of the vascular media, under conditions in which staining of platelets (along the surface of the vessel) is pronounced.

unable to detect immunoreactive TS in sections of rat aorta, even following a variety of protease treatments to reveal cryptic sites. TS can readily detected in the extracellular matrix of proliferative SMC cultured from the same source (Figure 5a). Although negative immunocytochemical results must be interpreted with caution, it appears that TS may not be a component of the extracellular matrix of medial SMC in uninjured, nonproliferative vessel walls. Consistent with this concept, we have found that cultured rat aortic SMC made quiescent by serum deprivation synthesize little or no TS (Figure 6a).

2. *TS synthesis by rat SMC is regulated by platelet-derived growth factor (PDGF), a potent SMC mitogen.* We have found that treatment of quiescent SMC with nanogram amounts of PDGF rapidly induces the synthesis and secretion of TS (Majack *et al.*, 1985), resulting in the elaboration of a TS-rich extracellular matrix (Figure 6). Other growth factors tested were not effective.

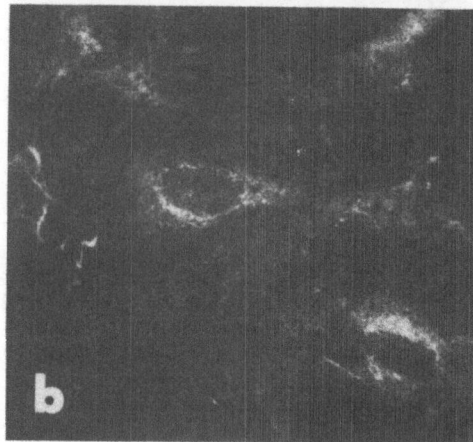

FIGURE 6. Immunolocation of thrombospondin in cultures of rat aortic SMC. Immunofluorescence was used to visualize the distribution of TS in quiescent SMC after-growth arrest in 10% plasma-derived serum (a), or SMC 24 hr after stimulation with 5 ng/ml PDGF (b). Note that PDGF, a potent mitogen for SMC, stimulates the synthesis and secretion of TS into the extracellular matrix (for details, see Majack *et al.*, 1985).

TABLE III. Effect of TS Antibodies on SMC Migration and Mitogenesis[a]

Assay	Effect
Migration	Distance/24 hr (μm)
Control	142.1 ± 9.4
Anti-TS	95.0 ± 8.6
Preimmune serum	148.2 ± 12.0
Mitogenesis	Percent labeled nuclei
Control	3.5 ± 1.8
PDGF, EGF, insulin	62.5 ± 9.3
PDGF, EGF, insulin, anti-TS	39.2 ± 6.6

[a] Data taken from Bornstein and Majack (1985).

The induction of TS synthesis by PDGF occurred within 1 hr and could be abolished by prior treatment of the cells with actinomycin D, an inhibitor of DNA-dependent RNA synthesis, suggesting a transcriptional control mechanism. Induction of TS occurred maximally at concentrations of PDGF (2.5 ng/ml) that were shown to be suboptimal for a mitogenic response in a serum-free mitogenesis assay. The elaboration of a TS-rich extracellular matrix in response to PDGF strongly suggests a role for TS in some aspect of the SMC growth response. TS synthesis can also be induced in quiescent Swiss 3T3 cells by treatment with PDGF (R. A. Majack, unpublished observations), suggesting that the effect may be a common property of all mesenchymal cells. Further studies are required to investigate the regulation of TS synthesis in other cell types.

3. Antibodies against TS block SMC migration and mitogenesis in response to purified growth factors. We have used an affinity-purified polyclonal antibody to TS (McPherson *et al.*, 1981; Raugi *et al.*, 1982; Wight *et al.*, 1985) to determine the involvement of TS in the SMC growth response to PDGF or serum. Cell migration experiments were performed as described previously (Majack and Clowes, 1984), using a two-dimensional *in vitro* assay. SMC migration was markedly reduced in the presence of antibodies against TS (Table III; Bornstein and Majack, 1985). Similarly, TS antibodies significantly reduced SMC mitogenesis in response to a panel of purified growth factors. These data establish an autocrine facilitative role for TS in the extracellular matrix; following growth stimulation, SMC express a TS-rich matrix that is, in some fashion, required for optimal growth and migration.

4. Addition of exogenous TS to SMC facilitates mitogenesis in response to epidermal growth factor (EGF). We have directly examined the role of TS in SMC proliferation using a serum-free mitogenesis assay, purified growth factors, and TS isolated from human platelets (Majack *et al.*, 1986). Addition of EGF (5 ng/ml) or TS (20 μg/ml) to quiescent rat vascular SMC did not substantially stimulate mitogenesis; the 30-hr nuclear labeling index increased from a mean of 7.3% in control cells to 19.8% or 16.6%, respectively. When TS and EGF were added together, however, a synergistic stimulation of DNA synthesis was observed, with 46.5% of the cells responding. The effect was specific for EGF, since TS did not potentiate growth induction in response to PDGF or insulinlike growth factors. The mechanism underlying the growth facilitative effect of TS is not yet known. We have hypothesized (Majack *et al.*, 1986) that TS produced by SMC in response to PDGF may enhance the response of SMC to EGF as a progression factor or may expand the proliferative response of SMC to neighboring cells by "sensitizing" them to respond to EGF as a complete mitogen.

  5. Heparin, which blocks the incorporation of TS into the extracellular matrix, prevents the growth-facilitative effects of TS. We have found (Figure 7) that SMC cultured in the presence of heparin do not incorporate newly synthesized

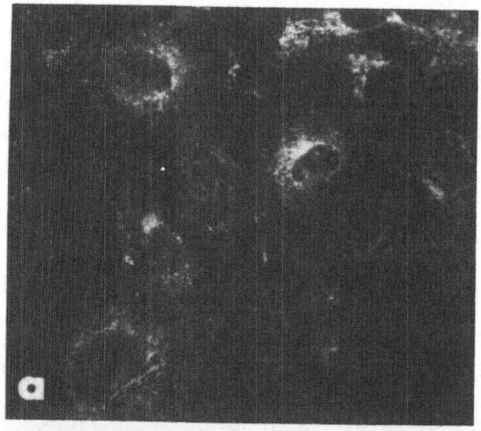

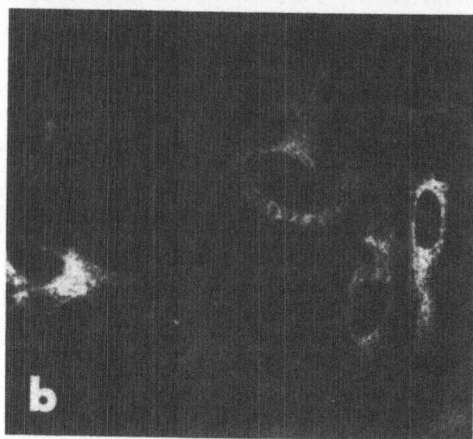

FIGURE 7. Immunolocation of TS in control cultures and in rat aortic SMC treated with heparin. Immunofluorescence was used to visualize the distribution of TS in cultures 24 hr after stimulation of the cells with 5 ng/ml PDGF. (a) Control stimulated cells showing intense staining in intracellular secretory granules and in the extracellular matrix. (b) Heparin-treated cells showing diminished incorporation of TS into the extracellular matrix. Heparin does not inhibit the PDGF-mediated stimulation of TS in these cells (for details, see Majack *et al.*, 1985).

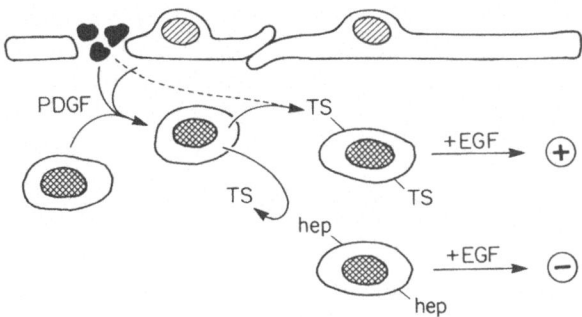

FIGURE 8. Possible role of TS as an extracellular integrator of SMC growth-regulatory signals. PDGF, derived from platelets (solid line), wounded endothelium (striped nuclei), or SMC (cross-hatched nuclei), stimulates SMC to synthesize and secrete TS, a protein also released from platelet alpha-granules (dashed line). The induction of TS can occur at levels of PDGF fourfold less than that required for a complete mitogenic response. The binding of TS to SMC may facilitate the growth-promoting effects of EGF, a growth factor that, by itself, is a weak mitogen for SMC. In the presence of endogenous heparinlike growth inhibitors (hep), TS is prevented from interacting appropriately with the SMC extracellular matrix or cell surface, and facilitation of growth will not occur (bottom). Thus, the extent of TS:SMC interactions, determined by the degree of stimulation of SMC by PDGF and by the concentration of inhibitory heparinlike molecules, may play a major regulatory role in the control of SMC proliferation. Such an integration of growth-regulatory signals at the SMC cell surface would allow graded proliferative responses to varying degrees of vascular injury.

TS into the extracellular matrix (Majack *et al.*, 1985). The inhibitory effect was specific for heparin and related glycosaminoglycans and occurred rapidly (maximal inhibition by 4 hr) and reversibly (by 4 hr). Heparin concentrations as low as 1 μg/ml completely inhibited the binding of SMC-derived TS into the Triton-X-100-insoluble matrix. Heparin is a potent inhibitor of SMC proliferation and migration *in vivo* (Clowes and Karnovsky, 1977) and *in vitro* (Castellot *et al.*, 1981; Majack and Clowes, 1984). Although the mechanism underlying this growth inhibition is not clear, the available evidence suggests that heparin acts extracellularly, possibly through an alteration in the extracellular matrix. In our serum-free mitogenesis assays (Majack *et al.*, 1986), heparin completely blocked TS-mediated mitogenic or growth-facilitative activities. Heparin did not block SMC mitogenesis in response to PDGF or EGF alone. McKeown-Longo *et al.* (1984) have previously demonstrated the ability of heparin to inhibit the incorporation of platelet TS into a fibroblast-derived extracellular matrix; others (Roberts *et al.*, 1985a) have demonstrated heparin-sensitive binding of TS to sulfated glycolipids, which may act as cell membrane receptors for the binding of TS to cell surfaces. Thus, the presence of heparinlike molecules in the extracellular milieu may alter or prevent the elaboration of a TS-rich matrix in response to the growth stimulation of mesenchymal cells. Failure to elaborate such a matrix may reduce cellular proliferation, as illustrated by our antibody data (Table III). A schematic representation of the postulated roles of TS and heparinlike glycosaminoglycans in the regulation of the growth of vascular SMC is presented in Figure 8.

6. SYNOPSIS

TS is a large, multifunctional glycoprotein, capable of a variety of interactions with components of the extracellular matrix and of the blood coagulation and fibrinolytic regulatory systems (Table I). The molecular characteristics of TS and a preliminary linear map of known binding domains along the chain have been summarized in Figures 1 and 2. Initial observations that TS was a major secretory product of activated platelets have led to the elucidation of several functional roles for TS in platelet aggregation, adhesion, and in fibrinolysis (Figure 3).

Recent studies have demonstrated that a variety of cultured cells synthesize and produce TS, and incorporate TS into their extracellular matrix (Table II). It is not known if cell-derived TS is identical to platelet-derived TS; like fibronectin, the two forms may differ in some domains. A function for cell-derived matrix TS has not been elucidated but appears to be related to cell growth. TS may also play a structural role in the matrix (e.g., acting as a focus for new matrix formation), may serve in cell–cell or cell–substrate interactions, or may serve in the formation of multimolecular complexes leading to more efficient activation of plasminogen (Figure 4).

We have used vascular SMC as a model to investigate the role of TS in cell proliferation and migration. Cultured SMC synthesize and incorporate TS into their extracellular matrix; quiescent cells, like untraumatized SMC in the vascular media *in situ*, do not appear to express a TS-containing matrix (Figures 5 and 6). We have shown that TS synthesis by SMC is regulated by PDGF, a potent SMC mitogen. The elaboration of a TS-rich matrix appears to be a necessary component of the SMC growth response, since antibodies against TS reduced SMC migration and mitogenesis (Table III). SMC exposed to TS responded to EGF as a mitogen, whereas control SMC did not, suggesting that the presence of TS in the SMC extracellular milieu may "sensitize" the cells to EGF. Heparin, a known inhibitor of SMC growth and migration, inhibited the deposition of TS into the SMC extracellular matrix (Figure 7) and blocked the growth-facilitative effects of TS. TS therefore appears to play a major regulatory role in the control of SMC proliferation and may act as an extracellular "integrator" of growth-regulatory signals present at the SMC cell surface (Figure 8).

Further investigation of the underlying mechanism whereby TS facilitates SMC growth is required.

Note added in proof: Since submission of this manuscript for publication, a useful review by Lawler (1986a) has appeared. Three groups have described cDNA clones for human TS (Dixit *et al.*, 1986; Lawler and Hynes, 1986b; Kobayashi *et al.*, 1986). The predicted amino acid sequence of the protein shows interesting internally repeating homologies. In addition, TS shares conserved amino acid sequences with several known proteins. Finally, recent data (Varani *et al.*, 1986; Roberts *et al.*, 1987) have established that TS may mediate the attachment and spreading of melanoma and carcinoma cells.

ACKNOWLEDGMENTS. We thank Monika Clowes for assistance with balloon catheterization of rat aortas and Donna Stewart for preparation of the manuscript. Original studies from this laboratory were supported by National Institute of Health grants HL 18645, AM 11248, and DE 02600 and by a grant from R. J. Reynolds Industries, Inc.

REFERENCES

Agam, G., Shohat, O., and Livne, A., 1984, Thrombospondin plays a role in platelet–platelet recognition during release-related aggregation, *Proc. Soc. Exp. Biol. Med.* **177**:482–486.

Baenziger, N., Brodie, G., and Majerus, P., 1971, A thrombin-sensitive protein of human platelet membranes, *Proc. Natl. Acad. Sci. USA* **68**:240–243.

Baenziger, N., Brodie, G., and Majerus, P., 1972, Isolation and properties of thrombin-sensitive protein of human platelets, *J. Biol. Chem.* **247**:2723–2731.

Bale, M. D., Westrich,L. G., and Mosher, D. F., 1985, Incorporation of thrombospondin into fibrin clots, *J. Biol. Chem.* **260**:7502–7508.

Booth, W. J., Castaldi, P. A., and Berndt, M. C., 1985, Platelet thrombospondin hemagglutinin activity is due to aggregate formation, *Thromb. Res.* **39**:29–42.

Bornstein, P., and Majack, R. A., 1985, Thrombospondin, an extracellular glycoprotein with possible growth-regulatory properties, in: *Proceedings, International Symposium on Biology and Chemistry of Basement Membranes* (S. Shibata, ed.), Elsevier Science Publishers, Amsterdam, pp. 301–308.

Castellot, J. J., Addonizio, M. L., Rosenberg, R. D., and Karnovsky, M. J., 1981, Cultured endothelial cells produce a heparin-like inhibitor of smooth muscle cell growth, *J. Cell Biol.* **90**:372–379.

Clowes, A. W., and Karnovsky, M. J., 1977, Suppression by heparin of smooth muscle cell proliferation in injured arteries, *Nature* **265**:625–626.

Coligan, J E., and Slayter, H. S., 1984, Structure of thrombospondin, *J. Biol. Chem.* **259**:3944–3948.

Danishefsky, K. J., Alexander, R. J., and Detwiler, T. C., 1984, Formation of a stable complex of thrombin and the secreted platelet protein glycoprotein G (thrombospondin, TSP) by thiol-disulfide exchange, *Biochemistry* **23**:4984–4990.

Dawes, J., Clemetson, K. J., Gogstad, G. O., McGregor, J., Clezardin, P., Prowse, C. V., and Peppers, D. S., 1983, A radioimmunoassay for thrombospondin, used in a comparative study of thrombospondin, β-thromboglobulin, and platelet factor 4 in healthy volunteers, *Thromb. Res.* **29:**569–581.

Dixit, V. M., Grant, G. A., Frazier, W. A., and Santoro, S. A., 1984a, Isolation of the fibrinogen-binding region of platelet thrombospondin, *Biochem. Biophys. Res. Commun.* **119:**1075–1081.

Dixit, V. M., Grant, G. A., Santoro, S. A., and Frazier, W. A., 1984b, Isolation and characterization of a heparin-binding domain from the amino terminus of platelet thrombospondin, *J. Biol. Chem.* **259:**10100–10105.

Dixit, V. M., Haverstick, D. M., O'Rourke, K. M., Hennessy, S. W., Grant, G. A., Santoro, S. A., and Frazier, W. A., 1985a, Effects of anti-thrombospondin monoclonal antibodies on the agglutination of erythrocytes and fixed, activated platelets by purified thrombospondin, *Biochemistry* **24:**4270–4275.

Dixit, V. M., Haverstick, D. M., O'Rourke, K. M., Hennessy, S. W., Grant, G. A., Santoro, J. A., and Frazier, W. A., 1985b, A monoclonal antibody against human thrombospondin inhibits platelet aggregation, *Proc. Natl. Acad. Sci. USA.* **82:**3472–3476.

Dixit, V. M., Hennessy, S. W., Grant, G. A., Rotwein, P., and Frazier, W. A., Characterization of a cDNA encoding the heparin and collagen binding domains of human thrombospondin, *Proc. Natl. Acad. Sci. USA.* **83:**5449–5453.

Galvin, N. J., Dixit, V. M., O'Rourke, K. M., Santoro, S. A., Grant, G. A., and Frazier, W. A., 1985, Mapping of epitopes for monoclonal antibodies against human platelet thrombospondin with electron microscopy and high sensitivity amino acid sequencing, *J. Cell Biol.* **101:**1434–1441.

Gartner, T. K., and Dockter, M. E., 1983, Secreted platelet thrombospondin binds monovalently to platelets and erythrocytes in the absence of free Ca^{2+}, *Thromb. Res.* **33:**19–30.

Gartner, T., Gerrard, J., White, J., and Williams, D., 1981a, The endogenous lectin of human platelets is an α-granule component, *Blood* **58:**153–157.

Gartner, T., Gerrard, J., White, J., and Williams, D., 1981b, Fibrinogen is the receptor for the endogenous lectin of human platelets, *Nature* **289:**688–690.

Gartner, T., Doyle, M., and Mosher, D., 1984a, Effect of thrombospondin antibodies on the hemagglutinating activities of the endogenous platelet lectin and thrombospondin, *Thromb. Haemost.* **52:**354–357.

Gartner, T. K., Wlaz, D. A., Aiken, M., Starr-Spires, L., and Ogilvie, M. L., 1984b, Antibodies against a 23Kd heparin binding fragment of thrombospondin inhibit platelet aggregation, *Biochem. Biophys. Res. Commun.* **124:**290–295.

George, N. J., Lyons, R. M., and Morgan, R. K., 1980, Membrane changes associated with platelet activation, *J. Clin. Invest* **66:**1–9.

Hagen, I., Gogstad, G. O., Brosstad, F., and Solum, N. O., 1983, Demonstration of ^{125}I-labeled thrombin binding platelet proteins by use of crossed immunoélectrophoresis and autoradiography, *Biochim. Biophys. Acta* **732:**600–606.

Haverstick, D. M., Dixit, V. M., Grant, G. A., Frazier, W. A., and Santoro, S. A., 1984, Localization of the hemagglutinating activity of platelet thrombospondin to a 140,000-dalton thermolytic fragment, *Biochemistry* **23:**5597–5603.

Haverstick, D. M., Dixit, V. M., Grant, G. A., Frazier, W. A., and Santoro, S. A., 1985, Characterization of the platelet agglutinating activity of thrombospondin, *Biochemistry* **24:**3128–3134.

Houdijk, W. P. M., de Groot, P. G., Nievelstein, P. F. E. M., Sakariassen, K. S., and Sixma, J. J., 1986, Subendothelial proteins and platelet adhesion. Von Willebrand factor and fibronectin, but not thrombospondin, are involved in platelet adhesion to extracellular matrix of human vascular endothelial cells, *Arteriosclerosis* **6:**24–33.

Hoylaerts, M., Rijken, D. C., Lijnen, H. R., and Collen, D., 1982, Kinetics of the activation of plasminogen by human tissue plasminogen activator: Role of fibrin, *J. Biol. Chem.* **257**:2912–2919.

Hunter, N. R., Dawes, J., MacGregor, I. R., and Pepper, D. S., 1984, Quantitation by radioimmunoassay of thrombospondin synthesized and secreted by human endothelial cells, *Thromb. Haemost.* **52**:288–291.

Hynes, R. O., and Yamada, K. M., 1982, Fibronectins: Multifunctional modular glycoproteins, *J. Cell Biol.* **95**:369–377.

Jaffe, E. A., Leung, L. L. K., Nachman, R., Levin, R., and Mosher, D. F., 1982, Thrombospondin is the endogenous lectin of human platelets, *Nature* **295**:246–248.

Jaffe, E. A., Ruggiero, J. T., Leung, L. L. K., Doyle, M. J.,McKeown-Longo, P. J., and Mosher, D. F., 1983, Cultured human fibroblasts synthesize and secrete thrombospondin and incorporate it into extracellular matrix, *Proc. Natl. Acad. Sci. USA.* **80**:998–1002.

Jaffe, E. A., Ruggiero, J. T., and Facone, D. J., 1985, Monocytes and macrophages synthesize and secrete thrombospondin, *Blood* **65**:79–84.

Kobayashi, S., Eden-McCutchan, F., Framson, P., and Bornstein, P., 1986, Partial amino acid sequence of human thrombospondin as determined by analysis of cDNA clones: Homology to malarial circumsporozite proteins, *Biochemistry* **25**:8418–8425.

Kramer, R. H., Fuh, G. M., Bensch, K. G., and Karasek, M. A., 1985, Synthesis of extracellular matrix glycoproteins by cultured microvascular endothelial cells isolated from the dermis of neonatal and adult skin, *J. Cell. Physiol.* **123**:1–9.

Lahav, J., Schwartz, M. A., and Hynes, R. O., 1982, Analysis of platelet adhesion with a radioactive chemical crosslinking reagent: Interaction of thrombospondin with fibronectin and collagen, *Cell* **31**:253–262.

Lahav, J., Lawler, J., and Gimbrone, M. A., 1984, Thrombospondin interactions with fibronectin and fibrinogen. Mutual inhibition in binding, *Eur. J. Biochem.* **145**:151–156.

Lawler, J., and Simons, E., 1983, Cooperative binding of calcium to thrombospondin: The effect of calcium on the circular dichroism and limited tryptic digestion of thrombospondin, *J. Biol. Chem.* **258**:12098–12101.

Lawler, J., and Slayter, H., 1981, The release of heparin binding peptides from platelet thrombospondin by proteolytic action of thrombin, plasmin, and trypsin, *Thromb. Res.* **22**:267–279.

Lawler, J. W., Chao, F. C., and Fang, P.-H., 1977, Observation of a high molecular weight platelet protein released by thrombin, *Thromb. Haemost.* **37**:355–357.

Lawler, J., Slayter, H., and Coligan, J., 1978, Isolation and characterization of high molecular weight glycoprotein from human blood platelets, *J. Biol. Chem.* **253**:8609–8616.

Lawler, J., Chao, F. C., and Cohen, C. M., 1982, Evidence for calcium-sensitive structure in platelet thrombospondin: Isolation and partial characterization of thrombospondin in the presence of calcium, *J. Biol. Chem.* **257**:12257–12265.

Lawler, J., Derick, L. H., Connolly, J. E., Chen, J.-H., and Chao, F. C., 1985, The structure of human platelet thrombospondin, *J. Biol. Chem.* **260**:3762–3772.

Lawler, J., 1986a, The structural and functional properties of thrombospondin, *Blood* **67**:1197–1209.

Lawler, J., and Hynes, R. O., 1986b, The structure of human thrombospondin, an adhesive glycoprotein with multiple calcium-binding sites and homologies with several different proteins, *J. Cell Biol.* **103**:1635–1648.

Leung, L. L. K., 1984, Role of thrombospondin in platelet aggregation, *J. Clin. Invest.* **74**:1764–1772.

Leung, L. L. K., and Nachman, R. L., 1982, Complex formation of platelet thrombospondin with fibrinogen, *J. Clin. Invest.* **70**:542–549.

Leung, L. L. K., Nachman, R.L., and Harpel, P. C., 1984, Complex formation of platelet thrombospondin with histidine rich glycoprotein, *J. Clin. Invest.* **73**:5–12.

Lijnen, H. R., Hoylaerts, M., and Collen, D., 1983, Heparin binding properties of human histidine rich glycoprotein, *J. Biol. Chem.* **255:**3803–3808.

McKeown-Longo, P. J., Hanning, R., and Mosher, D. F., 1984, Binding and degradation of platelet thrombospondin by cultured fibroblasts, *J. Cell Biol.* **98:**22–28.

McLaren, K. M., 1983, Immunohistochemical localization of thrombospondin in human megakaryocytes and platelets, *J. Clin. Pathol.* **36:**197–199.

McPherson, J., Sage, H., and Bornstein, P., 1981, Isolation and characterization of a glycoprotein secreted by aortic endothelial cells in culture: Apparent identity with platelet thrombospondin, *J. Biol. Chem.* **256:**11330–11336.

Majack, R. A., and Bornstein, P., 1984, Heparin and related glycosaminoglycans modulate the secretory phenotype of vascular smooth muscle cells, *J. Cell Biol.* **99:**1688–1695.

Majack, R. A., and Bornstein, P., 1985, Heparin regulates the collagen phenotype of vascular smooth muscle cells: Induced synthesis of an M_r 60,000 collagen, *J. Cell Biol.* **100:**613–619.

Majack, R. A., and Bornstein, P., 1987, Biosynthesis and modulation of extracellular matrix components by cultured vascular smooth muscle cells, in: *The Vascular Smooth Muscle Cell in Culture* (J. H. Chamley-Campbell and G. R. Campbell, eds.), CRC Press, Boca Raton, Florida (in press).

Majack R. A., and Clowes, A. W., 1984, Inhibition of vascular smooth muscle cell migration by heparin-like glycosaminoglycans, *J. Cell Physiol.* **118:**253–256.

Majack, R. A., Cook, S. C., and Bornstein, P., 1985, Platelet-derived growth factor and heparin-like glycosaminoglycans regulate thrombospondin synthesis and deposition in the matrix by smooth muscle cells, *J. Cell Biol.* **103:**1059–1070.

Majack, R. A., Cook, S. C., and Bornstein, P., 1986, Regulation of vascular smooth muscle cell growth by the extracellular matrix: An autocrine role for thrombospondin, *Proc. Natl. Acad. Sci. USA.* **83:**9050–9054.

Margossian, S., Lawler, J., and Slayter, H., 1981, Physical characterization of platelet thrombospondin, *J. Biol. Chem.* **256:**7495–7500.

Mosher, D. F., Doyle, M. J., and Jaffe, E. A., 1982, Synthesis and secretion of thrombospondin by cultured human endothelial cells, *J. Cell Biol.* **93:**343–348.

Mumby, S. M., 1984, The Role of Thrombospondin in the Extracellular Matrix, Ph.D. Dissertation, University of Washington, Seattle, Washington.

Mumby, S. M., Abbott-Brown, D., Raugi, G. R., and Bornstein, P., 1984a, Regulation of thrombospondin secretion by cells in culture, *J. Cell. Physiol.* **120:**280–288.

Mumby, S. M., Raugi, G. J., and Bornstein, P., 1984b, Interactions of thrombospondin with extracellular matrix proteins: Selective binding to type V collagen, *J. Cell Biol.* **98:**646–652.

Murphy-Ullrich, J. E., and Mosher, D. F., 1985, Localization of thrombospondin in clots formed *in situ*, *Blood* **66:**1098–1104.

Odermatt, E., Tamkun, J. W., and Hynes, R. O., 1985, Repeating modular structure of the fibronectin gene: Relationship to protein structure and subunit variation, *Proc. Natl. Acad. Sci. USA.* **82:**6571–6575.

Patthy, L., 1985, Evolution of the proteases of blood coagulation and fibrinolysis by assembly from modules, *Cell* **41:**657–663.

Phillips, D., Jennings, L., Hullahalli, R., 1980, Ca^{2+}-mediated association of glycoprotein G (thrombin-sensitive protein, thrombospondin) with human platelets, *J. Biol. Chem.* **255:**11629–11632.

Raugi, G. J., Mumby, S. M., Abbott-Brown, D., and Bornstein, P., 1982, Thrombospondin: Synthesis and secretion by cells in culture, *J. Cell Biol.* **95:**351–354.

Raugi, G. M., Mumby, S. M., Ready, C. A., and Bornstein, P., 1984, Location and partial characterization of the heparin-binding fragment of platelet thrombospondin, *Thromb. Res.* **36:**165–175.

Reich, E., 1978, Activation of plasminogen: A widespread mechanism for generating localized extracellular proteolysis, in: *Biological Markers of Neoplasia: Basic and Applied Aspects* (R. W. Rudden, ed.), Elsevier/North Holland, Amsterdam, pp. 491–500.

Reinders, J. H., deGroot, P. G., Dawes, J., Hunter, N. R., van Heugten, H. A. A., Zandbergen, J., Gonsalves, M. D., and van Mourik, J. A., 1985, Comparison of secretion and subcellular localization of von Willebrand protein with that of thrombospondin and fibronectin in cultured human vascular endothelial cells, *Biochim. Biophys. Acta* **844**:306–313.

Roberts, D. D., Haverstick, D. M., Dixit, V. M., Frazier, W. A., Santoro, S. A., and Ginsburg, V., 1985a, The platelet glycoprotein thrombospondin binds specifically to sulfated glycolipids, *J. Biol. Chem.* **260**:9405–9411.

Roberts, D. D., Sherwood, J. A., Spitalnik, S. L., Panton, L. J., Howard, R. J., Dixit, V. M., Frazier, W. A., Miller, L. H., and Ginsburg, V., 1985b, Thrombospondin binds falciparum malaria parasitized erythrocytes and may mediate cytoadherence, *Nature* **318**:64–66.

Roberts, D. D., Sherwood, J. A., and Ginsburg, V., 1987, Platelet thrombospondin mediates attachment and spreading of human melanoma cells, *J. Cell. Biol.* **104**:131–139.

Sage, H., and Bornstein, P., 1982, Endothelial cells from umbilical vein and an hemangioendothelioma secrete basement membrane largely to the exclusion of interstitial procollagens, *Arteriosclerosis* **2**:27–36.

Sage, H., Pritzl, P., and Bornstein, P., 1981, Characterization of cell matrix-associated collagens synthesized by aortic endothelial cells in culture, *Biochemistry* **20**:436–442.

Sage, H., Farin, F. M., Striker, G. E., and Fisher, A. B., 1983, Granular pneumocytes in primary culture secrete several major components of the extracellular matrix, *Biochemistry* **22**:2148–2155.

Saglio, S., and Slayter, H., 1982, Use of a radioimmunoassay to quantify thrombospondin, *Blood* **59**:162–166.

Silverstein, R. L., Leung, L. L. K., Harpel, P. C., and Nachman, R. L., 1984, Complex formation of platelet thrombospondin with plasminogen: Modulation of activation by tissue activator, *J. Clin. Invest.* **74**:1625–1633.

Silverstein, R. L., Leung, L. L. K., Harpel, P. C., Nachman, R.L., 1985a, Platelet thrombospondin forms a trimolecular complex with plasminogen and histidine-rich glycoprotein, *J. Clin. Invest.* **75**:2065–2073.

Silverstein, R. L., Nachman, R.L., Leung, L. L. K., and Harpel, P. C., 1985b, Activation of immobilized plasminogen by tissue activator: Multimolecular complex formation, *J. Biol. Chem.* **260**:10346–10352.

Silverstein, R. L., Leung, L. K., and Nachman, R. L., 1986, Thrombospondin: A versatile multifunctional glycoprotein, *Arteriosclerosis* **6**:245–253.

Sudhof, T. C., Goldstein, J. L., Brown, M. S., and Russell, D. W., 1985, The LDL receptor gene: A mosaic of exons shared with different proteins, *Science* **228**:815–822.

Takahashi, K., Aiken, M., Fenton, J. W., and Walz, D. A., 1984, Thrombospondin fragmentation by α-thrombin and resistance to γ-thrombin, *Biochem. J.* **224**:673–676.

Tuszynski, G. P., Srivastava, S., Switalska, H. I., Holt, J. C., Cierniewski, C. S., and Niewiarowski, S., 1985, The interaction of human platelet thrombospondin with fibrinogen, *J. Biol. Chem.* **160**:12240–12245.

Varani, J., Dixit, V. M., Fligiel, S. E. G., McKeever, P. E., and Carey, T. E., 1986, Thrombospondin-induced attachment an spreading of human squamous carcinoma cells, *Exp. Cell Res.* **167**:376–390.

Wencel-Drake, J. D., Plow, E. F., Zimmerman, T. S., Painter, R. G., and Ginsberg, M. H., 1984, Immunofluorescent localization of adhesive glycoproteins in resting and thrombin-stimulated platelets, *Am. J. Pathol.* **115**:156–164.

Wight, T. N., Raugi, G. J., Mumby, S. M., and Bornstein, P., 1985, Light microscopic immunolocation of thrombospondin in human tissues, *J. Histochem. Cytochem.* **33**:295–302.

3

NEURONAL ANTIGENS INVOLVED IN CELL ADHESION AND CELL RECOGNITION

Gregory J. Cole and Richard W. Bond

1. INTRODUCTION

With the introduction of immunological methods, particularly the hybridoma technology of Kohler and Milstein (1975), major advances have been made in our understanding of neuronal function. An important application of monoclonal antibodies in neurobiology has been the use of these probes to identify antigens that are restricted to specific neural cell types. Monoclonal antibodies have also been used to identify, purify, and then characterize antigens that are required for neuronal cell adhesion and cell recognition processes.

This review can be subdivided into two major sections, which focus on the application of monoclonal antibodies for identifying cell type-specific neural antigens and neuronal cell-adhesion molecules. Although the main emphasis of the review is on antigens that participate in neuronal cell adhesion, we have included a table summarizing the neuronal cell type-specific antigens that have been identified using immunological protocols. A number of these antigens may also be involved in neuronal cell recognition during neural development and have been discussed in the text. Due to the tremendous growth of this field and its literature in the last few years, we apologize for any omission from the table of neuronal antigens, which are presumed to be cell type or neural specific.

Gregory J. Cole ● Department of Anatomy and Cell Biology, Medical University of South Carolina, Charleston, South Carolina 29425. Richard W. Bond ● Department of Anatomy and Neurobiology, Washington University School of Medicine, St. Louis, Missouri 63110.

2. IDENTIFICATION OF NEURAL ANTIGENS USING IMMUNOLOGICAL APPROACHES

Following the introduction of hybridoma technology, it became apparent to neurobiologists that these reagents could serve as powerful tools for identifying minor neuronal antigens. Of particular interest was the fact that monoclonal antibodies could be employed to discriminate between different neural cell types and thus to obtain an understanding of the molecular composition of different neural cell classes. As shown in Tables I and II, we have compiled an extensive list of antigens that have been identified using immunological methods, primarily monoclonal antibodies. These antigens are restricted to individual cell types in the nervous system (Table I) or are components of the nervous system that are not detected in nonneural tissues (Table II). The antigens are listed in the tables according to the name of the antigen (or often the identifying monoclonal antibody) and are categorized according to the region of the nervous system that contains the antigen. We have also included a limited number of antigens (i.e., purpurin) that were identified using conventional antibodies; these antigens possess interesting properties or functions and are therefore included.

Owing to the number of antigens that have been described using immunological approaches, we have elected to discuss, in this review, only those antigens that may participate in cell recognition processes during neuronal development. Because there is considerable evidence describing the role of the

TABLE I. Cell Type-Specific Antigens

Antigen	Species	Cell type(s)	Identity	Other locations	Reference[a]
A. Retina					
HPC-1	Rat	Amacrine	35 Kd	Hippocampus	1
GAD1-4	Chick	Amacrine	GAD[b]	Brain	2
NCI/34HL	Chick	Amacrine	Substance P	CNS, PNS	3
YC5/45HL	Chick	Amacrine	Serotonin	CNS, PNS	3
NOC-1	Rat	Amacrine	Enkephalins	CNS, PNS	4
PNMT	Rat	Amacrine, ganglion	PNMT[c]	Brain, adrenals	5
94C2	Chick	Amacrine, ganglion	—	—	6
THY-1	Rat	Ganglion	25–29 Kd	Thymocytes, brain	7
D_1C_4	Chick	Ganglion	260 Kd	Brain, nonneural	8
C_3C_5	Chick	Ganglion	—	Brain, nonneural	8
Ret1,2,3	Chick	Ganglion	—	—	9
T61	Chick	Ganglion	—	—	10
AB5	Rabbit	Ganglion	—	—	11

TABLE I. (*Continued*)

Antigen	Species	Cell type(s)	Identity	Other locations	Reference[a]
23-4-C	Goldfish	Ganglion	140 Kd	—	12
HC-II.7	Carp	Horizontal	50 Kd	—	13
Ret 4,5	Chick	Inner retina	—	—	9
Ret-G1	Rat	Muller	—	Brain	14
R4	Mouse	Muller and horizontal	Filamentous	Many	15
Ret-P2	Rat	Photoreceptors	38 Kd	—	14
R3	Mouse	Several	185–200 Kd	—	15
R5	Mouse	Several	Filamentous	Many	15
pp60[c–src]	Chick	Several	60 Kd	Many	16

B. Cerebellum

CAT 101	Cat	Basket	—	Spinal cord	17
H9-H12	Rat	Cerebellar neurons	—	—	18
NSP-5	Mouse	Cerebellar neurons	180 Kd	DRG[d]	19
D1.1	Mouse	Germinal EGL[e]	Ganglioside	—	20
N1	Mouse	Granule	—	Cerebrum	21
7-8D2	Rat	Granule	—	None	22
8-20-1	Rat	Granule	48 Kd	Forebrain	22
69A1	Rat	Granule	—	—	23
M2	Mouse	Granule, Purkinje	—	—	24
12D5	Rat	Granular layer	270 Kd	—	25
CAT 301	Cat	Lugaro	—	Spinal cord	17
3G6.41	Rat	Molecular layer	140, 185 Kd	—	26
20D6	Rat	Purkinje	23, 240 Kd	—	25
PC1-PC4	Mouse	Purkinje	—	—	27
Spot 35	Rat	Purkinje	27 Kd	—	28
Q113	Rat	Purkinje	—	—	29
UCHT1	Human	Purkinje	—	T cells	30
CE5	Rat	Purkinje	60 Kd	PNS, heart	31
IF4	Rat	Purkinje	95, 130 Kd	Other brain areas	32
SMI.32	Rat	Purkinje, basket	200 Kd	PNS	33
CAT 201	Cat	Purkinje, others	—	Spinal cord	17

C. Olfactory system

BCL 1-2	Rat	Basal cell layer	—	—	34
GLA 1-9	Rat	Bowman's gland	—	—	34
LUM 1-7	Rat	Luminar surface	—	—	34
NEU 1-14	Rat	Neurons	—	—	34
2B8	Rat	Receptor neurons	163, 215 Kd	DRG, cerebellum	35
SBV-21	Rat	Receptor neurons	Vimentin	Many	36
RES 1-2	Rat	Respiratory	—	—	34
SUS 1-3	Rat	Sustentacular	—	—	34

(*continued*)

TABLE I. (Continued)

Antigen	Species	Cell type(s)	Identity	Other locations	Reference[a]
D. Glia					
M1	Mouse	Astrocytes	—	—	37
Ran 2	Rat	Astrocytes	140 Kd	—	38
S1	Mouse	Astrocytes, oligo[f]	—	—	39
C1	Mouse	Astrocytes, Schwann	—	—	37
H8, etc.	Rat	Glia	—	—	18
7D5	Rat	Glia	>30 bands	—	25
5E10	Chick	Muller, radial	250 Kd	Many	40
7G4	Chick	Muller, other	62 Kd	—	41
01–04	Mouse	Oligodendrocytes	—	—	42
217c	Rat	Schwann	—	—	43
MBP1-7	Human	Schwann	MBP[g]	—	44
224-58	Human	Schwann	SMGL[h]	—	45
MAG	Human	Schwann, oligo[f]	MAG[i]	—	46
E. Dorsal root ganglion					
RT97	Rat	Large light neurons	Neurofilament	CNS	47
SSEA-3,4	Mouse	Subsets of neurons	Carbohydrate	—	48
F. Nonvertebrate					
Mes-2	Grasshopper	Metathoracic	—	—	49
24B10	Fly	Photoreceptors	160 Kd	—	50
I5	Grasshopper	Subset of neurons	—	—	51
LAN 3-1	Leech	Two cells per ganglia	—	—	52

[a] (1) Barnstable *et al.* (1985), (2) Gottlieb (1986), (3) Cuello (1983), (4) Cuello *et al.* (1984), (5) Park *et al.* (1986), (6) Pessac *et al.* (1983), (7) Barnstable and Drager (1984), (8) Cole *et al.* (1986), (9) Lemmon and Gottlieb (1982), (10) Henke-Fahle and Bonheffer (1983), (11) Fry *et al.* (1985), (12) Schwartz and Eshhar (1984), (13) Young and Dowling (1984), (14) Barnstable (1980), (15) Drager *et al.* (1984), (16) Sorge *et al.* (1984), (17) McKay and Hockfield (1982), (18) Ghandour *et al.* (1984), (19) Rougon *et al.* (1984), (20) Levine *et al.* (1984), (21) Schnitzer *et al.* (1984a), (22) Webb and Woodhams (1984), (23) Pigott and Kelly (1984), (24) Lagenaur and Schachner (1981), (25) Hawkes *et al.* (1982), (26) Williams *et al.* (1985), (27) Weber and Schachner (1984), (28) Yamakuni *et al.* (1984), (29) Plioplys *et al.* (1985), (30) Caddy *et al.* (1982), (31) Wood *et al.* (1982), (32) Sajovic *et al.* (1986), (33) Sternberger *et al.* (1982), (34) Hempstead and Morgan (1985), (35) Allen and Akeson (1985), (36) Schwob *et al.* (1986), (37) Schachner *et al.* (1984), (38) Mirsky and Jessen (1984), (39) Schnitzer *et al.* (1984b), (40) Lemmon (1986), (41) Lemmon (1985), (42) Sommer and Schachner (1981), (43) Fields and Dammerman (1985), (44) Elfman *et al.* (1986), (45) Guerci *et al.* (1986), (46) Favilla *et al.* (1984), (47) Lawson *et al.* (1984), (48) Dodd *et al.* (1984), (49) Kotrla and Goodman (1984), (50) Zipursky *et al.* (1985), (51) Ho and Goodman (1982), (52) Zipser and McKay (1981).
[b] GAD, glutamic acid decarboxylase.
[c] PNMT, phenylethanolamine *N*-methyltransferase.
[d] DRG, dorsal root ganglia.
[e] EGL, external granule layer.
[f] Oligo, oligodendrocytes.
[g] MBP, myelin basic protein.
[h] SMGL, sulfomonogalactolipids.
[i] MAG, myelin-associated glycoprotein.

TABLE II. Neural Specific Antigens

Antigen	Species	Location	Identity	Reference[a]
NC-1	Avian	Neural crest	Three bands	1
GP130	Chick	CNS, PNS	130-Kd glycoprotein	2
A_2B_5	Chick	Brain, retina, PNS	Ganglioside	3
Purpurin	Chick	Retina	20 Kd	4
18B8	Chick	Retina	Ganglioside	5
TOP	Chick	Gradient in retina	Protein	6
H3	Chick	CNS, PNS	—	7
LAN3–8	Leech	All neural cells	63 Kd	8
A4	Rat	CNS	—	9
38/D7	Rat	PNS	—	10
G5	Rat	Brain	100 Kd	11
GP41	Rat	Brain	36, 38, 41 Kd	12
N-CAM	Many	CNS, PNS	140, 170 Kd	13
Ng-CAM	Many	CNS, PNS	200, 140 Kd	14

[a] (1) Vincent and Thiery (1984), (2) Ranscht et al. (1984), (3) Fredman et al. (1984), (4) Schubert et al. (1986), (5) Dubois et al. (1986), (6) Trisler et al. (1981), (7) Leah et al. (1984), (8) McKay et al. (1984), (9) Miller et al. (1984), (10) Vulliamy et al. (1981), (11) Akeson and Graham (1981), (12) Webb and Woodhams (1984), (13) Thiery et al. (1977), Grumet et al. (1984a).

neural cell-adhesion molecule or N-CAM, and other well-characterized cell-adhesion molecules in neuronal development, these antigens will be discussed separately in later sections. Initially, we will describe those antigens that have been identified by immunological protocols and that may play a role in neuronal cell recognition.

3. ROLE OF IDENTIFIED ANTIGENS IN NEURONAL CELL RECOGNITION

3.1. Cell Type-Specific Molecules

With the advent of hybridoma technology, it has been possible to identify antigens that are restricted to specific cell types in the developing nervous system. Although a majority of these antigens have not been well characterized (Table I), some appear to be promising candidates for cell-recognition molecules. In the dorsal root ganglia, monoclonal antibodies that recognize the stage-specific embryonal antigens SSEA-3 and SSEA-4 react with only a subset of sensory neurons (Dodd et al., 1984). These monoclonal antibodies recognize globosides, which are glycoconjugates that are usually associated with glycolipids. In light of evidence suggesting that interactions between carbohydrate and carbohydrate-binding proteins (lectins) may mediate cell interactions, these data raise the question of whether these antigens participate in cell-recognition events associated with development of the dorsal root gangion. In accord with these findings, Regan *et al.* (1986) have shown that endogenous lactose-binding lectins (RL-14.5 and RL-29) are restricted to subsets of rat dorsal root ganglia neurons. These lectins appear soon after formation of the dorsal root ganglion and are present in cells that produce lactoseries glycoconjugates. The data imply that these lectins, and the glycoconjugates they recognize, may be required for development and formation of primary sensory neurons.

A number of studies have identified antigens that are cell type specific and developmentally regulated; a likely function for these antigens, albeit as yet unconfirmed, is that they are involved in processes such as cell recognition associated with neural development.

In the goldfish retina, Schwartz and Eshhar (1984) used the 23-4-C monoclonal antibody to identify a 140-Kd glycoprotein that is specific for ganglion cells. This glycoprotein increases in amount following optic nerve injury, and if the 23-4-C monoclonal antibody is injected intraocularly after nerve injury, the regenerative response is augmented. These investigators propose that the monoclonal antibody either activates the retinal ganglion cells by interacting with their cell-surface molecules or augments regeneration by blocking an axonal growth inhibitor.

Our laboratory has also used monoclonal antibodies to identify a 260-Kd polypeptide that is restricted to chick retinal ganglion cells (Cole *et al.*, 1986a). The D_1C_4 antigen is developmentally regulated, and its expression is correlated with the period of ganglion cell axonal growth to the optic tectum. This polypeptide may therefore play a role in ganglion cell axonal growth or retinotectal synaptogenesis.

Barnstable and his co-workers (1985) have identified a 35-Kd polypeptide (using the HPC-1 monoclonal antibody) in rat retina that is a marker of amacrine and displaced amacrine cells. This molecule is expressed by amacrine cells in late embryonic retina during the period of amacrine cell migration and may be involved in this process.

Using a variety of experimental methods, other neuronal markers have been described that may play an important role in neuronal development. The Thy-1 glycoprotein is widely distributed in the central nervous system (CNS) but appears to show some cell-type specificity (Barnstable and Drager, 1984; Perry *et al.*, 1984). Recent studies conducted in the developing cerebellum have examined the developmental expression of Thy-1, and from these studies it was concluded that the appearance of Thy-1 is correlated with the period of synaptogenesis (Morris *et al.*, 1985a,b; Bolin and Rouse, 1986). Possible support for the hypothesis that Thy-1 participates in cell recognition has been obtained by analysis of the chromosome localization of the Thy-1 gene and the N-CAM gene (D'Eustachio *et al.*, 1985). The N-CAM gene has been localized to chromosome 9 in the mouse genome, which is also the chromosome that contains the Thy-1 gene. Thus, the gene encoding Thy-1 is in close proximity to the gene encoding a known cell-adhesion molecule. These data may indicate that the Thy-1 glycoprotein also has a cell-recognition function.

The TOP antigen, a 55-Kd polypeptide (Trisler *et al.*, 1981), is a molecule that appears to be a promising candidate for participating in neuronal cell recognition. It is well known that the retinotectal map exhibits a dorsoventral organization, and the TOP antigen is present in a gradient across the embryonic chicken neural retina; in this gradient, thirty-five-fold more antigen is detected in dorsoposterior retina than in ventroanterior retina. The antigen is also present in the synaptic layers of retina and has been proposed to be involved in synapse formation during retinogenesis. However, compelling experimental evidence for this hypothesis has been lacking.

Pigott and Kelly (1984) have used the 69A1 monoclonal antibody to identify an antigen that may be involved in neuronal cell recognition. The 69A1 antigen is restricted to the molecular layer of the developing rat cerebellum and is developmentally regulated; the antigen disappears from the cerebellum during the third postnatal week. This disappearance corresponds to the onset of synaptogenesis, so the antigen may be necessary for neurite extension or recognition between the growing axons and their target cells.

There is an intense interest in neurobiology to elucidate the mechanism underlying the process of myelin formation. The myelin-associated glycoprotein is a well-characterized neuronal antigen that has been proposed to be involved in interactions between axons and myelin (Quarles *et al.*, 1973; Sternberger *et al.*, 1979). Gulcher *et al.* (1986) have also identified two glycoproteins, with molecular weights of 150 and 225 Kd, that share epitopes with myelin-associated glycoprotein. These antigens are transiently expressed in the developing human CNS and are not detected in myelin. However, these antigens are present on processes of myelinating oligodendrocytes; the data may indicate that these antigens are involved in interactions between oligodendrocytes and axons during myelination.

A neural tissue-specific antigen, detected using conventional immunological protocols, has also been shown to participate in neuronal cell recognition. The antigen, named purpurin, is a chicken retina-specific molecule with a molecular weight of 20 Kd (Schubert and LaCorbiere, 1985b). Purpurin is secreted by neural retina cells *in vitro* and is a component of adherons. Adherons are particles of extracellular matrix-like material that promote retinal cell–substratum adhesion (Schubert *et al.*, 1983). Purpurin interacts with a heparin or heparan sulfate, and it appears that retinal cell attachment to purpurin occurs via a cell-surface heparan sulfate proteoglycan (Schubert and LaCorbiere, 1985a). Recent studies have shown that purpurin shares sequence homology with human serum retinol-binding protein, and serum retinol-binding protein promotes retinal cell attachment (Schubert *et al.*, 1986). Purpurin also promotes the survival of dissociated ciliary ganglion cells, raising the question of whether purpurin and the ciliary ganglion survival factor described by Collins (1985) are identical.

3.2. Monoclonal Antibodies as Probes for the Analysis of Axonal Guidance

Immunological protocols have been particularly useful for the study of axonal growth in the developing nervous system. Several different experimental approaches have been utilized, although most have utilized monoclonal antibodies. A number of monoclonal antibodies have been produced that inhibit neurite outgrowth *in vitro;* these data suggest that the antigens recognized by the monoclonal antibodies are necessary for cell interactions during axonal growth. Henke-Fahle and Bonhoeffer (1983) showed that the T61/3/12 monoclonal antibody inhibited ganglion cell axonal growth in the developing chick neural retina. This antigen is also restricted to the retinal ganglion cells. Another monoclonal antibody that inhibits neurite outgrowth and long-term neuronal survival has been described by Leah *et al.* (1984). The antigen, H3, has not been identified and is distributed throughout the peripheral nervous system (PNS) and the CNS. This antigen may therefore be involved in the more general mechanism of axonal growth, since the antigen described by Henke-Fahle and Bonhoeffer (1983) is a cell-specific molecule.

The insect nervous system has also served as an especially useful system for the analysis of axonal guidance. Monoclonal antibodies have been produced that identify antigens expressed by only small subpopulations of individual neurons in the grasshopper nervous system. These monoclonal antibodies can therefore be used in immunocytochemical studies to examine the outgrowth of identified axons. An example of a unique neuronal marker in the grasshopper nervous system is the Mes-2 antigen (Kotrla and Goodman, 1984). Mes-2 antigen is expressed by only four axons out of approximately 1000 neurons in each metathoracic hemisegment. Mes-2 is a cell-surface component that is transiently expressed during a short period of embryogenesis correlated with growth cone extension. Specifically, it appears that the Mes-2 antigen is required for fasciculation between the FETi and SETi axons, as fasciculation occurs only following expression of Mes-2 on each cell. Thus, a highly restricted neuronal marker appears to be required for axonal extension and cell recognition during grasshopper embryogenesis.

The I5 monoclonal antibody has also been employed to study the mechanisms underlying cell recognition during axonal growth in the grasshopper nervous system. The I5 monoclonal antibody recognizes a specific subset of grasshopper neurons (Ho and Goodman, 1982) and has been used to examine the pathways taken by the growth cones of pioneer axons. Using this monoclonal antibody it was possible to demonstrate that pioneer neurons interact with specific landmark cells during axonal growth and that these pathways are then followed by other neurons later in development.

3.3. Neural Cell Migration

Immunological approaches have been widely used to examine the role of various adhesive molecules in neural cell migration. Several of these studies have focused on the extracellular matrix molecule fibronectin, which is involved in fibroblast cell adhesion and migration (Hynes and Yamada, 1982). Hatten *et al.* (1982) used conventional immunological methods to demonstrate that granule cell migration in the developing cerebellum was dependent upon fibronectin. Their conclusions were based on observations that fibronectin was a component of the pathway followed by migrating granule cells. Although recent studies have demonstrated that astrocytes synthesize and secrete large amounts of a variant form of fibronectin (Price and Hynes, 1985), it is not known whether this fibronectin is involved in cell migration in the CNS. Likewise, the validity of the studies by Hatten *et al.* (1982) has been questioned, as Schachner *et al.* (1983) could not detect fibronectin along the path of migrating cerebellar granule cells.

Despite the uncertainty of the role of fibronectin in neural cell migration, a neuronal cell–cell adhesion molecule that participates in cerebellar granule cell migration has been identified using monoclonal antibodies. This antigen, named L1 (Rathjen and Schachner, 1984), will be described in more detail in Section 6 of this review.

Although it appears that fibronectin does not play a significant role in neuronal cell migration in the CNS, the molecule is an excellent candidate for regulating neural crest cell migration during embryogenesis. Several lines of evidence suggest that fibronectin is required for neural crest cell migration. These include studies that employed synthetic peptides containing the cell-recognition sequence of fibronectin. The synthetic peptide with the cell-recognition sequence inhibited gastrulation and neural crest cell migration in the avian embryo (Boucaut *et al.*, 1984). The role of fibronectin in neural crest cell migration has been confirmed by hybridoma methods. Monoclonal antibodies have been produced that recognize a 140-Kd glycoprotein complex and that inhibit muscle cell–substratum interactions (Greve and Gottlieb, 1982; Neff *et al.*, 1982). This glycoprotein complex has now been identified as a receptor for fibronectin (Horwitz *et al.*, 1985; Pytela *et al.*, 1985; Akiyama *et al.*, 1986) and has also been detected on the surface of neural crest cells, in close association with fibronectin (Duband *et al.*, 1986). In accord with these data, monoclonal antibodies against this antigen inhibit neural crest cell attachment to a substratum *in vitro* and perturb neural crest cell migration *in vivo* (Bronner-Fraser, 1985).

The studies described above have demonstrated how monoclonal antibodies can be used to identify and characterize novel neuronal antigens and how the monoclonal antibodies can then be employed as probes for analyzing the function of the antigen. In the following section we will describe several neuronal cell-adhesion molecules that have been identified using immunological approaches and that have been well characterized with respect to their biochemical and functional properties.

4. NEURONAL CELL-ADHESION MOLECULES

As described in the previous sections, monoclonal antibodies or conventional antibodies have proved to be useful probes for the identification of novel neural antigens. A primary aim of these studies has been to identify antigens that are critical to neural differentiation or development; however, an understanding of the function of antigens of this type has been facilitated for the most part by the application of *in vitro* assays that measure cell–cell adhesion between embryonal neural cells. Neural development can be viewed as a complex process that depends upon cell birth, migration, recognition and, ultimately, functional interaction. Cell–cell adhesion has been particularly well studied during neural development, since *in vitro* assay systems can be developed to identify molecules that are important for the adhesion process. Immunological techniques have played a vital role in identifying molecules that participate in neuronal cell–cell adhesion, and monoclonal antibodies have been particularly useful in recent studies as probes to identify neuronal cell adhesion molecules.

5. N-CAM

In this section we will describe primarily a cell-surface glycoprotein that has been directly implicated in neural cell adhesion. This molecule has been named N-CAM and is a well-characterized adhesive glycoprotein that has been described in detail in previous reviews (Edelman, 1983; Rutishauser, 1984). We will therefore provide limited background information on N-CAM and will focus attention primarily on recent developments regarding N-CAM-mediated adhesive processes. We will also describe other neuronal cell-adhesion molecules that have been identified using monoclonal antibodies, although the biochemical and functional properties of these molecules have been only partially characterized.

5.1. Chemistry of N-CAM

N-CAM was identified on the basis of an immunological analysis of cell adhesion, which was initially employed for the study of slime mold adhesion by Gerisch and co-workers (Beug *et al.*, 1973). To examine cell–cell adhesion in the embryonal chicken nervous system, an *in vitro* aggregation assay was employed that was based on the measurement of the rate of disappearance of single cells, which was coincident with the formation of cell aggregates. By using this assay, it was possible to investigate whether the formation of cell aggregates was capable of being inhibited by antibodies. An antiserum directed against the retinal cell surface (R10) inhibited cell–cell aggregation *in vitro*, and the antiserum recognized a 140-Kd protein that was later designated N-CAM (Brackenbury *et al.*, 1977; Thiery *et al.*, 1977). Support for the role of N-CAM in mediating cell–cell adhesion was provided by the ability of the 140-Kd protein to block the inhibitory effect of the R10 antiserum on cell adhesion.

Recent studies have produced a major advance in the understanding of the biochemistry of N-CAM. The N-CAM molecule is an integral membrane glycoprotein with a high content of sialic acid. Sialic acid represents 30% by weight of the embryonal N-CAM molecule (Hoffman *et al.*, 1982); but during development, the sialic acid content of the molecule is decreased (Edelman and Chuong, 1982; Rothbard *et al.*, 1982; Rougon *et al.*, 1982; Hirn *et al.*, 1983). The high and presumably variable sialic acid content of N-CAM results in a molecular-weight heterogeneity for the embryonal molecule, with the protein appearing as a heterogeneous 180- to 250-Kd band on polyacrylamide gels (Hoffman *et al.*, 1982). The heterogeneity in sialic acid content is responsible for the appearance of different N-CAM forms in various regions of the embryonic nervous system (Chuong and Edelman, 1984), and in adult nervous tissue the sialic acid-free N-CAM molecule exists as 140- and 170-Kd polypeptides that are similar in structure (Cunningham *et al.*, 1983) and probably arise from a common gene (Murray *et al.*, 1984; Goridis *et al.*, 1985; Murray *et al.*, 1986). However, presently, it is not known whether the two different N-CAM polypeptides mediate distinct functions. A recent study has described the production of a monoclonal antibody that specifically recognizes the higher molecular-weight form of N-CAM (Pollerberg *et al.*, 1985), and this N-CAM component appears to be selectively expressed in more differentiated neural cells. It is therefore apparent that the N-CAM is a heterogeneous protein, with various N-CAM forms being expressed in different cell types during embryogenesis (Williams *et al.*, 1985).

A tentative structure for the N-CAM polypeptide has been proposed (Cunningham *et al.*, 1983) and indicates that N-CAM has three carbohydrate attachment sites (Crossin *et al.*, 1984), a carboxyl terminus that is associated with the plasma membrane, and an aminoterminal region that has been suggested to be important in cell adhesion (Figure 1). Evidence for the participation of N-CAM in cell binding is derived from studies that demonstrate that anti-N-CAM antibodies inhibit cell–cell adhesion (Thiery *et al.*, 1977). In addition, the isolated N-CAM binds to neural cells (Rutishauser *et al.*, 1982), and liposomes containing N-CAM bind both to neural cells (Rutishauser *et al.*, 1982; Sadoul *et al.*, 1983) and other liposomes (Hoffman and Edelman, 1983). These data have provided the basis for the postulated homophilic binding mechanism involving N-CAM, which assumes that N-CAM on the surface of one cell binds to N-CAM on a

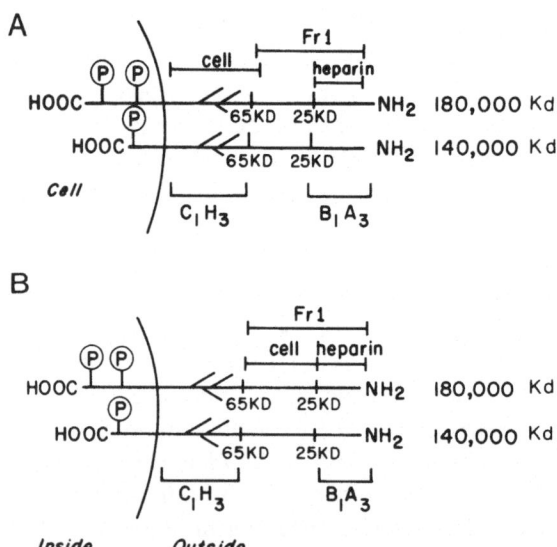

FIGURE 1. Linear model depicting structural and functional domains of N-CAM. Two models have been constructed, since the location of the cell (homophilic)-binding domain has not been established. Only the 140- and 180-Kd forms of N-CAM are shown, which contain the identical amino terminus but different carboxyl termini. Proposed locations of phosphorylation sites are also indicated, as well as carbohydrate chains. (A) The heparin-binding domain is aligned at the amino termini of N-CAM, and the cell-binding domain is proposed to be located at the region of the N-CAM that contains the carbohydrate chains. This location is based on data showing that the C_1H_3 monoclonal antibody inhibits cell adhesion but does not bind to fragment Fr1 (Cole *et al.*, 1986b,c; Fragment Fr1 has previously been shown to contain the cell-binding domain of N-CAM (Cunningham *et al.*, 1983). (B) The model is constructed based on data from studies in our laboratory and from Cunningham *et al.* (1983). In this model, both the cell- and heparin-binding domains are aligned in fragment Fr1. (For a more detailed discussion of this model, see Section 5.4.)

neighboring cell (Hoffman and Edelman, 1983). It can be presumed that the cell-binding region of one N-CAM would bind either to the cell-binding region of another N-CAM or to a receptor region, as yet unidentified, on N-CAM, although there is no direct experimental evidence to support these proposals. It can also be presumed that homophilic binding between N-CAM molecules does not occur as efficiently in solution, because immunopurified N-CAM is used for insertion into liposomes. If the N-CAM in solution was already bound to another N-CAM, then following insertion into liposomes, the N-CAM would no longer be capable of binding other N-CAMs.

The homophilic binding mechanism for N-CAM-mediated cell adhesion may at first appear to be an unattractive mechanism for promoting cell recognition, since any cell containing N-CAM will bind to a second N-CAM-bearing cell. Consistent with this view, neurons have been demonstrated to bind, via an N-CAM-mediated mechanism, to both glial (Keilhauer *et al.*, 1985) and muscle cells (Grumet *et al.*, 1982); these cell types also contain N-CAM on their surface. The binding between N-CAMs must be modulated during neural development if N-CAM is to play a role in neurogenesis. This is accomplished by varying the amounts of N-CAM on different neural cell types during embryogenesis (Chuong and Edelman, 1984), as well as modulating its binding affinity by the selective removal of sialic acid from the N-CAM (Hoffman and Edelman, 1983; Sadoul *et al.*, 1983). The removal of sialic acid from N-CAM has been termed embryonic to adult conversion and has been shown to increase the rate of N-CAM binding to neural cells (Hoffman and Edelman, 1983). However, embryonic-to-adult conversion is also characterized by a shift in the synthesis of the different forms of N-CAM, with more of the 140-Kd component being expressed in adult nervous tissue (Friedlander *et al.*, 1985). It has been proposed that the sialic acid on embryonal N-CAMs may produce a steric or ionic perturbation of cell–cell binding; thus, the removal of the sialic acid augments the binding affinity of the molecule. A recent study has provided strong support for the modulation of N-CAM binding affinity by the removal of sialic acid, since an endoneuraminidase that cleaves the polysialic acid from N-CAM produces an increase in neurite fasciculation of spinal ganglia neurites (Rutishauser *et al.*, 1985). By varying the amount of N-CAM on cells or the sialic acid content of the N-CAM, the interaction of distinct neural cell types can be partially regulated during neural development. A homophilic binding mechanism involving N-CAM may thus be capable of contributing to the complex cellular interactions that are required for tissue formation during neuronal development. It also remains possible, however, that N-CAM interacts with other cell ligands and that these interactions are also a requirement for pattern formation during development. As will be discussed in Section 6, other neuronal cell-adhesion molecules have also been identified, although their relationship to N-CAM function is unclear. For the present time, we will describe biological evidence for the role of N-

CAM in mediating neuronal cell interactions, with the early experimental data supporting a homophilic binding mechanism between N-CAMs.

5.2. Biological Function of N-CAM

Initial investigations into the biological function of N-CAM were conducted using *in vitro* model systems. These experiments provided the first evidence for N-CAM participating in particular aspects of neural morphogenesis and supported the hypothesis that N-CAM plays an integral role in neuronal cell interactions. Several different experimental protocols were employed to show that N-CAM was involved in neuronal cell adhesion. An initial experiment was based on the knowledge that there is a formation of distinct cell layers in the differentiated retina during retinal development, with the cell body layers being separated by plexiform layers that are composed of neurites and developing synapses. If Fab fragments of anti-N-CAM antibody are present during the growth of retinal explants *in vitro,* this developmental pattern is not observed (Buskirk *et al.,* 1980). Although these data were not particularly convincing, probably as a result of the nature of the *in vitro* experiment, it could be interpreted from the experimental observations that the sorting out of retinal neurons into cell body layers was dependent upon N-CAM.

Additional studies have been conducted *in vitro,* which support the proposal that N-CAM is an integral component of neural cell-adhesion processes. When dorsal root ganglia explants are maintained *in vitro,* growing neurites form fascicles. The formation of fascicles is not observed when dorsal root ganglia are incubated with anti-N-CAM Fab fragments (Rutishauser *et al.,* 1978); this suggests that the interaction between growing neurites, resulting in the formation of nerve bundles, is an N-CAM-mediated process. Similarly, when pioneering nerve fibers are incubated with myotube cultures, the growing neurites will interact with the muscle cell surface. However, in the presence of anti-N-CAM Fab, the interaction between neurites and myotubes is inhibited (Grumet *et al.,* 1982; Rutishauser *et al.,* 1983), suggesting that, at least under these experimental conditions, N-CAM is important for the formation of the neuromuscular junction; N-CAM may also serve to guide the growing axon to the myotube. Consistent with this view, N-CAM levels in muscle cells become elevated following denervation or paralysis (Covault and Sanes, 1985; Rieger *et al.,* 1985) and diminish following innervation. N-CAM is also detected by immunocytochemical methods in interstitial spaces surrounding denervated synaptic sites, which may imply that regenerating axons are guided to the synaptic site by N-CAM (Sanes *et al.,* 1986). Interestingly, fibronectin, heparan sulfate proteoglycan, and the cell-adhesion molecule J1 (Kruse *et al.,* 1985) are also localized in these interstitial spaces (Sanes *et al.,* 1986). It should be noted, however, that if N-CAM on the muscle cell surface or in interstitial spaces participates in the guidance of growing axons to the synaptic site, it is necessary to consider the role of N-CAM on the surface of the growing axons. Because N-CAM is involved in neurite fasciculation, it can be presumed that the growing axons will be tightly associated due to N-CAM–N-CAM interactions. There may therefore be secondary mechanisms that are necessary for neuron–muscle interaction. These may include a higher affinity reaction between neuronal N-CAM and muscle N-CAM, when compared with that of neuronal N-CAM–N-CAM binding, or the other cell-adhesion molecules that are localized at the muscle synaptic site may participate in the recognition process.

Recent studies by Schachner and her co-workers have implied that in addition to N-CAM participating in neuron–neuron and neuron–muscle adhesion, N-CAM may also regulate neuron–glial interactions (Keilhauer *et al.*, 1985). N-CAM has been demonstrated to be a cell-surface component of astrocytes (Keilhauer *et al.*, 1985; Noble *et al.*, 1985), and, accordingly, anti-N-CAM antibodies inhibit neuron–astrocyte adhesion (Keilhauer *et al.*, 1985). Although these data support the homophilic binding mechanism for N-CAM, in that astrocytes bearing N-CAM bind to neurons, these data raise the question as to why the majority of cell types in the nervous system contain N-CAM and use this ligand as a cell–cell linkage. A series of *in vivo* experiments, which are described in Section 5.3, have provided some insight into the mechanism by which N-CAM may mediate cell interactions during neural development and demonstrate how the restricted expression of N-CAM can control the interaction between these various cell types. Experiments from our laboratory that provide additional insights into the mechanism of N-CAM-mediated cell adhesion are described in Section 5.4.

5.3. *In Vivo* Analysis of N-CAM Function

Attempts to perturb N-CAM-mediated cell functions *in vivo* have focused primarily on the visual system, which offers the advantage of being readily accessible to experimental manipulation. In separate studies, Edelman's and Rutishauser's laboratories have demonstrated that the formation of the retinotectal projection depends upon N-CAM-mediated cell interactions. Using the the regenerating *Xenopus laevis* visual system, Fraser *et al.* (1984) showed that tectal implants of anti-NCAM antibodies markedly decreased the precision of the retinotectal projection. The anti-N-CAM antibody also produced a local distortion in the patterning of the projection in these experiments. These data therefore imply that the interaction between retinal ganglion cells and tectal target cells requires the N-CAM protein.

Rutishauser and his co-workers have utilized the embryonic chicken visual system to demonstrate that N-CAM plays an important role in establishing retinotectal connections. In their pioneering studies, Thanos *et al.* (1984) microinjected anti-N-CAM Fab fragments into the eye cup of embryonic day-3 chick embryos. Following the injection of rhodamine tracer, it was possible to monitor the growth of retinal ganglion cells to the optic tectum. Under normal conditions during neural development, local fiber order in the optic pathway is maintained during axonal growth (Scholes, 1979; Russoff and Easter, 1980; Reh *et al.*, 1983; Thanos and Bonhoeffer, 1984), and at the tectum the invading growth cones spread out over the surface of the tectum to establish normal retinotectal connections (Scholes, 1979; Constantine-Paton *et al.*, 1983). In the elegant studies of Thanos *et al.* (1984), it was demonstrated that anti-N-CAM Fab injected into the eye cup perturbed local fiber order in the optic fissure. Thus, fibers were misrouted leaving the eye cup but were able to grow in a normal optic tract that was free of anti-NCAM antibody. The majority of misrouted fibers were unable to reestablish order in the optic tract and therefore did not make appropriate connections in the tectum (Thanos *et al.*, 1984). The fibers that did make appropriate connections appeared to be capable of interacting with the tectal surface, which may imply that these fibers were responding to positional cues and could therefore correct their position. These experiments indicate either that N-CAM is required for the formation of retinotectal connections in the chick visual system or that the anti-N-CAM antibodies only disrupt the local order of fibers in the optic nerve, and that these misdirected fibers are unable to correct their position due to N-CAM-mediated neuron–neuron interactions (Rutishauser, 1984). Therefore, the strength of the axon–axon adhesion would cause the misdirected fibers to follow an incorrect pathway; it is also possible that once the local fiber order is perturbed, the majority of fibers are not capable of responding to tectal positional cues (which are distinct from N-CAM). Hence, the appropriate synaptic connection is not formed.

As discussed above, N-CAM has been detected on both neurons and glial cells and has been shown to participate in neuron–astrocyte adhesion *in vitro* (Keilhauer *et al.*, 1985). In studies conducted using the chicken visual system, Silver and Rutishauser (1984) have shown that a pathway of neuroepithelial cells lining the optic tract contain cell-surface N-CAM; these neuroepithelial cells are presumed to be precursors of glial cells. When anti-N-CAM Fab fragments are injected into the eye cups of developing chick embryos, the growth of retinal ganglion cell axons along the neuroepithelial cell pathway is perturbed, and an overall distortion in the optic pathway is observed. These data also indicated that a preformed pathway of N-CAM is involved in the growth of retinal ganglion cell axons to the optic tectum and, thus, provide strong support for earlier evidence obtained from *in vivo* experiments that N-CAM is critically important for the formation of retinotectal connections. Although it can be concluded from these studies that N-CAM is one component necessary for establishing retino-tectal connections, other mechanisms also appear to be important (Meyer, 1982; Schmidt and Edwards, 1983).

5.4. Role of Heparan Sulfate Proteoglycan in N-CAM Function

As described above, a homophilic binding mechanism for N-CAM-mediated cell adhesion has been proposed (Hoffman and Edelman, 1983). This mechanism states that cells containing N-CAM will interact only with other N-CAM-bearing cells and that the cell–cell linkage is formed by the direct binding between N-CAMs. In this respect, N-CAM appears to differ from other cell-adhesive proteins, such as fibronectin and laminin, which possess several distinct functional domains and do not exhibit homophilic binding (Hynes and Yamada, 1982; Yamada, 1983; Edgar et al., 1984). However, it can be envisioned that N-CAM function, particularly its homophilic binding properties, could be modulated by its interaction with other cell-surface molecules. Studies conducted in our laboratory have now provided support for this proposal and are described below.

Our laboratory has been interested in applying hybridoma technology for the analysis of cell interactions during neural development. We have generated monoclonal antibodies for the purpose of identifying unique cell-surface antigens in the nervous system, with the hypothesis that antigens of this type may participate in neuronal cell interactions. One monoclonal antibody of interest, designated C_1H_3, recognizes two polypeptides with molecular weights of 140 Kd and 170 Kd in the embryonic chicken nervous system (Cole and Glaser, 1984a). A putative function for these polypeptides was initially identified on the basis of the secretion of the 170-Kd protein by retinal cells in culture and its presence in adherons (Cole and Glaser, 1984b,c). Adherons are complexes of proteins and glycosaminoglycans that resemble extracellular matrix material and promote cell–substratum adhesion when adsorbed onto plastic surfaces (Schubert et al., 1983). The 170-Kd C_1H_3 antigen in adherons appears to participate in cell–substratum adhesion, since the C_1H_3 monoclonal antibody inhibits retinal cell attachment to adherons (Cole and Glaser, 1984b). The mechanism for the attachment of retinal cells to adherons is characterized by homophilic binding, as the C_1H_3 monoclonal antibody inhibits cell–adheron binding when incubated with either retinal cells or adheron-coated dishes. Thus, the mechanism for the binding of neural cells to adherons resembles N-CAM-mediated cell adhesion. Recent studies in our laboratory have documented that the C_1H_3 antigens are immunologically identical to N-CAM (Cole and Glaser, 1986), which indicates that N-CAM also participates in cell–substratum interactions in the developing nervous system. These data also raise the point that when using in vitro assay systems, it is possible to identify several distinct functions for the same molecule.

In the case of N-CAM, since it is present on neurons, muscle cells, and glia, it can be demonstrated that interactions between these cell types are capable of being inhibited by anti-N-CAM antibodies. Likewise, if a substratum is prepared that contains N-CAM, it can be presumed that N-CAM-containing cells will bind to the substratum. It is therefore necessary to interpret results from *in vitro* functional assays carefully, because if two ligands that are capable of interacting are present on two apposing surfaces, it is likely that the two surfaces (i.e., cell to cell or cell to substratum) will interact. However, as shown for N-CAM, the *in vivo* studies reported above strongly suggest that N-CAM is an important ligand in the adhesion of neural cells during development.

Prior to the demonstration that the C_1H_3 antigens were identical to N-CAM, it was shown that a homophilic binding mechanism could not adequately explain the experimental observations concerning the mechanism for cell–substratum interactions in the chicken nervous system. The studies of Schubert and La-Corbiere (1985a) initially indicated that a second molecular component, other than N-CAM, was involved in cell–substratum adhesion in the chick neural retina. These investigators showed that a retinal cell-surface heparan sulfate proteoglycan was also involved in the attachment of retinal cells to adherons. Accordingly, heparin, or heparan sulfate, and an antiserum that recognized the heparan sulfate proteoglycan inhibited cell–adheron binding (Schubert *et al.*, 1983; Schubert and LaCorbiere, 1985a). These data raised the possibility that retinal cell-surface heparan sulfate proteoglycan might be binding to N-CAM in adherons to promote cell attachment. It was therefore of interest to ascertain whether N-CAM could bind heparin (which was used as a functional analog for the physiologically relevant ligand heparan sulfate) and whether such an interaction was important for cell–substratum adhesion. Using a [^{3}H]heparin-binding assay, it was possible to demonstrate that N-CAM binds heparin (Cole *et al.*, 1985). When substrata were prepared using immunopurified N-CAM protein, retinal cell attachment to this substratum was inhibited by heparin, heparan sulfate, or an antiserum that binds retinal cell-surface heparan sulfate proteoglycan (Cole *et al.*, 1985). The ability of these molecules to inhibit cell attachment to an N-CAM substratum provides strong support for a mechanism where the interaction between heparan sulfate and N-CAM is required for cell–substratum adhesion. These data also suggest that homophilic binding between N-CAMs is just one facet of N-CAM-mediated cell adhesion. It can then be asked what is the mechanism by which heparan sulfate–N-CAM interactions promote cell adhesion? A second question that can be raised is whether heparan sulfate binding to N-CAM is also an important component of cell–cell adhesion in the embryonic nervous system.

The data indicating that N-CAM can bind both heparan sulfate and neural cells is the first evidence that N-CAM is a multifunctional protein. Since other multifunctional proteins, such as fibronectin and laminin, possess distinct functional domains that are present in the molecule as protease-resistant regions (Pierschbacher *et al.*, 1981; Ruoslahti *et al.*, 1981; Yamada *et al.*, 1980), our laboratory proposed that a heparin-binding domain from N-CAM could also be isolated. Using this isolated domain and monoclonal antibodies raised against it, it was possible to obtain valuable insight into the mechanism of N-CAM-mediated cell interactions. Proteolytic cleavage of N-CAM with subtilisin protease yielded a 25-Kd polypeptide fragment that was retained on a heparin–agarose column (Cole and Glaser, 1986). A monoclonal antibody (named B_1A_3) that binds N-CAM also recognized this fragment when using immunoblotting procedures, and this monoclonal antibody. The B_1A_3 monoclonal antibody inhibited [^{3}H]heparin binding to N-CAM and also disrupted retinal cell attachment to an N-CAM substratum (Cole and Glaser, 1986). In addition, the isolated 25-Kd heparin-binding fragment of N-CAM promoted cell–substratum adhesion when used as a substratum, and cell attachment to this fragment was capable of being inhibited by heparan sulfate, the B_1A_3 monoclonal antibody, and the anti–heparan sulfate proteoglycan antiserum. The isolated 25-Kd fragment was also capable of partially inhibiting cell attachment to adherons, which provides strong support that a heparin-binding domain from N-CAM is involved in cell–substratum adhesion.

Although these experiments suggest that the binding of heparan sulfate by N-CAM plays an integral role in cell–substratum adhesion, what is the mechanism for the modulation of cell adhesion by this interaction? The attachment of fibroblast-like cells to the extracellular matrix is mediated in part by fibronectin (for review, see Hynes and Yamada, 1982; Yamada, 1983), and the binding of fibronectin to heparan sulfate alters the affinity of fibronectin binding to the cell surface (Johannsen and Hook, 1984). Likewise, the binding of heparan sulfate to N-CAM appears to induce a conformational change in the protein (Cole *et al.*, 1985). It can thus be envisioned that the binding of heparan sulfate to N-CAM is an obligatory step in the attachment of neural cells to extracellular matrix material. A model depicting a possible mechanism for the attachment of neural cells to the extracellular matrix is shown in Figure 2. In this model it is proposed that homophilic binding between N-CAMs can only occur following a conformational change in N-CAM, which is induced by heparan sulfate binding. The binding of heparan sulfate to N-CAM occurs in the aminoterminal cell-binding region of the molecule (Figure 1), since the aminoterminal sequence of the isolated 25-Kd heparin-binding domain (Cole *et al.*, 1986c) is identical to the published aminoterminal sequence of intact N-CAM (Rougon and Marshak, 1986). The binding of heparan sulfate to N-CAM is proposed to be a low-affinity reaction, since N-CAM or the heparin-binding fragment bind only to heparin–agarose under low salt conditions. Scatchard analysis also suggests that the binding of heparan sulfate to N-CAM is weak (G.J. Cole and L. Glaser, unpublished observations). We have also analyzed the capacity of different heparin

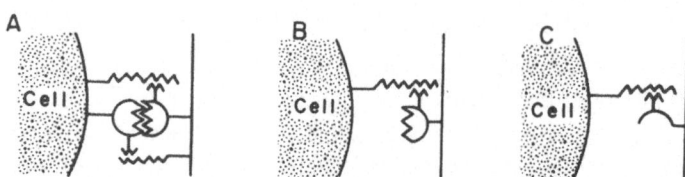

FIGURE 2. Schematic diagram depicting a possible mechanism for neural cell–substratum adhesion involving N-CAM and heparan sulfate proteoglycan. (A) The model shows that N-CAM undergoes a conformational change after binding of heparan sulfate proteoglycan, and homophilic binding between N-CAM molecules produces a stable cell adhesion to the matrix. Cell attachment does not occur in B (i.e., cell-surface N-CAM is unavailable for binding) since the interaction between N-CAM and heparan sulfate proteoglycan is weak. (C) Cell attachment does occur because the heparin-binding domain of N-CAM, when coupled to an inert surface, can bind heparan sulfate proteoglycan with higher affinity.

fragments to bind to N-CAM, and our data indicate that heparin fragments must be at least a decasaccharide to bind to N-CAM in the [³H]heparin assay. In addition, intact heparin molecules bind more efficiently to N-CAM than heparin fragments (Cole *et al.*, 1987). N-CAM is also capable of binding intact heparin molecules that either flow through or are retained on an antithrombin III affinity column.

From the model in Figure 2 it can be postulated that antibodies that block the conformational change (i.e., B_1A_3) or inhibit homophilic binding (i.e., C_1H_3) will prevent cell–substratum interactions. Although these data attempt to explain

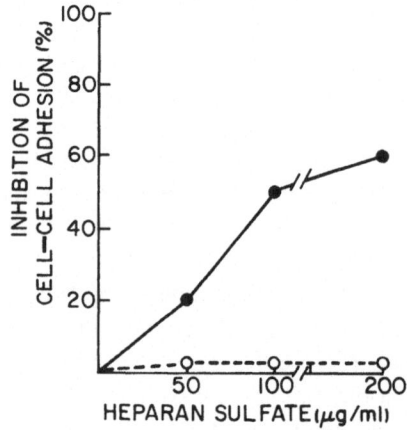

FIGURE 3. Effect of heparan sulfate glycosaminoglycan on retinal cell–cell adhesion. Monolayers of unlabeled embryonic retinal cells were coupled to glass minivials, and the adhesion of metabolically labeled retinal cells was measured in the presence of heparan sulfate (●) or chondroitin sulfate (○). Cell–cell adhesion was measured in a 60-min assay, and approximately 60% of input control cells adhered to the monolayers.

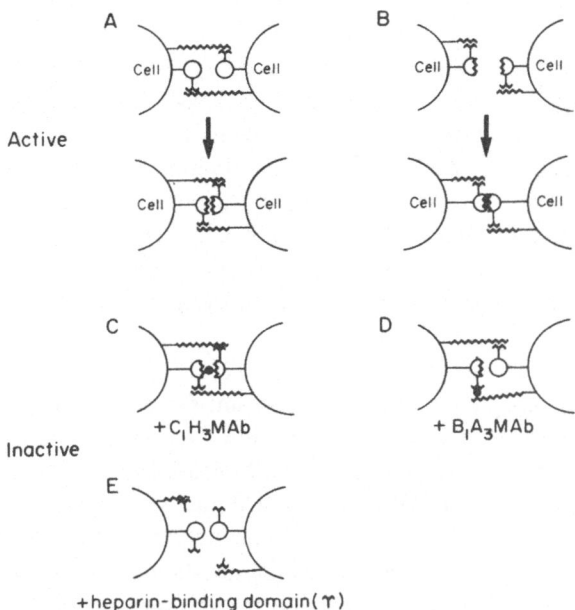

FIGURE 4. Schematic diagram depicting possible mechanisms for the role of heparan sulfate proteoglycan in N-CAM-mediated cell–cell adhesion. As described in Figure 2, it is proposed that N-CAM undergoes a conformational change following the binding of heparan sulfate. (A,B) Heparan sulfate binding to N-CAM depicted as *trans* (N-CAM binds to heparan sulfate on the opposing cell) or *cis;* we favor the *trans* mechanism at this time. (C–E) The inhibition of cell–cell adhesion by the C_1H_3 and B_1A_3 monoclonal antibodies and the heparin-binding domain. The C_1H_3 monoclonal antibody is proposed to interfere with homophilic binding, whereas the B_1A_3 monoclonal antibody and the heparin-binding domain derived from N-CAM prevent the conformational change in N-CAM and, hence, disrupt homophilic binding and stable cell attachment.

a mechanism for the mediation of cell–substratum adhesion by N-CAM and heparan sulfate proteoglycan, does cell–cell adhesion occur by a similar mechanism?

Because the interaction of N-CAM with the extracellular matrix is not well documented and the majority of studies indicate that N-CAM participates in cell–cell interactions, we conducted studies using a cell–cell adhesion system to determine whether heparan sulfate binding to N-CAM is involved in the adhesive process. The experiments conducted employed an assay system that measured the binding of metabolically labeled retinal cells to monolayers of retinal cells (Cole *et al.*, 1986b). When cell–cell adhesion was examined under conditions where chondroitin sulfate was present, adhesion was unaffected (Cole *et al.*, 1986b). However, when varying concentrations of heparin were included in the assay medium, cell–cell adhesion was inhibited in a dose-dependent manner.

The physiologically relevant ligand in the nervous system, heparan sulfate, also inhibited retinal cell–cell adhesion (Figure 3). It can therefore be concluded that the adhesion between neural cells in this assay system depends upon heparan sulfate binding to a cell-surface receptor. Evidence for N-CAM serving as this receptor was provided by experiments showing that the B_1A_3 monoclonal antibody inhibits cell–cell adhesion (Cole *et al.*, 1986b). Since the B_1A_3 monoclonal antibody recognizes the heparin-binding polypeptide fragment of N-CAM, these experiments strongly suggest that the binding of heparan sulfate by N-CAM is required for neuronal cell adhesion. A model depicting the dualistic role of N-CAM and heparan sulfate proteoglycan in neuronal cell–cell adhesion is shown in Figure 4.

In conclusion, it is apparent that the N-CAM molecule is an integral component of the neural cell-adhesion mechanism, since the homophilic binding of N-CAMs can mediate cell–cell adhesion. However, a second cell-surface ligand, heparan sulfate proteoglycan, is also vital to neuronal cell adhesion, in view of evidence that this molecule binds to N-CAM and may modulate the subsequent binding between N-CAM molecules. Other neuronal cell-adhesion molecules have also been identified, which raises the question as to how these molecules might be related to N-CAM. In Section 6, we describe some additional cell-adhesion molecules that were identified using immunological methods and how these molecules may act in tandem with N-CAM to regulate neuronal cell interactions.

6. RELATIONSHIP OF N-CAM TO OTHER NEURONAL CELL-ADHESION MOLECULES

Immunological screening procedures have been utilized to identify other cell-adhesion molecules in neural tissue. Several of these molecules, such as the D2 (Jorgensen *et al.*, 1980), BSP-2 (Rougon *et al.*, 1982), 2241A6 (Lemmon *et al.*, 1982), and C_1H_3 antigens (Cole and Glaser, 1984a,b), have been shown to be identical to N-CAM. The L1 cell-adhesion molecule, which is characterized as 200- and 140-Kd polypeptides, has been identified in mouse cerebellum with a monoclonal antibody (Rathjen and Schachner, 1984); this molecule is distinct from the N-CAM group of cell-adhesion molecules (Rathjen and Rutishauser, 1984; Faissner *et al.*, 1984a). To demonstrate a role for L1 in neuronal cell-recognition processes, a variety of *in vitro* assays were employed. When explants of developing mouse cerebellum were cultured in the presence of anti-L1 antibodies, at a time when granule cell migration was occurring, it was demonstrated that anti-L1 antibodies perturb granule cell migration (Lindner *et al.*, 1983). An *in vitro* cell–cell adhesion assay, which used a Coulter counter to measure the disappearance of single cells in aggregating cultures of neurons, also showed that the L1 molecule participated in the adhesion of mouse neural cells (Rathjen and Schachner, 1984). Although the L1 molecule is unrelated to N-CAM, except for a carbohydrate epitope recognized by the L2 monoclonal antibody (Kruse *et al.*, 1984), it appears that the N-CAM and L1 molecules may act synergistically to promote cell adhesion, because antibodies against both antigens increase the degree of inhibition markedly (Rathjen and Rutishauser, 1984). Like N-CAM, the L1 antigen is present on glial cells (Faissner *et al.*, 1984b), although the antibodies against L1 do not inhibit astrocyte–neuron adhesion (Keilhauer *et al.*, 1985). This observation is particularly surprising in light of recent data indicating that the L1 molecule is immunologically identical to the nerve growth factor–inducible large external (NILE) glycoprotein (Bock *et al.*, 1985). The NILE glycoprotein, which was initially identified in PC 12 cells (Greene and Tischler, 1976; McGuire *et al.*, 1978), appears to be involved in neurite fasciculation of neurons from embryonic rat brain (Stallcup and Beasely, 1985) and is expressed on neurons and Schwann cells (Stallcup *et al.*, 1983; Salton *et al.*, 1983). The NILE glycoprotein is also reported to be immunologically identical to Ng-CAM (Friedlander *et al.*, 1986), which is involved in the adhesion between neurons and glial cells. Ng-CAM was identified in chick brain (Grumet and Edelman, 1984; Grumet *et al.*, 1984a) and is characterized by three polypeptides with molecular weights of 200, 135, and 80 Kd (Grumet *et al.*, 1984b). Ng-CAM, is also identical in molecular weight to L1 antigen and appears to have a cellular distribution identical to L1 (Grumet *et al.*, 1984b); it is thus apparent that if L1 and Ng-CAM are identical, antibodies against L1 should inhibit neuron–glia adhesion. It remains possible that the anti-L1 antibodies recognize different epitopes, which are not involved in neuron–glia adhesion; however, it has also been reported that anti-NgCAM antibodies inhibit neuron–neuron interactions (Grumet *et al.*, 1984b). Additional studies will be necessary to resolve the discrepancies in the reported functions for these adhesion proteins.

The studies regarding the function of the NILE glycoprotein have also raised differences in the importance of N-CAM and NILE in mediating neuron–neuron interactions during different developmental periods. Stallcup and Beasley (1985) have demonstrated that anti-NILE antibodies inhibit neurite fasciculation in embryonic rat brain, but not in adult rat brain. They also noted the opposite effect for anti-N-CAM antibodies. It thus appears that the NILE glycoprotein has a more dominant role in neurite fasciculation in embryonic rat neural tissue than N-CAM does. With the loss of sialic acid from embryonal N-CAM, N-CAM may become the dominant cell-adhesion molecule in adult nervous tissue.

Because the Ng-CAM/L1/NILE molecule appears to participate in both neuron–neuron and neuron–glia adhesion, it can also be asked whether the interaction of this molecule with its cell-surface receptor undergoes some form of modulation. In light of the experimental evidence, which suggests that N-CAM also participates in several types of cellular interactions, neuronal cell adhesion can be viewed as a complex process that depends upon multiple mechanisms. As discussed above, it is also imperative to interpret results from these *in vitro* assays carefully, since several distinct functional activities for these cell-adhesion molecules have been described in different laboratories.

From current experimental observations in several laboratories, it can be concluded that there are two major cell-adhesion molecules in developing neural tissue. Using the nomenclature adopted by Edelman and his co-workers, the N-CAM and Ng-CAM polypeptides are unrelated proteins except for the L2 (or HNK-1) epitope, which is a shared determinant on the proteins (Kruse *et al.*, 1984). The L2 epitope appears to be a three-sulfated glucuronyl carbohydrate moiety (Chou *et al.*, 1986) and is proposed to be involved in neuronal cell adhesion (Kruse *et al.*, 1984). Consistent with this view, the HNK-1 monoclonal antibody inhibits neurite extension on neuronal extracellular matrix material (Riopelle *et al.*, 1986). However, since myelin-associated glycoprotein, N-CAM, and Ng-CAM possess this epitope, it is unclear which molecule is the donor of this structure during neurite extension. It appears that N-CAMs and Ng-CAMs employ different mechanisms for promoting cell adhesion, as N-CAM-mediated adhesion involves heparan sulfate binding to N-CAM, as well as homophilic binding between N-CAMs. Additionally, the role of the L2 epitope in N-CAM-mediated adhesion is not known. In contrast to N-CAM, Ng-CAM interacts with an unidentified receptor on glial cells (and presumably neurons) by a heterophilic binding mechanism (Grumet and Edelman, 1984). However, in view of recent studies describing a novel neural cell-adhesion molecule named cytotactin (Grumet *et al.*, 1985) or J1 (Kruse *et al.*, 1985), which is presumed to be localized in the neuronal extracellular matrix and binds to proteoglycans (Grumet *et al.*, 1985), it is of interest to determine whether Ng-CAM also binds heparan sulfate proteoglycan or other proteoglycans.

Because there is experimental evidence demonstrating that laminin and fibronectin possess both cell- and heparin-binding domains (Hynes and Yamada, 1982; Yamada, 1983; Edgar *et al.*, 1984), as well as the data indicating that N-CAM and cytotactin interact with proteoglycans, it will be of interest in future studies to examine whether these neuron cell-adhesion molecules can mediate functions currently assigned to fibronectin and laminin . For example, since it has been shown that neurite-outgrowth–promoting factors contain laminin (Lander *et al.*, 1985) that the heparin-binding domains of laminin and fibronectin promote neurite outgrowth (Carbonetto *et al.*, 1983; Edgar *et al.*, 1984; Rogers *et al.*, 1985), and that the HNK-1/L2 epitope is involved in neurite extension (Riopelle *et al.*, 1986), it can be asked whether N-CAM and cytotactin may possess similar activities. It will be of interest in future studies to examine in more detail the relationship between these cell-adhesion proteins and to determine the precise role heparin-binding proteins play in mediating neuronal cell interactions.

7. CONCLUSIONS

This review has described how the use of monoclonal antibodies, or other immunological methods, can facilitate the identification of novel neural antigens. The aim of many of these studies has been to identify antigens that are restricted to specific cell types in the developing nervous system, since antigens of this type may be involved in the differentiation of these cells, or may participate in cell-recognition processes. Although this review has demonstrated that many cell-specific antigens have been identified, the majority of these antigens are only cell specific in restricted regions of the nervous system. The function of only a few neural-specific antigens has been identified, and the function of these molecules was, for the most part, determined using *in vitro* cell-adhesion assays.

The best characterized neuronal cell-adhesion molecule is N-CAM, and the role of the Ng-CAM/L1/NILE polypeptide in mediating cell adhesion is also firmly established. Although the use of *in vitro* model systems has been useful in partially elucidating the mechanism by which N-CAM is involved in cell adhesion, it is apparent that the utilization of *in vivo* systems will provide greater insight into the specific role of these cell-adhesion molecules in neural development. For example, at present, a number of different functions for these cell-adhesion molecules have been described. Do these functions indicate that these molecules are required for only a general type of cell–cell interaction, or do these molecules also control the interaction between specific cell types in the nervous system? Also, what role do other neural antigens play in regulating neuronal cell-recognition processes? It is likely that the use of monoclonal antibodies will provide additional insight into the molecular mechanism of cell recognition during neural development. It will be of particular interest if monoclonal antibodies can permit the identification of antigens that are critically important to the initial recognition event between specific neural cell types, especially if molecules such as N-CAM are required at later stages of cell adhesion, which result in the formation of a "permanent" cell–cell linkage.

ACKNOWLEDGMENTS. This work was supported by grants GM-18405, EY-0566 and EY-05898.

REFERENCES

Akeson, R., and Graham, K. L., 1981, A new antigen common to the rat nervous and immune systems: 1. Detection with a hybridoma, *J. Neurosci. Res.* **6:**165–177.

Akiyama, S. K., Yamada, S. S., and Yamada, K. M., 1986, Characterization of a 140-kD avian cell surface antigen as a fibronectin-binding molecule, *J. Cell Biol.* **102:**442–448.

Allen W. K., and Akeson, R., 1985, Identification of a cell surface glycoprotein family of olfactory receptor neurons with a monoclonal antibody, *J. Neurosci.* **5:**284–296.

Barnstable, C. J., 1980, Monoclonal antibodies which recognize different cell types in the rat retina, *Nature* **286:**231–235.

Barnstable, C. J., and Drager, U. C., 1984, Thy-1 antigen: A ganglion cell specific marker in rodent retina, *Neuroscience* **11:**847–855.

Barnstable, C. J., Hofstein, R., and Akagawa, K., 1985, A marker of early amacrine cell development in rat retina, *Dev. Brain Res.* **20:**286–290.

Beug, H., Katz, F. E., and Gerisch, G., 1973, Dynamics of antigenic membrane sites relating to cell aggregation in *Dictyostelium discoideum*, *J. Cell Biol.* **56:**647–658.

Bock, E., Richter-Landsberg, C., Faissner, A., and Schachner, M., 1985, Demonstration of im-munochemical identity between the nerve growth factor–inducible large external (NILE) gly-coprotein and the cell adhesion molecule L1. *EMBO J.* **4:**2765–2768.

Bolin, L. M., and Rouse, R. V., 1986, Localization of Thy-1 expression during postnatal devel-opment of the mouse cerebellar cortex. *J. Neurocytol.* **15:**29–36.

Boucaut, J.-C., Darribere, T., Poole, T. J., Aoyama, H., Yamada, K. M., and Thiery, J. P., 1984, Biologically active synthetic peptides as probes of embryonic development: A competitive peptide inhibitor of fibronectin function inhibits gastrulation in amphibian embryos and neural crest cell migration in avian embryos. *J. Cell Biol.* **99:**1822–1830.

Brackenbury, R., Thiery, J. P., Rutishauser, U., and Edelman, G. M., 1977, Adhesion among neural cells of the chick embryo. I. An immunological assay for molecules involved in cell–cell binding. *J. Biol. Chem.* **252:**6835–6840.

Bronner-Fraser, M., 1985, Alterations in neural crest migration by a monoclonal antibody that affects cell adhesion. *J. Cell Biol.* **101:**610–617.

Buskirk, D. R., Thiery, J. P., Rutishauser, U., and Edelman, G. M., 1980, Antibodies to a neural cell adhesion molecule disrupt histogenesis in cultured chicken retinae. *Nature* **285:**488–489.

Caddy, K. W. T., Patterson, D. L., and Biscoe, T. J., 1982, Use of the UCHT1 monoclonal antibody to explore mouse mutants and development, *Nature* **300:**441–443.

Carbonetto, S., Gruver, M. M., and Turner, D. C., 1983, Nerve fiber growth in culture on fibronectin, collagen, and glycosaminoglycan substrates, *J. Neurosci.* **3:**2324–2335.

Chou, D. K. H., Schwarting, G. A., and Jungalwala, F. B., 1986, Sulfated glucuronyl glycolipids in the nervous system, *Trans. Am. Soc. Neurochem.* **17:**146.

Chuong, C.-M., and Edelman, G. M., 1984, Alterations in neural cell adhesion molecules during development of different regions of the nervous system, *J. Neurosci.* **4:**2354–2368.

Cole, G. J., and Glaser, L., 1984a, Identification of novel neural- and neural retina–specific antigens with a monoclonal antibody, *Proc. Natl. Acad. Sci. USA* **81:**2260–2264.

Cole, G. J., and Glaser, L., 1984b, Inhibition of embryonic neural retina cell–substratum adhesion with a monoclonal antibody, *J. Biol. Chem.* **259:**4031–4034.

Cole, G. J., and Glaser, L., 1984c, Cell–substratum adhesion in embryonic chick central nervous system is mediated by a 170,000-mol. wt. neural-specific polypeptide, *J. Cell Biol.* **99:**1605–1612.

Cole, G. J., and Glaser, L., 1986, A heparin-binding domain from N-CAM is involved in neural cell–substratum adhesion, *J. Cell Biol.* **102:**403–412.

Cole, G. J., Schubert, D., and Glaser, L., 1985, Cell–substratum adhesion in chick neural retina depends upon protein–heparan sulfate interactions, *J. Cell Biol.* **100:**1192–1199.

Cole, G. J., Bond, R., and Glaser, L., 1986a, Monoclonal antibodies specific for ganglion cells in the embryonic chicken neural retina, *Dev. Brain Res.* **26:**133–143.

Cole, G. J., Loewy, A., and Glaser, L., 1986b, Neuronal cell–cell adhesion depends on interactions on N-CAM with heparin-like molecules, *Nature* **320:**445–447.

Cole, G. J., Loewy, A., Cross, N. V., Akeson, R., and Glaser, L., 1986c, Topographic localization of the heparin-binding domain of the neural cell adhesion molecule N-CAM, *J. Cell Biol.* **103:**1739–1744.

Cole, G. J., Maimone, M., Tollefson, D., Loewy, A., and Glaser, L., 1987, Characterization of heparin binding to the neural cell adhesion molecule N-CAM, *Exp. Cell. Res.* (submitted).

Collins, F., 1985, Electrophoretic similarity of the ciliary ganglion survival factors from different tissues and species, *Dev. Biol.* **109**:255–258.

Constantine-Paton, M., Pitts, E. C., and Reh, T. A., 1983, The relationship between retinal axon ingrowth, terminal morphology, and terminal patterning in the optic tectum of the frog, *J. Comp. Neurol.* **218**:297–313.

Covault, J., and Sanes, J. R., 1985, Neural cell adhesion molecule (N-CAM) accumulates in denervated and paralyzed skeletal muscle, *Proc. Natl. Acad. Sci. USA* **82**:4544–4548.

Crossin, K. L., Edelman, G. M., and Cunningham, B. A., 1984, Mapping of three carbohydrate attachment sites in embryonic and adult forms of the neural cell adhesion molecule, *J. Cell Biol.* **99**:1848–1855.

Cuello, A. C., 1983, Monoclonal antibody immunohistochemistry. Applications in research and diagnosis, *Acta Histochem.* **28**:S9–S15.

Cuello, A. C., Milstein, C., Couture, R., Wright, B., Priestley, J. V., and Jarvis, J., 1984, Characterization and immunocytochemical application of monoclonal antibodies against en-kephalins, *J. Histochem. Cytochem.* **32**:947–957.

Cunningham, B. A., Hoffman, S., Rutishauser, U., Hemperly, J. J., and Edelman, G. M., 1983, Molecular topography of the neural cell adhesion molecule N-CAM: Surface orientation and location of sialic acid-rich and binding regions, *Proc. Natl. Acad. Sci. USA* **80**:3116–3120.

D'Eustachio, P., Owens, G. C., Edelman, G. M., and Cunningham, B. A., 1985, Chromosomal location of the gene encoding the neural cell adhesion molecule (N-CAM) in the mouse, *Proc. Natl. Acad. Sci. USA* **82**:7631–7635.

Dodd, J., Solter, D., and Jessell, T. M., 1984, Monoclonal antibodies against carbohydrate dif-ferentiation antigens identify subsets of primary sensory neurones, *Nature* **311**:469–472.

Drager, U. C., Edwards, D. L., and Barnstable, C. J., 1984, Antibodies against filamentous components in discrete cell types of the mouse retina, *J. Neurosci.* **4**:2025–2042.

Duband, J.-L., Rocher, S., Chen, W.-T., Yamada, K. M., and Thiery, J. P., 1986, Cell adhesion and migration in the early vertebrate embryo: Location and possible role of the putative fibro-nectin receptor complex, *J. Cell Biol.* **102**:160–178.

Dubois, C., Magnani, J. L., Grunwald, G. B., Spitalnik, S. L., Trisler, G. D., Nirenberg, M., and Ginsburg, V., 1986, Monoclonal antibody 18B8, which detects synapse-associated antigens, binds to ganglioside G_{T3} (II^3(NeuAc)$_3$LacCer), *J. Biol. Chem.* **261**:3826–3830.

Edelman, G. M., 1983, Cell adhesion molecules, *Science* **219**:450–457.

Edelman, G. M., and Chuong, C.-M., 1982, Embryonic to adult conversion of neural cell adhesion molecules in normal and staggerer mice, *Proc. Natl. Acad. Sci. USA* **79**:7036–7040.

Edgar, D., Timpl, R., and Thoenen, H., 1984, The heparin-binding domain of laminin is responsible for its effects on neurite outgrowth and neuronal survival, *EMBO J.* **3**:1463–1468.

Elfman, L., Kynoch, P. A. M., Siddle, K., and Thompson, R. J., 1986, Rat and mouse monoclonal antibodies to human myelin basic protein, *J. Neurosci.* **46**:509–515.

Faissner, A., Kruse, J., Goridis, C., Bock, E., and Schachner, M., 1984a, The neural cell adhesion molecule L1 is distinct from the N-CAM related group of surface antigens BSP-2 and D2, *EMBO J.* **3**:733–737.

Faissner, A., Kruse, J., Nieke, J., and Schachner, M., 1984b, Expression of neural cell adhesion molecule L1 during development, in neurological mutants, and in the peripheral nervous system, *Dev. Brain Res.* **15**:69–82.

Favilla, J. T., Frail, D. E., Palkovits, C. G., Stoner, G. L., Braun, P. E., and Webster, H., 1984, Myelin-associated glycoprotein (MAG) distribution in human central nervous tissue studied immunocytochemically with monoclonal antibody *J. Neuroimmunol.* **6**:19–30.

Fields, K. L., and Dammerman, M., 1985, A monoclonal antibody equivalent to anti-rat neural antigen-1 as a marker for Schwann cells, *Neuroscience* **15**:877–885.

Fraser, S. E., Murray, B. A., Chuong, C.-M., and Edelman, G. M., 1984, Alteration of the retinotectal map in *Xenopus* by antibodies to neural cell adhesion molecules, *Proc. Natl. Acad. Sci. USA* **81**:4222–4226.

Fredman, P., Magnani, J. L., Nirenberg, M., and Ginsburg, V., 1984, Monoclonal antibody A_2B_5 reacts with many gangliosides in neuronal tissue, *Arch. Biochem. Biophys.* **233**:661–666.

Friedlander, D. R., Brackenbury, R., and Edelman, G. M., 1985, Conversion of embryonic form to adult forms of N-CAM in vitro: Results from *de novo* synthesis of adult forms, *J. Cell Biol.* **101**:412–419.

Friedlander, D. R., Grumet, M., and Edelman, G. M., 1986, Nerve growth factor enhances expression of neuron–glia cell adhesion in PC12 cells, *J. Cell Biol.* **102**:413–419.

Fry, K. R., Tavella, D., Su, Y. Y. T., Peng, Y. W., Watt, C. B., and Lam, D. M. K., 1985, A monoclonal antibody specific for retinal ganglion cells of mammals, *Brain Res.* **338**:360–365.

Ghandour, M. S., Foucaud, B., and Gombos, G., 1984, Monoclonal antibodies specific for glial and neuronal antigens on the young rat cerebellum, *Neurosci. Lett.* **51**:119–125.

Goridis, C., Hirn, M., Santoni, M.-J., Gennarini, G., Deagostini-Bazin, H., Jordan, B. R., Kiefer, M., and Steinmetz, M., 1985, Isolation of mouse NCAM-related cDNA: Detection and cloning using monoclonal antibodies, *EMBO J.* **4**:631–635.

Gottlieb, D. I., Chang, Y.-C., and Schwob, J. E., 1986, Monoclonal antibodies to glutamic acid decarboxylase, *Proc. Natl. Acad. Sci. USA* **83**:8808–8812.

Greene, L. A., and Tischler, A. S., 1976, Establishment of a noradrenergic clonal line of rat adrenal pheochromocytoma cells which respond to nerve growth factor, *Proc. Natl. Acad. Sci. USA* **73**:2424–2428.

Greve, J. M., and Gottlieb, D. I., 1982, Monoclonal antibodies which alter the morphology of cultured chick myogenic cells, *J. Cell. Biochem.* **18**:221–229.

Grumet, M., and Edelman, G. M., 1984, Heterotypic binding between neuronal membrane vesicles and glial cells is mediated by a specific cell adhesion molecule, *J. Cell Biol.* **98**:1746–1756.

Grumet, M., Rutishauser, U., and Edelman, G. M., 1982, Neural cell adhesion molecule is on embryonic muscle cells and mediates adhesion to nerve cells *in vitro, Nature* **295**:693–695.

Grumet, M., Hoffman, S., and Edelman, G. M., 1984a, Two antigenically related neuronal cell adhesion molecules of different specificities mediate neuron–neuron and neuron–glia adhesion. *Proc. Natl. Acad. Sci. USA* **81**:267–271.

Grumet, M., Hoffman, S., Chuong, C.-M., and Edelman, G. M., 1984b, Polypeptide components and binding functions of neuron–glia cell adhesion molecules, *Proc. Natl. Acad. Sci. USA* **81**:7989–7993.

Grumet, M., Hoffman, S., Crossin, K. L., and Edelman, G. M., 1985, Cytotactin, an extracellular matrix protein of neural and non-neural tissues that mediates glia–neuron interactions, *Proc. Natl. Acad. Sci. USA* **82**:8075–8079.

Guerci, A., Monge, M., Baron-Van Evercooren, A., Lubetzki, C., Dancea, S., Boutry, J. M., Goujet-Zalc, C., and Zalc, B., 1986, Schwann cell marker defined by a monoclonal antibody (224-58) with species cross-reactivity. 1. Cellular localization, *J. Neurochem.* **46**:425–434.

Gulcher, J. R., Marton, L. S., and Stefansson, K., 1986, Two large glycosylated polypeptides found in myelinating oligodendrocytes but not in myelin, *Proc. Natl. Acad. Sci. USA* **83**:2118–2122.

Hatten, M. E., Furie, M. B., and Rifkin, D. B., 1982, Binding of developing mouse cerebellar cells to fibronectin: A possible mechanism for the foundation of the external granular layer, *J. Neurosci.* **2**:1195–1206.

Hawkes, R., Niday, E., and Matus, A., 1982, Monoclonal antibodies identify novel neural antigens, *Proc. Natl. Acad. Sci. USA* **79**:2410–2414.

Hempstead, J. L., and Morgan, J. I., 1985, A panel of monoclonal antibodies to the rat olfactory epithelium, *J. Neurosci.* **5**:438–449.

Henke-Fahle, S., and Bonhoeffer, F., 1983, Inhibition of axonal growth by a monoclonal antibody, *Nature* **303**:65–67.

Hirn, M., Ghandour, M. S., Deagostini-Bazin, H., and Goridis, C., 1983, Molecular heterogeneity and structural evolution during cerebellar ontogeny detected by monoclonal antibody of the mouse cell surface antigen BSP-2, *Brain Res.* **265**:87–100.

Ho, R. K., and Goodman, C. S., 1982, Peripheral pathways are pioneered by an array of central and peripheral neurones in grasshopper embryos, *Nature* **297**:404–406.

Hoffman, S., Sorkin, B. C., White, P. C., Brackenbury, R., Mailhammer, R., Rutishauser, U., Cunningham, B. A., and Edelman, G. M., 1982, Chemical characterization of a neural cell adhesion molecule purified from embryonic brain membrane, *J. Biol. Chem.* **257**:7720–7729.

Hoffman, S., and Edelman, G. M., 1983, Kinetics of homophilic binding by embryonic and adult forms of the neural cell adhesion molecule, *Proc. Natl. Acad. Sci. USA* **80**:5762–5766.

Horwitz, A., Duggan, K., Greggs, R., Decker, C., and Buck, C., 1985, The cell attachment (CSAT) antigen has properties of a receptor for laminin and fibronectin, *J. Cell Biol.* **101**:2134–2144.

Hynes, R. O., and Yamada, K. M., 1982, Fibronectin: Multifunctional modular glycoproteins, *J. Cell Biol.* **95**:369–377.

Johannsen, S., and Hook, M., 1984, Substrate adhesion of rat hepatocytes: On the mechanism of attachment to fibronectin, *J. Cell Biol.* **98**:810–817.

Jorgensen, O. S., Delouvee, A., Thiery, J. P., and Edelman, G. M., 1980, The nervous system specific protein D2 is involved in adhesion among neurites from cultured rat ganglia, *FEBS Lett.* **111**:39–42.

Keilhauer, G., Faissner, A., and Schachner, M., 1985, Differential inhibition of neurone–neurone, neurone–astrocyte, and astrocyte–astrocyte adhesion by L1, L2, and N-CAM antibodies, *Nature* **316**:728–730.

Kohler, G., and Milstein, C., 1975, Continuous cultures of fused cells secreting antibody of pre-defined specificity, *Nature* **256**:495–497.

Kotrla, K. J., and Goodman, C. S., 1984, Transient expression of a surface antigen on a small subset of neurones during embryonic development, *Nature* **311**:151–153.

Kruse, J., Mailhammer, R., Wernecke, H., Faissner, A., Sommer, I., Goridis, C., and Schachner, M., 1984, Neural cell adhesion molecules and myelin-associated glycoprotein share a carbo-hydrate moiety recognized by monoclonal antibodies L2 and HNK-1, *Nature* **311**:153–155.

Kruse, J., Keilhauer, G., Faissner, A., Timpl, R., and Schachner, M., 1985, The J1 glycoprotein— A novel nervous system cell adhesion molecule of the L2-HNK-1 family, *Nature* **316**:146–148.

Lagenaur, C., and Schachner, M., 1981, Monoclonal antibody (M2) to glial and neuronal cell surfaces, *J. Supramol. Struct. Cell. Biochem.* **15**:335–346.

Lander, A. D., Fujii, D. K., and Reichardt, L. F., 1985, Laminin is associated with the "neurite-promoting factors" found in conditioned medium. *Proc. Natl. Acad. Sci. USA* **82**:2183–2187.

Lawson, S. N., Harper, A. A., Harper, E. I., Garson, J. A., and Anderton, B. H., 1984, A monoclonal antibody against neurofilament protein specifically labels a subpopulation of rat sensory neurons, *J. Comp. Neurol.* **228**:263–272.

Leah, J., Gynther, B., and Kidson, C., 1984, A chick neural antigen identified by monoclonal antibodies, *Int. J. Dev. Neurosci.* **2**:517–527.

Lemmon, V., 1985, Monoclonal antibodies specific for glia in the chick nervous system, *Dev. Brain Res.* **23**:111–120.

Lemmon, V., 1986, Localization of a filamin-like protein in glia of the chick central nervous system, *J. Neurosci.* **6**:43–51.

Lemmon, V., and Gottlieb, D. I., 1982, Monoclonal antibodies selective for the inner portion of the chick retina, *J. Neurosci.* **2**:531–535.

Lemmon, V., Staros, E. B., Perry, H. E., and Gottlieb, D. I., 1982, A monoclonal antibody which binds to the surface of chick brain cells and myotubes: Cell selectivity and properties of the antigen, *Dev. Brain Res.* **3**:349–360.

Levine, J. M., Beasley, L., and Stallcup, W. B., 1984, The D1.1 antigen: A cell surface marker for germinal cells of the central nervous system, *J. Neurosci.* **4**:820–831.

Lindner, J., Rathjen, F. G., and Schachner, M., 1983, L1 mono- and polyclonal antibodies modify cell migration in early postnatal mouse cerebellum, *Nature* **305:**427–430.

McGuire, J., Greene, L., and Furano, A., 1978, NGF stimulates incorporation of fucose or glucosamine into an external glycoprotein in cultured rat PC12 pheochromocytoma cells, *Cell* **15:**357–365.

McKay, R. D. G., and Hockfield, S. J., 1982, Monoclonal antibodies distinguish antigenically discrete neuronal types in the vertebrate central nervous system, *Proc. Natl. Acad. Sci. USA* **79:**6747–6751.

McKay, R., Johansen, J., and Hockfield, S., 1984, Monoclonal antibody identifies a 63,000 dalton antigen found in all central neuronal cell bodies but in only a subset of axons in the leech, *J. Comp. Neurol.* **226:**448–455.

Meyer, R., 1982, Tetrodotoxin blocks the formation of ocular columns in goldfish, *Science* **218:**589–591.

Miller, R. H., Williams, J., and Raff, M. C., 1984, A4: An antigenic marker of neural tube-derived cells, *J. Neurocytol.* **13:**329–338.

Mirsky, R., and Jessen, K. R., 1984, A cell surface protein of astrocytes, Ran-2, distinguishes non-myelin-forming Schwann cells from myelin-forming Schwann cells, *Dev. Neurosci.* **6:**304–316.

Morris, R. J., Beech, J. N., Barber, P. C., and Raisman, G., 1985a, Early stages of Purkinje cell maturation demonstrated by Thy-1 immunohistochemistry on postnatal rat cerebellum, *J. Neurocytol.* **14:**427–452.

Morris, R. J., Beech, J. N., Barber, P. C., and Raisman, G., 1985b, Late emergence of Thy-1 on climbing fibres demonstrates a gradient of maturation from the fissures to the follial convexities in developing rat cerebellum, *J. Neurocytol.* **14:**453–467.

Murray, B. A., Hemperly, J. J., Gallin, W. J., MacGregor, J. S., Edelman, G. M., and Cunningham, B. A., 1984, Isolation of cDNA clones for the chicken neural cell adhesion molecule (N-CAM), *Proc. Natl. Acad. Sci. USA* **81:**5584–5588.

Murray, B. A., Hemperly, J. J., Prediger, E. A., Edelman, G. M., and Cunningham, B. A., 1986, Alternatively spliced mRNAs code for different polypeptide chains of the chicken neural cell adhesion molecule (N-CAM), *J. Cell Biol.* **102:**89–193.

Neff, N. T., Lowrey, C., Tovar, A., Decker, C., Damsky, C., Buck, C., and Horwitz, A., 1982, A monoclonal antibody detaches embryonic skeletal muscle from extracellular matrices, *J. Cell Biol.* **95:**654–666.

Noble, M., Albrechtsen, M., Moller, C., Lyles, J., Bock, E., Goridis, C., Watanabe, M., and Rutishauser, U., 1985, Glial cells express N-CAM/D2-CAM–like polypeptides *in vitro*, *Nature* **316:**725–728.

Park, D. H., Teitelman, G., Evinger, M. J., Woo, J. I., Ruggiero, D. A., Albert, V. R., Baetge, E. E., Pickel, V. M., Reis, D. J., and Joh, T. H., 1986, Phenylethanolamine *N*-methyltransferase-containing neurons in rat retina: Immunohistochemistry, immunochemistry, and molecular biology, *J. Neurosci.* **6:**1108–1113.

Perry, V. H., Morris, R. J., and Raisman, G., 1984, Is Thy-1 expressed only by ganglion cells and their axons in the retina and optic nerve? *J. Neurocytol.* **13:**809–824.

Pessac, B., Girard, A., Romey, G., Crisanti, P., Lorinet A., and Calothy, G., 1983, A neuronal clone derived from a Rous sarcoma virus–transformed quail embryo neuroretina established culture, *Nature* **302:**616–618.

Pierschbacher, M. D., Hayman, F. G., and Ruoslahti, E., 1981, Location of the cell-attachment site in fibronectin with monoclonal antibodies and proteolytic fragments of the molecule, *Cell* **26:**259–267.

Pigott, R., and Kelly, J. S., 1984, A cell surface antigen present on cultured cerebellar neurones appears to be transiently expressed during cerebellar development in the rat, *Neurosci. Lett.* **49:**105–110.

Plioplys, A. V., Thibault, J., and Hawkes, R., 1985, Selective staining of a subset of Purkinje cells in the human cerebellum with monoclonal antibody mabQ113, *J. Neurolog. Sci.* **70:**245–256.

Pollerberg, E. G., Sadoul, R., Goridis, C., and Schachner, M., 1985, Selective expression of the 180-kD component of the neural cell adhesion molecule N-CAM during development, *J. Cell Biol.* **101**:1921–1929.

Price, J., and Hynes, R. O., 1985, Astrocytes in culture synthesize and secrete a variant form of fibronectin, *J. Neurosci.* **5**:2205–2211.

Pytela, R., Pierschbacher, M., and Ruoslahti, E., 1985, Identification and isolation of a 140-kD cell surface glycoprotein with properties expected of a fibronectin receptor, *Cell* **40**:191–198.

Quarles, R. H., Everly, J. L., and Brady, R. O., 1973, Evidence for the close association of a glycoprotein with myelin in rat brain, *J. Neurochem.* **21**:1177–1191.

Ranscht, B., Moss, D. J., and Thomas, C., 1984, A neuronal surface glycoprotein associated with the cytoskeleton, *J. Cell Biol.* **99**:1803–1813.

Rathjen, F. G., and Rutishauser, U., 1984, Comparison of two cell surface molecules involved in neural cell adhesion, *EMBO J.* **3**:461–465.

Rathjen, G., and Schachner, M., 1984, Immunocytological and biochemical characterization of a newneuronal cell surface component (L1 antigen) which is involved in cell adhesion, *EMBO J.* **3**:1–10.

Regan, L. J., Dodd, J., Barondes, S. H., and Jessell, T.M., 1986, Selective expression of endogenous lactose-binding lectins and lactoseries glycoconjugates in subsets of rat sensory neurons, *Proc. Natl. Acad. Sci. USA* **83**:2248–2252.

Reh, T. A., Pitts, E. C., and Constantine-Paton, M., 1983, The organization of the fibers in the optic nerve of normal and tectum-less *Rana pipiens*, *J. Comp. Neurol.* **218**:297–313.

Rieger, F., Grumet, M., and Edelman, G. M., 1985, N-CAM at the vertebrate neuromuscular junction, *J. Cell Biol.* **101**:285–293.

Riopelle, R. J., McGarry, R. C., and Roder, J. C., 1986, Adhesion properties of a neuronal epitope recognized by the monoclonal antibody HNK-1, *Brain Res.* **367**:20–25.

Rogers, S. L., McCarthy, J. B., Palm, S. L., Furcht, L. T., and Letourneau, P. C., 1985, Neuron-specific interactions with two neurite-promoting fragments of fibronectin, *J. Neurosci.* **5**:369–378.

Rothbard, J. B., Brackenbury, R., Cunningham, B. A., and Edelman, G. M., 1982, Differences in the carbohydrate structures of neural cell adhesion molecules from adult and embryonic brains, *J. Biol. Chem.* **257**:11064–11069.

Rougon, G., and Marshak, D. R., 1986, Structural and immunological characterization of the amino-terminal domain of mammalian neural cell adhesion molecules, *J. Biol. Chem.* **261**:3396–3401.

Rougon, G., Deagostini-Bazin, H., Girsch, M., and Goridis, C., 1982, Tissue- and developmental stage-specific forms of a neural cell surface antigen linked to differences on glycosylation of a common polypeptide, *EMBO J.* **1**:1239–1244.

Rougon, G., Hirn, M., Hirsch, M. R., Guenet, J. L., and Goridis, C., 1984, Identification and immunolocalization by monoclonal antiobody of NSP-5, a surface polypeptide of neural cells, *J. Neuroimmunol.* **6**:411–426.

Ruoslahti, E., Hayman, E. G., Engvall, E., Cothran, W. C., and Butler, W. T., 1981, Alignment of biologically active domains in the fibronectin molecule, *J. Biol. Chem.* **256**:7277–7281.

Russoff, A. C., and Easter, S. S., 1980, Order in the optic nerve of goldfish, *Science* **208**:311–312.

Rutishauser, U., 1984, Developmental biology of a neural cell adhesion molecule, *Nature* **310**:549–554.

Rutishauser, U., Gall, W. E., and Edelman, G. M., 1978, Adhesion among neural cells of the chick embryo. IV. Relationship of the cell surface molecule CAM in the formation of neurite bundles in cultures of spoinal ganglia, *J. Cell Biol.* **79**:382–393.

Rutishauser, U., Hoffman, S., and Edelman, G. M., 1982, Binding properties of a cell adhesion molecule from neural tissue, *Proc. Natl. Acad. Sci. USA* **79**:685–689.

Rutishauser, U., Grumet, M., and Edelman, G. M., 1983, N-CAM mediates initial interactions between spinal cord neurons and muscle cells in culture, *J. Cell Biol.* **97**:145–152.

Rutishauser, U., Watanabe, M., Silver, J., Troy, F. A., and Vimr, E. R., 1985, Specific alteration of NCAM-mediated cell adhesion by an endoneuraminidase, *J. Cell Biol.* **101**:1842–1849.

Sadoul, R., Hirn, M., Deagostini-Bazin, H., Rougon, G., and Goridis, C., 1983, Adult and embryonic mouse neural cell adhesion molecules have different binding properties, *Nature* **304**:347–349.

Sajovic, P., Moraru, E., Greene, L. A., and Shelanski, M. L., 1986, Selective staining of large projection neurons by monoclonal antibody to a glycoprotein of PC12 cells, *J. Neurosci.* **6**:82–93.

Salton, S., Richter-Landsberg, C., Greene, L., and Shelanski, M., 1983, Nerve growth factor–inducible large external (NILE) glycoprotein: Studies of a central and peripheral neuronal marker, *J. Neurosci.* **3**:441–454.

Sanes, J., Schachner, M., and Covault, J., 1986, Expression of several adhesive macromolecules (N-CAM, L1, J1, NILE, uvomorulin, laminin, fibronectin, and a heparan sulfate proteoglycan) in embryonic, adult, and denervated adult skeletal muscle, *J. Cell Biol.* **102**:420–431.

Schachner, M., Faissner, A., Kruse, J., Lindner, J., Meier, D. H., Rathjen, F. G., and Wernecke, H., 1983, Cell type specificity and developmental expression of neural cell-surface components involved in cell interactions and of structurally related molecules, *Cold Spring Harbor Symp. Quant. Biol.* **48**:557–568.

Schachner, M., Sommer, I., and Lagenaur, C., 1984, Expression of glial antigens C1 and M1 in the peripheral nervous system during development and regeneration, *Dev. Brain Res.* **14**:165–178.

Schmidt, J. T., and Edwards, D. L., 1983, Activity sharpens the map during the regeneration of the retinotectal projection in goldfish, *Brain Res.* **269**:29–39.

Schnitzer, J., Kim, S. U., and Schachner, M., 1984a, Specificity of monoclonal antibody N1 for cell surfaces of mouse central nervous system neurons, *Dev. Brain Res.* **15**:21–32.

Schnitzer, J., Kim, S. U., and Schachner, M., 1984b, Some immature tetanus toxin-positive cells share antigenic properties with subclasses of glial cells. An immunofluorescence study in the developing nervous system of the mouse using a new monoclonal antibody S1, *Dev. Brain Res.* **16**:203–217.

Scholes, J. H., 1979, Nerve fibre topography in the retinal projection of the tectum, *Nature* **278**:620–624.

Schubert, D., and LaCorbiere, M., 1985a, Isolation of a cell surface receptor for chick neural retina adherons, *J. Cell Biol.* **100**:56–63.

Schubert, D., and LaCorbiere, M., 1985b, Isolation of an adhesion-mediating protein from chick neural retina adherons, *J. Cell Biol.* **101**:1071–1077.

Schubert, D., LaCorbiere, M., Klier, F. G., and Birdwell, C., 1983, A role for adherons in retinal cell adhesion, *J. Cell Biol.* **96**:990–998.

Schubert, D., LaCorbiere, M., and Esch, F., 1986, A chick neural retina adhesion and survival molecule is a retinol-binding protein, *J. Cell Biol.* **102**:2295–2301.

Schwartz, M., and Eshhar, N., 1984, Early regenerative responses induced by monoclonal antibodies directed against a cell surface glycoprotein of goldfish retinal ganglion cells, *EMBO J.* **3**:1287–1293.

Schwob, J. E., Farber, N. B., and Gottlieb, D. I., 1986, Neurons of the olfactory epithelial in adult rats contain vimentin, *J. Neurosci.* **6**:208–217.

Silver, J., and Rutishauser, U., 1984, Guidance of optic axons *in vivo* by a preformed adhesive pathway on neuroepithelial endfeet, *Dev. Biol.* **106**:485–499.

Sommer, I., and Schachner, M., 1981, Monoclonal antibodies (O1 to O4) to oligodendrocyte cell surfaces: An immunocytological study in the central nervous system, *Dev. Biol.* **83**:311–327.

Sorge, L. K., Levy, B. T., and Maness, P. F., 1984, pp60$^{c\text{-}src}$ is developmentally regulated in the neural retina, *Cell* **36**:249–257.

Stallcup, W. B., and Beasley, L., 1985, Involvement of the nerve growth factor–inducible large external glycoprotein (NILE) in neurite fasciculation in primary cultures of rat brain, *Proc. Natl. Acad. Sci. USA* **82**:1276–1280.

Stallcup, W. B., Arner, L., and Levine, J., 1983, An antiserum against the PC12 cell line defines cell surface antigens specific for neurons and Schwann cells, *J. Neurosci.* **3**:53–68.

Sternberger, L. A., Harwell, L. W., and Sternberger, N. H., 1982, Neurotypy: Regional individuality in rat brain detected by immunocytochemistry with monoclonal antibodies, *Proc. Natl. Acad. Sci. USA* **79:**1326–1330.

Sternberger, N. H., Quarles, R. H., Itoyama, Y., and Websteer, H. deF., 1979, Myelin-associated glycoprotein demonstrated immunocytochemically in myelin and myelin-forming cells of developing rat, *Proc. Natl. Acad. Sci. USA* **76:**1510–1514.

Thanos, S., and Bonhoeffer, F., 1984, Development of the transient ipsilateral retinotectal projection in the chick embryo: A numerical fluorescence–microscopic analysis, *J. Comp. Neurol.* **224:**407–414.

Thanos, S., Bonhoeffer, F., and Rutishauser, U., 1984, Fiber–fiber interactions and tectal cues influence the development of the chick retinotectal projection, *Proc. Natl. Acad. Sci. USA* **81:**1906–1910.

Thiery, J. P., Brackenbury, R., Rutishauser, U., and Edelman, G. M., 1977, Adhesion among neural cells of chick retina. II. Purification and characterization of cell adhesion molecule from neural retina, *J. Biol Chem.* **252:**6841–6845.

Trisler, G. D., Schneider, M. D., and Nirenberg, M., 1981, A topographic gradient of molecules in retina can be used to identify neuron position, *Proc. Natl. Acad. Sci. USA* **78:**2145–2149.

Vincent, M., and Thiery, J. P., 1984, A cell surface marker for neural crest and placodal cells: Further evolution in peripheral and central nervous system, *Dev. Biol.* **103:**468–481.

Vulliamy, T., Rattray, S., and Mirsky, R., 1981, Cell-surface antigen distinguishes sensory and autonomic peripheral neurones from central neurones, *Nature* **291:**418–420.

Webb, M., and Woodhams, P. L., 1984a, Monoclonal antibodies recognising cell surface molecules expressed by rat cerebellar interneurons, *J. Neuroimmunol.* **6:**283–300.

Webb, M., and Woodhams, P. L., 1984b, Recognition by a mouse monoclonal antibody of a glycoprotein antigen of rab brain which is expressed intracellularly by neurons, *Neuroscience* **13:**583–594.

Weber, A., and Schachner, M., 1984, Maintenance of immunocytologically identified Purkinje cells from mouse cerebellum in monolayer culture, *Brain Res.* **311:**119–130.

Williams, R. K., Goridis, C., and Akeson, R., 1985, Individual neural cell types express immunologically distinct N-CAM forms, *J. Cell Biol.* **101:**36–42.

Williams, R. K., Kelly, P. T., and Akeson, R. A., 1985, Cell-surface antigens of developing rat cerebellar neurons: Identification with monoclonal antibodies, *Dev. Brain Res.* **19:**253–266.

Wood, J. N., Hudson, L., Jessell, T. M., and Yamamoto, M., 1982, A monoclonal antibody defining antigenic determinants on subpopulations of mammalian neurones and *Trypanosoma cruzi* parasites, *Nature* **296:**34–38.

Yamada, K. M., 1983, Cell surface interactions with extracellular materials, *Annu. Rev. Biochem.* **52:**761–799.

Yamada, K. M., Kennedy, D. W., Kimata, K., and Pratt, R. M., 1980, Characterization of fibronectin interactions with glycosaminoglycans and identification of active proteolytic fragments, *J. Biol. Chem.* **255:**6055–6063.

Yamakuni, T., Usui, H., Iwanaga, T., Kondo, H., Odani, S., and Takahashi, Y., 1984, Isolation and immunohistochemical localization of a cerebellar protein, *Neurosci. Lett.* **45:**235–240.

Young, L. H. Y., and Dowling, J. E., 1984, Monoclonal antibodies distinguish subtypes of retinal horizontal cells, *Proc. Natl. Acad. Sci. USA* **81:**6255–6259.

Zipser, B., and McKay, R., 1981, Monoclonal antibodies distinguish identifiable neurones in the leech, *Nature* **289:**549–554.

Zipursky, S. L., Venkatesh, T. R., and Benzer, S., 1985, From monoclonal antibody to gene for a neuron-specific glycoprotein in *Drosophila, Proc. Natl. Acad. Sci. USA* **82:**1855–1859.

MECHANISMS OF α_1-ADRENERGIC AND RELATED RESPONSES

Roles of Calcium, Phosphoinositides, Guanine Nucleotides, Diacylglycerol, Calmodulin, and Changes in Protein Phosphorylation

John H. Exton

1. INTRODUCTION

Activation of the sympathetic nervous system leads to the release of catecholamines (principally epinephrine) from the adrenal medulla into the blood stream and of norepinephrine from adrenergic nerve endings throughout the body. The effects of these catecholamines are thus widespread and are mediated by four different types of adrenergic receptors, located on the plasma membranes of their target cells. Two of these receptors (β_1 and β_2) are linked positively to adenylate cyclase, and the effects of their activation can be attributed to an increase in cellular cAMP in almost all situations. The other two receptors (α_1 and α_2) elicit quite different responses. Activation of α_2-adrenergic receptors leads to inhibition of adenylate cyclase. Although most of the resulting physiological responses can be attributed to the decline in cAMP, there are some tissues (platelets, pancreatic islets) in which this is questionable. Activation of α_1-adrenergic receptors is linked to the breakdown of polyphosphoinositides in the plasma membrane with the generation of two intracellular messages, namely myoinositol 1,4,5-P_3 (IP_3) and 1,2-diacylglycerol (DAG) (Figure 1). The function of IP_3 is to release Ca^{2+} from intracellular stores, which are probably located

John H. Exton • Howard Hughes Medical Institute and Department of Molecular Physiology and Biophysics, Vanderbilt University School of Medicine, Nashville, Tennessee 37232.

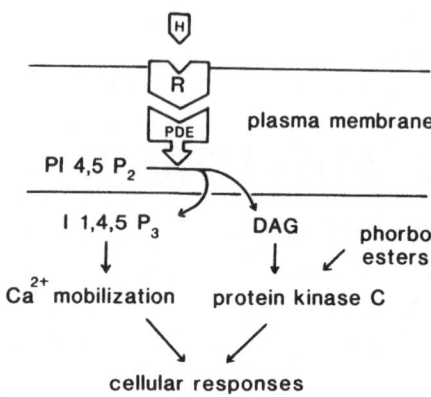

FIGURE 1. Mechanisms by which α_1-adrenergic and other Ca^{2+}-mobilizing agonists exert their physiological actions.

in the endoplasmic reticulum, thereby raising cytosolic Ca^{2+} and altering the activity of Ca^{2+} (calmodulin)-sensitive proteins; whereas that of DAG is to activate the Ca^{2+}-phospholipid-dependent protein kinase C.

The coupling of β- and α_2-adrenergic receptors to adenylate cyclase has been shown to involve guanine nucleotide–binding regulatory proteins that are stimulatory and inhibitory to adenylate cyclase, respectively. There is much

TABLE I. α_1-Adrenergic Agonists, Antagonists, Target Tissues, and Responses

Agonists	Antagonists	Tissue	Response
Epinephrine[a]	Prazosin	Smooth muscle	Contraction
Norepinephrine[a]	Corynanthine	(vascular, iris, radial,	
Phenylephrine	Ergot alkaloids[a]	ureter, pilomotor,	
Methoxamine	Phentolamine[a]	uterus, trigone, and	
	Phenoxybenzamine[a]	gastrointestinal and	
	Tolazoline[a]	bladder sphincters)	
		Smooth muscle	Relaxation
		(gastrointestinal)	
		Liver[b]	Glycogenolysis, gluconeogenesis
		Heart	Increased force[c], glycolysis
		Salivary glands	Secretion (K^+, H_2O)
		Adipose tissue	Glycogenolysis
		Sweat glands (localized)	Secretion
		Kidney (proximal tubule)	Gluconeogenesis, Na^+ reabsorption
		Brain	Neurotransmission

[a] Nonselective.
[b] Only important in some species; in man β_2-adrenergic responses predominate.
[c] β_1-adrenergic responses are greater.

evidence that the coupling of α_1-adrenergic receptors (and other receptors to Ca^{2+}-mobilizing agonists) to polyphosphoinositide breakdown also involves a guanine nucleotide–binding protein different from those involved in the regulation of adenylate cyclase.

α_1-adrenergic receptors are located in many tissues throughout the body and mediate many responses to catecholamines (Table I). Important α_1-adrenergic responses are the contraction of smooth muscle in a variety of tissues, e.g., blood vessels, uterus, ureter, and iris, relaxation of gastrointestinal smooth muscle, increased glycogen breakdown and gluconeogenesis in liver (also elicited through β_2-adrenergic receptors), increased force of contraction (inotropism) in heart (also elicited through β_1-adrenergic receptors), secretion of watery saliva, and neurotransmission in certain parts of the central nervous system. As shown in Table I, synthetic analogues of the natural catecholamines are also able to activate α_1-adrenergic receptors, and the activation of these receptors can be blocked by various ergot alkaloids and synthetic antagonists, the most specific of which is prazosin. These agonists and antagonists are used to define whether or not a given catecholamine response is mediated by α_1-adrenergic receptors and, when radioactively labeled, to identify and characterize these receptors. Since many α_1-adrenergic agonists and antagonists are nonselective, they are often used in combination with antagonists to other adrenergic receptors.

2. THE α_1-ADRENERGIC RECEPTOR

Radioactive ligands employed to identify the α_1-adrenergic receptor in tissues include ^{3}H-labeled prazosin, epinephrine, norepinephrine, dihydroergocryptine, phenoxybenzamine, and WB-4101, and ^{125}I-labeled HEAT (also called BE-2254), CP65,526, and APDQ (analogues of prazosin). Using these ligands, α_1-adrenergic receptors have been identified in brain, lung, liver, kidney, heart, uterus, adipose tissue, vas deferens, salivary glands, and certain blood vessels (Bylund and U'Prichard, 1983).

Molecular studies of the α_1-adrenergic receptor have been largely confined to rat liver and a smooth muscle cell line. Using ^{125}I-ADPQ as a photoaffinity probe to specifically label the α_1-adrenergic receptor of rat liver plasma membranes, a binding subunit of M_r 78,000–85,000 has been identified by polyacrylamide gel electrophoresis in sodium dodecylsulfate (Leeb-Lundberg et al., 1984). In the absence of protease inhibitors, species of lower relative molecular weight are observed. Photoaffinity labeling with another prazosin analogue (^{125}I-CP65,526) also identifies a 78-kD labeled protein in rat liver membranes, which is similarly degraded by endogenous proteases (Seidman et al., 1984; Lynch et al., 1986a). On the other hand, labeling of the membranes with low concentrations of [^{3}H]phenoxybenzamine has yielded labeled proteins of 58 and 80 kD (Kunos *et al.*, 1983) or 45 kD (Guellaen *et al.*, 1982). In these experiments, no specific precautions were taken to limit proteolysis. ^{125}I-ADPQ has been used to label the α_1-adrenergic receptor in other tissues e.g., spleen, lung, brain, and aortic smooth muscle cells (Leeb-Lundberg *et al.*, 1984). In all cases, a 78–79-kD protein is labeled, but in spleen, a species of lower relative molecular weight is also observed. Radiation inactivation analysis carried out in rat liver membranes indicates that the receptor exists as a dimer with subunits of approximately 85 kD (Venter *et al.*, 1984b). In summary, all these observations indicate that the native ligand–binding subunit of the α_1-adrenergic receptor has a M_r of approximately 80,000 but that it is very susceptible to proteolysis by endogenous membrane proteases.

Several efforts have been made to purify the α_1-adrenergic receptor from different tissues. Using a prazosin analogue (CP57,609) linked to agarose, a 72,000-fold purification of a protein that selectively binds [^{3}H]prazosin has been achieved in rat liver (Graham *et al.*, 1982). However this has a M_r of only 59,000, suggesting proteolytic degradation. Leeb-Lundberg *et al.* (1985) have purified the α_1-adrenergic receptor from DDT$_1$MF-2 cells, which are derived from vas deferens smooth muscle, using another prazosin analogue (A55414) linked to Affi-Gel. The purification was approximately 300-fold, and the resulting binding subunit had a M_r of 80,000. More recently, Lomasney *et al.* (1986) have purified the α_1-adrenergic receptor to apparent homogeneity from these cells using another prazosin analogue (A55453) linked to agarose, wheat germ agglutinin–agarose, and high-performance liquid chromatography. The purified ligand–binding subunit of the receptor had an apparent M_r of 80,000. Peptide maps of the receptor using several proteases revealed little structural homology with the α_2-adrenergic receptor.

The α_1-adrenergic receptor appears to be a glycoprotein based on its adsorption to wheat germ agglutinin–Sepharose (Meier *et al.*, 1984). Some monoclonal antibodies developed toward α_1-adrenergic receptors also precipitate muscarinic cholinergic (Venter *et al.*, 1984a) and α_2-adrenergic receptors (Shreeve *et al.*, 1985). This suggests the existence of common structural determinants, but these could be minor.

It is now clear that α_1-adrenergic receptors can exist in more than one agonist affinity state and that guanine nucleotides can influence the equilibrium between these states. There was some controversy about this initially. Some workers reported that agonist binding to these receptors was unaffected by guanosine triphosphate (GTP) and its analogues (Hoffman *et al.*, 1980; Stiles *et al.*, 1983), in contrast to the situations with α_2- and β-adrenergic receptors. On the other hand, others have observed effects of varying magnitude in liver, heart, and kidney (El-Refai *et al.*, 1979; Yamada *et al.*, 1980; Geynet *et al.*, 1980; Snavely and Insel, 1982; Goodhardt *et al.*, 1982; Boyer *et al.*, 1984; Lynch *et al.*, 1985b; Buxton and Brunton, 1986). The probable reason for the discrepancies is provided by the results of Geynet *et al.* (1980) and Lynch *et al.* (1985b, 1986a). These workers have shown that addition of proteases, or omission of metal ion chelators or protease inhibitors, leads to extensive proteolysis of the α_1-adrenergic receptor in liver plasma membranes and an associated loss of guanine nucleotide effects on agonist binding. Thus, varying degrees of proteolytic modification may account for the differences in the magnitude of nucleotide effects observed by various groups.

When endogenous proteases are inhibited, α_1-adrenergic receptors of rat liver plasma membrane can be shown to exist mainly in a form with high affinity for agonists (K_d for ($-$)epinephrine or ($-$)norepinephrine of 20–30 nM) (Lynch *et al.*, 1985b, 1986a). As illustrated in Figure 2, addition of micromolar or higher concentrations of GTP and its nonhydrolyzable analogues causes the receptors to change entirely to a form with low agonist affinity [K_d for ($-$)epinephrine or ($-$)norepinephrine greater than 1 μM]. These data are similar to those observed with receptors linked positively or negatively to adenylate cyclase, e.g., β- and α_2-adrenergic receptors, and suggest that α_1-adrenergic receptors may also couple to a guanine nucleotide-binding regulatory protein. Further evidence for this is presented in Section 5.

Although α_1-adrenergic receptors are linked to Ca^{2+} mobilization in most tissues, they may become coupled to cAMP accumulation under certain circumstances. For example, in the livers of aging rats, β_2-adrenergic receptor–mediated cAMP accumulation declines, whereas α_1-adrenergic receptor–induced cAMP elevation appears (Blair *et al.*, 1979; Morgan *et al.*, 1983a). The subtype of α_1-adrenergic receptor responsible for the cAMP response appears to be the same as that which mobilizes Ca^{2+} (Morgan *et al.*, 1983e). Calcium depletion of hepatocytes enhances the cAMP accumulation elicited by α_1-adrenergic stimulation (Chan and Exton, 1977; Morgan *et al.*, 1983a), but the mechanism of the enhancement is unknown. The elevation of cAMP induced by α_1-adrenergic

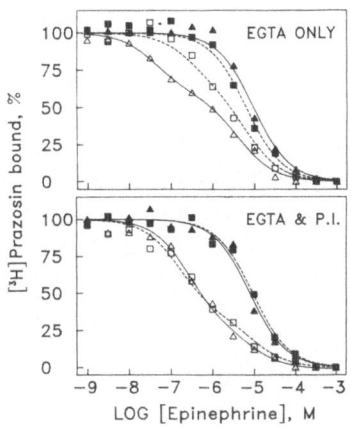

FIGURE 2. Effect of a GTP analogue on epinephrine binding to liver α_1-adrenergic receptor. Liver plasma membranes were prepared from young ($\square$) or old ($\triangle$) rats in the presence of ethyleneglycoltetraacetic acid (EGTA) alone or EGTA plus protease inhibitors (P.I.). Epinephrine displacement of 1 nM [^{3}H]prazosin binding was assayed without (open symbols) or with 0.4 mM GTPγS (closed symbols).

agonists is not large, but it probably accounts for reports that these agonists have two mechanisms of action in liver (Hernandez-Sotomayor and Garcia-Sainz, 1984; Pushpendran *et al.*, 1984; Corvera *et al.*, 1984; Garcia-Sainz and Hernandez-Sotomayor, 1985). Other Ca^{2+}-mobilizing agonists do not induce cAMP accumulation in calcium-depleted hepatocytes or hepatocytes from aging rats (Morgan *et al.*, 1983a). Elevation of cAMP in response to α-adrenergic stimulation has also been reported in brain and spinal cord (Perkins and Moore, 1973; Schultz and Daly, 1973; Jones and McKenna, 1980). A possible explanation for the phenomenon is that in liver cells depleted of Ca^{2+} or obtained from aged rats, α_1-adrenergic receptors may aberrantly couple to N_s or G_s (the stimulatory guanine nucleotide–binding protein of the adenylate cyclase system) in addition to the binding protein involved in polyphosphoinositide breakdown (see Section 5).

In cerebral cortex, α_1-adrenergic and other Ca^{2+}-mobilizing agonists augment the cAMP accumulation induced by adenosine and β-adrenergic agonists (Hollingsworth and Daly, 1985). Since the effect is mimicked by phorbol esters, it has been proposed to be due to activation of protein kinase C (Hollingsworth *et al.*, 1985). α_1-adrenergic agonists also potentiate the accumulation of cAMP induced by β-adrenergic agonists in pinealocytes (Vanecek *et al.*, 1985). The potentiation is associated with stimulation of Ca^{2+} influx and may also involve activation of protein kinase C (Sugden *et al.*, 1986).

In a thyroid cell line exposed to thyrotropin, there is evidence that activation of α_1-adrenergic receptors induces the release of arachidonic acid, which is metabolized to prostaglandin E_2 and smaller amounts of other eicosanoids (Burch *et al.*, 1986a). These are believed to mediate the observed stimulation of cell replication. Interestingly, the receptors appear to be coupled to both phospholipase C and phospholipase A_2 through separate guanine nucleotide–binding regulatory proteins (Burch *et al.*, 1986b), as shown by the differential effects of neomycin, extracellular Ca^{2+}, and islet-activating protein on inositol phosphate formation and arachidonic acid release.

In addition to age, other factors alter α_1-adrenergic responses in liver and other tissues. These include thyroid hormones, glucocorticoids, hepatectomy, cell culture, sex, and chronic exposure to agonists. In liver, thyroidectomy decreases α_1-adrenergic responses but increases β-adrenergic responses (Malbon *et al.*, 1978; Preiksaitis and Kunos, 1979; Preiksaitis *et al.*, 1982; Storm *et al.*, 1984). These alterations are accompanied by corresponding changes in the density of α_1- and β-adrenergic receptors (Malbon, 1980; Preiksaitis *et al.*, 1982; cf. Malbon and Lo Presti, 1981). In contrast, hypothyroidism decreases β-adrenergic responses in adipose tissue (Malbon *et al.*, 1978), apparently because of an impairment in the coupling of the β-adrenergic receptor to N_s (Malbon *et al.*, 1984) but does not affect α-adrenergic responses (Garcia-Sainz and Fain, 1980; Garcia-Sainz *et al.*, 1981).

Adrenalectomy also alters α- and β-adrenergic responses in liver. There is an enhancement of β-adrenergic responses, which can be attributed to an increased number of β-adrenergic receptors (Chan *et al.*, 1979; Wolfe *et al.*, 1976; Guellaen *et al.*, 1978; Studer and Borle, 1984). On the other hand, α_1-adrenergic responses are diminished (Chan *et al.*, 1979; Studer and Borle, 1982) due to a decrease in high-affinity α_1-adrenergic receptors (El-Refai and Chan, 1986). Interestingly, adrenalectomy also causes an increase in α_2-adrenergic receptors, which offsets the physiological effects of the increase in β-adrenergic receptors (El-Refai and Chan, 1986).

Hepatectomy causes a marked decrease in α_1-adrenergic responsiveness in the liver and an increase in β-adrenergic responsiveness. The latter is associated with an increase in β-adrenergic receptors, whereas there seems to be no change in α_1-adrenergic receptors (Huerta-Bahena *et al.*, 1983). There is also a loss of responsiveness to vasopressin, angiotensin II, and ionophore A23187, although phosphatidylinositol turnover is apparently unchanged (Huerta-Bahena and Garcia-Sainz, 1983, 1984). These findings suggest that hepatectomy causes a defect in intracellular Ca^{2+} action in response to Ca^{2+}-mobilizing agonists in liver.

Primary culture of rat hepatocytes leads to a gradual loss of α_1-adrenergic responses and enhancement of β_2-adrenergic responses (Okajima and Ui, 1982; Itoh *et al.*, 1984; Kunos *et al.*, 1984). These changes are associated with a progressive diminution in the ADP ribosylation of a 41-kD membrane protein under the influence of islet-activating protein, which is a *Bordetella pertussis* toxin (Itoh *et al.*, 1984). The changes in the 41-kD protein (which is assumed to be N_i or G_i, the inhibitory guanine–nucleotide binding protein of the adenylate cyclase system) could explain the observed increase in β-adrenergic receptor–mediated cAMP accumulation, but its relationship to the loss of α_1-adrenergic responses is uncertain. Kunos *et al.* (1984) have observed no change in α_1- or β-adrenergic receptors during hepatocyte culture for 4 hr and believe that the altered adrenergic responses are due to increased membrane phospholipase A_2 activity. This conclusion is based on studies with the phospholipase A_2 inhibitors melittin and lipomodulin; however, these agents exert effects on other cell components.

Livers of female rats display greater β-adrenergic responses compared to those of male rats (Studer and Borle, 1982, 1983; Morgan *et al.*, 1983b) and also show different cellular Ca^{2+} changes in response to epinephrine (Studer and Borle, 1982, 1983). These differences in Ca^{2+} fluxes may be partly due to the difference in the levels of cAMP induced by the catecholamine (Morgan *et al.*, 1983b).

There have been relatively few studies of the effects of chronic agonist exposure on α_1-adrenergic receptors. Incubation of MDCK-D-1 (cloned renal epithelial) cells with a high concentration of epinephrine for 21 hr caused an 81% loss of α_1-adrenergic receptors (Meier *et al.*, 1985). The loss occurred more slowly than the loss of β_2-adrenergic receptors and was due to a decrease in maximal binding capacity (B_{max}) without change in K_d for epinephrine. Likewise, prolonged exposure of cultured aortic smooth muscle cells to norepinephrine did not alter the binding affinity of the α_1-adrenergic receptors but decreased their B_{max} and norepinephrine-stimulated Ca^{2+} efflux (Colucci and Alexander, 1986). A detailed analysis of these changes indicated that an alteration in receptor occupancy–response coupling was primarily responsible for the desensitization of Ca^{2+} mobilization.

There have also been studies of desensitization in BC3H-1 muscle cells that possess both α_1-adrenergic and H1-histaminergic receptors. Prolonged exposure of these cells to norepinephrine reduced Ca^{2+} efflux induced by either norepinephrine and histamine by 30–40% (Brown et al., 1986), apparently because of altered intracellular Ca^{2+} stores. However, long exposure of the cells to histamine caused a complete loss of responsiveness to histamine while reducing the norepinephrine response by about 30%. This suggests that desensitization to histamine occurred at the level of the receptor itself. Other in vivo studies of the effects of chemical sympathectomy, epinephrine treatment, or pheochromocytoma have also given evidence of down-regulation of α_1-adrenergic receptors (Colucci et al., 1981; Snavely et al., 1983).

As noted above, activation of α_1-adrenergic receptors can lead to an increase in cAMP in certain situations, apparently due to activation of adenylate cyclase. There is also evidence that α_1-adrenergic and other Ca^{2+}-mobilizing agonists can decrease cAMP in liver or heart (Assimacopoulos-Jeannet et al., 1982; Morgan et al., 1983c,d; Buxton and Brunton, 1985, 1986). Since these agonists can also inhibit the actions of exogenous cAMP (Assimacopoulos-Jeannet et al., 1982) and antagonize forskolin (Morgan et al., 1983d), and since inhibitors of cyclic nucleotide phosphodiesterase eliminate the effect in cardiac myocytes (Buxton and Brunton, 1985), the effect appears to be due to activation of cAMP phosphodiesterase. A similar effect of muscarinic cholinergic agonists has been reported (Meeker and Harden, 1982; Evans et al., 1985; A.R. Hughes, et al., 1984; Masters et al., 1984). The mechanisms by which the phosphodiesterase is activated by Ca^{2+}-mobilizing agonists are unknown but could involve, in part at least, the Ca^{2+}-calmodulin-activated cyclic nucleotide phosphodiesterase (see Section 8).

3. CHANGES IN CELL Ca^{2+} INDUCED BY α_1-ADRENERGIC AGONISTS

During the 1970s, evidence began to accumulate that epinephrine and norepinephrine did not always exert their effects by increasing cAMP and that the actions not involving an elevation of cAMP were mediated by α-adrenergic receptors (e.g., Tolbert *et al.*, 1973; Hutson *et al.*, 1976). It was also becoming clear that α-adrenergic receptors were comprised of two subtypes, α_1 and α_2 (Langer, 1974, 1977; Starke, 1977; Berthelson and Pettinger, 1977). Subsequent work demonstrated that these subtypes were functionally distinct, i.e., it was shown that activation of α_2 receptors decreased cAMP, whereas the stimulation of α_1 receptors was usually without effect on cAMP but altered cellular Ca^{2+} fluxes (reviewed by Exton, 1980, 1981, 1985).

The initial experiments showing effects of α_1-adrenergic agonists on Ca^{2+} fluxes were performed in liver and smooth muscle and utilized $^{45}Ca^{2+}$. Stimulation of both cellular uptake and efflux of $^{45}Ca^{2+}$ was observed (reviewed by Bolton, 1979; Exton, 1980, 1981; Williamson *et al.*, 1981; Reinhart *et al.*, 1984c,d). The stimulation of $^{45}Ca^{2+}$ influx gave rise to the view that the agonists acted by opening a plasma membrane Ca^{2+} channel. However, studies of agonist-induced cellular responses known to be mediated by Ca^{2+}, e.g., liver glycogen breakdown, K^+ permeability in parotid, tonic smooth muscle contraction, showed that they were initially unimpaired by removal of extracellular Ca^{2+} or inhibition of its entry (Deth and Van Breeman, 1974; Putney, 1976; Assimacopoulos-Jeannet *et al.*, 1977; Weiss and Putney, 1978; Blackmore *et al.*, 1978; Parod and Putney, 1978; Casteels and Raeymaekers, 1979; Blackmore *et al.*, 1982; Reinhart *et al.*, 1984a). This finding suggested that a major initial change in cell Ca^{2+} induced by α_1-adrenergic agonists was the mobilization of Ca^{2+} from intracellular stores. This idea was confirmed by measurements of Ca by atomic absorption or Ca^{2+} electrode, which showed that the agonists caused a rapid release of Ca^{2+} from hepatocytes or perfused livers (Figure 3; Blackmore *et al.*, 1978, 1979, 1983a; Studer and Borle, 1983; Reinhart *et al.*, 1982). It was also supported by observations that $^{45}Ca^{2+}$ previously accumulated into the internal stores of liver, smooth muscle, and other tissues could be released rapidly by agonists (Assimacopoulos-Jeannet *et al.*, 1977; Casteels and Raeymaekers, 1979; Chen *et al.*, 1978; Deth and Casteels, 1977; Blackmore *et al.*, 1978; Haylett, 1976; Jenkinson *et al.*, 1978; Smith *et al.*, 1984; Ambler *et al.*, 1984; Parod and Putney, 1979; Haddas *et al.*, 1979; Miller and Nelson, 1977; R.D. Brown *et al.*, 1984; Amitai *et al.*, 1984; Connolly *et al.*, 1984). More direct proof of internal mobilization came when measurements of the Ca content of liver subcellular fractions revealed that the fractions enriched in mitochondria and microsomes showed large decreases in response to α_1-adrenergic and other Ca^{2+}-mobilizing agonists (Blackmore *et al.*, 1979; Babcock *et al.*, 1979; Murphy *et al.*, 1980; Reinhart *et al.*, 1982).

The concept that Ca^{2+}-mobilizing agonists initially release Ca^{2+} from an intracellular pool is supported by other studies in which livers were perfused with $^{45}Ca^{2+}$ and the $^{45}Ca^{2+}$ content of subcellular fractions was measured (Barritt *et al.*, 1981; Kimura *et al.*, 1982; Studer and Borle, 1983), or in which chlortetracycline fluorescence was measured in hepatocytes (Babcock *et al.*, 1979). Although early studies suggested that mitochondria represented the major pool

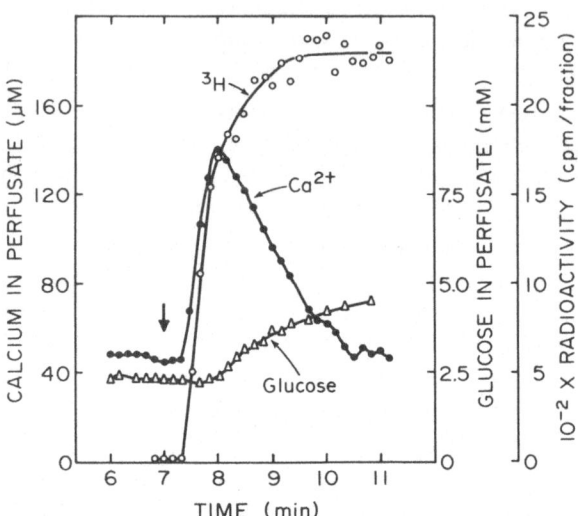

FIGURE 3. Time course of epinephrine action on Ca^{2+} and glucose release from the perfused rat liver. To determine precisely the time at which epinephrine reached the liver, [^3H]epinephrine was included in the epinephrine infusion, which was started at 7 min. (From Blackmore *et al.*, 1982, with permission of the authors and the publisher.)

from which Ca^{2+} was mobilized (Blackmore *et al.*, 1979; Babcock *et al.*, 1979; Murphy *et al.*, 1980; Barritt *et al.*, 1981; Studer and Borle 1983; Reinhart *et al.*, 1982), more recent investigations indicate another source (Althaus-Salzmann *et al.*, 1980; Poggioli *et al.*, 1980; Berthon *et al.*, 1981; Kimura *et al.*, 1982; Shears and Kirk 1984a,b; Kleineke and Soling, 1985). This is most likely the endoplasmic reticulum, as shown by subcellular fractionation (Blackmore *et al.*, 1979; Berthon *et al.*, 1981; Joseph and Williamson, 1983) and electron probe X-ray microanalysis (Bond *et al.*, 1984; A.P. Somylo *et al.*, 1985), but other sources have not been excluded. Several studies have indicated that only a functionally discrete portion of the total endoplasmic reticulum is involved (Dawson and Irvine, 1984; Joseph *et al.*, 1984b; Prentki *et al.*, 1984b).

The intracellular Ca^{2+} pool that is mobilized by agonists does not refill until the agonists are removed or antagonists added (Putney 1977; Morgan *et al.*, 1982; Breant *et al.*, 1981; Aub *et al.*, 1982; Dewitt and Putney, 1983; Reinhart *et al.*, 1984b; Joseph *et al.*, 1985), and refilling does not occur if extracellular Ca^{2+} is absent or its entry is blocked (Marier *et al.*, 1978; Aub *et al.*, 1982; Putney, 1976; Weiss and Putney, 1978; Reynolds and Dubyak, 1985; Joseph *et al.*, 1985). As will be discussed in Section 6, the refilling of the pool is apparently prevented by continuing production of IP_3 (Prentki *et al.*, 1985). When this declines after agonist removal, Ca^{2+} reaccumulates into the stores and readdition of agonists can produce a further response (Putney, 1977; DenHertog, 1981; Parod and Putney, 1978; Joseph et al., 1985). During the reaccumulation phase, cytosolic Ca^{2+} levels and the associated physiological responses decline, indicating the rate of reuptake of Ca^{2+} by internal organelles exceeds the rate of net Ca^{2+} influx (Charest et al., 1983; Joseph et al., 1985; Morgan et al., 1982; Poggioli and Putney, 1982; Blackmore et al., 1982; Casteels and Droogmans, 1981). The alternative view that the internal pool fills directly from the extracellular space seems less likely now that it is known that IP_3 can control the Ca^{2+} content of the internal pool.

With the introduction of the fluorescent Ca^{2+} probe quin 2 by Tsien and co-workers (Tsien, 1980; Tsien et al., 1982, 1984), measurements of cytosolic Ca^{2+} have been possible in many cells. These show a very rapid rise in cytosolic Ca^{2+} (within a few seconds) in response to α_1-adrenergic and other Ca^{2+}-mobilizing agonists in many cells (Pozzan et al., 1982; Charest et al., 1983; Hesketh et al., 1983; Tsien et al., 1984; Korchak et al., 1984; Capponi et al., 1985; Nabika et al., 1985; Berthon et al., 1984; Smith et al., 1984; Reynolds and Dubyak, 1985; Figure 4). In confirmation of earlier predictions, the increase in cytosolic Ca^{2+} is mainly independent of extracellular Ca^{2+} initially, but beyond 0.5–1 min the increase declines rapidly unless Ca^{2+} of 0.5 mM or higher

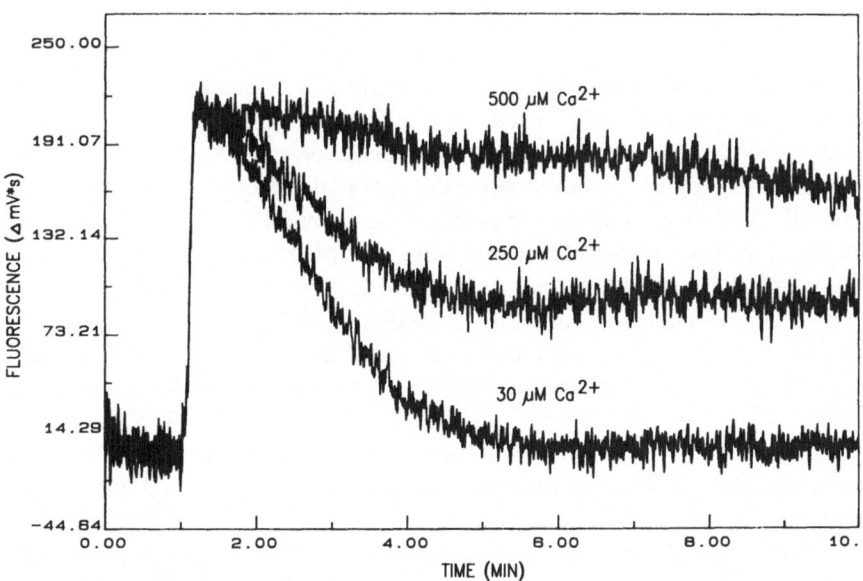

FIGURE 4. Increase in cystosolic Ca^{2+} induced by vasopressin in hepatocytes incubated in the presence of different extracellular Ca^{2+} concentrations (30–500 μM). Cytosolic Ca^{2+} was measured fluorimetrically in quin 2–loaded hepatocytes. Vasopressin (10 nM) was added at 1 min. (From Charest et al., 1985, with permission of the authors and publisher.)

is present in the medium (Figure 4; Charest et al., 1985; Joseph et al., 1985; cf. Berthon et al., 1984; Binet et al., 1985).

The intracellular stores of Ca^{2+} in most cells are limited and rapidly become depleted with continued agonist stimulation (Exton, 1985; Charest et al., 1985; Joseph et al., 1985). Calcium released from the stores into the cytosol is extruded from the cell through the action of the plasma membrane Ca^{2+} pump or the Na^+/Ca^{2+} exchanger, or is taken up by organelles not sensitive to IP_3. In the absence of extracellular Ca^{2+}, this results in a rapid decline in cytosolic Ca^{2+} and a waning of any physiological responses dependent upon cytosolic Ca^{2+} (Charest et al., 1985; Joseph et al., 1985; Binet et al., 1985). However, in the presence of normal levels of extracellular Ca^{2+}, agonists cause a persisting increase in cytosolic Ca^{2+} (Charest et al., 1985; Binet et al., 1985) and continuing physiological responses (Exton, 1985; Joseph et al., 1985). This implies that α_1-adrenergic and other Ca^{2+}-mobilizing agonists also affect the processes by which Ca^{2+} is transferred across the plasma membrane.

There have been several reports of agonist effects on both Ca^{2+} uptake and efflux at the level of the plasma membrane in several tissues. Evidence for a stimulation of Ca^{2+} entry is based on measurements of $^{45}Ca^{2+}$ uptake into hepatocytes measured 15–105 sec after agonist addition (Mauger et al., 1984, 1985; Poggioli et al., 1985). However, as discussed in detail elsewhere (Williamson et al., 1981; Blackmore et al., 1982), part of the increase in $^{45}Ca^{2+}$ in the cells could be secondary to the mobilization of internal unlabeled Ca^{2+}, which occurs within a few seconds. A more detailed analysis of $^{45}Ca^{2+}$ fluxes in hepatocytes has been carried out by Barritt and co-workers (Barritt et al., 1981; Parker et al., 1983). These workers concluded that epinephrine causes both a mobilization of Ca^{2+} from an intracellular compartment that includes mitochondria and endoplasmic reticulum and a stimulation of Ca^{2+} influx into the cell. Additional evidence for agonist stimulation of Ca^{2+} entry comes from studies using Ca^{2+}-depleted cells and quin 2 to measure the influx of extracellular Ca^{2+} into the cytosol (Joseph et al., 1985; R. Charest, P.F. Blackmore, and J.H. Exton, unpublished material). In addition, high concentrations of Ca^{2+}-channel blockers, such as diltiazem, can block the influx of Ca^{2+} observed in the presence of agonists and accelerate the decline in phosphorylase activity observed 1 min after addition of agonists (Joseph et al., 1985; R. Charest, P.F. Blackmore, and J.H. Exton, unpublished material). The molecular mechanism(s) by which Ca^{2+}-mobilizing agonists stimulates the influx of Ca^{2+} into cells is presently unknown.

The increased influx of Ca^{2+} caused by Ca^{2+}-mobilizing agonists is eventually balanced by increased efflux of Ca^{2+}, since the cytosolic Ca^{2+} concentration becomes steady after a few minutes and there is no net uptake of Ca^{2+} by organelles until agonists are removed (Morgan et al., 1982; Charest et al., 1983). The increased Ca^{2+} efflux may simply be attributable to stimulation of the plasma membrane Ca^{2+} pump resulting from the elevated concentration of cytosolic Ca^{2+}. This would result in increased bidirectional flux of Ca^{2+} across the plasma membrane in the presence of agonists compared with basal conditions. Reinhart et al. (1984b) have presented some studies of $^{45}Ca^{2+}$ uptake by perfused rat livers, which suggest such increased cycling.

Another means of producing a sustained increase in cytosolic Ca^{2+} is to alter the kinetics of the plasma membrane Ca^{2+} pump so that a higher cytosolic Ca^{2+} concentration is required to produce the same rate of Ca^{2+} efflux in the presence of agonists as observed under basal conditions. In the absence of any concurrent stimulation of Ca^{2+} influx, this would result in an unchanged cycling of Ca^{2+} across the plasma membrane. Evidence for inhibition of the plasma membrane Ca^{2+} pump by several agonists in liver has been presented by Prpic et al. (1984). In addition, there have been reports of the inhibition of the plasma membrane $(Ca^{2+} + Mg^{2+})$-ATPase in liver by vasopressin and phenylephrine (Lin et al., 1983) and in myometrium by oxytocin (Soloff and Sweet, 1982). The mechanism by which Ca^{2+}-mobilizing agonists inhibit the plasma membrane Ca^{2+} pump is unknown but could be due to the changes in phosphoinositides produced by these agonists (Buckley and Hawthorne, 1972; Penniston, 1983; Prpic et al., 1984; Charest et al., 1985).

4. ROLE OF PHOSPHOINOSITIDE CHANGES

Michell (1975, 1979) first emphasized the association between the actions of Ca^{2+}-mobilizing hormones and the turnover of phosphoinositides in a variety of tissues. In particular, this was pointed out for α-adrenergic agonists in brain, parotid, pineal, iris, liver, vas deferens, aorta, and submaxillary gland (Jones and Michell, 1978). The initial studies demonstrated that the agonists increased both the synthesis and breakdown of phosphatidylinositol (PI), as measured by labeling studies with $^{32}P_i$ (Jones and Michell, 1978). These observations were confirmed using [^{3}H]myoinositol (Tolbert et al., 1980; Prpic et al., 1982). However, it was observed that the turnover of PI induced by α_1-adrenergic agonists or other Ca^{2+}-mobilizing agents, although rapid, was not fast enough to be responsible for the physiological responses to these agents, which occurred within seconds (Canessa de Scarnatti and Lapetina, 1974; Kirk et al., 1977, 1981; Billah and Michell, 1979; Uchida et al., 1982; Prpic et al., 1982).

Early observations by Schacht and Agranoff (1972) and Abdel-Latif et al. (1977) indicated that Ca^{2+}-mobilizing agonists stimulated the phosphodiesteratic breakdown of phosphatidylinositol 4,5-P$_2$ (PIP$_2$) in addition to PI in neural and smooth muscle tissue. The group working with Kirk and Michell then demonstrated that the breakdown of the polyphosphoinositide induced by these agonists in liver occurred much more rapidly than that of PI (Kirk *et al.*, 1981; Michell *et al.*, 1981; Creba *et al.*, 1983). This has now been confirmed by others in liver (Rhodes *et al.*, 1983; Thomas *et al.*, 1983; Litosch *et al.*, 1983), parotid (Weiss *et al.*, 1982; Downes and Wusteman, 1983), platelets (Billah and Lapetina, 1982; Agranoff *et al.*, 1983; Mauco *et al.*, 1983), kidney cortex (Wirthensohn *et al.*, 1984), exocrine pancreas (Putney *et al.*, 1983), neutrophils (Volpi *et al.*, 1983; Yano *et al.*, 1983; Dougherty *et al.*, 1984), and pituitary (Martin, 1983; Rebecchi and Gershengorn, 1983).

The significance of the enhanced breakdown of PIP_2 induced by Ca^{2+}-mobilizing agonists was made clear when Berridge and associates measured the changes in the concentration of one of the products, IP_3, in various tissues (Berridge, 1983; Berridge *et al.*, 1983) and when Streb *et al.* (1983) showed that this compound released Ca^{2+} from internal stores when added to permeabilized pancreatic acinar cells. Since that time, α_1-adrenergic and other Ca^{2+}-mobilizing agonists have been shown to increase IP_3 rapidly in many tissues, including liver (Thomas *et al.*, 1984; Charest *et al.*, 1985), brain (Berridge *et al.*, 1983), platelets (Agranoff *et al.*, 1983; Vickers *et al.*, 1984; Rittenhouse and Sasson, 1985), salivary glands (Berridge *et al.*, 1983; Berridge, 1983; Downes and Wustemann, 1983; Aub and Putney, 1984, 1985; Irvine *et al.*, 1985), pituitary (Martin, 1983; Rebecchi and Gershengorn, 1983; Enjalbert *et al.*, 1986), exocrine pancreas (Rubin *et al.*, 1984), endocrine pancreas (Morgan *et al.*, 1985), Swiss 3T3 cells (Berridge *et al.*, 1984), adrenal cortex (Gallo-Payet *et al.*, 1986), smooth muscle cells (Akhtar and Abdel-Latif, 1984; Smith *et al.*, 1984), lymphocytes (Imboden and Stobo, 1985), gastric mucosal cells (Baudiere *et al.*, 1986), astrocytoma cells (Masters *et al.*, 1985b), PC 12-pheochromocytoma cells (Vincentini *et al.*, 1985a), and adipocytes (Nanberg and Putney, 1986). In contrast to all these reports, there has been one study indicating an inhibitory effect of an agonist (dopamine) on inositol phosphate production (Enjalbert *et al.*, 1986).

The increase in IP_3 induced by agonists is detectable within a few seconds and is approximately coincident with the rise in cytosolic Ca^{2+}. The concentrations of agonists that produce half-maximal decreases in PIP_2 or increases in IP_3 are similar to their K_ds for binding to their receptors in plasma membranes (Creba *et al.*, 1983; Lynch *et al.*, 1985a). In addition, the maximum generation of IP_3 by agonists is proportional to their maximum number of plasma membrane–binding sites (Lynch *et al.*, 1985a). These findings imply a close relationship between receptor occupancy and phosphoinositide breakdown. However, because of the presence of spare receptors in most cells, the concentrations of agonists required to half-maximally elevate cytosolic Ca^{2+} and elicit physiological responses are lower than those required to half-maximally increase IP_3 or decrease PIP_2, i.e., small increases in IP_3 can elicit large physiological responses (Lynch *et al.*, 1985a; Creba *et al.*, 1983; Rhodes *et al.*, 1983; Charest *et al.*, 1985; Thomas *et al.*, 1984; Aub and Putney, 1985). However, it should be recognized that in some cells, the receptor reserve for α_1-adrenergic agonists may be small or nonexistent compared to that for other agonists (El-Refai *et al.*, 1979; Ambler *et al.*, 1984; Amitai *et al.*, 1984; Lynch *et al.*, 1985a).

The metabolism of the phosphoinositides is depicted in Figure 5. It is generally agreed that the reaction primarily stimulated by α_1-adrenergic agonists and other Ca^{2+}-mobilizing agents is the breakdown of PIP_2 to IP_3 and DAG under the influence of a Ca^{2+}-requiring phosphodiesterase (phospholipase C). This hormone-sensitive phosphodiesterase may also have some effect on phosphatidylinositol 4-P (PIP) but apparently does not act on PI (Uhing *et al.*, 1985, 1986; Aub and Putney, 1984; Downes and Wustemann, 1983; Martin, 1983). The loss of PI that was observed in early experiments was presumably due to the accelerated conversion of PI to PIP and the PIP_2 to replace that broken down by the phosphodiesterase. The kinases that catalyze the conversion of PI to the polyphosphoinositides are very active and are located in the plasma membrane, as are the phosphomonoesterases that reverse their actions (Berridge, 1984). To date, there is no evidence that the kinases and phosphatases involved in PIP_2 metabolism (Figure 5) are regulated by agonists, and the increased conversion of PI to PIP_2 is thought to occur through mass action. IP_3 produced by cleavage of PIP_2 is released into the cytosol where it acts to release Ca^{2+} from internal stores (see Section 6). As shown in Figure 5, it is also degraded rapidly to myoinositol $1,4$-P_2 (IP_2) by a specific phosphomonoesterase found in the plasma membrane or soluble phase (Downes *et al.*, 1982; Seyfred *et al.*, 1984; Storey

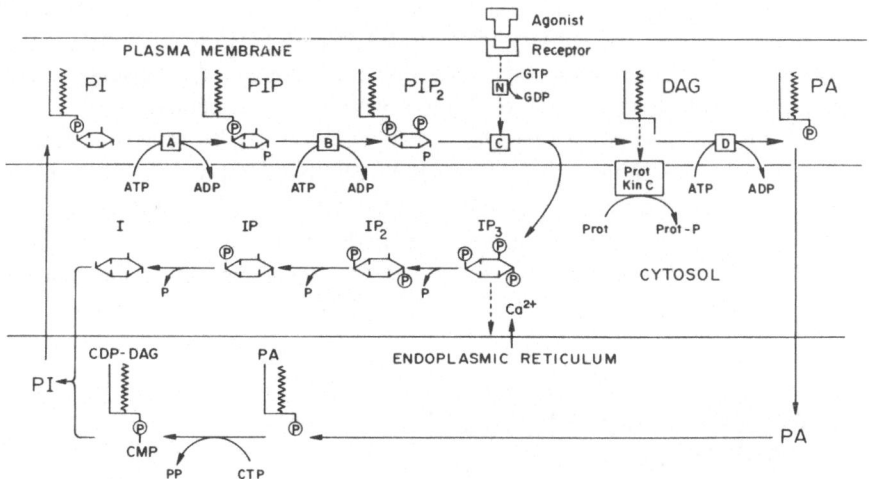

FIGURE 5. Pathways of cellular phosphoinositide metabolism. Abbreviations not defined in the text as follows: (N) guanine nucleotide–binding regulatory protein; (CDP–DAG) CDP–diacylglycerol; (A) phosphatidylinositol kinase; (B) phosphatidylinositol 4-P kinase; (C) phosphatidylinositol 4,5-P_2 phosphodiesterase.

et al., 1984; Joseph and Williams, 1985; Connolly et al., 1985). IP_2 is then sequentially degraded to myoinositol 1-P (IP) or myoinositol 4-P and myoinositol by soluble phosphomonoesterases (Joseph and Williams, 1985; Storey et al., 1984). The myoinositol can then be reincorporated into PI through the action of cytosine 3',5'-diphosphate (CDP)-diacylglycerol:inositol transferase in the endoplasmic reticulum. Synthesized PI is then transferred to the plasma membrane by a specific phospholipid carrier protein (Michell, 1975). The other product of PIP_2 breakdown, namely DAG, can activate protein kinase C in the membrane or be converted to phosphatidic acid (PA) through the action of DAG kinase (Figure 5). The PA can then be used for the resynthesis of PI and the synthesis of other phospholipids and triacylglycerol. Other fates of DAG are degradation to monoacylglycerol and fatty acids plus glycerol by lipases.

In addition to IP_3, myoinositol $1,3,4-P_3$ and myoinositol $1,3,4,5-P_4$ have been found in tissues exposed to Ca^{2+}-mobilizing agonists (Irvine et al., 1984b, 1985, 1986b; Batty et al., 1985). Myoinositol $1,3,4,5-P_4$ arises from IP_3 through the action of a kinase and is then dephosphorylated to myoinositol $1,3,4-P_3$ by a phosphatase (Irvine et al., 1986a; Hansen et al., 1986; Downes et al., 1986; Stewart et al., 1986; Biden and Wollheim, 1986). Myoinositol $1,3,4,5-P_4$ has been reported to activate sea urchin eggs, apparently by opening a plasma membrane Ca^{2+} channel (Irvine and Moor, 1986), but a physiological role for myoinositol $1,3,4-P_3$ has not yet been defined. The 3-phosphokinase that converts IP_3 to myoinositol $1,3,4,5-P_4$ is stimulated by Ca^{2+} (Biden and Wollheim, 1986). This may account for the transiency of the elevation in IP_3 following agonist stimulation of cells and the delayed formation of myoinositol $1,3,4-P_3$ (Lew et al., 1986). The phosphomonoesterase specific for IP_3 has also been reported to be phosphorylated and stimulated by protein kinase C (Connolly et al., 1986), and activators of the kinase have been shown to stimulate the conversion of IP_3 to IP_2 in permeabilized platelets (Molina y Vedia and Lapetina, 1986).

The formation of myoinositol 1,2-(cyclic)-4,5-trisphosphate has been reported during phosphoinositide cleavage by phospholipase C in vitro (Wilson et al., 1985) and thrombin stimulation of platelets (Ishii et al., 1986). Although this cyclic compound is more potent than IP_3 in increasing Ca^{2+} in Limulus photoreceptors (Wilson et al., 1985), it is of similar potency to the noncyclic compound in other systems (Irvine et al., 1986b).

5. ROLE OF GUANINE NUCLEOTIDE–BINDING REGULATORY PROTEIN

As described above, the ability of GTP and its nonhydrolyzable analogues to alter the agonist affinity of the α_1-adrenergic receptor and of the receptors for other Ca^{2+}-mobilizing agonists implies that these receptors can couple to a guanine nucleotide–binding regulatory protein analogous to those involved in the regulation of adenylate cyclase. Further evidence for the involvement of such a protein in the actions of these agonists comes from a variety of studies. For example, in permeabilized mast cells and platelets, nonhydrolyzable analogues of GTP elicit Ca^{2+}-dependent exocytotic secretion (Gomperts, 1983; Haslam and Davidson 1984b, 1984c), and some Ca^{2+}-mobilizing agonists stimulate a low K_m membrane GTPase activity (Hinkle and Phillips, 1984) or the binding/exchange of a GTP analogue to membranes (Lad et al., 1985). In liver cells, NaF stimulates the breakdown of PIP_2 to IP_3 and DAG, with resultant increase in cytosolic Ca^{2+}, phosphorylase activation, and glycogen synthase inactivation (Blackmore et al., 1985). All these effects are potentiated by $AlCl_3$, implying that AlF_4^- is the active molecule. AlF_4^- is known to modulate the activity of other guanine nucleotide–binding regulatory proteins (Sternweis and Gilman, 1982; Katada et al., 1984; Kanaho et al., 1985).

More direct evidence for a role of a guanine nucleotide–binding protein is provided by studies showing stimulatory effects of GTP and its analogues on PIP_2 and PIP breakdown in isolated plasma membranes from liver (Uhing et al., 1985, 1986; Wallace and Fain, 1985; Guillon et al., 1986), polymorphonuclear leukocytes (Cockcroft and Gomperts, 1985), salivary glands (Litosch et al., 1985), astrocytoma cells (Hepler and Harden, 1986), pituitary cells (Lucas et al., 1985; Martin et al., 1986; Straub and Gershengorn, 1986), and platelets (Baldassare and Fisher, 1986). The effect is greater with the nonhydrolyzable analogues than with GTP and is not seen with other nucleoside triphosphates or with guanosine diphosphate (GDP) or guanosine monophosphate (GMP) (Figure

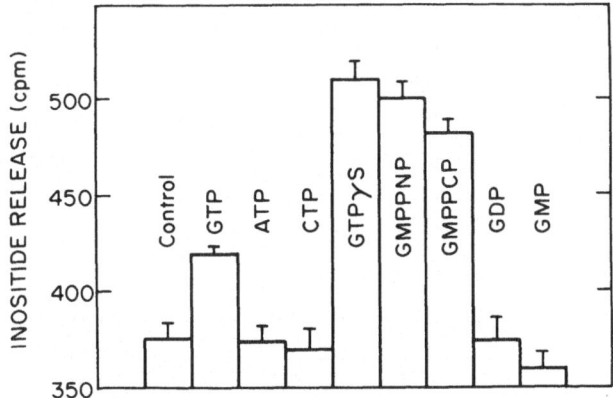

FIGURE 6. Stimulatory effects of GTP and its analogues on inositide release from rat liver plasma membranes. Liver membranes prepared from rats injected previously with [^{3}H]myoinositol were incubated with the nucleotides shown at 100 µM concentration, and the release of labeled inositides (IP_3, IP_2, IP plus myoinositol) was measured. (From Uhing et al., 1985, with permission of the authors and publisher.)

6; Uhing *et al.*, 1985; Wallace and Fain, 1985; Cockcroft and Gomperts, 1985; Litosch *et al.*, 1985). It is observed with micromolar concentrations of GTP analogues, is Mg^{2+}-dependent, and is inhibited by GDPβS (Uhing *et al.*, 1986). The breakdown of PIP_2 is greater than that of PIP and requires the presence of free Ca^{2+} of 10 nM or higher (Uhing *et al.*, 1985, 1986; Cockcroft and Gomperts, 1985; cf. Wallace and Fain, 1985). IP_3 is the product formed initially but is degraded to IP_2 by the IP_3 phosphatase activity of the membranes (Uhing *et al.*, 1985; Wallace and Fain, 1985).

Recently, many workers have observed direct effects of Ca^{2+}-mobilizing agonists on the breakdown of polyphosphoinositides in membranes from liver, salivary glands, polymorphonuclear leukocytes, GH_3 pituitary cells, astrocytoma cells, WRK1 cells, and platelets (C.D. Smith *et al.*, 1985; Litosch *et al.*, 1985; Lucas *et al.*, 1985; Uhing *et al.*, 1986; Martin *et al.*, 1986; Straub and Gershengorn, 1986; Guillon *et al.*, 1986; Hepler and Harden, 1986; Baldassare and Fisher, 1986). In all cases, the effect is dependent upon the presence of GTP or its analogues (Figure 7). The major inositol phosphate formed is IP_3, but this breaks down rapidly to IP_2 (Uhing *et al.*, 1985). The rate of formation of IP_3 is maximal within 1 min (Uhing *et al.*, 1986). The concentration dependence

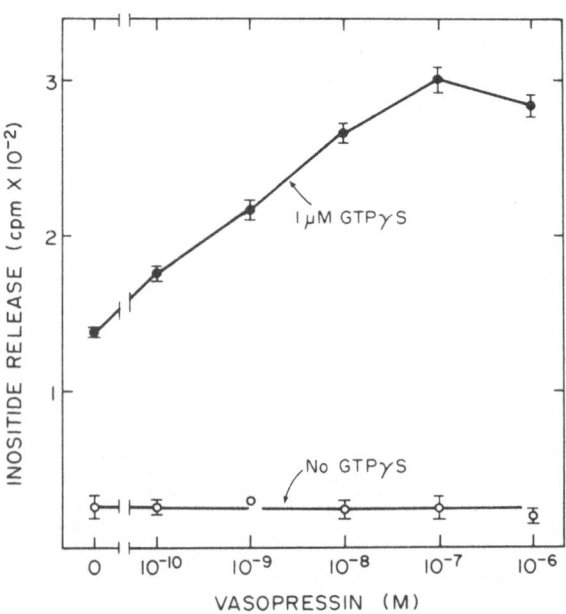

FIGURE 7. Stimulatory effect of arginine–vasopressin on inositide release from rat liver plasma membranes. Conditions were as described for Figure 6. (From Uhing *et al.*, 1986, with permission of the authors and publisher.)

for agonist-induced inositide formation in liver or salivary gland membranes is similar to that for IP_3 formation in the intact tissue (Uhing et al., 1986; Litosch et al., 1985). These observations differ from earlier reports of direct effects of catecholamines and vasopressin on phosphoinositide breakdown in isolated plasma membranes, which were observed in the absence of guanine nucleotides (Lin and Fain, 1981; Wallace et al., 1982, 1983; Harrington and Eichberg, 1983; Seyfred and Wells, 1984).

The formation of a hormone receptor–guanine nucleotide–binding protein complex of high relative molecular weight has been observed in liver plasma membranes treated with vasopressin and then solubilized with digitonin or another detergent (Fitzgerald et al., 1986; Bojanic and Fain, 1986). The complex is dissociated by GTP analogues and exhibits GTPase activity. The presence in the complex of a 35-kD β subunit common to several other guanine nucleotide–binding proteins has been demonstrated immunologically (Fitzgerald et al., 1986).

There is no general agreement about the nature of the guanine nucleotide–binding protein involved in polyphosphoinositide breakdown and Ca^{2+} mobilization. In neutrophils and mast cells, the breakdown of PIP_2 and the associated physiological events induced by 48/80 and chemotactic peptide, respectively, are inhibited by islet-activating protein (Okajima et al., 1985; Volpi et al., 1985; Verghese et al., 1985; Nakamura and Ui, 1983, 1985; Bokoch et al., 1984; Okajima and Ui, 1984). The toxin can also ADP ribosylate transducin, a protein involved in coupling rhodopsin to cGMP phosphodiesterase in rod outer segments of the retina, and N_o or G_o, a guanine nucleotide–binding protein isolated from brain and certain other tissues (Manning et al., 1984; Van Dop et al., 1984; Watkins et al., 1984; Sternweis and Robishaw, 1984). Thus, the inhibitory effects of the toxin on neutrophils and mast cells may be due to the involvement of one of these guanine nucleotide–binding proteins or to a novel protein that is also a substrate for the toxin. In NG108-15 hybrid cells, addition of purified N_o or N_i to membranes prepared from cells treated with islet-activating protein restored bradykinin activation of GTPase activity (Higashida et al., 1986).

In liver, islet-activating protein is without effect on the stimulation of PIP_2 breakdown and Ca^{2+} mobilization induced by agonists in either intact hepatocytes or isolated plasma membranes, under conditions in which N_i is ADP ribosylated and its functions are blocked (Uhing et al., 1986; Lynch et al., 1986b). Furthermore, the ability of GTP analogues to decrease high-affinity binding of epinephrine, vasopressin, or angiotensin II to liver plasma membranes is unaffected by treatment with the toxin (Lynch et al., 1986b). Likewise, the toxin does not affect muscarinic cholinergic effects on phosphoinositide hydrolysis in cardiac myocytes and pancreatic acinar cells (Masters et al., 1985a; Merritt et al., 1986), thrombin action on inositol release in 3T3 fibroblasts (Murayama and Ui, 1985), α_1-adrenergic agonist binding to kidney cortex membranes (Boyer *et al.*, 1984), carbachol binding to 1321N1 astrocytoma cells (Martin *et al.*, 1985), or α_1-adrenergic agonist stimulation of respiration in brown adipocytes (Schimmel *et al.*, 1985). It also does not inhibit agonist-induced PIP_2 hydrolysis in plasma membranes from liver, GH_3 pituitary cells, and astrocytoma cells (Uhing *et al.*, 1986; Martin *et al.*, 1986; Hepler and Harden, 1986). Therefore, the guanine nucleotide–binding regulatory protein involved in the actions of Ca^{2+}-mobilizing agonists in these tissues does not appear to be ADP ribosylated by islet-activating protein, and the reasons for the difference in neutrophils and mast cells are unknown.

6. ROLE OF MYOINOSITOL TRISPHOSPHATE AND Ca^{2+} RELEASE

Streb *et al.* (1983) originally demonstrated that IP_3 added to saponin-permeabilized pancreatic acinar cells caused the release of Ca^{2+} from a nonmitochondrial store. Since that time, IP_3 has been shown to release Ca^{2+} from internal stores in permeabilized liver cells (Figure 8; Burgess *et al.*, 1984a, b; Joseph *et al.*, 1984a; Joseph and Williamson, 1986), insulin-secreting cells (Joseph *et al.*, 1984b; Biden *et al.*, 1984; Prentki *et al.*, 1985; B.A. Wolf *et al.*, 1985), smooth muscle cells (Suematsu *et al.*, 1984; A.V. Somlyo *et al.*, 1985), vesicles (dense tubular system) from platelets (O'Rourke *et al.*, 1985; Authi and Crawford, 1985; Brass and Joseph, 1985), neutrophils (Prentki *et al.*, 1984b), 3T3 fibro-

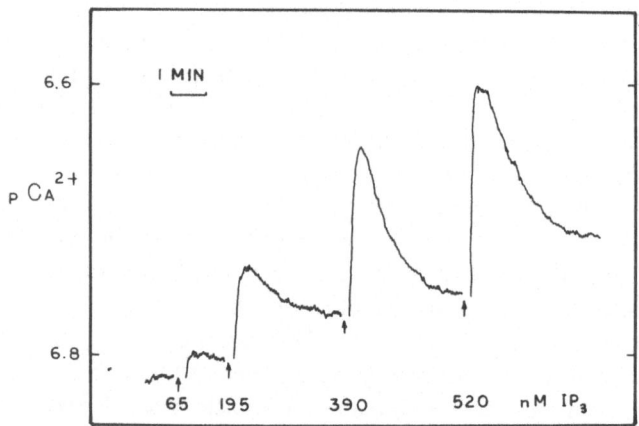

FIGURE 8. Effects of IP_3 added in increasing concentrations on intracellular Ca^{2+} release in digitonin-permeabilized rat hepatocytes (P. Thiyagarajah, R. Charest, P.F. Blackmore, and J.H. Exton, unpublished data).

blasts (Irvine *et al.*, 1984a), macrophages (Hirata *et al.*, 1985), pituitary cells (Gershengorn *et al.*, 1984; Biden *et al.*, 1986), leukocytes (Burgess *et al.*, 1984c), N1E-115 neuronal cells (Chueh and Gill, 1986; Ueda *et al.*, 1986), adipocytes (Delfert *et al.*, 1986), kidney cortex cells (Thevenod *et al.*, 1986), and adrenal chromaffin cells (Stoehr *et al.*, 1986). The action of IP_3 is extremely rapid and is observed with submicromolar concentrations, i.e., those calculated or measured to exist intracellularly (Charest *et al.*, 1985; Thomas *et al.*, 1984; Rittenhouse and Sasson, 1985). It is transient, due to the rapid degradation of IP_3 (Streb *et al.*, 1985; Prentki *et al.*, 1985). Both rapid action and rapid breakdown are desirable properties for a molecule involved in the regulation of intracellular Ca^{2+}. Myoinositol 2,4,5-P_2 and myoinositol 4,5-P_2 also release Ca^{2+} but are, respectively, approximately ten- and one-hundred-fold less potent than IP_3. IP_2 and IP are ineffective (Burgess *et al.*, 1984b; Streb *et al.*, 1983; Irvine *et al.*, 1984a; B.A. Wolf *et al.*, 1985). Myoinositol 1,2-(cyclic)-4,5-trisphosphate is of similar potency to IP_3, whereas myoinositol 1,3,4-trisphosphate is about thirtyfold less potent and myoinositol 1,3,4,5-tetrakisphosphate is ineffective (Irvine *et al.*, 1986b).

The intracellular pool from which Ca^{2+} is released by IP_3 cosediments with mitochondria and microsomes during centrifugation of liver or other tissue homogenates (Dawson and Irvine, 1984; Prentki *et al.*, 1984b; Streb *et al.*, 1984; Delfert *et al.*, 1986). However, there is much evidence that it is not mitochondrial (Streb *et al.*, 1983, 1984; Gershengorn *et al.*, 1984; Joseph *et al.*, 1984a,b; Thevenod *et al.*, 1986; Biden *et al.*, 1986). It is probably a component of the endoplasmic reticulum, based on studies with uncouplers and other inhibitors of mitochondrial energy production and with ruthenium red, an inhibitor of mitochondrial Ca^{2+} transport (Streb *et al.*, 1983; Dawson and Irvine, 1984; Gershengorn *et al.*, 1984; Joseph *et al.*, 1984a, b; A.V. Somlyo *et al.*, 1985). The uptake of Ca^{2+} into the IP_3-releasable pool requires MgATP and is inhibited by orthovanadate (Biden *et al.*, 1984; Muallem *et al.*, 1985). Enzyme marker measurements in subcellular fractions of rat exocrine pancreas and platelets indicate codistribution of NADPH cytochrome *c* reductase and RNA with the IP_3-sensitive pool (Streb *et al.*, 1984; Authi and Crawford, 1985), which is consistent with its location in the rough endoplasmic reticulum. However, it is clear that only a small fraction of the endoplasmic reticulum responds to IP_3 (Prentki *et al.*, 1984a; Joseph *et al.*, 1984b; Dawson and Irvine, 1984; Biden *et al.*, 1986).

Addition of IP_3 to microsomal fractions isolated from insulinoma cells or liver rapidly releases Ca^{2+} (Prentki *et al.*, 1984a; Dawson and Irvine, 1984; Muallem *et al.*, 1985; Joseph *et al.*, 1984b). Similar effects are obtained with membrane vesicles from platelets thought to correspond to the endoplasmic reticulum (O'Rourke *et al.*, 1985; Authi and Crawford, 1985). The action of IP_3 appears to be exerted on Ca^{2+} efflux rather than on Ca^{2+} uptake, but the mechanism remains unknown. It is relatively insensitive to temperature (J.B.

Smith *et al.*, 1985; Chueh and Gill, 1986; Henne and Soling, 1986; Joseph and Williamson, 1986), which suggests that it involves a Ca^{2+} channel rather than a carrier. It also requires the countermovement of K^+ or another monovalent cation (Muallem *et al.*, 1985; Joseph and Williamson, 1986) and is inhibited by high concentrations of Cl^- or other anions. These findings indicate that the Ca^{2+} release process is electrogenic, but it is unlikely that an anion exchange mechanism or a cation/anion cotransport system is involved since the release is insensitive to 4,4'-dithiocyanostilbene-2,2'-disulfonic acid or furosemide (Joseph and Williamson, 1986). Voltage-dependent Ca^{2+} channel blockers and agonists are also without effect (Henne and Soling, 1986).

The effects of IP_3 on isolated organelles are small relative to those observed in permeabilized cells or require higher concentrations (e.g., Joseph *et al.*, 1984b). Dawson (1985) has reported that GTP (but not its nonhydrolyzable analogues) enhances the effect of IP_3 on Ca^{2+} release from liver microsomes, but the enhancement depends upon the presence of polyethylene glycol. GTP has also been found to stimulate the release of Ca^{2+} from microsomes prepared from two neuronal cell lines under these conditions (Ueda *et al.*, 1986; Jean and Klee, 1986). The Ca^{2+} pool acted on by GTP and the mechanism of action of the nucleotide appear to be different from those for IP_3 (Chueh and Gill, 1986; Henne and Soling, 1986; Jean and Klee, 1986). Nonhydrolyzable analogues of GTP are ineffective by themselves but block the action of GTP. The physiological significance of the effect of GTP on intracellular Ca^{2+} mobilization is presently unclear.

There have been some recent reports of IP_3 binding to microsomal fractions from liver and adrenal cortex (Baukal *et al.*, 1985; Spat *et al.*, 1986). Some high-affinity (k_d, 5–10 nM), low-capacity binding sites have been identified, but it is unclear whether these are responsible for Ca^{2+} mobilization.

In addition to mediating the effects of certain hormones on Ca^{2+} mobilization, IP_3 may act as a chemical messenger between transverse-tubular membrane depolarization and Ca^{2+} release from sarcoplasmic reticulum in skeletal muscle (Vergara *et al.*, 1985). For example, it has been reported that electrical stimulation of frog muscles increases IP_3 (Vergara *et al.*, 1985) and that IP_3 releases Ca^{2+} from skinned skeletal muscles or sarcoplasmic reticulum (Vergara *et al.*, 1985; Volpe *et al.*, 1985) or enhances contraction of skinned muscle fibers (Nosek *et al.*, 1986; Thieleczek and Heilmeyer, 1986). There is also evidence that it is involved in light-induced excitation and adaptation in *Limulus* photoreceptors (Fein *et al.*, 1984; J.E. Brown *et al.*, 1984; Brown and Rubin, 1984). For example, light causes a decrease in labeled PIP_2 and an increase in [^{3}H]-IP_3 in photoreceptors labeled previously with [^{3}H]myoinositol (J.E. Brown *et al.*, 1984; Vandenberg and Montal, 1984). Light also induces an increase in IP_3 within 200 ms in squid retina (Szuts *et al.*, 1986). When injected into photoreceptors, IP_3 produces electrical changes identical to those caused by light

(J.E. Brown et al., 1984; Fein et al., 1984). In addition to the above evidence that IP$_3$ may be involved in phototransduction, there are some experiments indicating that it may play a role in adaptation. Injection of IP$_3$ into photoreceptors reduces their response to subsequent flashes of light (Fein et al., 1984), and, conversely, light exposure reduces their response to repetitive injections of IP$_3$ (J.E. Brown et al., 1984). IP$_3$ has also been shown to increase intracellular Ca^{2+} in Limulus photoreceptors (Brown and Rubin, 1984).

IP$_3$ has also been proposed to have a role in fertilization in sea urchins and Xenopus. For example, microinjection of IP$_3$ into sea urchin eggs or Xenopus oocytes produces changes in membrane potential similar to those seen during fertilization or exposure to muscarinic cholinergic agonists (Oron et al., 1985; Busa et al., 1985; Slack et al., 1986). In sea urchin eggs, there is also an elevation of the fertilization envelope and an increase in intracellular pH (Whitaker and Irvine, 1984), as observed during fertilization. These changes have been attributed to the release of intracellular Ca^{2+}, and IP$_3$ has been shown to mobilize Ca^{2+} in Xenopus oocytes and sea urchin eggs (Busa et al., 1985; Clapper and Lee, 1985; Nadler et al., 1986). During fertilization of sea urchin eggs, there is an increase of IP$_3$ and DAG (Ciapa and Whitaker, 1986).

7. ROLE OF DIACYLGLYCEROL AND PROTEIN KINASE C

With the discovery of the Ca^{2+}-phospholipid-dependent protein kinase, now commonly known as protein kinase C (for references, see Nishizuka, 1984), a second mechanism of intracellular signaling for α_1-adrenergic and other Ca^{2+}-mobilizing agonists was revealed. This enzyme has a requirement for Ca^{2+} and a phospholipid for activity. Phosphatidylserine is the most effective phospholipid, but PI, phosphatidylethanolamine, and PA are also active, whereas phosphatidylcholine is inactive by itself and inhibitory in the presence of phosphatidylserine (Takai et al., 1979a; Kaibuchi et al., 1981). The enzyme is present in particulate and soluble fractions in all tissues examined but is highest in brain, spleen, platelets, and lymphocytes (Kikkawa et al., 1982; Kuo et al., 1980). As will be discussed later, the distribution of the enzyme between membrane and cytosol phases is apparently under the control of Ca^{2+} and diacylglycerol and, hence, hormones that alter these.

Protein kinase C is activated by sn-1,2-diacylglycerols (DAGs), with the forms containing at least one unsaturated fatty acid being more effective than the saturated forms, unless the latter contain symmetrically two C$_6$–C$_{10}$ fatty acids (Takai et al., 1979a, 1979b; Kishimoto et al., 1980; Mori et al., 1982; Lapetina et al., 1985). The sn-1,3- and sn-2,3-DAG isomers are inactive (Boni and Rando, 1985). The naturally occurring DAGs can be replaced by synthetic

DAGs or by tumor-promoting phorbol esters, which have a structure similar to DAG (Castagna *et al.*, 1982; Davis *et al.*, 1985; Ebeling *et al.*, 1985; Niedel *et al.*, 1983). The phorbol esters appear to bind to the same "receptor" on protein kinase C as the DAGs (Kikkawa *et al.*, 1983; Ebeling *et al.*, 1985; Sharkey *et al.*, 1984). DAGs and phorbol esters increase the activity of protein kinase C at maximum Ca^{2+} but, more importantly, decrease the concentration of Ca^{2+} for half-maximal activity to the submicromolar range found in the cytosol (Takai *et al.*, 1979b; Kishimoto *et al.*, 1980; Kuo *et al.*, 1980). In the absence of phospholipid, DAGs and phorbol esters have little effect. Although it is often assumed that protein kinase C is the sole cellular target of DAGs and phorbol esters, other possible mechanisms of action should be kept in mind (e.g., Gonzatti-Haces and Traugh, 1986).

Using a mixed micellar assay, it has been shown that a single molecule of 1,2-dioleoylglycerol and of Ca^{2+} and four molecules of phosphatidylserine are required to activate monomeric protein kinase C (Hannun *et al.*, 1986; Ganong *et al.*, 1986; Hannun and Bell, 1986). The four phospholipid molecules are believed to bind Ca^{2+} through the four carboxyl groups in the serine head groups, and protein kinase C binds to this surface structure but is inactive (Hannun *et al.*, 1985; Ganong *et al.*, 1986). The complex then binds active phorbol esters or diacylglycerols resulting in activation of the kinase (Hannun *et al.*, 1985; Ganong *et al.*, 1986). The diacylglycerol or phorbol ester is thought to have at least three attachment points to the complex, including the kinase and Ca^{2+} (Ganong *et al.*, 1986; Hannun and Bell, 1986). This model also explains the translocation of the enzyme to membranes induced by phorbol esters and Ca^{2+}.

Since the phosphoinositides contain predominantly stearic acid at the *sn*-1 position of glycerol and arachidonic acid at the *sn*-2 position (Holub and Kuksis, 1978), their hydrolysis by phosphodiesterase yields stearoyl arachidonoylglycerol, which would activate protein kinase C. Thus, PIP_2 breakdown induced by Ca^{2+}-mobilizing agonists is associated with protein kinase C activation (Nishizuka, 1984). However, all the major phospholipids contain some unsaturated fatty acids, predominantly in the *sn*-2 position of glycerol (Holub and Kuksis, 1978), and their breakdown by phosphodiesterase could therefore yield DAGs capable of activating protein kinase C.

α_1-Adrenergic and other Ca^{2+}-mobilizing agonists have been shown to increase DAG in liver, platelets, and exocrine pancreas (Figure 9; Bocckino *et al.*, 1985; Banschbach *et al.*, 1981; B.P. Hughes *et al.*, 1984; Kawahara *et al.*, 1980; Rink *et al.*, 1983; Thomas *et al.*, 1983; Haslam and Davidson, 1984a), and it is likely that they do so in other tissues. In platelets labeled with [^{3}H]arachidonic acid, the increase in [^{3}H]-DAG in response to activating factors is very rapid and transient (Kawahara *et al.*, 1980; Rink *et al.*, 1983; Rittenhouse-Simmons, 1979), but chemical measurements of DAG in hepatocytes and other cells show a slower and more stable increase (Bocckino *et al.*, 1985; Griendling

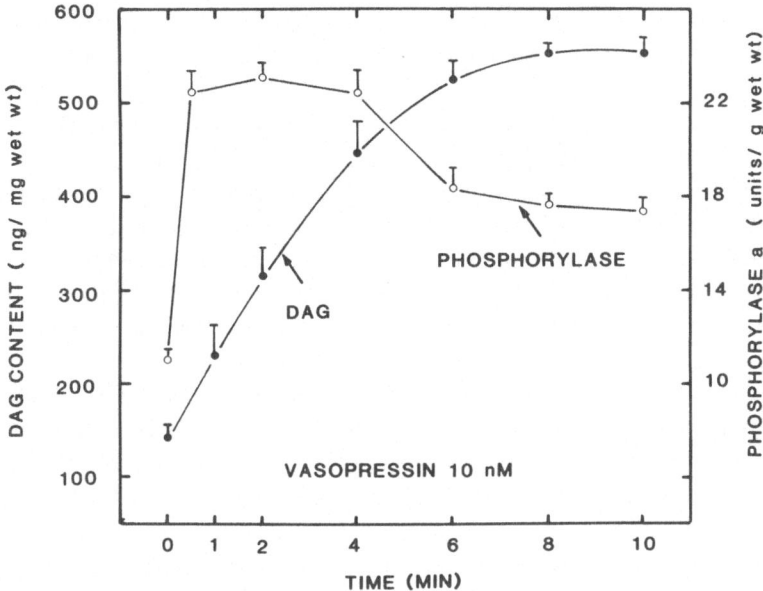

FIGURE 9. Time course of vasopressin (10 nM) action on DAG and phosphorylase a in insolate rat hepatocytes. (From Bocckino *et al.*, 1985, with permission of the authors and publisher.)

et al., 1986; Preiss *et al.*, 1986). The studies of Bocckino *et al.* (1985) show, furthermore, that at least two species of DAG are generated by Ca^{2+}-mobilizing agonists in liver, namely one enriched in arachidonic and stearic acids, which presumably arises from the phosphoinositides, and another enriched in palmitic and linoleic acids, which is of unknown origin.

The time course of DAG generation in hepatocytes, vascular smooth muscle cells, and pancreatic acini differs markedly from that for IP_3 (Bocckino *et al.*, 1985; Griendling *et al.*, 1986; Preiss *et al.*, 1986; Pandol and Schoeffield, 1986), consistent with the idea that DAG is formed from other sources besides PIP_2. This is reinforced by the fact that DAG release is quantitatively much greater than IP_3 release. However, the agonist concentration dependence of DAG formation resembles that for IP_3 accumulation (Bocckino *et al.*, 1985; Griendling *et al.*, 1986), suggesting that they both utilize the same receptor signaling mechanism, at least in part. GTP-dependent agonist stimulation of phosphatidylcholine breakdown in liver plasma membranes has been reported recently (Irving and Exton, 1987) and may account for some of the DAG formed. Relatively low concentrations (10^{-6} M) of the ionophore A23187 can increase DAG in hepatocytes without raising IP_3 detectably, whereas higher concentrations (10^{-5} M) increase both DAG and IP_3 (Bocckino *et al.*, 1985). Calcium depletion causes a partial reduction in the effects of Ca^{2+}-mobilizing agonists on IP_3 and DAG (Charest *et al.*, 1985; Bocckino *et al.*, 1985), suggesting that changes in cellular Ca^{2+} may account for part, but not all, of the effect of these agonists on DAG and IP_3.

There have been no *direct* demonstrations that Ca^{2+}-mobilizing agonists activate protein kinase C in cells. However, there have been several reports showing that these agonists increase the phosphorylation of several substrates in platelets, liver cells, and mast cells, which are also selectively affected by active phorbol esters or synthetic DAGs (Kaibuchi *et al.*, 1983; Katakami *et al.*, 1984; Haslam and Davidson, 1984a; Garrison *et al.*, 1984). Some of these substrates have been shown to be phosphorylated by protein kinase C *in vitro* (Kawahara *et al.*, 1980; Sano *et al.*, 1983; Cooper *et al.*, 1984).

There have been reports showing that phorbol esters or Ca^{2+}-mobilizing agonists induce the translocation of protein kinase C from the soluble phase to the plasma membrane in several cells (Kraft and Anderson, 1983; Kraft *et al.*, 1982; Drust and Martin, 1985; Wooten and Wrenn, 1984). The data indicate that protein kinase C present in the soluble phase is inactive due to the absence of lipid. However, it is postulated that when agonists induce a rise in DAG in the plasma membrane and in cytosolic Ca^{2+}, the enzyme becomes associated with the membrane where it becomes activated by the accumulated DAG (M. Wolf *et al.*, 1985; May *et al.*, 1985). These events have been demonstrated in intact GH_3 pituitary cells treated with phorbol ester, using [^{35}S]methionine-labeled protein kinase C and antisera to the enzyme (Ballester and Rosen, 1985). They imply that soluble protein substrates for protein kinase C can only be phosphorylated at the plasma membrane or at other membranes where there might be a rise in DAG.

DAG is further metabolized to phosphatidic acid in the plasma membrane due to the action of diacylglycerol kinase. Other routes of metabolism are hydrolysis by diacylglycerol and monoglycerol lipases. Diacylglycerol lipase is present in the plasma membrane of some cells (Mauco *et al.*, 1984; Authi *et al.*, 1985), but it is unclear to what extent membrane-associated DAG is metabolized by this enzyme. PA can be reconverted to DAG by phosphatidate phosphohydrolyase, but it is not known whether this enzyme is present in the plasma membrane. The major fate of PA generated in the plasma membrane appears to be its transfer to the endoplasmic reticulum for phospholipid and triacylglycerol synthesis. This transfer presumably involves a phospholipid exchange protein.

Protein kinase C has been shown to phosphorylate a large number of proteins *in vitro*, but it is unclear to what extent these serve as substrates in intact cells. Addition of active phorbol esters to liver cells increases the phosphorylation of several soluble proteins of unknown function (Garrison *et al.*, 1984; Cooper *et al.*, 1984). They also cause inactivation of glycogen synthase in these cells

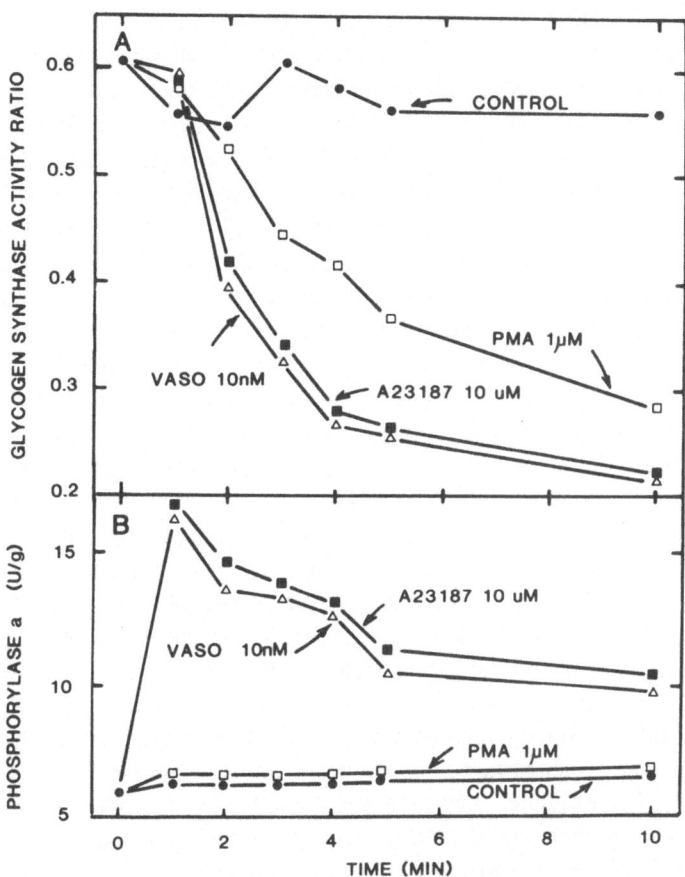

FIGURE 10. Effects of vasopressin, ionophore A23187, and 4β-phorbol 12β-myristate 13α-acetate (PMA) on glycogen synthase activity ratio (−Glc6P/ + Glc6P) and phosphorylase *a* in isolated rat hepatocytes. (From Blackmore *et al.*, 1986, with permission of the authors and publisher.)

(Figure 10; Roach and Goldman, 1983; Blackmore *et al.*, 1986). The inactivation of this enzyme caused by Ca^{2+}-mobilizing agonists is also better correlated with changes in DAG than in cytosolic Ca^{2+} (Bouscarel and Exton, 1986) and is also seen in the absence of changes in cell Ca^{2+} (Blackmore *et al.*, 1986). However, it is unclear whether or not the inactivation is due to a direct effect of protein kinase C in this enzyme (Imazu *et al.*, 1984; Ahmad *et al.*, 1984). Phorbol esters and synthetic DAGs also activate a phosphatase for IP_3 (Molina y Vedia and Lapetina, 1986) and phosphorylate a 40- to 47-kD protein in platelets (Kawahara *et al.,* 1980; Sano *et al.,* 1983; Kaibuchi *et al.,* 1983). This protein appears to be the same as that phosphorylated in response to platelet-activating factors.

There is also evidence that phorbol esters induce phosphorylation and/or alter the function of several plasma membrane hormone receptors, including α_1-adrenergic receptors (Corvera and Garcia-Sainz, 1984; Labarca *et al.,* 1984; Danthuluri and Deth, 1984; Lynch *et al.,* 1985c; Cooper *et al.,* 1985; Baraban *et al.,* 1985a; Van de Werve *et al.,* 1985; Leeb-Lundberg *et al.,* 1985), epidermal growth factor receptors (Lee and Weinstein, 1978, 1979; Shoyab *et al.,* 1979; Moon *et al.,* 1984; Davis and Czech, 1984; Davis *et al.,* 1985; Cochet *et al.,* 1984; Beguinot *et al.,* 1985), insulin receptors (Jacobs *et al.,* 1983; Thomopoulos *et al.,* 1982; Grunberger and Gorden, 1982), somatostatin receptors (Matozaki *et al.,* 1986), and transferrin receptors (May *et al.,* 1984). Inhibition by phorbol esters of the actions of other agonists has been reported, e.g., the chemotactic peptide formyl Met-Leu-Phe in neutrophils (White *et al.,* 1984; Naccache *et al.,* 1985), thyrotropin-releasing hormone in pituitary cells (Albert and Tashjian, 1985), muscarinic cholinergic agonists in hippocampus and astrocytoma and pheochromocytoma cells (Labarca *et al.,* 1984; Orellana *et al.,* 1985; Vincentini *et al.,* 1985b), and several activating factors in platelets (MacIntyre *et al.,* 1985). Although phosphorylation of membrane receptors probably underlies the inhibitory effects of phorbol esters in most cases, there is also evidence that they may affect guanine nucleotide–binding proteins (Blackmore and Exton, 1986; Jakobs *et al.,* 1985; Katada *et al.,* 1985).

In addition to their inhibitory actions on some receptor responses, there is accumulating evidence that phorbol esters can increase the responses to other receptors. For example, they increase β-adrenergic or adenosine responses in brain, S49 lymphoma cells, and pinealocytes (Hollingsworth *et al.,* 1985; Bell *et al.,* 1985; Sugden *et al.,* 1985). The effects of the esters may be exerted at the level of the guanine nucleotide–binding regulatory proteins (Bell *et al.,* 1985). This may also be true in part for the receptor systems that are inhibited by DAG and its analogues (Blackmore and Exton, 1986).

Phorbol esters and synthetic DAGs have been shown to have effects on cells that are not directly related to modification of receptor functions. For example, they can induce serotonin secretion in platelets (Yamanishi *et al.,* 1983; Rink *et al.,* 1983), stimulate amylase secretion in pancreatic acini (Wooten and Wrenn, 1984), induce superoxide generation or O_2 consumption in neutrophils (Dale and Penfield, 1984; De Virgilio *et al.,* 1984; Sha'afi *et al.,* 1983), stimulate protein secretion in parotid gland (Putney *et al.,* 1984), regulate ionic conductance in hippocampal pyramidal neurons (Baraban *et al.,* 1985b) and in bag cell neurons of *Aplysia* (De Riemer *et al.,* 1985), stimulate insulin release from islets (Hutton *et al.,* 1984; Malaisse *et al.,* 1985; Zawalich *et al.,* 1983), stimulate prolactin release by pituitary cells (Osborne and Tashjian, 1981; Delbeke *et al.,* 1984), induce contraction in certain smooth muscles (Baraban *et al.,* 1985a; Rasmussen *et al.,* 1984), stimulate the Na^+/H^+ antiporter in several cells (Besterman and Cuatrecasas, 1984; Volpi et al., 1985), cause histamine release from mast cells (Katakami et al., 1984), stimulate the Na^+ pump in liver (Lynch et al., 1986c), and increase the phosphorylation and activity of tyrosine hydroxylase and catecholamine secretion in adrenal chromaffin cells (Pocotte et al., 1985; Pocotte and Holz, 1986). In most cases, the effects of the DAG analogues are synergistic with those of Ca^{2+} ionophores, and addition of both types of agent is necessary to mimic the effects of natural agonists completely (Nishizuka, 1984). However, this synergism is not seen for all actions, indicating that some agonist effects are mediated by an increase in Ca^{2+} or DAG alone (Lynch et al., 1985c; Cooper et al., 1985; cf. Fain et al., 1984; Kimura et al., 1984).

In Swiss 3T3 fibroblasts, Ca^{2+} ionophores increase expression of the c-myc gene, and this is enhanced by a tumor-promoting phorbol ester (Kaibuchi et al., 1986). These results suggest that activation of both protein kinase C and Ca^{2+} mobilization are involved in the stimulatory effects of platelet-derived growth factor and fibroblast growth factor on c-myc mRNA levels in these cells (Kaibuchi et al., 1986). The protein product of the c-myc protooncogene is functionally involved in DNA synthesis and is believed to be important in the regulation of the cell cycle. In BALB/c 3T3 and NIH 3T3 cells, induction of transcription of the c-myc and c-fos protooncogenes by phorbol esters alone has been reported (Greenberg and Ziff, 1984; Kelly et al., 1983; Kruijer et al., 1984). Thus, one mechanism of platelet-derived growth factor–induced gene expression in these cells could be via activation of protein kinase C, since the growth factor is known to activate PIP_2 phospholipase C.

The mechanisms involved in the synergistic interactions of Ca^{2+} and DAG have not been defined. They could be due to the effects of these agents on protein kinase C per se, but this explanation seems inadequate in some cases. Alternative explanations are that some responses require the phosphorylation of single proteins by both DAG- and Ca^{2+}-sensitive protein kinases, that some proceses require the separate phosphorylation of two or more proteins by these kinases, and that some effects involve a phosphorylation cascade in which protein kinase C phosphorylates a Ca^{2+}-dependent protein kinase or vice versa. Protein kinase C has been shown to phosphorylate myosin light-chain kinase in vitro, but this leads to inactivation (Nishikawa et al., 1985; Ikebe et al., 1985).

8. ROLE OF Ca^{2+}-CALMODULIN-REGULATED ENZYMES AND OTHER PROTEINS

When Ca^{2+} ions were demonstrated to play a pivotal role in the actions of α_1-adrenergic and other agonists, the search for intracellular targets of these ions intensified. Troponin C was well known as the Ca^{2+}-responsive protein involved in contraction in skeletal muscle, but most of the proteins involved in Ca^{2+} actions in smooth muscle and other cells were unknown. A major breakthrough came with the discovery of the 17-kD Ca^{2+}-dependent regulatory protein calmodulin, by Kakiuchi, Cheung, Wang, and their associates (for reviews, see Cheung, 1980; Klee and Vanaman, 1982). This protein was soon shown to be involved in a large number of Ca^{2+}-mediated cellular responses and to be distributed widely in the soluble and particulate fractions of many tissues from animal and plant species.

Calmodulin is homologous to troponin C and has four nonidentical Ca^{2+}-binding sites of high affinity (K_d between 10^{-7} and 10^{-5} M). Thus, a rise in cytosolic Ca^{2+} within the physiological range leads to increased formation of Ca^{2+}-calmodulin complexes. Ca^{2+} binding to calmodulin results in a conformational change in the protein that increases its reversible interaction with certain target proteins, thereby altering their activities. Some of these proteins are a form of cyclic nucleotide phosphodiesterase, a form of adenylate cyclase, a plasma membrane Ca^{2+}-ATPase, and a specific phosphoprotein phosphatase termed calcineurin (Klee and Vanaman, 1982). In addition to these proteins, the Ca^{2+}-calmodulin complex interacts with and activates certain protein kinases leading to the phosphorylation of a diversity of proteins.

A major target of Ca^{2+} calmodulin formed in response to a rise in cytosolic Ca^{2+} in smooth muscle and platelets is myosin light-chain kinase. This 105- to 125-kD enzyme phosphorylates the regulatory 20-kD light chains of myosin, thereby increasing actin-stimulated myosin ATPase activity and the increased cross-bridge cycling associated with contraction in smooth muscle (Figure 11; Chacko et al., 1977; Dabrowska et al., 1978; Adelstein and Eisenberg, 1980; Driska et al., 1981; Ruegg, 1982) or shape change in platelets (Adelstein and Conti, 1975; Daniel et al., 1981, 1984). The enzyme is also present in brain, heart, and skeletal muscle, but its function in the latter is not to play a role in the initiation of contraction but to augment force generation (Stull et al., 1980).

Another Ca^{2+}-dependent protein kinase with high substrate specificity is phosphorylase b kinase. This differs from most calmodulin-responsive enzymes in that it contains calmodulin as one of its subunits (Cohen et al., 1978; Chan and Graves, 1984). It has a M_r of approximately 1.3 million and consists of a tetramer of α or α', β, γ, and δ subunits (Chan and Graves, 1984). The α and β subunits are regulatory and can be autophosphorylated or phosphorylated by cAMP-dependent protein kinase, whereas the γ subunit contains the catalytic domain (although there is a suggestion that the β subunit may also have catalytic activity). The δ subunit is virtually identical with calmodulin, which means that Ca^{2+} can interact directly with the enzyme (Shenolikar et al., 1979). In addition, the α and β subunits of most tissue forms of the enzyme can bind additional Ca^{2+} calmodulin, and this binding leads to increased activity (Picton et al., 1980; Cohen, 1980). Thus, a common response to a rise in cytosolic Ca^{2+},

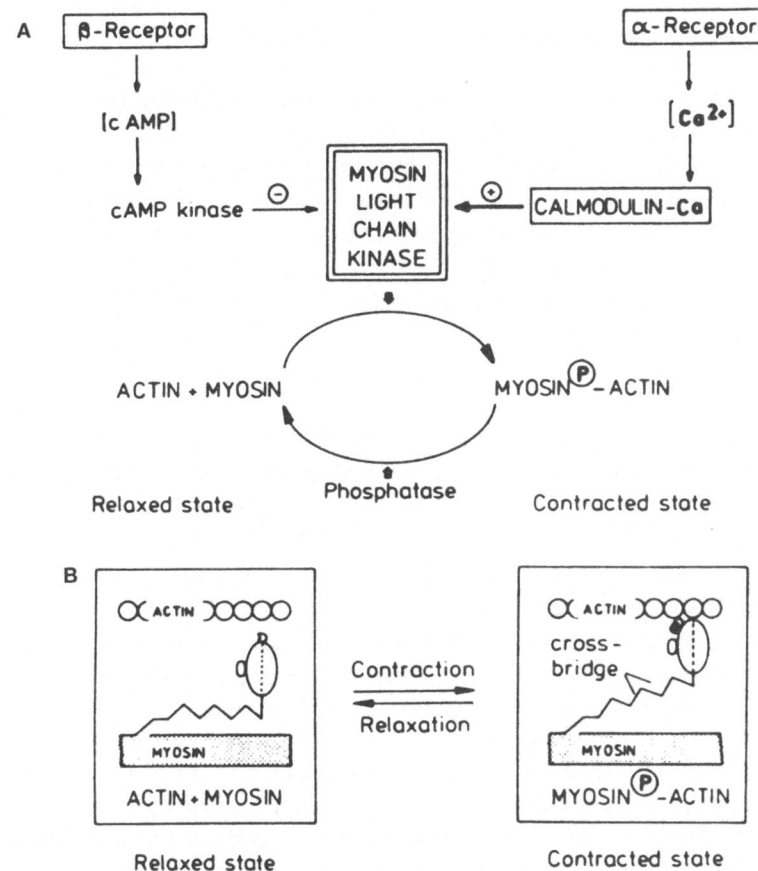

FIGURE 11. (A) Pathways of α-adrenergic stimulation and β-adrenergic inhibition of smooth muscle contraction. (B) Mechanism of smooth muscle contraction involving phosphorylation of myosin. (Reproduced from Ruegg, 1982, with permission of the author and publisher.)

induced by α_1-adrenergic agonists and other hormones or neurotransmitters in many tissues, is activation of phosphorylase b kinase. This, in turn, leads to phosphorylation of phosphorylase b, converting it to the more active form, phosphorylase a. This enzyme is rate limiting for glycogen breakdown, and its activation leads to enhanced formation of glucose-6-phosphatase for energy production via glycolysis in most tissues. In the case of liver, there is also production of glucose due to the presence of glucose 6-phosphatase. Phosphorylase b kinase has also been shown to phosphorylate and inactivate liver muscle glycogen synthase (Roach *et al.*, 1978), but it is unclear whether this is important in the inhibition of glycogen synthase by Ca^{2+}-mobilizing agonists (Strickland *et al.*,

1983). It is also uncertain whether the rise in sarcoplasmic Ca^{2+} occurring during skeletal muscle contraction alters glycogen synthase activity.

Another protein kinase of importance in the actions of Ca^{2+}-mobilizing agonists is a multifunctional Ca^{2+}-calmodulin-dependent protein kinase that is found widely distributed in mammalian tissues. This kinase is not as selective in its substrate specificity as myosin light-chain kinase or phosphorylase kinase and exists in more than one isozymic form. It was originally discovered as a Ca^{2+}-dependent protein kinase in brain (Schulman and Greengard, 1978a,b) and as a glycogen synthase kinase in liver (Payne and Soderling, 1980), and has now also been purified from skeletal muscle (Woodgett *et al.*, 1983; 1984). There is also evidence that it is present in adipose tissue (Landt and McDonald, 1984; *Torpedo* electric organ (Palfrey *et al.*, 1983), and pancreatic islets (Landt *et al.*, 1982).

The Ca^{2+}-calmodulin-dependent protein kinases of brain have been subdivided into isozymic forms. Type II is found in both particulate and cytosolic fractions and phosphorylates a unique site on the neuron-specific protein, called synapsin 1 (Kennedy and Greengard, 1981). Another (type I) is present mainly in the cytosol and phosphorylates synapsin 1 on a site that is also phosphorylated by cAMP-dependent protein kinase (Kennedy and Greengard, 1981). The type II isozyme from brain appears to be identical with the Ca^{2+}-calmodulin-dependent glycogen synthase kinase of skeletal muscle (McGuinness *et al.*, 1983; Woodgett *et al.*, 1984) and liver (Schworer and Soderling, 1983).

Type II Ca^{2+}-calmodulin-dependent protein kinase has a wider substrate specificity and is more abundant than type I. It is a 550- to 650-kD polymer, with subunits of approximately 50 and 60 kD weight, which undergo autophosphorylation (Bennett *et al.*, 1983; McGuinness *et al.*, 1985; Kuret and Shulman, 1984). The isozymes contain different ratios of the subunits (McGuinness *et al.*, 1985). A large number of *in vitro* substrates of type II Ca^{2+}-calmodulin-dependent protein kinase have been identified. These include glycogen synthase, synapsin 1, microtubule-associated protein 2, myosin light chains, tyrosine hydroxylase, phenylalanine hydroxylase, ATP-citrate lyase, acetyl-CoA carboxylase, pyruvate kinase (Schworer and Soderling, 1983; McGuinness *et al.*, 1983; Woodgett *et al.*, 1983, 1984; Doskeland *et al.*, 1984; Vulliet *et al.*, 1984; Schulman, 1984a,b). Many of these proteins are phosphorylated when neuronal and other cells are stimulated by nervous or hormonal signals that induce a rise in cytosolic Ca^{2+} (Nestler *et al.*, 1984; Exton, 1985; Nestler and Greengard, 1983; Schulman, 1984a,b; Garrison and Wagner, 1982; Blackmore and Exton, 1985; Garrison *et al.*, 1984). However, although it is likely that a Ca^{2+}-calmodulin-dependent protein kinase is responsible for these phosphorylations, it has not been clearly established, because many of the proteins are also substrates for protein kinase C and/or cAMP-dependent protein kinase. The case that Ca^{2+}-calmodulin-dependent protein kinase is important in the phosphorylations induced by Ca^{2+}-

mobilizing agonists appears to be strongest for synapsin 1, tyrosine hydroxylase, and microtubule proteins in brain, and phenylalanine hydroxylase and pyruvate kinase in liver.

Phosphorylation of the synaptic vesicle–associated protein synapsin 1 induces neurotransmitter release in the giant squid synapse, and there is evidence that it produces an analogous effect in the mammalian nervous system (Nestler et al., 1984). Tyrosine hydroxylase converts tyrosine to dihydroxyphenylalanine (Dopa) and is rate controlling for epinephrine and norepinephrine synthesis in adrenal medulla and, presumably, brain. Phosphorylation of this enzyme leads to an increase in its activity when assayed in the presence of an "activator protein" (Yamauchi et al., 1981). Type II Ca^{2+}-calmodulin-dependent protein kinase phosphorylates microtubule-associated protein 2, α- and β-tubulin, and τ factor from brain (Burke and Lorenzo, 1981; Yamamoto et al., 1983; Schulman, 1984a,b). This suggests that some aspect of microtubule function (state of polymerization, treadmilling, or interaction with other cell components) may be regulated by Ca^{2+}-mobilizing agonists through this enzyme.

Glycogen synthase was the substrate utilized initially to identify Ca^{2+}-calmodulin-dependent protein kinase in liver (Payne and Soderling, 1980; Ahmad et al., 1982; Payne et al., 1983). It phosphorylates the enzyme on site 2, which is serine 7 near the amino terminus, and also on site 1b toward the carboxyl terminus (Payne et al., 1983; Juhl et al., 1983). Site 2 is also phosphorylated by cAMP-dependent protein kinase and phosphorylase b kinase (Juhl et al., 1983), and its phosphorylation is associated with inactivation of the enzyme. Although Ca^{2+}-calmodulin-dependent protein kinase is a good candidate for mediating the effects of Ca^{2+}-mobilizing agonists on glycogen synthase in liver (Strickland et al., 1980), it should be noted that the enzyme is also a substrate for phosphorylase b kinase and protein kinase C (Imazu et al., 1984; Ahmad et al., 1984), which are also involved in the actions of these agonists.

There is much evidence that α_1-adrenergic agonists and other Ca^{2+}-mobilizing hormones induce phosphorylation and inactivation of pyruvate kinase in liver and that this is important in the stimulation of gluconeogenesis by these agents (Chan and Exton, 1978; Garrison et al., 1979; Nagano et al., 1980). The kinase involved does not appear to be phosphorylase kinase, since the phosphorylation occurs in animals lacking this enzyme, or protein kinase C, since phorbol esters do not induce phosphorylation of pyruvate kinase (Garrison et al., 1984). It is probably type II Ca^{2+}-calmodulin-dependent protein kinase, since this can phosphorylate and inactivate the enzyme in vitro (Schworer et al., 1985), although the rate of phosphorylation is slow relative so that of glycogen synthase (Schworer and Soderling, 1983; Schworer et al., 1985). Phenylalanine hydroxylase converts phenylalanine to tyrosine and is controlled by both cAMP- and Ca^{2+}-dependent stimuli in liver (Fisher and Pogson, 1984; Fisher et al., 1984). Phosphorylation and activation of the enzyme occurs in hepatocytes in response to several Ca^{2+}-mobilizing agonists (Garrison and Wagner, 1982; Garrison et al., 1984; Fisher et al., 1984), and there is evidence that neither phosphorylase b kinase nor protein kinase C is involved (Garrison et al., 1984). On the other hand, the enzyme is phosphorylated and activated in vitro by a Ca^{2+}-calmodulin-dependent protein kinase (Doskeland et al., 1984), suggesting that this is the kinase responsible.

Although the focus of the preceding paragraphs has been on the specific or multifunctional Ca^{2+}-calmodulin-dependent protein kinases involved in α_1-adrenergic and other Ca^{2+}-mobilizing agonist actions, other proteins are sensitive to Ca^{2+} calmodulin. Microtubules, which are key cytoskeletal elements associated with cell movement, flagellar and ciliary motility, chromosome movement, and axonal transport, are targets of calmodulin (Means and Dedman, 1980). Polymerization of α- and β-tubulin to form microtubules is inhibited by Ca^{2+} in the presence of calmodulin (Marcum et al., 1978; Kamagai and Nishida, 1979), and there is evidence that nucleation rather than elongation may be inhibited (Berkowitz and Wolff, 1981). In addition to its direct effects on tubulin, Ca^{2+} calmodulin can also influence microtubule assembly/disassembly through phosphorylation of microtubule components by a Ca^{2+}-calmodulin-dependent protein kinase, as noted above.

Ca^{2+}-calmodulin can activate a form of cyclic nucleotide phosphodiesterase found in brain, heart, liver, and most other tissues (Klee and Vanaman, 1982). This has a subunit M_r of approximately 60,000 and exists as a dimer in the native state. Ca^{2+} calmodulin binds stoichiometrically to a specific site on the enzyme to form a complex that hydrolyzes cGMP with a low K_m (5–10 μM) and cAMP with a high K_m (approximately 100 μM). Despite the well-demonstrated effects of Ca^{2+}-calmodulin on this enzyme in vitro, there are no clear-cut examples of Ca^{2+} regulation of cAMP or cGMP by this mechanism in intact cells. Perhaps this relates to the high K_ms of the enzyme for its two substrates relative to their cellular concentrations.

Ca^{2+}-calmodulin activates a form of adenylate cyclase in brain, pancreatic islets, adrenal medulla, and kidney cells (Klee and Vanaman, 1982). The effect does not appear to involve a guanine nucleotide–binding protein and is exerted directly on the catalytic subunit of the enzyme (Coussen et al., 1985). Presently, there are no unequivocal examples of Ca^{2+} regulation of the enzyme under physiological conditions, although such would be difficult to demonstrate in the intact brain.

The plasma membrane Ca^{2+}-pump ATPase is calmodulin sensitive in most tissues, with an important exception being the liver (Carafoli, 1984). This pump is responsible for most of the Ca^{2+} extracted from nonexcitable cells and from excitable cells during rest. In the latter, the lower-affinity Na^+/Ca^{2+} exchanger is responsible for most of the Ca^{2+} ejected during excitation. Addition of Ca^{2+} calmodulin to the ATPase lowers its K_m for Ca^{2+} and increases its maximum velocity. The enzyme appears to function as a $Ca^{2+}-2H^+$ exchanger (Niggli *et al.*, 1979; Waisman *et al.*, 1981). In consort with other calmodulin-responsive proteins, the 138-kD ATPase has a specific Ca^{2+}-calmodulin-binding domain that is approximately 25 kD, as defined by proteolytic fragmentation (Zurini *et al.*, 1984).

In contrast to the plasma membrane Ca^{2+}-pump ATPase, that of the endoplasmic (sarcoplasmic) reticulum is not *directly* controlled by Ca^{2+} calmodulin. In heart, the sarcoplasmic reticular Ca^{2+} ATPase is regulated by phospholamban, a 22-kD proteolipid that can be phosphorylated by both cAMP-dependent and Ca^{2+}-calmodulin-dependent protein kinases, leading to increased uptake of Ca^{2+} by the sarcoplasmic reticulum (Tada *et al.*, 1979; LePeuch *et al.*, 1979; Tada and Katz, 1982; Davis *et al.*, 1983). Ca^{2+}-calmodulin-dependent phosphorylation of phospholamban increases the maximum rate of Ca^{2+} transport by isolated cardiac sarcoplasmic reticulum vesicles with a small decrease in K_m for Ca^{2+} (Davis *et al.*, 1983), but efforts to demonstrate that physiological increases in cytosolic Ca^{2+} increase phospholamban phosphorylation in intact myocardium have not been successful to date. Ca^{2+}-calmodulin-dependent protein kinase and phosphorylase *b* kinase phosphorylate several proteins in skeletal muscle sarcoplasmic reticulum (Varsanyi and Heilmeyer, 1981; Campbell and MacLennan, 1982), but there is no evidence that phospholamban is present or is phosphorylated in this tissue. The functional significance of these phosphorylations is uncertain, but it has been suggested that the phosphorylation/dephosphorylation cycle of a 60-kD protein may control the Ca^{2+}-release channel (Campbell and MacLennan, 1982).

Ca^{2+} ions can also regulate cellular processes by interacting with proteins other than calmodulin. As alluded to above, troponin C is a major target in skeletal and cardiac muscle and is fundamentally involved in contraction. Other Ca^{2+} targets are a group of proteins that alter aspects of actin filament assembly and severance and, thus, may be important in cell architecture, cytoplasmic flow, and exocytosis (Stossel, 1984). These include gelsolin, profilin, villin, and fragmin, which act on actin in various ways, e.g., by sequestering actin monomers and by nucleating, end-blocking, and severing actin filaments. Ca^{2+} binds to gelsolin with high affinity, and this causes shortening of actin filaments, contributing to the collapse of their three-dimensional lattice (Yin and Stossel, 1982). This gel–sol transformation may be involved in the regulation of cell motility. Other gelsolin-related proteins can bind Ca^{2+} and may contribute to the effects of this ion on actin filament assembly/disassembly.

Denton, McCormack, and others (reviewed by Denton and McCormack, 1981; Hansford, 1985) have identified another group of Ca^{2+}-responsive, calmodulin-independent enzymes in liver, heart, and adipose tissue. These are all mitochondrial and include pyruvate dehydrogenase phosphate phosphatase, α-oxoglutarate dehydrogenase, and NAD^+-isocitrate dehydrogenase. The pyruvate dehydrogenase complex is under elaborate control by allosteric effectors and phosphorylation/dephosphorylation mechanisms. The component that is phosphorylated by pyruvate dehydrogenase kinase is the α subunit of the pyruvate decarboxylase moiety. Phosphorylation of a specific serine residue in the subunit causes inactivation, and dephosphorylation by pyruvate dehydrogenase phosphate phosphatase leads to activation. Low concentrations (0.1–10 μM) of Ca^{2+} stimulate the phosphatase (Denton *et al.*, 1972; McCormack, 1985a) and activate pyruvate dehydrogenase in isolated mitochondria (McCormack *et al.*, 1982; McCormack and Denton, 1984). Thus, Ca^{2+} has been implicated in the stimulatory effects of glucagon, epinephrine, angiotensin II, vasopressin, and A23187 ionophore on pyruvate dehydrogenase in liver (Figure 12; Hems *et al.*, 1978; Assimacopoulos-Jeannet *et al.*, 1983; Blackmore *et al.*, 1983b; Sies *et al.*, 1983; Oviasu and Whitton, 1984; McCormack, 1985b,c) and of inotropic agents in heart (McCormack and Denton, 1981a,b, 1984; McCormack *et al.*, 1982). These stimulatory effects can be observed in tissue extracts (Assimacopoulos-Jeannet *et al.*, 1983; Blackmore *et al.*, 1983b; Sies *et al.*, 1983; Oviasu and Whitton,

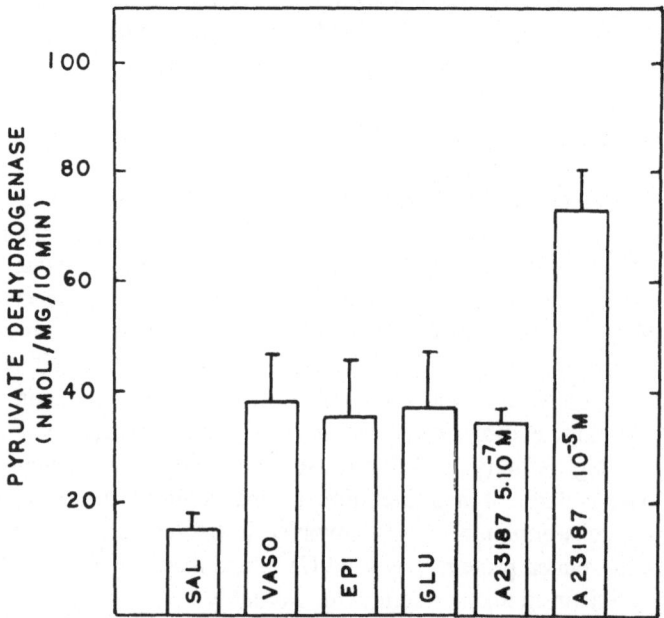

FIGURE 12. Effects of vasopressin (10 nM), epinephrine (10 μM), glucagon (10 nM), and ionophore A23187 (0.5 or 10 μM) on pyruvate dehydrogenase activity in isolated rat hepatocytes. Hepatocytes were incubated with the agonists shown for 5 min and pyruvate dehydrogenase assayed in the frozen/thawed extracts according to Assimacopoulos-Jeannet *et al.* (1983). (B.P. Hughes, P.F. Blackmore, and J.H. Exton, unpublished material.)

1984; McCormack and Denton, 1981a,b) or in mitochondria isolated from livers or hearts exposed to the agonists (McCormack, 1985b,c; McCormack and Denton, 1984; Assimacopoulos-Jeannet et al., 1986).

There is some controversy regarding the effects of Ca^{2+}-mobilizing agonists on pyruvate dehydrogenase activity in intact liver when this is assayed indirectly by measuring CO_2 production from isotopically labeled pyruvate. However, this approach is complicated by intracellular changes in precursor specific radioactivity and by entry of the label into the citric acid cycle via pyruvate carboxylation. Thus, some workers have reported that α_1-adrenergic agonists and vasopressin decrease the production of $^{14}CO_2$ from [1-^{14}C]pyruvate in isolated hepatocytes or the perfused rat liver (Sies et al., 1983; Fisher et al., 1985). However, in these experiments there was concomittant breakdown of glycogen with production of unlabeled pyruvate, which would lead to a decrease in pyruvate specific radioactivity.

Increased α-oxoglutarate dehydrogenase activity has been observed in liver mitochondria isolated from rats or livers treated with epinephrine or glucagon (McCormack, 1985b,c; Assimacopoulos-Jeannet et al., 1986). It has also been deduced from measurements of $^{14}CO_2$ and [^{14}C]glucose production from labeled glutamine, glutamate, or proline and from α-oxoglutarate levels in livers perfused with glucagon (Ui et al., 1973) or with α_1-adrenergic agonists (Haussinger and Sies, 1984; Ochs, 1984; Verhoeven et al., 1985), or in hepatocytes incubated with vasopressin (Staddon and McGivan, 1985). Evidence that the increase, like that of pyruvate dehydrogenase, is due to an elevation of intramitochondrial Ca^{2+} has been presented by McCormack (1985a,b,c). Thus, when Ca^{2+} influx into the mitochondria is prevented during their isolation by ethyleneglycoltetraacetic acid (EGTA) and when Ca^{2+} efflux is minimized by the use of Na^+-free media, the hormone effect is preserved (McCormack, 1985b,c). Furthermore, manipulation of the extramitochondrial Ca^{2+} concentration within the physiological range and examination of the effects of ruthenium red (an inhibitor of mitochondrial Ca^{2+} uptake) and of Na^+ and diltiazem (an inhibitor of Na^+-induced mitochondrial Ca^{2+} efflux) strongly implicate intramitochondrial Ca^{2+} as a major regulator of both pyruvate dehydrogenase and α-oxoglutarate dehydrogenase in liver (McCormack, 1985a). There is strong evidence that a similar regulation occurs in heart (Denton et al., 1980; Hansford, 1981, 1985; Hansford and Castro, 1981; McCormack et al., 1982; McCormack and Denton, 1981b, 1984).

Activation of pyruvate dehydrogenase, α-oxoglutarate dehydrogenase, and isocitrate dehydrogenase is probably largely responsible for the increase in respiration induced by α_1-adrenergic agonists, vasopressin, and angiotensin II in perfused rat liver or isolated hepatocytes (Jakob and Diem, 1975; Sugano et al., 1980; Dehaye et al., 1981; Reinhart et al., 1982; Taylor et al., 1983; Blackmore et al., 1983a) and the increased reduction state of NAD(P) (Figure 13; Sugano

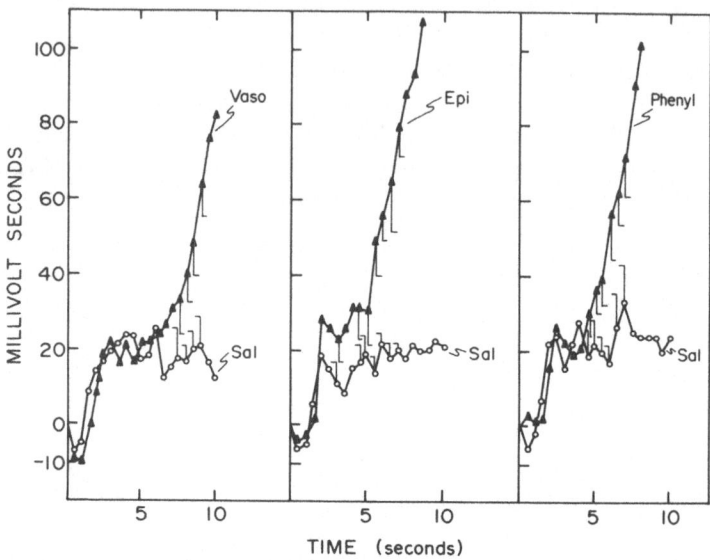

FIGURE 13.   Effects of vasopressin (10 nM), epinephrine (1 μM), and phenylephrine (10 μM) on NAD(P) reduction in isolated hepatocytes, as determined fluorimetrically. (From Blackmore *et al.*, 1983a.)

et al., 1980; Balaban and Blum, 1982; Buxton *et al.*, 1982; Blackmore *et al.*, 1983a). The activation may also account for the observation that Ca^{2+}-mobilizing agonists increase the oxidation of fatty acids to CO_2 and inhibit ketogenesis in hepatocytes (Sugden *et al.*, 1980; Williamson *et al.*, 1980; Sugden and Watts, 1983), since both effects can be attributed to increased turnover of the citric acid cycle.

   An initial problem with the hypothesis that a rise in intramitochondrial Ca^{2+} is responsible for the effects of α_1-adrenergic agonists, vasopressin, and angiotensin II on pyruvate dehydrogenase and α-oxoglutarate dehydrogenase was the observation from many laboratories that these agonists caused a loss of Ca^{2+} from mitochondria-enriched subcellular fractions of liver (for references, see Williamson *et al.*, 1981; Exton, 1981; Reinhart *et al.*, 1984a,b). However, recent work has indicated that components of the endoplasmic reticulum, rather than mitochondria, are the organelles from which Ca^{2+} is mobilized by these agonists (see Sections 3 and 7). Thus, it is now probable that mitochondria take up Ca^{2+} in response to the elevation of cytosolic Ca^{2+} induced by the agonists and that the concept that mitochondria are a site from which Ca^{2+} is released by the agonists (Exton, 1980, 1981; Williamson *et al.*, 1981; Reinhart *et al.*, 1984a,b) is incorrect. Furthermore, the idea that the mitochondrial Ca^{2+} cycle controls the concentration of cytosolic Ca^{2+} within the physiological range (Nicholls, 1978; Nicholls and Akerman, 1982) seems unlikely now that it is known that the cytosolic Ca^{2+} level in unstimulated cells is approximately 0.2 μM (Charest et al., 1983, 1985; Murphy et al., 1980; Joseph et al., 1985) and that the mitochondrial Ca^{2+} content is low in situ (Bond et al., 1984; A.P. Somlyo et al., 1985; Hansford, 1985). It now appears that the function of the mitochondria in regulating cell Ca^{2+} is to take up cytosolic Ca^{2+} when it rises to high levels and thus protect the cell from damage.

Consistent with the view that mitochondria take up Ca^{2+} in response to Ca^{2+}-mobilizing agonists in liver are the observations that the increases in respiration, NAD(P) reduction state, and pyruvate dehydrogenase activity induced by these agonists lag significantly (5–20 sec) behind the increase in cytosolic Ca^{2+} and its associated activation of phosphorylase and initiation of cellular Ca^{2+} efflux (Figure 13; Blackmore et al., 1983a,b; Charest et al., 1983, 1985). Stable increases in Ca^{2+} uptake by mitochondria isolated from livers perfused with α_1-adrenergic agonists or glucagon have also been reported (Taylor et al., 1980), and similar effects have been observed with mitochondria from hearts exposed to α_1-adrenergic agonists (Kessar and Crompton, 1981). It is unclear whether or not these changes occur in the intact cell, but if they do, they would be expected to contribute to the increase in intramitochondrial Ca^{2+} induced by these agents and the corresponding metabolic changes.

Although the foregoing account indicates that much is now known about the biochemical reactions underlying α_1-adrenergic phenomena, it should be noted that the molecular bases of many effects of α_1-adrenergic stimulation remain to be defined. These include the increase in plasma membrane K^+ permeability and other ion fluxes in salivary and lacrimal glands; the increase in K^+ efflux and thermogenesis in brown adipose tissue; the stimulation of K^+ fluxes, ureogenesis, and pyruvate carboxylation in liver; the stimulation of gluconeogenesis in kidney; the alterations in contractility and glycolysis in heart; and the hyperpolarization and relaxation of gastrointestinal muscle (Exton, 1985).

9. SUMMARY

α_1-Adrenergic receptors mediate many actions of the sympathetic nervous system by interacting with its component transmitters epinephrine and norepinephrine. Some important α_1-adrenergic responses are the contraction of smooth muscle in vascular and other tissues, the secretion of certain glands, alterations in carbohydrate metabolism in certain tissues, and neurotransmission. α_1-Adrenergic receptors have a ligand-binding subunit of approximately 80 kD and can exist in low- and high-agonist affinity states. The interconversion between these states can be controlled by GTP and its analogues, implying that the receptors can couple to a GTP-binding regulatory protein.

As illustrated in Figure 14, the primary effect of α_1-adrenergic receptor activation is the breakdown of PIP_2 in the plasma membrane to yield IP_3 and DAG. There is much evidence that coupling of the receptor to the phosphodiesterase (phospholipase C) responsible for the breakdown involves a GTP-binding regulatory protein. For example, the stimulation of PIP_2 breakdown and formation of IP_3 by α_1-adrenergic and other Ca^{2+}-mobilizing agonists in isolated plasma membranes is dependent upon GTP and its nonhydrolyzable analogues, and micromolar concentrations of GTP analogues can stimulate IP_3 formation in a Mg^{2+}-dependent manner. In some tissues, islet-activating protein inhibits agonist-induced PIP_2 breakdown, implying the involvement of N_i or a related protein. However, in most tissues, neither cholera toxin nor islet-activating protein modifies this breakdown, indicating that a novel GTP-binding protein is involved.

The formation of IP_3 in response to Ca^{2+}-mobilizing agonists is very rapid and is proportional to receptor occupancy. IP_3 releases Ca^{2+} rapidly from nonmitochondrial stores in permeabilized cells and from isolated microsomes (Figure 14). It appears to act by stimulating Ca^{2+} efflux from a component of the

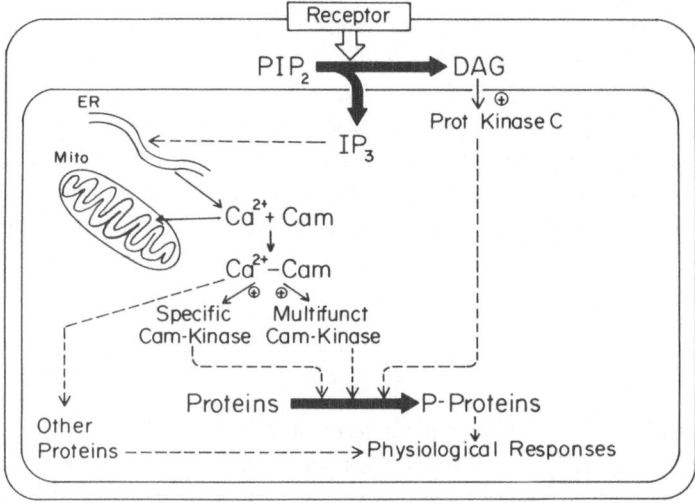

FIGURE 14.   General scheme by which α_1-adrenergic and other Ca^{2+}-mobilizing agonists alter cellular Ca^{2+} and the phosphorylation state of proteins leading to physiological responses. Abbreviations not given in the text are as follows: (Cam) calmodulin; (Cam-kinase) specific or multifunctional calmodulin-dependent protein kinase; (ER) endoplasmic reticulum; (Mito) mitochondrion.

endoplasmic reticulum and not by inhibiting Ca^{2+} uptake. IP_3 is rapidly degraded by phosphomonoesterases present in the plasma membrane and soluble phase. This leads to a cessation of Ca^{2+} efflux from the endoplasmic reticulum, unless IP_3 generation continues. IP_3 is also converted by a kinase to myoinositol 1,3,4,5-tetrakisphosphate, which is then broken down by a phosphomonoesterase to myoinositol 1,3,4-trisphosphate, which is degraded further. The inositol tetrakisphosphate has been reported to stimulate Ca^{2+} entry in sea urchin eggs, but the role of myoinositol 1,3,4-trisphosphate remains unclear. IP_3 is almost certainly the intracellular messenger responsible for Ca^{2+} mobilization.

The formation of DAG in response to Ca^{2+}-mobilizing agonists is of rapid onset, but the increase to maximum is generally slower than that of IP_3. The accumulation of DAG appears to cause the translocation of the Ca^{2+}-phospholipid-dependent protein kinase C from the cytosol to the plasma membrane. There it is activated by unsaturated DAG (Figure 14), which reduces its Ca^{2+} requirement to the cytosolic range. Although many enzymes and other proteins have been shown to be phosphorylated by protein kinase C *in vitro,* few intracellular targets of the enzyme have been characterized. Protein kinase C is presumed to be the cellular target of tumor-promoting phorbol esters, which activate the enzyme in a manner analogous to DAG. Protein kinase C phosphorylates the α_1-adrenergic receptor and certain other membrane receptors, thereby altering agonist binding.

In addition to promoting the release of intracellular Ca^{2+} through IP_3 generation, α_1-adrenergic agonists alter Ca^{2+} fluxes across the plasma membrane. These effects are responsible for prolonging the physiological responses to these agonists since the intracellular Ca^{2+} stores become depleted rapidly. Evidence exists that there is stimulation of Ca^{2+} influx, presumably due to the opening of Ca^{2+} channels. This could be due to the action of myoinositol 1,3,4,5-tetrakisphosphate. In some tissues, there is inhibition of Ca^{2+} efflux due to altered kinetics of the plasma membrane Ca^{2+}-ATPase/pump, and there is indirect evidence that this is due to alterations in plasma membrane inositol phospholipids.

The Ca^{2+}-dependent regulatory protein calmodulin is a major target of intracellular Ca^{2+} (Figure 14) and is involved in many physiological responses. It has four high-affinity-binding sites for Ca^{2+} and is present in all mammalian tissues. It is a subunit of phosphorylase *b* kinase and thus mediates the effects of increased cytosolic Ca^{2+} on glycogen breakdown and perhaps glycogen synthesis, since phosphorylase *b* kinase phosphorylates and activates glycogen phosphorylase *b* and phosphorylates and inactivates glycogen synthase *a*. More commonly, calmodulin does not exist as an enzyme subunit but forms a complex with Ca^{2+}, which then binds to a variety of enzymes and other proteins, thereby altering their functions. Ca^{2+} calmodulin interacts with myosin light-chain kinase, leading to increased phosphorylation of the regulatory 20-kD light chains of myosin. In smooth muscles and platelets, this promotes actin-stimulated myosin ATPase and increased cross-bridge cycling, resulting in contraction and shape change.

Another major target of Ca^{2+} calmodulin is a calmodulin-dependent protein kinase that exists in isozymic forms. One multifunctional isozyme is distributed widely and acts on many substrates (Figure 14) and, thus, its involvement in Ca^{2+}-mediated physiological responses is important. It acts on several substrates, including glycogen synthase, synapsin 1, tubulin, microtubule-associated proteins, tyrosine hydroxylase, phenylalanine hydroxylase, and pyruvate kinase. Ca^{2+}-calmodulin-dependent phosphorylation of the synaptic vesicle–associated protein synapsin 1 probably controls synaptic neurotransmitter release, and the phosphorylation of microtubular and associated proteins probably affects functions associated with microtubules, e.g., motility, chromosome movement, and axonal transport. Tyrosine hydroxylase is rate controlling for epinephrine and norepinephrine synthesis in the nervous system, and its Ca^{2+}-dependent phosphorylation leads to increased activity in the presence of an activator protein. Phenylalanine hydroxylase converts phenylalanine to tyrosine in liver, and pyruvate kinase is a regulatory enzyme in the glycolytic/gluconeogenic pathway. Phosphorylation by Ca^{2+}-calmodulin-dependent protein kinase stimulates the former enzyme and inhibits the latter.

Cells contain many other protein targets of Ca^{2+} calmodulin that are not protein kinases but may be involved in the actions of α_1-adrenergic and other Ca^{2+}-mobilizing agonists. These include microtubules, whose assembly is inhibited by Ca^{2+} calmodulin. Other targets are a form of cyclic nucleotide phosphodiesterase and of adenylate cyclase, and the plasma membrane Ca^{2+}-pump ATPase of most cells.

Ca^{2+} can regulate cellular processes independently of calmodulin. It can interact directly with troponin C to initiate contraction in skeletal and cardiac muscle and with gelsolin and other proteins, which alter actin filament assembly/disassembly and thus affect cell architecture, cytoplasmic flow and, perhaps, exocytosis. An increase in cytosolic Ca^{2+} also leads to an increase in mitochondrial Ca^{2+} in liver, heart, and probably other tissues (Figure 14). This results in stimulation of respiration and the citric acid cycle because of increased activity of α-oxoglutarate dehydrogenase and NAD^+-isocitrate dehydrogenase, and activation of pyruvate dehydrogenase due to stimulation of pyruvate dehydrogenase phosphate phosphatase.

Other α_1-adrenergic responses have been shown to be Ca^{2+} dependent, but the mechanisms involved remain obscure. These include (1) altered plasma membrane fluxes of K^+ and other ions and related membrane potential changes in salivary and lacrimal glands and brown adipose tissue, (2) stimulation of certain gluconeogenic reactions in liver and kidney, (3) alterations in contractility, glucose uptake, and glycolysis in heart, and (4) hyperpolarization and relaxation of gastrointestinal muscle.

ACKNOWLEDGMENT. The author wishes to thank Carolyn Sielbeck for her superlative assistance in the preparation of this chapter.

REFERENCES

Abdel-Latif, A. A., Akhtar, R. A., and Hawthorne, J. N., 1977, Acetylcholine increases the breakdown of triphosphoinositide of rabbit iris muscle prelabelled with [^{32}P]phosphate, *Biochem. J.* **162:**61–73.

Adelstein, R. S., and Conti, M. A., 1975, Phosphorylation of platelet myosin increases actin-activated myosin ATPase activity, *Nature* **256:**597–598.

Adelstein, R. S., and Eisenberg, E., 1980, Regulation and kinetics of the actin–myosin–ATP interaction, *Annu. Rev. Biochem.* **49:**921–956.

Agranoff, B. W., Murthy, P., and Seguin, E. B., 1983, Thrombin-induced phosphodiesteratic cleavage of phosphatidylinositol bisphosphate in human platelets, *J. Biol. Chem.* **258:**2076–2078.

Ahmad, Z., DePaoli-Roach, A. A., and Roach, P. J., 1982, Purification and characterization of a rabbit liver calmodulin-dependent protein kinase able to phosphorylate glycogen synthase, *J. Biol. Chem.* **257:**8348–8355.

Ahmad, Z., Lee, F. T., DePaoli-Roach, A., and Roach, P. J., 1984, Phosphorylation of glycogen synthase by the Ca^{2+}- and phospholipid-activated protein kinase (protein kinase C), *J. Biol. Chem.* **259:**8743–8747.

Akhtar, R. A., and Abdel-Latif, A. A., 1984, Carbachol causes rapid phosphodiesteratic cleavage of phosphatidylinositol 4,5-bisphosphate and accumulation of inositol phosphates in rabbit iris smooth muscle; prazosin inhibits noradrenaline- and ionophore A23187–stimulated accumulation of inositol phosphates, *Biochem. J.* **224:**291–300.

Albert, P. R., and Tashjian, A. H., Jr., 1985, Dual actions of phorbol esters on cytosolic free Ca^{2+} concentrations and reconstitution with ionomycin of acute thyrotropin-releasing hormone responses, *J. Biol. Chem.* **260:**8746–8759.

Althaus-Salzmann, M. Carafoli, E., and Jakob, A, 1980, Ca^{2+}, K^+ redistributions and α-adrenergic activation of glycogenolysis in perfused rat livers, *Eur. J. Biochem.* **106:**241–248.

Ambler, S. K., Brown, R. D., and Taylor, P., 1984, The relationship between phosphoinositol metabolism and mobilization of intracelluar calcium elicited by alpha$_1$-adrenergic receptor stimulation in BC3H-1 muscle cells, *Mol. Pharmacol.* **26:**405–413.

Amitai, G., Brown, R. D., and Taylor, P., 1984, The relationship between α_1-adrenergic receptor occupation and the mobilizatiron of intracellular calcium, *J. Biol. Chem.* **259:**12519–12527.

Assimacopoulos-Jeannet, F. D., Blackmore, P. F., and Exton, J. H., 1977, Studies on α-adrenergic activatiron of hepatic glucose output: Studies on role of calcium in α-adrenergic activation of phosphorylase, *J. Biol. Chem.* **252:**2662–2669.

Assimacopoulos-Jeannet, F. D., Blackmore, P. F., and Exton, J. H., 1982, Studies on the interaction between glucagon and α-adrenergic agonists in the control of hepatic glucose output, *J. Biol. Chem.* **257:**3759–3765.

Assimacopoulos-Jeannet, F., McCormack, J. G., and Jeanrenaud, B., 1983, Effect of phenylephrine on pyruvate dehydrogenase activity in rat hepatocytes and its interaction with insulin and glucagon, *FEBS Lett.* **159:**83–88.

Assimacopoulos-Jeannet, F., McCormack, J. G., and Jeanrenaud, B., 1986, Vasopressin and/or glucagon rapidly increases mitochondrial calcium and oxidative enzyme activities in the perfused rat liver, *J. Biol. Chem.* **261:**8799–8804.

Aub, D. L., and Putney, J. W., Jr., 1984, Metabolism of inositol phosphates in parotid cells: Implications for the pathway of the phosphoinositide effect and for the possible messenger role of inositol trisphosphate, *Life Sci.* **34:**1347–1355.

Aub, D. L., and Putney, J. W., Jr., 1985, Properties of receptor-controlled inositol trisphosphate formation in parotid acinar cells, *Biochem. J.* **225**:263–266.

Aub, D. L., McKinney, J. S., and Putney, J. W., Jr., 1982, Nature of the receptor-regulated calcium pool in the rat parotid gland, *J. Physiol.* **331**:557–565.

Authi, K. S., and Crawford, N., 1985, Inositol 1,4,5-trisphosphate-induced release of sequestered Ca^{2+} from highly purified human platelet intracellular membranes, *Biochem. J.* **230**:247–253.

Authi, K. S., Lagarde, M., and Crawford, N., 1985, Diacylglycerol lipase activity in human platelet intracellular and surface membranes *FEBS Lett.* **180**:95–101.

Babcock, D. F., Chen, J.-L. J., Yip, B. P., and Lardy, H. A., 1979, Evidence for mitochondrial localization of the hormone-responsive pool of Ca^{2+} in isolated hepatocytes, *J. Biol. Chem.* **254**:8117–8120.

Balaban, R. S., and Blum, J. J., 1982, Hormone-induced changes in NADH fluorescence and O2 consumption of rat hepatocytes, *Am. J. Physiol.* **242**:C172–C177.

Baldassare, J. J., and Fisher, G. J., 1985, Regulation of membrane associated and cytosolic phospholipase C activities in human platelets by guanosine triphosphate, *J. Biol. Chem.* **261**:11942–11944.

Ballester, R., and Rosen, O. M., 1985, Fate of immuno-precipitable protein kinase C in GH3 cells treated with phorbol 12-myristate 13-acetate, *J. Biol. Chem.* **260**:15194–15199.

Banschbach, M. W., Geison, R. L., and Hokin-Neaverson, M., 1981, Effects of cholinergic stimulation on levels and fatty acid composition of diacylglycerols in mouse pancreas, *Biochim. Biophys. Acta* **663**:34–45.

Baraban, J. M., Gould, R. J., Peroutka, S. J., and Snyder, S. H., 1985a, Phorbol ester effects on neurotransmission: Interaction with neurotransmitters and calcium in smooth muscle, *Proc. Natl. Acad. Sci. USA* **82**:604–607.

Baraban, J. M., Snyder, S. H., and Alger, B. E., 1985b, Protein kinase C regulates ionic conductance in hippocampal pyramidal neurons: Electrophysiological effects of phorbol esters, *Proc. Natl. Acad. Sci. USA* **82**:2538–2542.

Barritt, G. J., Parker, J. C., and Wadsworth, J. C., 1981, A kinetic analysis of effects of adrenaline on calcium distribution in isolated rat liver parenchymal cells, *J. Physiol.* **312**:29–55.

Batty, I. R., Nahorski, S. R., and Irvine, R. F., 1985, Rapid formation of inositol (1,3,4,5) tetrakisphosphate following muscarinic receptor stimulation of rat cerebral corticol slices, *Biochem. J.* **232**:211–215.

Baudiere, B., Guillon, G., Bali, J.-P., and Jard, S., 1986, Muscarinic stimulation of inositol phosphate accumulation and acid secretion in gastric fundic mucosal cells, *FEBS Lett.* **198**:321–325.

Baukal, A. J., Guillemette, G., Rubin, R., Spat, A., and Catt, K. J., 1985, Binding sites for inositol trisphosphate in the bovine adrenal cortex, *Biochem. Biophys. Res. Commun.* **133**:532–538.

Beguinot, L., Hanover, J. A., Ito, S., Richert, N. D., Willingham, M. C., and Pastan, I., 1985, Phorbol esters induce transient internalization without degradation of unoccupied epidermal growth factor receptors, *Proc. Natl. Acad. Sci. USA* **82**:2774–2778.

Bell, J. D., Buxton, I. L. O., and Brunton, L. L., 1985, Enhancement of adenylate cyclase activity in S49 lymphoma cells by phorbol esters, *J. Biol. Chem.* **260**:2625–2628.

Bennett, M. K., Erondu, N. E., and Kennedy, M. B., 1983, Purification and characterization of a calmodulin-dependent protein kinase that is highly concentrated in brain, *J. Biol. Chem.* **258**:12735–12744.

Berkowitz, S. A., and Wolff, J., 1981, Intrinsic calcium sensitivity of tubulin polymerization: The contributions of temperature, tubulin concentration, and associated proteins, *J. Biol. Chem.* **256**:11216–11223.

Berridge, M. J., 1983, Rapid accumulation of inositol trisphosphate reveals that agonists hydrolyse polyphosphoinositides instead of phosphatidylinositol. *Biochem. J.* **212**:849–858.

Berridge, M. J., 1984, Inositol trisphosphate and diacylglycerol as second messengers, *Biochem. J.* **220**:345–360.

Berridge, M. J., Dawson, R. M. C., Downes, C. P., Heslop, J. P., and Irvine, R. F., 1983, Changes in the levels of inositol phosphates after agonist-dependent hydrolysis of membrane phosphoinositides, *Biochem. J.* **212:**473–482.

Berridge, M. J., Heslop, J. P., Irvine, R. F., and Brown, K. D., 1984, Inositol trisphosphate formation and calcium mobilization in Swiss 3T3 cells in response to platelet-derived growth factor, *Biochem. J.* **222:**195–201.

Berthelson, S., and Pettinger, W. A., 1977, A functional basis for classification of α-adrenergic receptors, *Life Sci.* **21:**595–606.

Berthon, B., Poggioli, J., Capiod, T., and Claret, M., 1981, Effect of the α-agonist noradrenaline on total and $^{45}Ca^{2+}$ movements in mitochondria of rat liver cells, *Biochem. J.* **200:**177–180.

Berthon, B., Binet, A., Mauger, J. P., and Claret, M., 1984, Cytosolic free Ca^2 in isolated rat hepatocytes as measured by quin2, *FEBS Lett.* **167:**19–24.

Besterman, J. M., and Cuatrecasas, P., 1984, Phorbol esters rapidly stimulate amiloride-sensitive Na^+/H^+ exchange in a human leukemic cell line, *J. Cell Biol.* **99:**340–343.

Biden, T. J., and Wollheim, C. B., 1985, Ca^{2+} regulates the inositol tris/tetrakisphosphate pathway in intact and broken preparations of insulin-secreting R1Nm5F cells, *J. Biol. Chem.* **261:**11931–11934.

Biden, T. J., Prentki, M., Irvine, R. F., Berridge, M. J., and Wollheim, C. B., 1984, Inositol 1,4,5-trisphosphate mobilizes intracellular Ca^{2+} from permeabilized insulin-secreting cells, *Biochem. J.* **223:**467–473.

Biden, T. J., Wollheim, C. B., and Schlegel, W., 1986, Inositol 1,4,5-trisphosphate and intracellular Ca^{2+} homeostasis in clonal pituitary cells, *J. Biol. Cehm.* **261:**7223–7229.

Billah, M. M., and Lapetina, E. G., 1982, Rapid decrease of phosphatidylinositol 4,5-bisphosphate in thrombin-stimulated platelets, *J. Biol. Chem.* **257:**12705–12708.

Billah, M. M., and Michell, R. H., 1979, Phosphatidylinositol metabolism in rat hepatocytes stimulated by glycogenolytic hormones, *Biochem. J.* **182:**661–668.

Binet, A., Berthon, B., and Claret, M., 1985, Hormone-induced increase in free cytosolic calcium and glycogen phosphorylase activation in rat hepatocytes incubated in normal and low-calcium media, *Biochem. J.* **228:**565–574.

Blackmore, P. F., and Exton, J. H., 1985, Mechanisms involved in the actions of calcium dependent hormones, in : *Biochemical Actions of Hormones*, Vol. 12 (G. Litwak, ed.), Academic Press, New York, pp. 215–235.

Blackmore, P. F., and Exton, J. H., 1985, Studies on the hepatic calcium-mobilizing activity of aluminum fluoride and glucagon. Modulation by cAMP and phorbol myristate acetate, *J. Biol. Chem.* **261:**11056–11063.

Blackmore, P. F., Brumley, F. T., Marks, J. L., and Exton, J. H., 1978, Studies on α-adrenergic activation of hepatic glucose output: Relationship between α-adrenergic stimulation of calcium efflux and activation of phosphorylase in isolated rat liver parenchymal cells, *J. Biol. Chem.* **253:**4851–4858.

Blackmore, P. F., Dehaye, J.-P., Exton, J. H., 1979, Studies on α-adrenergic activation of hepatic glucose output: The role of mitochondrial calcium release in α-adrenergic activation of phosphorylase in perfused rat liver, *J. Biol. Chem.* **254:**6945–6950.

Blackmore, P. F., Hughes, B. P., Shuman, E. A., and Exton, J. H., 1982, α-adrenergic activation of phosphorylase in liver cells involves mobilization of intracellular calcium without influx of extracellular calcium, *J. Biol. Chem.* **257:**190–197.

Blackmore, P. F., Hughes, B. P., Charest, R., Shuman, E. A., IV, and Exton, J. H., 1983a, Time course of α₁-adrenergic and vasopressin actions on phosphorylase activation, calcium efflux, pyridine nucleotide reduction, and respiration in hepatocytes, *J. Biol. Chem.* **258:**10488–10494.

Blackmore, P. F., Hughes, B. P., and Exton, J. H., 1983b, Time course of α-adrenergic and vasopressin effects in isolated hepatocytes, in *Isolation, Characterization, and Use of Hepatocytes* (R. A. Harris and N. W. Cornell, eds.), Elsevier, New York, pp. 433–438.

Blackmore, P. F., Bocckino, S. B., Waynick, L. E., and Exton, J. H., 1985, Role of a guanine nucleotide–binding regulatory protein in the hydrolysis of hepatocyte phosphatidylinositol 4,5-bisphosphate by calcium-mobilizing hormones and the control of cell calcium. Studies utilizing aluminum fluoride, *J. Biol. Chem.* **260:**14477–14483.

Blackmore, P. F., Strickland, W. G., Bocckino, S. B., and Exton, J. H., 1986, Mechanism of hepatic glycogen synthase inactivation induced by Ca^{2+}-mobilizing hormones, *Biochem. J.* **237:**235–242.

Blair, J. B., James, M. E., and Foster, J. L., 1979, Adrenergic control of glucose output and adenosine 3′ : 5′-monophosphate levels in hepatocytes from juvenile and adult rats, *J. Biol. Chem.* **254:**7579–7584.

Bocckino, S. B., Blackmore, P. F., and Exton, J. H., 1985, Stimulation of 1,2-diacylglycerol accumulation in hepatocytes by vasopressin, epinephrine, and angiotensin II, *J. Biol. Chem.* **260:**14201–14207.

Bojanic, D., and Fain, J. N., 1986, Guanine nucleotide regulation of [³H]vasopressin binding to liver plasma membranes and solubilized receptors. Evidence for the involvement of a guanine nucleotide regulatory protein, *Biochem. J.* **240:**361–365.

Bokoch, G. M., Katada, T., Northup, J. K., Ui, M., and Gilman, A. G., 1984, Purification and properties of the inhibitory guanine nucleotide–binding regulatory component of adenylate cyclase, *J. Biol. Chem.* **259:**3560–3567.

Bolton, T. B., 1979, Mechanisms of action of transmitters and other substances on smooth muscle, *Physiol. Rev.* **59:**606–718.

Bond, M., Kitazawa, T., Somlyo, A. P., and Somlyo, A. V., 1984, Release and recycling of calcium by the sarcoplasmic reticulum in guinea-pig portal vein smooth muscle, *J. Physiol.* **355:**677–695.

Boni, L. T., and Rando, R. R., 1985, The nature of protein kinase C activation by physically defined phospholipid vesicles and diacylglycerols, *J. Biol. Chem.* **260:**10819–10825.

Bouscarel, B., and Exton, J. H., 1986, Regulation of hepatic glycogen phosphorylase and glycogen synthase by calcium and diacylglycerol, *Biochim. Biophys. Acta* **888:**126–134.

Boyer, J. L., Garcia, A., Posadas, C., and Garcia-Sainz, J. A., 1984, Differential effect of pertussis toxin on the affinity state for agonists of renal α_1- and α_2-adrenoceptors, *J. Biol. Chem.* **259:**8076–8079.

Brass, L. F., and Joseph, S. K., 1985, A role for inositol triphosphate in intracellular Ca^{2+} mobilization and granule secretion in platelets, *J. Biol. Chem.* **260:**15172–15179.

Breant, B., Keppens, S., and DeWulf, H., 1981, Desensitization of the cAMP-independent glycogenolytic response in rat hepatocytes, *Arch. Int. Physiol. Biochim.* **89:**B90–B91.

Brown, J. E., and Rubin, L. J., 1984, A direct demonstration that inositol trisphosphate induces an increase in intracellular calcium in Limulus photoreceptors, *Biochem. Biophys. Res. Commun.* **125:**1137–1142.

Brown, J. E., Rubin, L. J., Ghalayini, A. J., Tarver, A. P., Irvine, R. F., Berridge, M. J., and Anderson, R. E., 1984, *myo*-inositol polyphosphate may be a messenger for visual excitation in Limulus photoreceptors, *Nature* **311:**160–163.

Brown, R. D., Berger, K. D., and Taylor, P., 1984, α_1-Adrenergic receptor activation mobilizes cellular Ca^{2+} in a muscle cell line, *J. Biol. Chem.* **260:**7554–7562.

Brown, R. D., Prendiville, P., and Cann, C., 1986, α_1-Adrenergic and H1-histamine receptor control of intracellular Ca^{2+} in a muscle cell line: The influence of prior agonist exposure on receptor responsiveness, *Mol. Pharmacol.* **29:**531–539.

Buckley, J. T., and Hawthorne, J. N., 1972, Erythrocyte membrane polyphosphoinositide metabolism and the regulation of calcium binding, *J. Biol. Chem.* **247:**7218–7223.

Burch, R. M., Luini, A., Mais, D. E., Corda, D., Vanderhoek, J. Y., Kohn, L. D., and Axelrod, J., 1986a, α_1-Adrenergic stimulation of arachidonic acid release and metabolism in a rat thyroid cell line, *J. Biol. Chem.* **261:**11236–11241.

Burch, R. M., Luini, A., and Axelrod, J., 1986b, Phospholipase A_2 and phospholipase C are activted by distinct GTP-binding proteins in resposne to α_1-adrenergic stimulation in FRTL5 thyroid cells, *Proc. Natl. Acad. Sci. USA* **33**:7201–7205.

Burgess, G. M., Godfrey, P. P., McKinney, J. S., Berridge, M. J., Irvine, R. F., and Putney, J. W., Jr., 1984a, The second messenger linking receptor activation to internal Ca release in liver, *Nature* **309**:63–66.

Burgess, G. M., Irvine, R. F., Berridge, M. J., McKinney, J. S., and Putney, J. W., Jr., 1984b, Actions of inositol phosphates on Ca^{2+} pools in guinea-pig hepatocytes, *Biochem. J.* **224**:741–746.

Burgess, G. M., McKinney, J. S., Irvine, R. F., Berridge, M. J., Hoyle, P. C., and Putney, J. W., Jr., 1984c, Inositol 1,4,5-trisphosphate may be a signal for f-Met-Leu-Phe- induced intracellar calcium mobilization in human leucocytes (HL-60 cells), *FEBS Lett.* **176**: 193–196.

Burke, B. E., and Lorenzo, R. J., 1981, Ca^{2+}- and calmodulin-stimulated endogenous phosphorylation of neurotubulin, *Proc. Natl. Acad. Sci. USA* **78**:991–995.

Busa, W. B., Ferguson, J. E., Joseph, S. K., Williamson, J. R., and Nuccitelli, R., 1985, Activation of frog (*Xenopus laevis*) eggs by inositol trisphosphate. 1. Characterization of Ca^{2+} release from intracellular stores, *J. Cell. Biol.* **101**:677–682.

Buxton, I. L. O., and Brunton, L. L., 1985, Action of the cardiac α_1-adrenergic receptor: Activation of cyclic AMP degradation, *J. Biol. Chem.* **260**:6733–6737.

Buxton, I. L. O., and Brunton, L. L., 1986, α-Adrenergic receptors on rat ventricular myocyte: Characteristics and linkage to cAMP metabolism, *Am. J. Physiol.* **251**:H307–H313.

Buxton, D., Barron, L. L., and Olson, M. S., 1982, The effects of α-adrenergic agonists on the regulation of the branched chain α-ketoacid oxidation in the perfused rat liver, *J. Biol. Chem.* **257**:14318–14323.

Bylund, D. B., and U'Prichard, D. C., 1983, Characterization of α_1- and α_2-adrenergic receptors, *Int. Rev. Neurobiol.* **24**:343–431.

Campbell, K. P., and MacLennan, D. H., 1982, A calmodulin-dependent protein kinase system from skeletal muscle sarcoplasmic reticulum: Phosphorylation of a 60,000 dalton protein, *J. Biol. Chem.* **257**:1238–1246.

Canessa de Scarnatti, O., and Lapetina, E., 1974, Adrenergic stimulation of phosphatidylinositol labelling in rat vas deferens, *Biochim. Biophys. Acta* **360**:298–305.

Capponi, A. M., Lew, P. D., and Vallotton, M. B., 1985, Cytosolic free calcium levels in monolayers of cultured rat aortic smooth muscle cells, *J. Biol. Chem.* **260**:7836–7842.

Carafoli, E., 1984, Calmodulin-sensitive calcium-pumping ATPase of plasma membranes: Isolation, reconstitution, and regulation, *Fed. Proc.* **43**:3005–3010.

Castagna, M., Takai, Y., Kaibuchi, K., Sano, K., Kikkawa, U., and Nishizuka, Y., 1982, Direct activation of calcium-activated, phospholipid-dependent protein kinase by tumor-promoting phorbol esters, *J. Biol. Chem.* **257**:7847–7851.

Casteels, R., and Droogmans, G., 1981, Exchange characteristics of the noradrenaline-sensitive calcium store in vascular smooth muscle cells of rabbit ear artery, *J. Physiol.* **317**:263–279.

Casteels, R., and Raeymaekers, L., 1979, The action of acetylcholine and catecholamines on an intracellular calcium store in smooth muscle cells of guinea-pig taenia coli, *J. Physiol.* **294**:51–68.

Chacko, S., Conti, M. A., and Adelstein, R. S., 1977, Effect of phosphorylation of smooth muscle myosin on actin activation and Ca^{2+} regulation, *Proc. Natl. Acad. Sci. USA* **74**:129–133.

Chan, K.-F., and Graves, D. J., 1984, Molecular properties of phosphorylase kinase, in: *Calcium and Cell Function*, Vol. 5 (W. Y. Cheung, ed.), Academic Press, New York, pp. 1–31.

Chan, T. M., and Exton, J. H., 1977, α-Adrenergic-mediated accumulation of adenosine $3':5'$-monophosphate in calcium-depleted hepatocytes, *J. Biol. Chem.* **252**:8645–8651.

Chan, T. M., and Exton, J. H., 1978, Studies on α-adrenergic activatin of hepatic glucose output: Studies on α-adrenergic inhibition of hepatic pyruvate kinase and activation of gluconeogenesis, *J. Biol. Chem.* **253**:6393–6400.

Chan, T. M., Blackmore, P. F., Steiner, K. E., and Exton, J. H., 1979, Effects of adrenalectomy on hormone action on hepatic glucose metabolism, *J. Biol. Chem.* **254**:2428–2433.

Charest, R., Blackmore, P. F., Berthon, B., and Exton, J. H., 1983, Changes in free cytosolic Ca^{2+} in hepatocytes following α_1-adrenergic stimulation, *J. Biol. Chem.* **258**:8769–8773, 1983.

Charest, R., Prpic, V., Exton, J. H., and Blackmore, P. F., 1985, Stimulation of inositol tris-phosphate formation in hepatocytes by vasopressin, epinephrine, and angiotensin II and its relationship to changes in cytosolic free Ca^2, *Biochem. J.* **227**:79–90.

Chen, J.-L. J., Babcock, D. F., and Lardy, H. A., 1978, Norepinephrine, vasopressin, glucagon, and A23187 induce efflux of calcium from an exchangeable pool in isolted rat hepatocytes, *Proc. Natl. Acad. Sci. USA* **75**:2234–2238.

Cheung, W. Y., 1980, Calmodulin plays a pivotal role in cellular regulation, *Science* **207**:19–27.

Chueh, S.-H., and Gill, D. L., 1986, Inositol 1,4,5-trisphosphate and guanine nucleotides activate calcium release from endoplasmic reticulum via distinct mechanisms, *J. Biol. Chem.* **261**:13883–13886.

Ciapa, B., and Whitaker, M., 1986, Two phases of inositol polyphosphate and diacylglycerol production at fertilization, *FEBS Lett.* **195**:347–351.

Clapper, D. L., and Lee, H. C., 1985, Inositol trisphosphate induces calcium release from non-mitochondrial stores in sea urchin egg homogenates, *J. Biol. Chem.* **260**:13947–13954.

Cochet, C., Gill, G. N., Meisenhelder, J., Coper, J. A., and Hunter, T., 1984, C-kinase phos-phorylates the epidermal growth factor receptor and reduces its epidermal growth fac-tor–stimulated tyrosine protein kinase activity, *J. Biol. Chem.* **259**:2553–2558.

Cockcroft, S., and Gomperts, B. D., 1985, Role of guanine nucleotide binding protein in the activation of polyphosphoinositide phosphodiesterase, *Nature* **314**:534–536.

Cohen, P., 1980, The role of calcium ions, calmodulin, and troponin in regulation of phosphorylase kinase from rabbit skeletal muscle, *Eur. J. Biochem.* **111**:563–574.

Cohen, P., Burchell, A., Foulkes, J. G., and Cohen, P. T. W., 1978, Identification of the Ca^{2+}-dependent modulator protein as the fourth subunit of rabbit skeletal muscle phosphorylase kinase, *FEBS. Lett.* **92**:287–293.

Colucci, W. S., and Alexander, R. W., 1986, Norepinephrine-induced alteration in the coupling of α_1-adrenergic receptor occupancy to calcium efflux in rabbit aortic smooth muscle cells, *Proc. Natl. Acad. Sci USA* **83**:1743–1746.

Colucci, W. S., Gimbrone, M. A., Jr., and Alexander, R. W., 1981, Regulation of postsynaptic α-adrenergic receptor in rat mesentery artery: Effects of chemical sympathectomy and epi-nephrine treatment, *Circ. Res.* **48**:104–111.

Connolly, E., Nanberg, E., and Nedergaard, J., 1984, Na^+-dependent, α-adrenergic mobilization of intracellular (mitochondrial) Ca^{2+} in brown adipocytes, *Eur. J. Biochem.* **141**:187–193.

Connolly, T. M., Bross, T. E., and Majerus, P. W., 1985, Isolation of a phosphomonoesterase from human platelets that specifically hydrolyzes the 5-phosphate of inositol 1,4,5-trisphosphate, *J. Biol. Chem.* **260**:7868–7874.

Connolly, T. M., Lawing, W. J., Jr., and Majerus, P. W., 1986, Protein kinase C phosphorylates human platelet inositol trisphosphate 5′-phosphomonoesterase increasing the phosphatase ac-tivity, *Cell* **49**:951–958.

Cooper, R. H., Kobayashi, K., and Williamson, J. R., 1984, Phosphorylation of a 16-kDa protein by diacylglycerol-activated protein kinase C *in vitro* and by vasopressin in intact hepatocytes, *FEBS Lett.* **166**:125–130.

Cooper, R. H., Coll, K. E., and Williamson, J. R., 1985, Differential effects of phorbol ester on phenylephrine and vasopressin-induced Ca^{2+} mobilization in isolated hepatocytes, *J. Biol. Chem.* **260**:3281–3288.

Corvera, S., and Garcia-Sainz, J. A., 1984, Phorbol esters inhibit alpha_1 adrenergic stimulation of glycogenolysis in isolated rat hepatocytes, *Biochem. Biophys. Res. Commun.* **119**:1128–1133.

Corvera, S., Hernandez-Sotomayor, S. M. T., and Garcia-Sainz, J. A., 1984, Modulation by thyroid status of cyclic AMP-dependent and Ca^{2+}-dependent mechanisms of hormone action in rat liver cells, *Biochim. Biophys. Acta* **803**:95–105.

Coussen, F., Haiech, J., d'Alayer, J., and Monneron, A., 1985, Identification of the catalytic subunit of brain adenylate cyclase: A calmodulin binding protein of 135kDa, *Proc. Natl. Acad. Sci. USA* **82**:6736–6740.

Creba, J. A., Downes, C. P. K., Hawkins, P. T., Brewster, G., Michell, R. H., and Kirk, C. J., 1983, Rapid breakdown of phosphatidylinositol 4-phosphate and phosphatidylinositol 4,5-bis-phosphate in rat hepatocytes stimulated by vasopressin and other Ca^{2+}-mobilizing hormones, *Biochem. J.* **212**:733–747.

Dabrowska, R., Sherry, J. M. F., Aromatorio, D. K., and Hartshorne, D. J., 1978, Modulator protein as a component of the myosin light chain kinase from chicken gizzard, *Biochemistry* **17**:253–258.

Dale, M. M., and Penfield, A., 1984, Synergism between phorbol ester and A23187 in superoxide productin by neutrophils, *FEBS. Lett.* **175**:170–172.

Daniel, J. L., Molish, I. R., and Holmsen, H., 1981, Myosin phosphorylation in intact platelets, *J. Biol. Chem.* **256**:7510–7514.

Daniel, J. L., Molish, I. R., Rigmaiden, M., and Stewart, G., 1984, Evidence for a role of myosin phosphorylatin in the initiation of the placelet shape change resposne, *J. Biol. Chem.* **259**:9826–9831.

Danthuluri, N. R., and Deth, R. C., 1984, Phorbol ester–induced contraction of arterial smooth muscle and inhibition of α-adrenergic response, *Biochem. Biophys. Res. Commun.* **125**:1103–1109.

Davis, R. J., and Czech, M. P., 1984, Tumor-promoting phorbol diesters mediate phosphorylation of the epidermal growth factor receptor, *J. Biol. Chem.* **259**:8545–8549.

Davis, B. A., Schwartz, A., Samaha, F. J., and Kranias, E. G., 1983, Regulation of cardiac sarcoplasmic reticulum calcium transport by calcium–calmodulin-dependent phosphorylation, *J. Biol. Chem.* **258**:13587–13591.

Davis, R. J., Ganong, B. R., Bell, R. M., and Czech, M. P., 1985, Structural requirements for diacylglycerols to mimic tumor-promoting phorbol diester action on the epidermal growth factor receptor, *J. Biol. Chem.* **260**:5315–5322.

Dawson, A. P., 1985, GTP enhances inositol trisphosphate–stimulated Ca^{2+} release from rat liver microsomes, *FEBS Lett.* **185**:147–150.

Dawson, A. P., and Irvine, R. F., 1984, Inositol(1,4,5)trisphosphate–promoted Ca^{2+} release from microsomal fractions of rat liver, *Biochem. Biophys. Res. Commun.* **120**:858–864.

Dehaye, J.-P., Hughes, B. P., Blackmore, P. F., and Exton, J. H., 1981, Insulin inhibition of α-adrenergic actions in liver, *Biochem. J.* **194**:949–956.

Delbeke, D., Kojima, I., Dannies, P. S., and Rasmussen, H., 1984, Synergistic stimulation of prolactin release by phorbol ester, A23187, and forskolin, *Biochem. Biophys. Res. Commun.* **123**:735–741.

Delfert, D. M., Hill, S., Pershadsingh, H. A., Sherman, W. R., and McDonald, J. M., 1986, *myo*Inositol 1,4,5-trisphosphate mobilizes Ca^{2+} from isolated adipocyte endoplamsic reticulum but not from plasma membranes, *Biochem. J.* **236**:37–44.

DenHertog, A., 1981, Calcium and the α-action of catecholamines on guinea-pig taenia caeci, *J. Physiol.* **316**:109–125.

Denton, R. M., and McCormack, J. G., 1981, Calcium ions, hormones, and mitochondrial metabolism, *Clin. Sci.* **61**:135–140.

Denton, R. M., Randle, P. J., and Martin, B. R., 1972, Stimulation by calcium ions of pyruvate dehydrogenase phosphate phosphatase, *Biochem. J.* **128**:161–163.

Denton, R. M., McCormack, J. G., and Edgell, N. J., 1980, Role of calcium ions in the regulation of intramitochondrial metabolism: Effects of Na^+, Mg^{2+}, and Ruthenium Red on the Ca^{2+}-

stimulated oxidatin of oxoglutarate and on pyruvate dehydrogenase activity in intact rat heart mitochondria, *Biochem. J.* **190**:107–117.

DeRiemer, S. A., Strong, J. A., Albert, K. A., Greengaard, P., and Kaczmarek, L. K., 1985, Enhancement of calcium current in Aplysia neurones by phorbol ester and protein kinase C, *Nature* **313**:313–316.

Deth, R., and Casteels, R., 1977, A study of releasable Ca fractions in smooth muscle cells of rabbit aorta, *J. Gen. Physiol.* **69**:401–416.

Deth, R., and van Breemen, C., 1974, Relative contributions of Ca^{2+} influx and cellular Ca^{2+} release during drug induced activation of the rabbit aorta, *Pfluegers Arch.* **348**:13–22.

DeVirgilio, F., Lew, D. P., and Pozzan, T., 1984, Protein kinase C activation of physiological processes in human neutrophils at vanishingly small cytosolic Ca^{2+} levels, *Nature* **310**:691–693.

DeWitt, L. M., and Putney, J. W., 1983, α-Adrenergic stimulation of potassium efflux in guinea pig hepatocytes may involve calcium influx and calcium release, *J. Physiol.* **346**:395–407.

Doskeland, A. P., Schworer, C. M., Doskeland, S. O., Chrisman, T. D., Soderling, T. R., Corbin, J. D., and Flatmark, T., 1984, Some aspects of phosphorylation of phenylalanine 4-monooxygenase by a calcium-dependent and calmodulin-dependent protein kinase, *Eur. J. Biochem.* **145**:31–37.

Dougherty, R. W., Godfrey, P. P., Hoyle, P. C., Putney, J. W., Jr., and Freer, R. J., 1984, Secretagogue-induced phosphoinositide metabolism in human leucocytes, *Biochem. J.* **222**:307–314.

Downes, C. P., and Wuseman, M. M., 1983, Breakdown of polyphosphoinositides and not phosphatidylinositol accounts for muscarinic agonist-stimulated inositol phospholipid metabolism in rat parotid glands, *Biochem. J.* **216**:633–640.

Downes, C. P., Mussat, M. C., and Michell, R. H., 1982, The inositol triphosphate phosphomonoesterase of the human erythrocyte membrane, *Biochem. J.* **203**:169–177.

Downes, C. P., Hawkins, P. T., and Irvine, R. F., 1986, Inositol 1,3,4,5-tetrakisphosphate is the probable precursor of inositol 1,3,4-trisphosphate in agonist-stimulated parotid gland, *Biochem. J.* **238**:501–506.

Driska, S. P., Aksoy, M. O., and Murphy, R. A., 1981, Myoson light chain phosphorylation associated with contraction in arterial smooth muscle, *Am. J. Physiol.* **240**:C222–C233.

Drust, D. W., and Martin, T. F. J., 1985, Protein kinase C translocates from cytosol to membrane upon hormone activation: Effects of thyrotropin-releasing hormone in GH3 cells, *Biochem. Biophys. Res. Commun.* **128**:531–537.

Ebeling, J. G., Vandenbark, G. R., Kuhn, L. J., Ganong, B. R., Bell, R. M., and Niedel, J. E., 1985, Diacylglycerols mimic phorbol diester induction of leukemic cell differentiation, *Proc. Natl. Acad. Sci. USA* **82**:815–819.

El-Refai, M. F., and Chan, T. M., 1986, Effects of adrenalectomy on binding to and actions of adrenergic receptors, *Biochem. J.* **237**:527–531.

El-Refai, M. F., Blackmore, P. F., and Exton, J. H., 1979, Evidence for two α-adrenergic binding sites in liver plasma membranes. Studies with [³H]epinephrine and [³H]dihydroergocryptine, *J. Biol. Chem.* **254**:4375–4386.

Enjalbert, A., Sladeczek, F., Guilon, G., Bertrand, P., Shu, C., Epelbaum, J., Garcia-Sainz, A., Jard, S., Lombard, C., Kordon, C., and Bockaert, J., 1986, Angiotensin II and dopamine modulate both cAMP and inositol phosphate production in anterior pituitary cells, *J. Biol. Chem.* **261**:4071–4075.

Evans, T., Martin, M. W., Hughes, A. R., and Harden, T. K., 1985, Guanine nucleotide–sensitive, high affinity binding of carbachol to muscarinic cholinergic receptors of 1321N1 astrocytoma cells is insensitive to pertussis toxin, *Mol. Pharmacol.* **27**:32–37.

Exton, J. H., 1980, Mechanisms involved in α-adrenergic phenomena: Role of calcium ions in actions of catecholamines in liver and other tissues, *Am. J. Physiol.* **238**:E3–E12.

Exton, J. H., 1981, Molecular mechanisms involved in α-adrenergic responses, *Mol. Cell. Endocrinol.* **23**:233–264.

Exton, J. H., 1985, Mechanisms involved in α-adrenergic phenomena, *Am. J. Physiol.* **248**:E633–E647.

Fain, J. N., Li, S. Y., Litosch, I., and Wallace, M., 1984, Synergistic activation of rat hepatocyte glycogen phosphorylase by A23187 and phorbol ester, *Biochem. Biophys. Res. Commun.* **119**:88–94.

Fein, A., Payne, R., Corson, D. W., Berridge, M. J., and Irvine, R. F., 1984, Photoreceptor excitation and adaptation by inositol 1,4,5-trisphosphate, *Nature* **311**:157–160.

Fisher, M. J., and Pogson, C. I., 1984, Phenylalanine hydroxylase in liver cells: Correlation of glucagon-stimulated enzyme phosphorylation with expressed activity, *Biochem. J.* **219**:79–85.

Fisher, M. J., Santana, M. A., and Pogson, C. I., 1984, Effects of adrenergic agents, vasopressin, and ionophore A23187, on the phosphorylation of, and flux through, phenylalanine hydroxylase in rat liver cells, *Biochem. J.* **219**:87–90.

Fisher, R. A., Tanabe, S., Buxton, D. B., and Olson, M. S., 1985, The effects of α-adrenergic stimulation on the regulation of the pyruvate dehydrogenase complex in the perfused rat liver, *J. Biol. Chem.* **260**:9223–9229.

Fitzgerald, T. J., Uhing, R. J., and Exton, J. H., 1986, Solubilization of the vasopressin receptor from liver plasma membranes. Evidence for a receptor·GTP·binding protein complex, *J. Biol. Chem.* **261**:16871–16877.

Gallo-Payet, N., Guillon, G., Balestre, M. N., and Jard, S., 1986, Vasopressin induces breakdown of membrane phosphoinositides in adrenal glomerulosa and fasciculata cells, *Endocrinology* **119**:1042–1047.

Ganong, B. R., Loomis, C. R., Hannun, Y. A., and Bell, R. M., 1986, Specific and mechanism of protein kinase C activation by *sn*-1,2-diacylglycerols, *Proc. Natl. Acad. Sci. USA* **83**:1184–1188.

Garcia-Sainz, J. A., and Fain, J. N., 1980, Effect of adrenergic amines on phosphatidylinositol labeling and glycogen synthase activity in fat cells from euthyroid and hypothyroid rats, *Mol. Pharmacol.* **28**:72–77.

Garcia-Sainz, J. A., and Hernandez-Sotomayor, S. M. T., 1985, Adrenergic regulation of gluconeogenesis: Possible involvement of two mechanisms of signal transduction in α_1-adrenergic action, *Proc. Natl. Acad. Sci. USA* **82**:6727–6730.

Garcia-Sainz, J. A., Litosch, I., Hoffman, B. B., Lefkowitz, R. J., and Fain, J. N., 1981, Effect of thyroid status on α- and β-catecholamine responsiveness of hamster adipocytes, *Biochim. Biophys. Acta* **678**:334–341.

Garrison, J. C., and Wagner, J. D., 1982, Glucagon and the Ca^{2+}-linked hormones angiotensin II, norepinephrine, and vasopressin stimulate the phosphorylation of distinct substrates in intact hepatocytes, *J. Biol. Chem.* **257**:13135–13143.

Garrison, J. C., Borland, M. K., Florio, V. A., and Twible, D. A., 1979, The role of calcium ion as a mediator of the effects of angiotensin II, catecholamines, and vasopressin on the phosphorylation and activity of enzymes in isolated hepatocytes, *J. Biol. Chem.* **254**:7147–7156.

Garrison, J. C., Johnsen, D. E., and Campanile, C. P., 1984, Evidence for the role of phosphorylase kinase, protein kinase C, and other Ca^{2+}-sensitive protein kinases in the response of hepatocytes to angiotensin II and vasopressin, *J. Biol. Chem.* **259**:3283–3292.

Gershengorn, M. C., Geras, E., Purrello, V. S., and Rebecchi, M. J., 1984, Inositol trisphosphate mediates thyrotropin-releasing hormone mobilization of non-mitochondrial calcium in rat mammotropic pituitary cells, *J. Biol. Chem.* **259**:10675–10681.

Geynet, P., Borsodi, A., Ferry, N., and Hanoune, J., 1980, Proteolysis of rat liver plasma membranes cancels the guanine nucleotide sensitivity of agonist binding to the alpha-receptor, *Biochem. Biophys. Res. Commun.* **97**:947–954.

Gomperts, B. D., 1983, Involvement of guanine nucleotide-binding protein in the gating of Ca^{2+} by receptors, *Nature* **306**:64–66.

Gonzatti-Haces, M. I., and Trauch, J. A., 1986, Ca^{2+}-independent activation of protease-activated kinase II by phospholipids/diolein and comparison with the Ca^{2+}/phospholipid-dependent protein kinase, *J. Biol. Chem.* **261**:15266–15272.

Goodhardt, M., Ferry, N., Geynet, P., and Hanoune, J., 1982, Hepatic α_1-adrenergic receptors show agonist-specific regulation by guanine nucleotides. Loss of nucleotide effect after adrenalectomy, *J. Biol. Chem.* **257**:11577–11583.

Graham, R. M., Hess, H.-J., and Homcy, C. J., 1982, Biophysical characterization of the purified α_1-adrenergic receptor and identification of the hormone binding subunit, *J. Biol. Chem.* **257**:15174–15181.

Greenberg, M. E., and Ziff, E. B., 1984, Stimulation of 3T3 cells induces transcription of the *c-fos* proto-oncogene, *Nature* **311**:433–438.

Griendling, K. K., Rittenhouse, S. E., Brock, T. A., Ekstein, L. S., Gimbrone, M. A., Jr., and Alexander, R. W., 1986, Sustained diacylglycerol formation from inositol phospholipids in angiotension II–stimulated vascular smooth muscle cells, *J. Biol. Chem.* **261**: 5901–5906.

Grunberger, G., and Gorden, P., 1982, Affinity alteration of insulin receptor induced by a phorbol ester, *Am. J. Physiol.* **243**:E319–E324.

Guellaen, G., Yates-Aggerbeck, M., Vauquelin, G., Strosberg, D., and Hanoune, J., 1978, Characterization with [^{3}H]dihydroergocryptine of the α-adrenergic receptor of the hepatic plasma membrane, *J. Biol. Chem.* **253**:1114–1120.

Guellaen, G., Goodhardt, M., Barouki, R., and Hanoune, J., 1982, Subunit structure of rat liver α_1-adrenergic receptor, *Biochem. Pharmacol.* **31**:2817–2820.

Guillon, G., Balestre, M.-N., Mouillac, B., and Devilliers, G., 1986, Activation of membrane phospholipase C by vasopressin. A requirement for guanyl nucleotides, *FEBS Lett.* **196**:155–159.

Haddas, R. A., Landis, C. A., and Putney, J. W., Jr., 1979, Relationship between calcium release and potassium release in rat parotid gland, *J. Physiol.* **291**:457–465.

Hannun, Y. A., and Bell, R. M., 1986, Phorbol ester binding and activation of protein kinase C on Triton X-100 mixed micelles containing phosphatidylserine, *J. Biol. Chem.* **261**: 9341–9347.

Hannun, Y. A., Loomis, C. R., and Bell, R. M., 1985, Activation of protein kinase C by Triton X-100 mixed micelles containing diacylglycerol and phosphatidylserine, *J. Biol. Chem.* **260**:10039–10043.

Hannun, Y. A., Loomis, C. R., and Bell, R. M., 1986, Protein kinase C activation in mixed micelles. Mechanistic implications of phospholipid, diacylglycerol, and calcium interdependencies, *J. Biol. Chem.* **261**:7184–7190.

Hansen, C. A., Mah, S., and Williamson, J. R., 1986, Formation and metabolism of inositol 1,3,4,5-tetrakisphosphate in liver, *J. Biol. Chem.* **261**:8100–8103.

Hansford, R. G., 1981, Effect of micromolar concentrations of free Ca^{2+} ions on pyruvate dehydrogenase interconversion in intact rat heart mitochondria, *Biochem. J.* **194**:721–732.

Hansford, R. G., 1985, Relation between mitochondrial calcium transport and control of energy metabolism, *Rev. Physiol. Biochem. Pharmacol.* **102**:1–72.

Hansford, R. G., and Castro, F., 1981, Effect of micromolar concentrations of free calcium ions on the reduction of heart mitochondrial NAD(P) by 2-oxoglutarate, *Biochem. J.* **198**:525–533.

Harrington, C. A., and Eichberg, J., 1983, Norepinephrine causes α_1-adrenergic receptor-mediated decrease of phosphoinositide in isolated rat liver plasma membranes supplemented with cytosol, *J. Biol. Chem.* **258**:2087–2090.

Haslam, R. J., and Davidson, M. M. L., 1984a, Potentiation by thrombin of the platelets equilibrated with Ca^{2+} buffers. Relationship to protein phosphorylation and diacylglycerol formation, *Biochem J.* **222**:351–361.

Haslam R. J., and Davidson, M. M. L., 1984b, Guanine nucleotides decrease the free $[Ca^{2+}]$ required for secretion of serotonin from permeabilized blood platelets. Evidence of a role for a GTP-binding protein in platelet activation, *FEBS Lett.* **174**:90–95.

Haslam, R. J., and Davidson, M. M. L., 1984c, Receptor-induced diacylglycerol formation in permeabilized platelets; possible role for a GTP-binding protein, *J. Receptor Res.* **4**:605–629.

Haussinger, D., and Sies, H., 1984, Effect of phenylephrine on glutamate and glutamine metaboilsm in isolated perfused rat liver, *Biochem. J.* **221**:651–658.

Haylett, D. G., 1976, Effects of sympathomimetic amines on ^{45}Ca efflux from liver slices, *Brit. J. Pharmacol.* **57**:158–160.

Hems, D. A., McCormack, J. G., and Denton, R. M., 1978, Activation of pyruvate dehydrogenase in the purified rat liver by vasopressin, *Biochem. J.* **176**:627–629.

Henne, V., and Soling, H.-D., 1986, Guanosine 5'-triphosphate releases calcium from rat liver and guinea pig parotid gland endoplasmic reticulum independently of inositol 1,4,5-trisphosphate, *FEBS Lett.* **202**:267–273.

Hepler, J. R., and Harden, T. K., 1986, Guanine nucleotide–dependent pertussis toxin–insensitive stimulation of inositol phosphate formation by carbachol in a membrane preparation from human astrocytoma cells, *Biochem. J.* **239**:141–146.

Hernandez-Sotomayor, S. M. T., and Garcia-Sainz, J. A., 1984, Adrenergic regulation of ureogenesis in hepatocytes from adrenalectomized rats, *FEBS Lett.* **166**:385–388.

Hesketh, T. R., Smith, G. A., Moore, J. P., Taylor, M. V., and Metcalfe, J. C., 1983, Free cytoplasmic calcium concentration and the mitogenic stimulation of lymphocytes, *J. Biol. Chem.* **258**:4876–4882.

Higashida, H., Streaty, R. A., Klee, W., and Nirenberg, M., 1986, Bradykinin-activated transmembrane signals are coupled via N_o or N_i to production of inositol 1,4,5-trisphosphate, a second messenger in NG108-15 neuroblastoma glioma hybrid cells, *Proc. Natl. Acad. Sci. USA* **83**:942–946.

Hinkle, P. M., and Phillips, W. J., 1984, Thyrotropin-releasing hormone stimulates GTP hydrolysis by membranes from GH_4C_1 rat pituitary tumor cells, *Proc. Natl. Acad. Sci. USA* **81**:6183–6187.

Hirata, M., Kukita, M., Sasaguri, T., Suematsu, E., Hashimoto, T., and Koga, T., 1985, Increase in Ca^{2+} permeabilization of intracellular Ca^{2+} store membrane of saponin-treated guinea pig peritoneal macrophages by inositol 1,4,5-trisphosphate, *J. Biochem.* **97**:1575–1582.

Hoffman, B. B.,Mullikin-Kilpatrick, D., and Lefkowitz, R. J., 1980, Heterogeneity of radioligand binding to α-adrenergic receptors, *J. Biol. Chem.* **255**:4645–4652.

Hollingsworth, E. B., and Daly, J. W., 1985, Accumulation of inositol phosphates and cyclic AMP in guinea-pig cerebral cortical preparations. Effects of norepinephrine, histamine, carbamylcholine, and 2-chloroadenosine, *Biochim. Biophys. Acta* **847**:207–216.

Hollingsworth, E. B., Sears, E. B., and Daly, J. W., 1985, An activator of protein kinase C (phorbol-12-myristate-13-acetate) augments 2-chloroadenosine-elicited accumulation of cyclic AMP in guinea pig cerebral cortical particulate preparations, *FEBS Lett.* **184**:339–342.

Holub, B. J., and Kuksis, A., 1978, Metabolism of molecular species of diacylglycerophospholipids, *Adv. Lipid Res.* **16**:1–125.

Huerta-Bahena, J., and Garcia-Sainz, J. A., 1983, Inositol administration restores the sensitivity of liver cells formed during liver regeneration to alpha$_1$-adrenergic amines, vasopressin, and angiotensin II, *Biochim. Biophys. Acta* **763**:125–128.

Huerta-Bahena, J., and Garcia-Sainz, J. A., 1984, Effect of inositol and tri-iodothyronine on the hormonal responsiveness of hepatocytesobtained from partially hepatectomized rats, *Biochem. J.* **223**:925–928.

Huerta-Bahena, J., Vallalobos-Molina, R., and Garcia-Sainz, J. A., 1983, Roles of alpha$_1$- and beta-adrenergic receptors in adrenergic responsiveness of liver cells formed after partial hepatectomy, *Biochim. Biophys. Acta* **763**:112–119.

Hughes, A. R., Martin, M. W., and Harden, T. K., 1984, Pertussis toxin differentiates between two mechanisms of attenuation of cyclic AMP accumulation by muscarinic cholinergic receptors, *Proc. Natl. Acad. Sci. USA* **81**:5680–5684.

Hughes, B. P., Rye, K.-A., Pickford, L. B., Barritt, G. J., and Chalmers, A. H., 1984, A transient increase in diacylglycerols is associated with the action of vasopressin on hepatocytes, *Biochem. J.* **222**:535–540.

Hutson, N. J., Brumley, F. T., Assimacopoulos, F. D., Harper, S. C., and Exton, J. H., 1976, Studies on the α-adrenergic activation of hepatic glucose output, *J. Biol. Chem.* **251**:5200–5208.

Hutton, J. C., Peshavaria, M., and Brocklehurst, K. W., 1984, Phorbol ester stimulation of insulin release and secretory-granule protein phosphorylation in a transplantable rat insulinoma, *Biochem. J.* **224**:483–490.

Ikebe, M., Inagaki, M., Kanamuaru, K., and Hidaka, H., 1985, Phosphorylation of smooth muscle myosin light chain kinase by Ca^{2+}-activted, phospholipid-dependent protein kinase, *J. Biol. Chem.* **260**:4547–4550.

Imazu, M., Strickland, W. G., Chrisman, T. D., and Exton, J. H., 1984, Phosphorylation and inactivation of liver glycogen synthase by liver protein kinases, *J. Biol. Chem.* **259**:1813–1821.

Imboden, J. B., and Stobo, J. D., 1985, Transmembrane signalling by the T cell antigen receptor, *J. Exper. Med.* **161**:446–456.

Irvine, R. F., and Moor, R. M., 1986, Microinjection of inositol 1,3,4,5-tetrakisphosphate activates sea urchin eggs by a mechanism dependent upon external Ca^{2+}, *Biochem. J.* **240**:917–920.

Irvine, R. F., Brown, K. D., and Berridge, M. J., 1984a, Specificity of inositol trisphosphate-induced calcium release from permeabilized Swiss-mouse 3T3 cells, *Biochem. J.* **221**:269–272.

Irvine, R. F., Letcher, A. J., Lander, D. J., and Downes, C. P., 1984b, Inositol trisphosphates in carbachol-stimulated rat parotid glands, *Biochem. J.* **223**:237–243.

Irvine, R. F., Anggard, E. E., Letcher, A. J., and Downes, C. P., 1985, Metabolism of inositol 1,4,5-trisphosphate and inositol 1,3,4-trisphosphate in rat parotid glands, *Biochem. J.* **229**:505–511.

Irvine, R. F., Letcher, A. J., Heslop, J. P., and Berridge, M. J., 1986a, The inositol tris/tetrakisphosphate pathway—Demonstration of Ins(1,4,5)P₃ 3-kinase activity in animal tissues, *Nature* **320**:631–634.

Irvine, R. F., Letcher, A. J., Lander, D. J., and Berridge, M. J., 1986b, Specificity of inositol phosphate–stimulated Ca^{2+} mobilization from Swiss-mouse 3T3 cells, *Biochem. J.* **240**:301–304.

Irving, H. R., and Exton, J. H., 1987, Phosphatidylcholine breakdown in rat liver plasma membranes: roles of guanine nucleotides and P₂-purinergic agonists, *J. Biol. Chem.* **262**:3440–3443.

Ishii, H., Connolly, T. M., Bross, T. E., and Majerus, P. W., 1986, Inositol cyclic trisphosphate [inositol 1,2-(cyclic)-4,5-trisphosphate] is formed upon thrombin stimulation of human platelets, *Proc. Natl. Acad. Sci. USA* **83**:6397–6401.

Itoh, H., Okajima, F., and Ui, M., 1984, Conversion of adrenergic mechanism from an α- to a β-type during primary culture of rat hepatocytes, *J. Biol. Chem.* **259**:15464–15473.

Jacobs, S., Sahyoun, N. E., Saltiel, A. R., and Cuatrecasas, P.,1983,Phorbol esters stimulate the phosphorylation of receptors for insulin and somatomedin C, *Proc. Natl. Acad. Sci. USA* **80**:6211–6213.

Jakob, A., and Diem, S., 1975, Metabolic responses of perfused rat livers to alpha- and beta-adrenergic agonists, glucagon, and cyclic AMP, *Biochim. Biophys. Acta* **404**:57–66.

Jakobs, K. H., Bauer, S., and Watanabe, Y., 1985, Modulation of adenylate cyclase of human platelets by phorbol ester. Impairment of the hormone-sensitive inhibitory pathway, *Eur. J. Biochem.* **151**:425–430.

Jean, T., and Klee, C. B., 1986, Calcium modulation of inositol 1,4,5-trisphosphate-induced calcium release from neuroblastoma × glioma hybrid (NG108-15) microsomes, *J. Biol. Chem.* **261**:16414–16420.

Jenkinson, D. H., Haylett, D. G., Koller, K., and Burgess, G., 1978, Classification and actions of liver cell adrenoceptors, in: *Recent Advances in the Pharamcology of Adrenoceptors* (E. Szabadi, ed.), Elsevier/North-Holland, pp. 23–33.

Jones, D. J., and McKenna, L. F., 1980, Alpha-adrenergic receptor mediated formation of cyclic AMP in rat spinal cord, *J. Cyclic Nucleotide Res.* **6:**133–141.

Jones, L. M., and Michell, R. H., 1978, Stimulus–response coupling at α-adrenergic receptors, *Biochem. Soc. Trans.* **6:**673–688.

Joseph, S. K., and Williams, R. J., 1985, Subcellular localization and some properties of the enzymes hydrolysing inositol polyphosphates in rat liver, *FEBS Lett.* **180:**150–154.

Joseph, S. K., and Williamson, J. R., 1983, The origin, quantitation, and kinetics of intracellular calcium mobilization by vasopressin and phenylephrine in hepatocytes, *J. Biol. Chem.* **258:**10425–10432.

Joseph, S. K., and Williamson, J. R., 1986, Characteristics of inositol trisphosphate-mediated Ca^{2+} release from permeabilized hepatocytes, *J. Biol. Chem.* **261:**14658–14664.

Joseph, S. K., Thomas, A. P., Williams, R. J., Irvine, R. F., and Williamson, J. R., 1984a, *myo*-inositol 1,4,5-trisphosphate: A second messenger for the hormonal mobilization of intracellular Ca^{2+} in liver, *J. Biol. Chem.* **259:**3077–3081.

Joseph, S. K., Williamson, R. J., Corkey, B. E., Matschinsky, F. M., and Williamson, J. R., 1984b, The efect of inositol trisphosphate on Ca^{2+} fluxes in insulin-secreting tumor cells, *J. Biol. Chem.* **259:**12952–12955.

Joseph, S. K., Coll, K. E., Thomas, A. P., Rubin, R., and Williamson, J. R., 1985, The role of extracellular Ca^{2+} in the response of the hepatocyte to Ca^{2+}-dependent hormones, *J. Biol. Chem.* **260:**12508–12515.

Juhl, H., Sherorain, V. S., Schworer, C. M., Jett, M. F., and Soderling, T. R., 1983, Phosphorylatin site specificites of glycogen synthase kinases: Determination by peptide mapping using high performance liquid chromatography, *Arch. Biochem. Biophys.* **222:**518–526.

Kaibuchi, K., Takai, Y., and Nishizuka, Y., 1981, Cooperative roles of various membrane phospholipids in the activation of calcium-activated, phospholipid-dependent protein kinase, *J. Biol. Chem.* **256:**7146–7149.

Kaibuchi, K., Takai, Y., Sawamura, M., Hoshijima, M., Fujikura, T., and Nishizuka, Y., 1983, Synergistic functions of protein phosphorylation and calcium mobilization in platelet activation, *J. Biol. Chem.* **258:**6701–6704.

Kaibuchi, K., Tsuda, T., Kikuchi, A., Tanimoto, T., Yamashita, T., and Takai, Y., 1986, Possible involvement of protein kinase C and calcium ion in growth factor–induced expression of *c-myc* oncogene in Swiss 3T3 fibroblasts, *J. Biol. Chem.* **261:**1187–1192.

Kamagai, H., and Nishida, E., 1979, The interactions between calcium-dependent regulator protein of cyclic nucleotide phosphodiesterase and microtubule proteins, *J. Biochem. (Tokyo)* **85:**1267–1274.

Kanaho, Y., Moss, J., and Vaughan, M., 1985, Mechanism of inhibition of transducin GTPase activity by fluoride and aluminum, *J. Biol. Chem.* **260:**11493–11497.

Katada, T., Bokoch, G. M., Northup, J. K., Ui, M., and Gilman, A. G., 1984, The inhibitory guanine nucleotide-binding regulatory component of adenylate cyclase. Properties and function of the purified protein, *J. Biol. Chem.* **259:**3568–3577.

Katada, T., Gilman, A. G., Watanabe, Y., Buer, S., and Jakobs, K. H., 1985, Protein kinase C phosphorylates the inhibitory guanine–nucleotide-binding regulatory component and apparently suppresses its function in hormonal inhibition of adenylate cyclase, *Eur. J. Biochem.* **151:**431–437.

Katakami, Y., Kaibuchi, K., Sawamura, M., Takai, Y., and Nishizuka, Y., 1984, Synergistic action of protein kinase C and calcium for histamine release from rat peritoneal mast cells, *Biochem. Biophys. Res. Commun.* **121:**573–578.

Kawahara, Y., Takai, Y., Minakuchi, R., Sano, K., and Nishizuka, Y., 1980, Phospholipid turnover as a possible transmembrane signal for protein phosphorylation during human platelet activation by thrombin, *Biochem. Biophys. Res. Commun.* **97:**309–317.

Kelly, K., Cochran, B. H., Stiles, C. D., and Leder, P., 1983, Cell-specific regulation of the *c-myc* gene by lymphocyte mitogens and platelet derived growth factor, *Cell* **35:**603–610.

Kennedy, M. B., and Greengard, P., 1981, Two calcium/calmodulin-dependent protein kinases, which are highly concentrated in brain, phosphorylate protein I at distinct sites, *Proc. Natl. Acad. Sci. USA* **78:**1293–1297.

Kessar, P., and Crompton, M., 1981, The α-adrenergic-mediated activation of Ca^{2+} influx into cardiac mitochondria, *Biochem. J.* **200:**379–388.

Kikkawa, U., Takai, Y., Minakuchi, R., Inohara, S., and Nishizuka, Y., 1982, Calcium-activated, phospholipid-dependent protein kinase from rat brain, *J. Biol. Chem.* **257:**13341–13348.

Kikkawa, U., Takai, Y., Tanaka, Y., Miyake, R., and Nishizuka, Y., 1983, Protein kinase C as a possible receptor protein of tumor-promoting phorbol esters, *J. Biol. Chem.* **258:**11442–11445.

Kimura, S., Kugai, N., Tada, R., Kojima, I., Abe, K., and Ogata, E., 1982, Sources of calcium mobilized by α-adrenergic stimulation in perfused rat liver, *Horm. Metab. Res.* **14:**133–138.

Kimura, S., Nagasaki, K., Adachi, I., Yamaguchi, K., Fujiki, H., and Abe, K., 1984, Stimulation of hepatic glycogenolysis by 12-0-tetradecanoylphorbol-13-acetate (TPA) via a calcium requiring process, *Biochem. Biophys. Res. Commun.* **122:**1057–1064.

Kirk, C. J., Verrinder, T. R., and Hems, D. A., 1977, Rapid stimulation, by vasopressin and adrenaline, or inorganic phosphate incorporation into phosphatidylinositol in isolated hepatocytes, *FEBS Lett.* **83:**267–271.

Kirk, C. J., Creba, J. A., Downes, C. P., and Michell, R. H., 1981, Hormone-stimulated metabolism of inositol lipids and its relationship to hepatic receptor function, *Biochem. Soc. Trans.* **9:**377–379.

Kishimoto, A., Takai, Y., Mori, T., Kikkawa, U., and Nishizuka, Y., 1980, Activation of calcium and phospholipid-dependent protein kinase by diacylglycerol, its possible relation to phosphatidylinositol turnover, *J. Biol. Chem.* **255:**2273–2276.

Klee, C. B., and Vanaman, T. C., 1982, Calmodulin, *Adv. Protein Chem.* **35:**213–321.

Kleineke, J., and Soling, H. D., 1985, Mitochondrail and extramitochondrial Ca^{2+} pools in the perfused rat liver: Mitochondria are not the origin of calcium mobilized by vasopressin, *J. Biol. Chem.* **260:**1040–1045.

Korchak, H. M., Rutherford, L. E., and Weissman, G., 1984, Stimulus response coupling in the human neutrophil. I. Kinetic analysis of changes in calcium permeability, *J. Biol. Chem.* **259:**4070–4075.

Kraft, A. S., and Anderson, W. B., 1983, Phorbol esters increase the amount of Ca^{2+}, phospholipid-dependent protein kinase associated with plasma membrane, *Nature* **301:**621–623.

Kraft, A. S., Anderson, W. B., Cooper, H. L., and Sando, J. J., 1982, Decrease in cytosolic calcium/phospholipid-dependent protein kinase activity following phorbol ester treatment of EL4 thymoma cells, *J. Biol. Chem.* **257:**13193–13196.

Kruijer, W., Cooper, J. A., Hunter, T., and Verma, I. M., 1984, Platelet-derived growth factor induces rapid but transient expression of the *c-fos* gene and protein, *Nature* **312:**711–716.

Kunos, G., Kan, W. H., Greguski, R., and Venter, J. C., 1983, Selective affinity labeling and molecular characterization of hepatic α_1-adrenergic receptors with [^{3}H]phenoxybenzamine, *J. Biol. Chem.* **258:**326–332.

Kunos, G., Hirata, F., Ishac, E. J. N., and Tchakarov, L., 1984, Time-dependent conversion of α_1- to β-adrenoceptor-mediated glycogenolysis in isolated rat liver cells: Role of membrane phospholipase A2, *Proc. Natl. Acad. Sci. USA* **81:**6178–6182.

Kuo, J. F., Andersson, R. G. G., Wise, B. C., Mackerlova, L., Salomonsson, I., Brackett, N. L., Katoh, N., Shoji, M., and Wrenn, R. W., 1980, Calcium-dependent protein kinase: Widespread occurrence in various tissues and phyla of the animal kingdom and comparison of effects of phospholipid, calmodulin, and trifluoperazine, *Proc. Natl. Acad. Sci. USA* **77:**1039–1043.

Kuret, J., and Schulman, H., 1984, Purification and characterization of a Ca2 + /calmodulin-dependent protein kinase from rat brain, *Biochemistry* **23:**5495–5504.

Kuret, J., and Schulman, H., 1985, Mechanism of autophosphorylation of the multifunctional Ca^{2+} calmodulin-dependent protein kinase, *J. Biol. Chem.* **260:**6427–6433.

Labarca, R., Janowsky, A., Patel, J., and Paul, S. M., 1984, Phorbol esters inhibit agonist-induced [^{3}H]inositol-1-phosphate accumulation in rat hippocampal slices, *Biochem. Biophys. Res. Commun.* **123**:703–709.

Lad, P. M., Olson, C. V., and Smiley, P. A., 1985, Association of the N-formyl-Met-Leu-Phe receptor in human neutrophils with a GTP-binding protein sensitive to pertussis toxin, *Proc. Natl. Acad. Sci. USA* **82**:869–873.

Landt, M., and McDonald, J. M., 1984, Characterization of calmodulin-activated protein kinase activity of rat adipocyte endoplasmic reticulum fraction, *Int. J. Biochem.* **16**:161–169.

Landt, M., McDaniel, M. L., Bry, C. G., Kotagal, N., Colca, J. R., Lacy, P. E., and McDonald, J. M., 1982, Calmodulin-activated protein kinase activity in rat pancreatic islet cell membranes, *Arch. Biochem. Biophys.* **213**:148–154.

Langer, S. Z., 1974, Presynaptic regulation of catecholamine release (Commentary), *Biochem. Pharmacol.* **23**:1793–1800.

Langer, S. Z., 1977, Presynaptic receptors and their role in the regulation of transmitter release, *Br. J. Pharmacol.* **60**:481–497.

Lapetina, E. G., Reep, B., Ganong, B. R., and Bell, R. M., 1985, Exogenous sn-1,2-diacylglycerols containing saturated fatty acids function as bioregulators of protein kinase C in human platelets, *J. Biol. Chem.* **260**:1358–1361.

Lee, L. S., and Weinstein, I. B., 1978, Tumor-promoting phorbol esters inhibit binding of epidermal growth factor to cellular receptors, *Science* **202**:313–315.

Lee, L. S., and Weinstein, I. B., 1979, Mechanism of tumor promoter inhibition of cellular binding of epidermal growth factor, *Proc. Natl. Acad. Sci. USA* **76**:5168–5172.

Leeb-Lundberg, L. M. F., Dickinson, K. E. J., Heald, S. L., Wikberg, J. E. S., Lefkowitz, R. J., and Caron, M. G., 1984, Photoaffinity labeling of mammalian α_1-adrenergic receptors, *J. Biol. Chem.* **259**:2579–2587.

Leeb-Lundberg, L. M. F., Cotecchia, S., Lomasney, J. W., Debernadis, J. F., Lefkowitz, R. J., and Caron, M. G., 1985, Phorbol esters promote α_1-adrenergic receptor phosphorylation and receptor uncoupling from inositol phospholipid metabolism, *Proc. Natl. Acad. Sci. USA* **82**:5651–5655.

LePeuch, C. J., Haiech, J., and Demaille, J. G., 1979, Concerted regulation of cardiac sarcoplasmic reticulum calcium transport by cAMP dependent and calcium-calmodulin-dependent phosphorylation, *Biochemistry* **18**:5150–5157.

Lew, P. D., Monod, A., Krause, K.-H., Waldvogel, F. A., Biden, T. J., and Schlegel, W., 1986, The role of cytosolic free calcium in the generation of inositol 1,4,5-trisphosphate and inositol 1,3,4-trisphosphate in HL-60 cells, *J. Biol. Chem.* **261**:13121–13127.

Lin, S. H., and Fain, J. N., 1981, Vasopressin and epinephrine stimulation of phosphatidylinositol breakdown in the plasma membrane of rat hepatocytes, *Life Sci.* **29**:1905–1912.

Lin, S. H., Wallace, M. A., and Fain, J. N., 1983, Regulation of Ca^{2+}-Mg^{2+}-ATPase activity in hepatocyte plasma membranes by vasopressin and phenylephrine, *Endocrinology* **113**:2268–2275.

Litosch, I., Lin, S. H., and Fain, J. N., 1983, Rapid changes in hepatocyte phosphoinositides induced by vasopressin, *J. Biol. Chem.* **258**:13727–13732.

Litosch, I., Wallis, C., and Fain, J. N., 1985, 5-Hydroxytryptamine stimulates inositol phosphate production in a cell-free system from blowfly salivary glands, *J. Biol. Chem.* **260**:5464–5471.

Lomasney, J. W., Leeb-Lundberg, L. M. F., Cotecchia, S., Regan, J. W., DeBernadis, J. F., Caron, M. G., and Lefkowitz, R. J., 1986, Mammalian α_1-adrenergic receptor. Purification and characterization of the native receptor ligand binding subunit, *J. Biol. Chem.* **261**:7710–7716.

Lucas, D. O., Bajjalich, S. M., Kowalchyk, J. A., and Martin, T. F. J., 1985, Direct stimulation by thyrotropin-releasing hormone of polyphosphoinositide hydrolysis in GH3 cell membranes by a guanine nucleotide–modulated mechanism, *Biochem. Biophys. Res. Commun.* **132**:721–728.

Lynch, C. J., Blackmore, P. F., Charest, R., and Exton, J. H., 1985a, The relationships between receptor binding capacity for norepinephrine, angiotensin II, and vasopressin and release of inositol trisphosphate, Ca^{2+} mobilization, and phosphorylase activation in rat liver, *Mol. Pharmacol.* **28:**93–99.

Lynch, C. J., Charest, R., Blackmore, P. F., and Exton, J. H., 1986b, Studies on the hepatic α_1-adrenergic receptor. Modulation of guanine nucleotide effects by calcium temperature and age, *J. Biol. Chem.* **260:**1593–1600.

Lynch, C. J., Chraest, R., Bocckino, S. B., Exton, J. H., and Blackmore, P. F., 1985c, Inhibition of hepatic α_1-adrenergic effects and binding by phorbol myristate acetate, *J. Biol. Chem.* **260:**2844–2851.

Lynch, C. J., Sobo, G. E., and Exton, J. H., 1986a, Studies on the hepatic α_1-adrenergic receptor. An endogenous Ca^{2+}-sensitive protease converts the α_1-adrenergic receptor to a guanine nucleotide insensitive form, *Biochim. Biophys. Acta* **885:**110–120.

Lynch, C. J., Prpic, V., Blackmore, P. F.,and Exton, J. H., 1986b, Effect of islet activating pertussis toxin on the binding characteristics of Ca^{2+}-mobilizing hormones and on agonist activation of phosphorylase in hepatocytes, *Mol. Pharmacol.* **29:**196–203.

Lynch, C. J., Wilson, P. B., Blackmore, P. F., and Exton, J. H., 1986c, The hormone-sensitive hepatic Na^+ pump. Evidence for regulation by diacylglycerol and tumor promotors, *J. Biol. Chem.* **261:**14551–14556.

MacIntyre, D. E., McNicol, A., and Drummond, A. H., 1985, Tumour-promoting phorbol esters inhibit agonist-induced phosphatidate formation and Ca^{2+} flux in human platelets, *FEBS Lett.* **180:**160–164.

Malaisse, W. J., Dunlop, M. E., Mathias, P. C. F., Malaisse-Lagae, F., and Sener, A., 1985, Stimulation of protein kinase C and insulin release by 1-oleoyl-2-acetyl-glycerol, *Eur. J. Biochem.* **149:**23–27.

Malbon, C. C., 1980, Liver cell adenylate cyclase and β-adrenergic receptors, *J. Biol. Chem.* **255:**8692–8699.

Malbon, C. C., and Lo Presti, J. J., 1981, Hyperthyroidism impairs the activation of glycogen phosphorylase by epinephrine in rat hepatocytes, *J. Biol. Chem.* **256:**12199–12204.

Malbon, C. C., Li, S. Y., and Fain, J. N., 1978, Hormonal activation of glycogen phosphorylase in hepatocytes from hypothyroid rats, *J. Biol. Chem.* **253:**8820–8825.

Malbon, C. C., Graziano, M. P., and Johnson, G. L., 1984, Fat cell β-adrenergic receptor in the hypothyroid rat, *J. Biol. Chem.* **259:**3254–3260.

Manning, D. R., Fraser, B. A., Kahn, R. A., and Gilman, A. G., 1984, ADP-ribosylation of transducin by islet-activating protein. Identification of asparagine as the site of ADP-ribosylation, *J. Biol. Chem.* **259:**749–756.

Marcum, J. M., Dedman J. R., Brinkley, B. R., and Means, A. R., 1978, Control of microtubule assembly–disassembly by calcium-dependent regulator protein, *Proc. Natl. Acad. Sci. USA* **75:**3771–3775.

Marier, S. H., Putney, J. W., Jr., and Van De Walle, C. M., 1978, Control of calcium channels by membranen receptors in rat parotid gland, *J. Physiol.* **279:**141–151.

Martin, M. W., Evans, T., and Harden, T. K., 1985, Further evidence that muscarinic cholinergic receptors of 1321N1 astrocytoma cells couple to a guanine nucleotide regulatory protein that is not N_i, *Biochem. J.* **229:**539–544.

Martin, T. F. J., 1983, Thyrotropin-releasing hormone rapidly activates the phosphodiesterase hydrolysis of polyphosphoinositides in GH3 pituitary cells, *J. Biol. Chem.* **258:**14816–14822.

Martin, T. F. J., Bajjalieh, S. M., Lucas, D. O., and Kowalchyk, J. A., 1986, Thyrotropin-releasing hormone stimulation of polyphosphoinositide hydrolysis in GH_3 cell membranes is GTP-dependent but insensitive to cholera or pertussis toxin, *J. Biol. Chem.* **261:** 10041–10049.

Masters, S. B., Harden, T. K., and Brown, J. H., 1984, Relationships between phosphoinositide and calcium responses to muscarinic agonists in astrocytoma cells, *Mol. Pharmacol.* **26**:149–155.

Masters, S. B., Martin, M. W., Harden, T. K., and Brown, J. H., 1985a, Pertussis toxin does not inhibit muscarinic receptor–mediated phosphoinositide hydrolysis or calcium mobilization, *Biochem. J.* **227**:933–937.

Masters, S. B., Quinn, M. T., and Brown, J. H., 1985b, Agonist induced desenitization of muscarinic receptor–mediated calcium efflux without concomitant desensitization of phosphoinositide hydrolysis, *Mol. Pharmacol.* **27**:325–332.

Matozaki, T., Sakamoto, C., Nagao, M., and Baba, S., 1986, Phorbol ester or diacylglycerol modulates somatostatin binding to its receptors on rat pancreatic acinar cell membranes, *J. Biol. Chem.* **261**:1414–1420.

Mauco, G., Chap, H., and Douste-Blazy, L., 1983, Platelet activating factor (PAF-acether) promotes an early degradation of phosphatidylinositol-4,5-bisphosphate in rabbit platelets, *FEBS Lett.* **153**:361–365.

Mauco, G., Fauvel, J., Chap, H., and Douste-Blazy, L., 1984, Studies on enzymes related to diacylglycerol production in activated platelets, *Biochim. Biophys. Acta* **796**:169–177.

Mauger, J. P., Poggioli, J., Guesdon, F., and Claret, M., 1984, Noradrenaline, vasopressin, and angiotensin increase Ca^{2+} influx by opening a common pool of Ca^{2+} channels in isolated rat liver cells, *Biochem. J.* **221**:121–127.

Mauger, J. P., Poggioli, J., and Claret, M., 1985, Synergistic stimulation of the Ca^{2+} influx in rat hepatocytes by glucagon and the Ca^{2+}-linked hormones vasopressin and angiotensin II, *J. Biol. Chem.* **260**:11635–11642.

May, W. S., Jacobs, S., and Cuatrecasas, P., 1984, Association of phorbol ester-induced hyperphosphorylation and reversible regulation of transferrin membrane receptors in HL60 cells, *Proc. Natl. Acad. Sci. USA* **81**:2016–2020.

May, W. S., Jr., Sahoyn, N., Wolf, M., and Cuatrecasas, P., 1985, Role of intracellular calcium mobilization in the regulation of protein kinase C–mediated membrane processes, *Nature* **317**:549–551.

McCormack, J. G., 1985a, Characterization of the effects of Ca^{2+} on the intramitochondrial Ca^{2+}-sensitive enzymes from rat liver and within intact rat liver mitochondria, *Biochem. J.* **231**:581–595.

McCormack, J. G., 1985b, Studies on the activation of rat liver pyruvate dehydrogenase and 2-oxoglutarate dehydrogenase by adrenaline and glucagon, *Biochem. J.* **231**:597–608.

McCormack, J. G., 1985c, Evidence that adrenaline activates key oxidative enzymes in rat liver by increasing intramitochondrial $[Ca^{2+}]_i$, *FEBS Lett.* **180**:259–264.

McCormack, J. G., and Denton, R. M., 1981a, Comparative study of regulation by Ca^{2+} of activities of 2-oxoglutarate dehydrogenase complex and NAD^+-isocitrate dehydrogenase from a variety of sources, *Biochem. J.* **196**:619–624.

McCormack, J. G., and Denton, R. M., 1981b, The activation of pyruvate dehydrogenase in perfused rat heart by adrenaline and other inotropic agents, *Biochem. J.* **194**:639–643.

McCormack, J. G., and Denton, R. M., 1984, Role of Ca^{2+} ions in the regulation of intramitochondrial metaoblism in rat heart, *Biochem. J.* **218**:235–247.

McCormack, J. G., Edgell, N. J., and Denton, R. M., 1982, Studies on the interactions of Ca^{2+} and pyruvate in regulation of rat heart pyruvate dehydrogenase activity, *Biochem. J.* **202**:419–427.

McGuinness, T. L., Lai, Y., Greengard, P., Woodgett, J. R., and Cohen, P., 1983, A multifunctional calmodulin-dependent protein kinase: Similarities between skeletal muscle glycogen synthase kinase and a brain synapsin I kinase, *FEBS Lett.* **163**:329–334.

McGuinness, T. L., Lai, Y., and Greengard, P., 1985, Ca^{2+}/Calmodulin-dependent protein kinase II. Isozymic forms from rat forebrain and cerebellum, *J. Biol. Chem.* **260**:1969–1704.

Means, A. R., and Dedman, J. R., 1980, Calmodulin—An intracellular calcium receptor, *Nature* **285**:73–77.

Meeker, R. B., and Harden, T. K., 1982, Muscarinic cholinergic receptor–mediated activation of phosphodiesterase, *Mol. Pharmacol.* **22**:310–319.

Meier, K. E., Sternfeld, D. R., and Insel, P. A., 1984, Alpha$_1$- and beta$_2$-adrenergic receptors co-expressed on cloned MDCK cells are distinct glycoprotens, *Biochem. Biophys. Res. Commun.* **118**:73–81.

Meier, K. E., Sperling, D. M., and Insel, P. A., 1985, Agonist-mediated regulation of α_1- and β_2-adrenergic receptors in cloned MDCK cells, *Am. J. Physiol.* **249**:C69–C77.

Merritt, J. E., Taylor, C. W., Rubin, R. P., and Putney, J. W., Jr., 1986, Evidence suggesting that a novel guanine nucleotide regulatory protein couples receptors to phospholipase C in exocrine pancreas, *Biochem. J.* **236**:337–343.

Michell, R. H., 1975, Inositol phospholipids and cell surface receptor function, *Biochim. Biophys. Acta* **20**:339–344.

Michell, R. H., 1979, Inositol phospholipids in membrane function, *Trends Biochem. Sci.* **40**:128–131.

Michell, R. H., Kirk, C. J., Jones, L. M., Downes, C. P., and Creba, J. A., 1981, Stimulation of inositol lipid metabolism that accompanies calcium mobilization in stimulated cells: Defined characteristics and unanswered questions, *Philos. Trans. R. Soc. Lond. B* **296**:123–38.

Miller, B. E., and Nelson, D. L., 1977, Calcium fluxes in isolated acinar cells from rat parotid: Effect of adrenergic and cholinergic stimulation, *J. Biol. Chem.* **252**:3629–3636.

Molina y Vedia, L. M., and Lapetina, E. G., 1986, Phorbol 12,13-dibutyrate and 1-oleyl-2-acetyldiacylglycerol stimulate inositol trisphosphate dephosphorylation in human platelets, *J. Biol. Chem.* **261**:10493–10495.

Moon, S. O., Palfrey, H. C., and King, C., 1984, Phorbol esters potentiate tyrosine phosphorylation of epidermal growth factor receptors in A431 membranes by a calcium-independent mechanism, *Proc. Natl. Acad. Sci. USA* **81**:2298–2302.

Morgan, N. G., Shuman, E. A., Exton, J. H., and Blackmore, P. F., 1982, Stimulation of hepatic glycogenolysis by α_1- and α_2-adrenergic agonists, *J. Biol. Chem.* **257**:13907–13910.

Morgan, N. G., Blackmore, P. F., and Exton, J. H., 1983a, Age-related changes in the control of hepatic cyclic AMP levels by α_1- and β_2-adrenergic receptors in male rats, *J. Biol. Chem.* **258**:5103–5109.

Morgan, N. G., Blackmore, P. F., and Exton, J. H., 1983b, Modulation of the α_1-adrenergic control of hepatocyte calcium redistribution by increases in cyclic AMP, *J. Biol. Chem.* **258**:5110–5116.

Morgan, N. G., Exton, J. H., and Blackmore, P. F., 1983c, Angiotensin II inhibits hepatic cAMP accumulation induced by glucagon and epinephrine and their metabolic effects, *FEBS Lett.* **153**:77–80.

Morgan, N. G., Shipp, C. C., and Exton, J. H., 1983d, Studies on the mechanism of inhibition of hepatic cAMP accumulation by vasopressin, *FEBS Lett.* **163**:277–281.

Morgan, N. G., Waynick, L. E., and Exton, J. H., 1983e, Characterisation of the α_1-adrenergic control of hepatic cAMP in male rats, *Eur. J. Pharmacol.* **96**:1–10.

Morgan, N. G., Rumford, G. M., and Montague, W., 1985, Studies on the role of inositol trisphosphate in the regulation of insulin secretion from isolated rat islets of Langerhans, *Biochem. J.* **228**:713–718.

Mori, T., Takai, Y., Yu, B., Takahashi, J., Nishizuka, Y., and Fujikura, T., 1982, Specificity of the fatty acyl moieties of diacylglycerol for the activation of calcium-activated, phospholipid-dependent protein kinase, *J. Biochem.* **91**:427–431.

Muallem, S., Schoeffield, M., Pandol, S., and Sachs, G., 1985, Inositol trisphosphate modification of ion transport in rough endoplasmic reticulum, *Proc. Natl. Acad. Sci. USA* **82**:4433–4437.

Murayama, T., and Ui, M., 1985, Receptor-mediated inhibition of adenylate cyclase and stimulation of arachidonic acid release in 3T3 fibroblasts, *J. Biol. Chem.* **260**:7226–7233.

Murphy, E., Coll, K., Rich, T. L., and Williamson, J. R., 1980, Hormonal effects on calcium homeostasis in isolated hepatocytes, *J. Biol. Chem.* **255**:6600–6608.

Nabika, T., Velletri, P. A., Lovenberg, W., and Beaven, M. A., 1985, Increase in cytosolic calcium and phosphoinositide metabolism induced by angiotensin II and [arg]vasopressin in vascular smooth muscle cells, *J. Biol. Chem.* **260**:4661–4670.

Naccache, P. H., Molski, T. F. P., Borgeaut, P., White, J. R., and Sha'afi, R. I., 1985, Phorbol esters inhibit the fMet-Leu-Phe- and leukotriene B4-stimulated calcium mobilization and enzyme secretion in rabbit neutrophils, *J. Biol. Chem.* **260**:2125–2131.

Nadler, E., Gillo, B., Lass, Y., and Oron, Y., 1986, Acetylcholine- and inositol 1,4,5-trisphosphate–induced calcium mobilization in *Xenopus laevis* oocytes, *FEBS Lett.* **199**:208–212.

Nagano, M., Ishibashi, H., McCully, V., and Cottam, G. L., 1980, Epinephrine-stimulated phosphorylation of pyruvate kinase in hepatocytes, *Arch. Biochem. Biophys.* **203**:271–281.

Nakamura, T., and Ui, M., 1983, Suppression of passive cutaneous anaphylaxis by pertussis toxin, an islet-activating protein, as a result of inhibition of histamine release from mast cells, *Biochem. Pharmacol.* **32**:3435–3441.

Nakamura, T., and Ui, M., 1985, Simultaneous inhibitions of inositol phospholipid breakdown, arachidonic acid release, and histamine secretion in mast cells by islet-activating protein, pertussis toxin, *J. Biol. Chem.* **260**:3584–3593.

Nanberg, E., and Putney, J. W., Jr., 1986, α_1-Adrenergic activation of brown adipocytes leads to an increased formation of inositol polyphosphates, *FEBS Lett.* **195**:319–322.

Nestler, E. J., and Greengard, P., 1983, Protein phosphorylation in the brain, *Nature* **305**:583–588.

Nestler, E. J., Walaas, S. I., and Greegard, P., 1984, Neuronal phosphoproteins: Physiological and clinical implications, *Science* **225**:1357–1364.

Nicholls, D. G., 1978, The regulation of extramitochondrial free calcium ion concentration by rat liver mitochondria, *Biochem. J.* **176**:463–474.

Nicholls, D., and Akerman, K., 1982, Mitochondrial calcium transport, *Biochim. Biophys. Acta* **683**:57–88.

Niedel, J. E., Kuhn, L. J., and Vandenbark, G. R., 1983, Phorbol diester receptor copurifies with protein kinase C, *Proc. Natl. Acad. Sci. USA* **80**:36–40.

Niggli, V., Penniston, J. T., and Carafoli, E., 1979, Purification of the (Ca^{2+}-Mg^{2+})-ATPase from human erythrocyte membranes using a calmodulin affinity column, *J. Biol. Chem.* **254**:9955–9958.

Nishikawa, M., Shirakawa, S., and Adelstein, R. S., 1985, Phosphorylation of smooth muscle myosin light chain kinase by protein kinase C: Comparative study of the phosphorylated sties, *J. Biol. Chem.* **260**:8978–8983.

Nishizuka, Y., 1984, The role of protein kinase C in cell surface signal transduction and tumour promotion, *Nature* **308**:693–698.

Nosek, T. M., Williams, M. F., Zeigler, S. T., and Godt, R. E., 1986, Inositol trisphosphate enhances calcium release in skinned cardiac and skeletal muscle, *Am. J. Physiol.* **250**:C807–C811.

Ochs, R., 1984, Glutamine metabolism of isolated rat hepatocytes: Evidence for catecholamine activation of α-ketoglutarate dehydrogenase, *J. Biol. Chem.* **259**:13004–13010.

Okajima, F., and Ui, M., 1982, Conversion of adrenergic regulation of glycogen phosphorylase and synthase from an α to β type during primary culture of rat hepatocytes, *Arch. Biochem. Biophys.* **213**:658–668.

Okajima, F., and Ui, M., 1984, ADP-ribosylation of the specific membrane protein by islet-activating protein, pertussis toxin, associated with inhibition of a chemotactic peptide-induced arachidonate release in neutrophils, *J. Biol. Chem.* **259**:13863–13871.

Okajima, F., Katada, T., and Ui, M., 1985, Coupling of guanine nucleotide regulatory protein to chemotactic peptide receptors in neutrophil membranes and its uncoupling by islet-activating protein, pertussis toxin, *J. Biol. Chem.* **260**:6761–6768.

Orellana, S. A., Solski, P. A., and Brown, J. H., 1985, Phorbol ester inhibits phosphoinositide hydrolysis and calcium mobilization in cultured astrocytoma cells, *J. Biol. Chem.* **2650**:5236–5239.

174   JOHN H. EXTON

Oron, Y., Dascal, N., Nadler, E., and Lupu, M., 1985, Inositol 1,4,5-trisphosphate mimics muscarinic response in Xenopus oocytes, *Nature* **313**:141–143.

O'Rourke, F. A., Halenda, S. P., Zavoico, G. B., and Feinstein, M. B., 1985, Inositol 1,4,5-trisphosphate releases Ca^{2+} from a Ca^{2+}-transporting membrane vesicle fraction derived from human platelets, *J. Biol. Chem.* **260**:956–962.

Osborne, R., and Tashjian, A. H., Jr., 1981, Tumor-promoting phorbol esters affect production of prolactin and growth hormone by rat pituitary cells, *Endocrinology* **108**:1164–1170.

Oviasu, O. A., and Whitton, P. D., 1984, Hormonal control of pyruvate dehydrogenase activity in rat liver, *Biochem. J.* **224**:181–186.

Palfrey, H. C., Rothlein, J. E., and Greengard, P., 1983, Calmodulin-dependent protein kinase α associated substrates in Torpedo electric organ, *J. Biol. Chem.* **256**:496–503.

Pandol, S. J., and Schoeffield, M. S., 1986, 1,2-diacylglycerol, protein kinase C, and pancreatic enzyme secretion, *J. Biol. Chem.* **261**:4438–4444.

Parker, J. C., Barritt, G. J., and Wadsworth, J. C., 1983, A kinetic investigation of the effects of adrenaline on $^{45}Ca^{2+}$ exchange in isolated hepatocytes at different Ca^{2+} concentrations. A kinetic investigation of the effects of adrenaline on $^{45}Ca^{2+}$ exchange in isolated hepatocytes at different Ca^{2+} concentrations, at 20°C and in the presence of inhibitors of mitochondrial Ca^{2+} transport, *Biochem. J.* **216**:51–62.

Parod, R. J., and Putney, J. W., Jr., 1978, The role of calcium in the receptor mediated control of potassium permeability in rat lacrimal gland, *J. Physiol.* **281**:371–382.

Parod, R. J., and Putney, J. W., Jr., 1979, Stimulation of ^{45}Ca efflux from rat lacrimal gland slices by carbachol and epinephrine, *Life Sci.* **25**:2211–2215.

Payne, M. E., and Soderling, T. R., 1980, Calmodulin-dependent glycogen synthase kinase, *J. Biol. Chem.* **255**:8054–8056.

Payne, M. E., Schworer, C. M., and Soderling, T. R., 1983, Purification and characterization of rabbit liver calmodulin-dependent glycogen synthase kinase, *J. Biol. Chem.* **258**:2376–2382.

Penniston, J., 1983, Plasma membrane Ca^{2+}-ATPases as active Ca^{2+} pumps, in: *Calcium and Cell Function*, Vol. IV (W. Y. Cheung, ed.), Academic Press, New York, pp. 99–149.

Perkins, J. P., and Moore, M. M., 1973, Characterization of the adrenergic receptors mediating a rise in cyclic 3',5'-adenosine monophosphate in rat cerebral cortex, *J. Pharmacol. Exp. Ther.* **185**:371–378.

Picton, C., Klee, C. B., and Cohen, P., 1980, Phosphorylase kinase from rabbit skeletal muscle: Identifiction of calmodulin-binding subunits, *Eur. J. Biochem.* **111**:553–561.

Pocotte, S. L., and Holz, R. W., 1986, Effects of phorbol ester on tyrosine hydroxylase phosphorylation and activation in cultured bovine adrenal chromaffin cells, *J. Biol. Chem.* **261**:1873–1877.

Pocotte, S. L., Frye, R. A., Senter, R. A., Terbush, D. R., Lee, S. A., and Holz, R. W., 1985, Effects of phorbol ester on catecholamine secretion and protein phosphorylation in adrenal medullary cell cultures, *Proc. Natl. Acad. Sci. USA* **82**:930–934.

Poggioli, J., and Putney, J. W., Jr., 1982, Net calcium fluxes in rat parotid acinar cells: Evidence for a hormone-sensitive calcium pool in or near the plasma membrane, *Pfluegers Arch.* **392**:239–243.

Poggioli, J., Berthon, B., and Claret, M., 1980, Calcium movements in *in situ* mitochondria following activation of α-adrenergic receptors in rat liver cells, *FEBS Lett.* **115**:243–246.

Poggioli, J., Mauger, J.-P., Guesdon, F., and Claret, M., 1985, A regulatory calcium-binding site for calcium channel in isolated rat hepatocytes, *J. Biol. Chem.* **260**:3289–3294.

Pozzan, T., Arslan, P., Tsien, R. Y., and Rink, T. J., 1982, Anti-immunoglobulin, cytoplasmic free calcium, and capping in B lymphocytes, *J. Cell Biol.* **94**:335–340.

Preiksaitis, H. G., and Kunos, G., 1979, Adrenoceptor-mediated activation of liver glycogen phosphorylase: Effects of thyroid state, *Life Sci.* **24**:35–41.

Preiksaitis, H. G., Kan, W. H., and Kunos, G., 1982, Decreased α_1-adrenoceptor responsiveness and density in liver cells of thyroidectomized rats, *J. Biol. Chem.* **257:**4321–4327.

Preiss, J., Loomis, C. R., Bishop, W. R., Stein, R., Niedel, J. E., and Bell, R. M., 1986, Quantitative measurement of *sn*-1,2-diacylglycerols present in platelets, hepatocytes, and *ras*- and *sis*-transformed normal rat kidney cells, *J. Biol. Chem.* **261:**8597–8600.

Prentki, M. M., Biden, T. J., Janjic, D., Irvine, R. F., Berridge, M. J., and Wollheim, C. B., 1984a, Rapid mobilization of Ca^{2+} from rat insulinoma microsomes by inositol-1,4,5-trisphosphate, *Nature* **309:**562–564.

Prentki, M., Wollheim, C. B., and Lew, P. D., 1984b, Ca^{2+} homeostasis in permeabilized human neutrophils. Characterization of Ca^{2+}-sequestering pools and the action of inositol 1,4,5-trisphosphate, *J. Biol. Chem.* **259:**13777–13782.

Prentki, M., Corkey, B. E., and Matschinsky, F. M., 1985, Inositol 1,4,5-trisphosphate and the endoplamsic reticulum Ca^{2+} cycle of a rat insulinoma cell line, *J. Biol. Chem.* **260:**9185–9190.

Prpic, V., Blackmore, P. F., and Exton, J. H., 1982, Phosphatidylinositol breakdown induced by vasopressin and epinephrine in hepatocytes is calcium-dependent, *J. Biol. Chem.* **257:**11323–11331.

Prpic, V., Green, K. C., Blackmore, P. F., and Exton, J. H., 1984, Vasopressin-, angiotensin II-, and α_1-adrenergic-induced inhibition of Ca^{2+} transport by rat liver plasma membrane vesicles, *J. Biol. Chem.* **259:**1382–1385.

Pushpendran, C. K., Corvera, S., and Garcia-Sainz, J. A., 1984, Effect of insulin on alpha$_1$-adrenergic actions in hepatocytes from euthyroid and hypothyroid rats, *Biochem. Biophys. Res. Commun.* **118:**451–459.

Putney, J. W., Jr., 1976, Biphasic modulation of potassium release in rat parotid gland by carbachol and phenylephrine, *J. Pharmacol. Exp. Ther.* **198:**375–384.

Putney, J. W., Jr., 1977, Muscarinic, alpha-adrenergic, and peptide receptors regulate the same calcium influx sites in the parotid gland, *J. Physiol.* **268:**139–149.

Putney, J. W., Jr., Burgess, G. M., Halenda, S. P., McKinney, J. S., and Rubin, R. P., 1983, Effects of secretagogues on [^{32}P]phosphatidylinositol 4,5-bisphosphate metabolism in the exocrine pancreas, *Biochem. J.* **212:**483–488.

Putney, J. W., Jr., McKinney, J. S., Aub, D. L., and Leslie, B. A., 1984, Phorbol ester–induced protein secretion in rat parotid gland, *Mol. Pharmacol.* **26:**261–66.

Rasmussen, H., Forder, J., Kojima, I., and Scriabine, A., 1984, TPA-induced contraction of isolated rabbit vascular smooth muscle, *Biochem. Biophys. Res. Commun.* **122:**776–784.

Rebecchi, M. J., and Gershengorn, M. C., 1983, Thyroliberin stimulates rapid hydrolysis of phosphatidylinositol 4,5-bisphosphate by a phosphodiesterase in rat mammotropic pituitary cells, *Biochem. J.* **216:**287–294.

Reinhart, P. H., Taylor, W. M., and Bygrave, F. L., 1982, Calcium ion fluxes induced by the action of α-adrenergic agonists in perfused rat liver, *Biochem. J.* **208:**619–630.

Reinhart, P. H., Taylor, W. M., and Bygrave, F. L., 1984a, The contribution of both extracellular and intracellular calcium to the action of α-adrenergic agonists in perfused rat liver, *Biochem. J.* **220:**35–42.

Reinhart, P. H., Taylor, W. M., and Bygrave, F. L., 1984b, The action of α-adrenergic agonists on plasma-membrane calcium fluxes in perfused rat liver, *Biochem. J.* **220:**43–50.

Reinhart, P. H., Taylor, W. M., and Bygrave, F. L., 1984c, The role of calcium ions in the mechanism of action of α-adrenergic agonists in rat liver, *Biochem. J.* **223:**1–13.

Reinhart, P. H., Taylor, W. M., and Bygrave, F. L., 1984d, The mechanism of α-adrenergic agonist action in liver, *Biol. Rev.* **59:**511–557.

Reynolds, E. E., and Dubyak, G. R., 1985, Activation of calcium mobilization and calcium influx by alpha$_1$-adrenergic receptors in a smooth muscle cell line, *Biochem. Biophys. Res. Commun.* **130:**627–632.

Rhodes, D., Prpic, V., Exton, J. H., and Blackmore, P. F., 1983, Stimulation of phosphatidylinositol 4,5-bisphosphate hydrolysis in hepatocytes by vasopressin, *J. Biol. Chem.* **258**:2770–2773.

Rink, T. J., Sanchez, A., and Hallam, T. J., 1983, Diacylglycerol and phorbol ester stimulate secretion without raising cytoplasmic free calcium in human platelets, *Nature* **305**:317–319.

Rittenhouse, S. E., and Sasson, J. P., 1985, Mass changes in myoinositol trisphosphate in human platelets stimulated by thrombin: Inhibitory effects of phorbol ester, *J. Biol. Chem.* **260**:8657–8660.

Rittenhouse-Simmons, S., 1979, Production of diglyceride from phosphatidylinositol in activated human platelets, *J. Clin. Invest.* **63**:580–587.

Roach, P. J., and Goldman, M., 1983, Modification of glycogen synthase activity in isolated rat hepatocytes by tumor-promoting phorbol esters: Evidence for differential regulation of glycogen synthase and phosphorylase, *Proc. Natl. Acad. Sci. USA* **80**:7170–7172.

Roach, P. J., DePaoli-Roach, A. A., and Larner, J., 1978, Ca^{2+}-stimulated phosphorylation of muscle glycogen synthase by phosphorylase *b* kinase, *J. Cyclic Nucleotide Res.* **4**:245–257.

Rubin, R. P., Godfrey, P. P., Chapman, D. A., and Putney, J. W., Jr., 1984, Secretagogue-induced formation of inositol phosphates in rat exocrine pancreas, *Biochem. J.* **219**:655–659.

Ruegg, J. C., 1982, Vascular smooth muscle: Intracellular aspects of adrenergic receptor contraction coupling, *Experientia* **38**:1400–1404.

Sano, K., Takai, Y., Yamanishi, J., and Nishizuka, Y., 1983, A role of calcium-activated phospholipid-dependent protein kinase in human platelet activation, *J. Biol. Chem.* **258**:2010–2013.

Schacht, J., and Agranoff, B. W., 1972, Effects of acetylcholine on labeling of phosphatidate and phosphoinositides by [^{32}P]orthophosphate in nerve, *J. Biol. Chem.* **247**:774–777.

Schimmel, R. J., McCarthy, L, and Dzierzanowski, D., 1985, Effects of pertussis toxin treatment on metabolism in hamster brown adipocytes, *Am. J. Physiol.* **249**:C456–C463.

Schulman, H., 1984a, Phosphorylation of microtubule-associated proteins by a Ca^{2+}/calmodulin dependent protein kinase, *J. Cell Biol.* **99**:11–19.

Schulman, H., 1984b, Calcium-dependent protein kinases and neuronal function, *Trends Pharmacol. Sci.* **5**:188–192.

Schulman, H., and Greengard, P., 1978a, Stimulation of brain membrane protein phosphorylation by calcium and an endogenous heat-stable protein, *Nature* **271**:478–479.

Schulman, H., and Greengard, P., 1978b, Ca^{2+}-dependent protein phosphorylation system in membranes from various tissues, and its activation, by "calcium-dependent regulator," *Proc. Natl. Acad. Sci. USA* **75**:5432–5436.

Schultz, J., and Daly, J. W., 1973, Adenosine 3',5'-monophosphate in guinea pig cerebral cortical slices: Effects of α- and β-adrenergic agents, histamine, serotonin, and adenosine, *J. Neurochem.* **21**:573–579.

Schworer, C. M., and Soderling, T. R., 1983, Substrate specificity of liver calmodulin-dependent glycogen synthase kinase, *Biochem. Biophys. Res. Commun.* **116**:412–416.

Schworer, C. M., El-Maghrabi, M. R., Pilkis, S. J., and Soderling, T. R., 1985, Phosphorylation of L-type pyruvate kinase by a Ca^{2+}/calmodulin-dependent protein kinase, *J. Biol. Chem.* **260**:13018–13022.

Seidman, C. E., Hess, H. J., Homcy, C. J., and Graham, R. M., 1984, Photoaffinity labeling of the α_1-adrenergic receptor using an ^{125}I-labeled aryl azide analogue of prazosin, *Biochemistry* **23**:3765–3770.

Seyfred, M. A., and Wells, W. W., 1984, Subcellular site and mechanism of vasopressin-stimulated hydrolysis of phosphoinositides in rat hepatocytes, *J. Biol. Chem.* **259**:7666–7672.

Seyfred, M. A., Farrell, L. E., and Wells, W. W., 1984, Characterization of D-myo-inositol 1,4,5-trisphosphate phosphatase in rat liver plasma membranes, *J. Biol. Chem.* **259**:13204–13208.

Sha'afi, R. I., White, J. R., Molski, T. F. P., Shefcyk, J., Volpi, M., Naccache, P. H., and Feinstein, M. B., 1983, Phorbol 12-myristate 13-acetate activates rabbit neutrophils without an apparent rise in the level of intracellular free calcium, *Biochem. Biophys. Res. Commun.* **114:**638–645.

Sharkey, N. A., Leach, K. L., and Blumberg, P. M., 1984, Competitive inhibition by diacylglycerol of specific phorbol ester binding, *Proc. Natl. Acad. Sci. USA* **81:**607–610.

Shears, S. B., and Kirk, C. J., 1984a, Determination of mitochondrial calcium content in hepatocytes by a rapid cellular-fractionation technique. α-Adrenergic agonists do not mobilize mitochondrial Ca^{2+}, *Bichem. J.* **219:**383–89.

Shears, S. B., and Kirk, C. J., 1984b, Determination of mitochondrial calcium content in hepatocytes by a rapid cellular fractionation technique. Vasopressin stimulates mitochondrial Ca^{2+} uptake, *Biochem. J.* **220:**417–421.

Shenolikar, S., Cohen, P. T. W., Cohen, P., Nairn, A. C., and Perry, S. V., 1979, The role of calmodulin in the structure and regulation of phosphorylase kinase from rabbit skeletal muscle, *Eur. J. Biochem.* **100:**329–337.

Shoyab, M., De Larco, J. E., and Todaro, G. J., 1979, Biologically active phorbol esters specifically alter affinity of epidermal growth factor membrane receptors, *Nature* **279:**387–391.

Shreeve, S. M., Fraser, C. M., and Venter, J. C., 1985, Molecular comparison of α_1- and α_2-adrenergic receptors suggsts that these proteins are structurally related "isoreceptors," *Proc. Natl. Acad. Sci. USA* **82:**4842–4846.

Sies, H., Graf, P., and Crane, D., 1983, Decreased flux through pyruvate dehydrogenase during calcium ion movements induced by vasopressin, α-adrenergic agonists, and the ionophore A23187 in perfused rat liver, *Biochem. J.* **212:**271–278.

Slack, B. E., Bell, J. E., and Benos, D. J., 1986, Inositol-1,4,5-trisphosphate injection mimics fertilization potentials in sea urchin eggs, *Am. J. Physiol.* **250:**C340–C344.

Smith, C. D., Lane, B. C., Kusaka, I., Verghese, M. W., and Snyderman, R., 1985, Chemoattractant receptor-induced hydrolysis of phosphatidylinositol 4,5-bisphosphate in human polymorphonuclear leukocyte membranes. *J. Biol. Chem.* **260:**5875–5878.

Smith, J. B., Smith, L., Brown, E. R., Barnes, D., Sabir, M. A., Davis, J. S., and Farese, R. V., 1984, Angiotension II rapidly increases phosphatidate–phosphoinositide hydrolysis and mobilizes intracellular calcium in cultured arterial muscle cells, *Proc. Natl. Acad. Sci. USA* **81:**7812–7816.

Smith, J. B., Smith, L., and Higgins, B. L., 1985, Temperature and nucleotide dependence of calcium release by *myo*inositol 1,4,5-trisphosphate in cultured vascular smooth muscle cells, *J. Biol. Chem.* **260:**14413–14416.

Snavely, M. D., and Insel, P. A., 1982, Characterization of alpha-adrenergic receptor subtypes in the rat renal cortex, *Mol. Pharmacol.* **22:**532–546.

Snavely, M. D., Mahan, L. C., O'Connor, D. T., and Insel, P. A., 1983, Selective down regulation of adrenergic receptor subtypes in tissues from rats with pheochromocytoma, *Endocrinology* **113:**354–360.

Soloff, M. S., and Sweet, P., 1982, Oxytocin inhibition of $(Ca^{2+} + Mg^{2+})$-ATPase activity in rat myometrial plasma membranes, *J. Biol. Chem.* **257:**10687–10693.

Somlyo, A. P., Bond, M., and Somlyo, A. V., 1985, Calcium content of mitochondria and endoplasmic reticulum in liver frozen rapidly *in vivo, Nature* **314:**622–625.

Somlyo, A. V., Bond, M., Somlyo, A. P., and Scarpa, A., 1985, Inositol trisphosphate–induced calcium release and contraction in vascular smooth muscle, *Proc. Natl. Acad. Sci. USA* **82:**5231–5235.

Spat, A., Fabiato, A., and Rubin, R. P., 1985, Binding of inositol trisphosphate by a liver microsomal fraction, *Biochem. J.* **223:**929–932.

Staddon, J. M., and McGivan, J. D., 1985, Ca^{2+}-dependent activation of oxoglutarate dehydrogenase by vasopressin in isolated hepatocytes, *Biochem. J.* **225:**327–333.

Starke, K., 1977, Regulation of noradrenaline release by presynaptic receptor systems, *Rev. Physiol. Bicohem. Pharmacol.* **77:**1–124.

Sternweis, P. C., and Gilman, A. G., 1982, Aluminum: A requirement for activation of the regulatory component of adenylate cyclase by fluoride, *Proc. Natl. Acad. Sci. USA* **79:**4888–4891.

Sternweis, P. C., and Robishaw, J. D., 1984, Isolation of two proteins with high affinity for guanine nucleotides from membranes of bovine brain, *J. Biol. Chem.* **259:**13806–13813.

Stewart, S. J., Prpic, V., Powers, F. S., Bocckino, S. B., Isaacks, R. E., and Exton, J. H., 1986, Perturbation of the human T-cell antigen receptor–T_3 complex leads to the production of inositol tetrakisphosphate: Evidence for conversion from inositol trisphosphate, *Proc. Natl. Acad. Sci. USA* **83:**6098–6102.

Stiles, G. L., Hoffman, B. B., Hubbard, M., Caron, M. G., and Lefkowitz, R. J., 1983, Guanine nucleotides and alpha$_1$-adrenergic receptors in the heart, *Biochem. Pharmacol.* **32:**69–71.

Stoehr, S. J., Smolen, J. E., Holz, R. W., and Agranoff, B. W., 1986, Inositol trisphosphate mobilizes intracellular calcium in permeabilized adrenal chromaffin cells, *J. Neurochem.* **46:**637–640.

Storey, D. J., Shears, S. B., Kirk, C. J., and Michell, R. H., 1984, Stepwise enzymatic dephosphorylation of inositol 1,4,5-trisphosphate to inositol in liver, *Nature* **312:**374–376.

Storm, H., Van Hardeveld, C., and Kassenaar, A. A. H., 1984, The influence of hypothyroidism on the adrenergic stimulation of glycogenolysis in perfused rat liver, *Biochim. Biophys. Acta* **798:**350–360.

Stossel, T. P., 1984, Contribution of actin to the structure of the cytoplasmic matrix, *J. Cell Biol.* **99:**15s–21s.

Straub, R. E., and Gershengorn, M. C., 1986, Thyrotropin-releasing hormone and GTP activate inositol trisphosphate formation in membranes isolated from rat pituitary cells, *J. Biol. Chem.* **261:**2712–2717.

Streb, H., Irvine, R. F., Berridge, M. J., and Schulz, I., 1983, Release of Ca^{2+} from a nonmitochondrial intracellular store in pancreatic acinar cells by inositol 1,4,5-triphosphate, *Nature* **306:**67–69.

Streb, H., Bayerdorffer, E., Haase, W., Irvine, R. F., and Schulz, I., 1984, Effect of inositol-1,4,5-trisphosphate in isolated subcellular fractions of rat pancreas, *J. Membr. Biol.* **81:** 241–253.

Streb, H., Heslop, J. P., Irvine, R. F., Schulz, I., and Berridge, M. J., 1985, Relationship between secretagogue-induced Ca^{2+} release and inositol polyphosphate production is permeabilized pancreatic acinar cells, *J. Biol. Chem.* **260:**7309–7315.

Strickland, W. G., Blackmore, P. F., and Exton, J. H., 1980, The role of calcium in alpha-adrenergic inactivation of glycogen synthase in rat hepatocytes and its inhibition by insulin, *Diabetes* **29:**617–622.

Strickland, W. G., Imazu, M., Chrisman, T. D., and Exton, J. H., 1983, Regulation of rat liver glycogen synthase: Roles of Ca^{2+}, phosphorylase kinase, and phosphorylase *a*, *J. Biol. Chem.* **258:**5490–5497.

Studer, R. K., and Borle, A. B., 1982, Differences between male and female rats in the regulation of hepatic glycogenolysis: The relative role of calcium and cAMP in phosphorylase activation by catecholamines, *J. Biol. Chem.* **257:**7987–7993.

Studer, R. K., and Borle, A. B., 1983, Sex difference in cellular calcium metabolism of rat hepatocytes and in α-adrenergic activation of glycogen phosphorylase, *Biochim. Biophys. Acta* **762:**302–314.

Studer, R. K., and Borle, A. B., 1984, Effect of adrenalectomy on cellular calcium metabolism and on the response to adrenergic stimulation of hepatocytes isolated from male and female rats, *Biochim. Biophys. Acta* **804:**377–385.

Stull, J. T., Manning, D. R., High, C. W., and Blumenthal, D. K., 1980, Phosphorylation of contractile proteins in heart and skeletal muscle, *Fed. Proc.* **39:**1552–1557.

Suematsu, E., Hirata, M., Hashimoto, T., and Kuriyama, H., 1984, Inositol 1,4,5-trisphosphate releases Ca^{2+} from intracellular store sites in skinned single cells of procine coronary artery, *Biochem. Biophys. Res. Commun.* **120**:481–485.

Sugano, T., Shiota, M., Khono, H., Shimada, M., and Oshino, N., 1980, Effects of calcium ions on the activation of gluconeogenesis by norepinephrine in perfused rat liver, *J. Biochem. (Tokyo)* **87**:465–472.

Sugden, A. L., Sugden, D., and Klein, D. C., 1986, Essential role of calcium influx in the adrenergic regulation of cAMP and cGMP in rat pinealocytes, *J. Biol. Chem.* **261**:11608–11612.

Sugden, D., Vanecek, J., Klein, D. C., Thomas, T. P., and Anderson, W. B., 1985, Activation of protein kinase C potentiates isoprenaline-induced cyclic AMP accumulation in rat pinealocytes, *Nature* **314**:359–361.

Sugden, M. C., and Watts, D. I., 1983, Stimulation of [1-^{14}C]oleate oxidation to $^{14}CO_2$ in isolated rat hepatocytes by the catecholamines, vasopressin, and angiotensin, *Biochem. J.* **212**:85–91.

Sugden, M. C., Tordoff, A. F. C., Ilic, V., and Williamson, D. H., 1980, α-Adrenergic stimulation of [1-^{14}C]oleate oxidation to $^{14}CO_2$ in isolated rat hepatocytes, *FEBS Lett.* **120**:80–84.

Szuts, E. Z., Wood, S. F., Reid, M. S., and Fein, A., 1986, Light stimulates the rapid formation of inositol trisphosphate in squid retinas, *Biochem. J.* **240**:929–932.

Tada, M., and Katz, A. M., 1982, Phosphorylation of the sarcoplasmic reticulum and sarcolemma, *Annu. Rev. Physiol.* **44**:401–423.

Tada, M., Ohmori, F., Yamada, M., and Abe, H., 1979, Mechanism of the stimulation of Ca^{2+}-dependent ATPase of cardiac sarcoplasmic reticulum by adenosine $3' : 5'$-monophosphate-dependent protein kinase, *J. Biol. Chem.* **254**:319–326.

Takai, Y., Kishimoto, A., Iwasa, Y., Kawahara, Y., Mori, T., and Nishizuka, Y., 1979a, Calcium-dependent activation of a multifunctional protein kinase by membrane phospholipids, *J. Biol. Chem.* **254**:3692–3695.

Takai, Y., Kishimoto, A., Kikkawa, U., Mori, T., and Nishizuka, Y., 1979b, Unsaturated diacylglycerol as a possible messenger for activation of calcium-activated, phospholipid-dependent protein kinase system, *Biochem. Biophys. Res. Commun.* **91**:1218–1224.

Taylor, W. M., Prpic, V., Exton, J. H., and Bygrave, F. L., 1980, Stable changes to calcium fluxes in mitochondria isolated from rat livers perfused with α-adrenergic agonists α with glucagon, *Biochem. J.* **188**:443–450.

Taylor, W. M., Reinhart, P. H., and Bygrave, F. L., 1983, Stimulation by α-adrenergic agonists of Ca^{2+} fluxes, mitochondrial oxidation, and gluconeogenesis in perfused rat liver, *Biochem. J.* **212**:555–565.

Thevenod, F., Streb, H., Ullrich, K. J., and Schulz, I., 1986, Inositol trisphosphate releases Ca^{2+} from a nonmitochondrail store site in permeabilized rat cortical kidney cells, *Kidney Int.* **29**:695–702.

Thieleczek, R., and Heilmeyer, L. M. G., Jr., 1986, Inositol 1,4,5-trisphosphate enhances Ca^{2+}-sensitivity of the contractile mechanism of chemically skinned rabbit skeletal muscle fibres, *Biochem. Biophys. Res. Commun.* **135**:662–669.

Thomas, A. P., Marks, J. S., Coll, K. E., and Williamson, J. R., 1983, Quantitation and early kinetics of inositol lipid changes induced by vasopressin in isolated and cultured hepatocytes, *J. Biol. Chem.* **258**:5716–5725.

Thomas, A. P., Alexander, J., and Williamson, J. R., 1984, Relationship between inositol polyphosphate production and the increase of cytosolic free Ca^{2+} induced by vasopressin in isolated hepatocytes, *J. Biol. Chem.* **259**:5574–5584.

Thomopoulos, P., Testa, U., Gourdin, M. F., Hervy, C., Titeaux, M., and Vaincheaker, W., 1982, Inhibition of insulin receptor binding by phorbol esters, *Eur. J. Biochem.* **129**:389–393.

Tolbert, M. E. M., Butcher, F. R., and Fain, J. N., 1973, Lack of correlation between catecholamine effects on cyclic adenosine $3' : 5'$-monophosphate and glyconeogenesis in isolated rat liver cells, *J. Biol. Chem.* **248**:5686–5692.

Tolbert, M. E. M., White, A. C., Aspry, K., Cutts, J., and Fain, J. N., 1980, Stimulation by vasopressin and α-catecholamines of phosphatidylinositol formation in isolated rat liver parenchymal cells, *J. Biol. Chem.* **255**:1938–1944.

Tsien, R. Y., 1980, New calcium indicators and buffers with high selectivity against magnesium and protons: Design, synthesis, and properties of prototype structures, *Biochemistry* **19**:2396–2404.

Tsien, R. Y., Pozzan, T., and Rink, T. J., 1982, Calcium homeostasis in intact lymphocytes: Cytoplasmic free calcium monitored with a new intracellularly trapped fluorescent indicator, *J. Cell Biol.* **94**:325–334.

Tsien, R. Y., Pozzan, T., and Rink, T. J., 1984, Measuring and manipulating cytosolic Ca^{2+} with trapped indicators, *Trends Biochem. Sci.* **9**:263–266.

Uchida, T., Ito, H., Baum, B. J., Roth, G. S., Filburn, C.R., and Sacktor, B., 1982, Alpha$_1$-adrenergic stimulation of phosphatidylinositol–phosphatidic acid turnover in rat parotid cells, *Mol. Pharmacol.* **21**:128–132.

Ueda, T., Chueh, S.-H., Noel, M. W., and Gill, D. L., 1986, Influence of inositol 1,4,5-trisphosphate and guanine nucleotides on intracellular calcium release within the N1E-115 neuronal cell line, *J. Biol. Chem.* **261**:3184–3192.

Uhing, R. J., Jiang, H., Prpic, V., and Exton, J. H., 1985, Regulation of a liver plasma membrane phosphoinositide phosphodiesterase by guanine nucleotides and calcium, *FEBS Lett.* **188**:317–320.

Uhing, R. J., Prpic, V., Jiang, H., and Exton, J. H., 1986, Hormone stimulated polyphosphoinositide breakdown in rat liver plasma membranes: Roles of guanine nucleotides and calcium, *J. Biol. Chem.* **261**:2140–2146.

Ui, M., Exton, J. H., and Park, C. R., 1973, Effects of glucagon on glutamate metabolism in the perfused rat liver, *J. Biol. Chem.* **248**:5350–5359.

Vandenberg, C. A., and Montal, M., 1984, Light-regulated biochemical events in invertebrate photoreceptors. 2. Light-regulated phosphorylation of rhodopsin and phosphoinositides in squid photoreceptor membranes, *Biochemistry* **23**:2347–2353.

Van de Werve, G., Proietto, J., and Jeanrenaud, B., 1985, Control of glycogen phosphorylase interconversion by phorbol esters, diacylglycerols, Ca^{2+}, and hormones in isolated rat hepatocytes, *Biochem. J.* **231**:511–516.

Van Dop, C., Yamanaka, G., Steinberg, F., Sekura, R. D., Manclark, C. R., Stryer, L., and Bourne, H. R., 1984, ADP-ribosylation of transducin by pertussis toxin blocks the light-stimulated hydrolysis of GTP and cGMP in retinal photoreceptors, *J. Biol. Chem.* **259**:23–26.

Vanecek, J., Sugden, D., Weller, J. L., and Klein, D. C., 1985, Atypical synergistic α$_1$- and β-adrenergic regulation of adenosine 3',5'-monophosphate and guanosine 3',5'-monophosphate in rat pinealocytes, *Endocrinology* **116**:2167–2173.

Varsanyi, M., and Heilmeyer, L. M. G., Jr., 1981, Phosphorylation of the 100,000 Mr Ca^{2+}-transport ATPase by Ca^{2+} or cyclic AMP-dependent and -independent protein kinases, *FEBS Lett.* **131**:223–228.

Venter, J. C., Eddy, B., Hall, L. M., and Fraser, C. M., 1984a, Monoclonal antibodies detect the conservation of muscarinic cholinergic receptor structure from Drosophila to human brain and detect possible structural homology with α$_1$-adrenergic receptors, *Proc. Natl. Acad. Sci. USA* **81**:272–276.

Venter, J. C., Horne, P., Eddy, B., Gregusta, R., and Fraser, C. M., 1984b, Alpha$_1$-adrenergic receptor structure, *Mol. Pharmacol.* **26**:196–205.

Vergara, J., Tsien, R. Y., and Delay, M., 1985, Inositol 1,4,5-trisphosphate: A possible chemical link in excitation–contraction coupling in muscle, *Proc. Natl. Acad. Sci. USA* **82**:6352–6356.

Verghese, M. W., Smith, C. D., and Snyderman, R., 1985, Potential role for a guanine nucleotide regulatory protein in chemoattractant receptor mediated polyphosphoinositide metabolism, Ca^{2+} mobilization, and cellular respiration by leukocytes, *Biochem. Biophys. Res. Commun.* **127**:450–457.

Verhoeven, A. J., Estrela, J. M., and Meijer, A. J., 1985, α-Adrenergic stimulation of glutamine metabolism in isolated rat hepatocytes, *Biochem. J.* **230**:457–463.

Vickers, J. D., Kinlough-Rathbone, R. L., and Mustard, J. F., 1984, Changes in the platelet phosphoinositides during the first minute after stimulation of washed rabbit platelets with thrombin, *Biochem. J.* **219**:25–31.

Vincentini, L. M., Ambrosini, A., DiVirgilio, F., Pozzan, T., and Meldolesi, J., 1985a, Muscarinic receptor–induced phosphoinositide hydrolysis at resting cytosolic Ca^{2+} concentration in PC12 cells, *J. Cell. Biol.* **100**:1330–1333.

Vincentini, L. M., DiVirgilio, F., Ambrosini, A., Pozzan, T., and Meldolesi, J., 1985b, Tumor promoter phorbol 12-myristate, 13-acetate inhibits phosphoinositol hydrolysis and cytosolic Ca^{2+} rise induced by activation of muscarinic receptors in PC12 cells, *Biochem. Biophys. Res. Commun.* **127**:310–317.

Volpe, J., Salviati, G., DiVirgilio, R., and Pozzan, T., 1985, Inositol 1,4,5-trisphosphate induces calcium release from sarcoplasmic reticulum of skeletal muscle, *Nature* **316**:347–349.

Volpi, M., Yassin, R., Naccache, P. H., and Sha'afi, R. I., 1983, Chemotactic factors cause rapid decreases in phosphatidylinositol, 4,5-bisphosphate, and phosphatidylinositol 4-monophosphate in rabbit neutrophils, *Biochem. Biophys. Res. Commun.* **112**:957–964.

Volpi, M., Naccache, P. H., Molski, T. F. P., Shefcyk, J., Huang, C.-K., Marsh, M. L., Munoz, J., Becker, E. L., and Sha'afi, R. I., 1985, Pertussis toxin inhibits fMet-Leu-Phe but not phorbol ester stimulated changes in rabbit neutrophils, *Proc. Natl. Acad. Sci. USA* **82**:2708–2712.

Vulliet, P. R., Woodgett, J. R., and Cohen, P., 1984, Phosphorylation of tyrosine hydroxylase by calmodulin-dependent multiprotein kinase, *J. Biol. Chem.* **259**:13680–13683.

Waisman, D. M., Gimble, J. M., Goodman, D. B. P., and Rasmussen, H., 1981, Studies on the Ca^{2+} transport mechanism of human erythrocyte inside–out plasma membrane vesicles J. Biol. Chem. **256**:415–424.

Wallace, M. A., and Fain, J. N., 1985, Guanosine 5'-0-thiotriphosphate stimulates phospholipase C activity in plasma membranes of rat hepatocytes, *J. Biol. Chem.* **260**:9527–9530.

Wallace, M. A., Randazzo, P., Li, S. Y., and Fain, J. N., 1982, Direct stimulation of phosphatidylinositol degradation by addition of vasopressin to purified rat liver plasma membranes, *Endocrinology* **111**:341–343.

Wallace, M. A., Poggioli, J., Giraud, F., and Claret, M., 1983, Norepinephrine-induced loss of phosphatidylinositol from isolated rat liver plasma membrane, *FEBS Lett.* **156**:239–243.

Watkins, P. A., Moss, J., Burns, D. L., Hewlett, E. L., and Vaughan, M., 1984, Inhibition of bovine outer rod segment GTPase by *Bordetella pertussis* toxin, *J. Biol. Chem.* **259**:1378–1381.

Weiss, S. J., and Putney, J. W., Jr., 1978, Does calcium mediate the increase in potassium permeability due to phenylephrine or angiotension II in the liver, *J. Pharmacol. Exp. Ther.* **207**:669–676.

Weiss, S. J., McKinney, J. S., and Putney, J. W., Jr., 1982, Receptor-mediated net breakdown of phosphatidylinositol 4,5-bisphosphate in parotid acinar cells, *Biochem. J.* **206**:555–560.

Whitaker, M., and Irvine, R. F., 1984, Inositol 1,4,5-trisphosphate microinjection activates sea urchin eggs, *Nature* **312**:636–639.

White, J. R., Huang, C. K., Hill, J. M., Jr., Naccache, P. H., Becker, E. L., and Sha'afi, R. I., 1984, Effect of phorbol 12-myristate 13-acetate and its analogue 4α-phorbol 12,13-didecanoate on protein phosphate and lysosomal enzyme release in rabbit neutrophils, *J. Biol. Chem.* **259**:8605–8611.

Williamson, D. H., Ilic, V., Tordoff, A. F. C., and Ellington, E. V., 1980, Interactions between vasopressin and glucagon on ketogenesis and oleate metabolism in isolated hepatocytes from fed rats, *Biochem. J.* **186**:621–624.

Williamson, J. R., Cooper, R. H., and Hoek, J. B., 1981, Role of calcium in the hormonal regulation of liver metabolism, *Biochim. Biophys. Acta* **639**:243–295.

Wilson, D. B., Connolly, T. M., Bross T. E., Majerus, P. W., Sherman, W. R., Tyler, A. N.,

Rubin, L. J., and Brown, J. E., 1985, Isolation and characterization of the inositol cyclic phosphate products of polyphosphoinositide cleavage by phospholipase C, *J. Biol. Chem.* **260:**13496–13501.

Wirthensohn, G., Lefrank, S., and Guder, W. G., 1984, Phospholipid metabolism in rat kidney cortical tubules. II. Effects of hormones on ^{32}P incorporation, *Biochim. Biophys. Acta* **795:**401–410.

Wolf, B. A., Comens, P. G., Ackermann, K. E., Sherman, W. R., and McDaniel, M. L., 1985, The digitonin-permeablized pancreatic islet model, *Biochem. J.* **227:**965–969.

Wolf, M., LeVine, H., III, May, W. S., Jr., Cuatrecasas, P., and Sahyoun, N., 1985, A model for intracellular translocation of protein kinase C involving synergism between Ca^{2+} and phorbol esters,, *Nature* **317:**546–551.

Wolfe, B. B., Harden, K., and Molinoff, P. B., 1976, β-Adrenergic receptors in rat liver: Effects of adrenalectomy, *Proc. Natl. Acad. Sci. USA* **73:**1343–1347.

Woodgett, J. R., Davison, M. T., and Cohen, P., 1983, The calmodulin-dependent glycogen synthase kinase from rabbit skeletal muscle: Purification, subunit structure, and substrate specificity, *Eur. J. Biochem.* **136:**481–487.

Woodgett, J. R., Cohen, P., Yamauchi, T., and Fujisawa, H., 1984, Comparison of calmodulin-dependent glycogen synthase kinase from skeletal muscle and calmodulin-dependent protein kinase-II from brain, *FEBS Lett.* **163:**329–334.

Wooten, M. W., and Wrenn, R. W., 1984, Phorbol ester induces intracellular translocation of phospholipid/Ca^{2+}-dependent protein kinase and stimulates amylase secretion in isol pancreatic acini, *FEBS Lett.* **171:**183–186.

Yamada, S., Yamamura, H. I., and Roeske, W. R., 1980, The regulation of cardiac α_1-adrenergic receptors by guanine nucleotides and by muscarinic cholinergic agonists, *Eur. J. Pharmacol.* **63:**239–241.

Yamamoto, H., Fukunaga, K., Tanaka, E., and Miyamoto, E., 1983, Ca^{2+}- and calmodulin-dependent prosphorylation of microtubule-associated protein 2 and factor τ, and inhibition of microtubule assembly, *J. Neurochem* **41:**1119–1125.

Yamanishi, J., Takai, Y., Kaibuchi, K., Sano, K., Castagna, M., and Nishizuka, Y., 1983, Synergistic functions of phorbol ester and calcium in serotonin release from human platelets, *Biochem. Biophys. Res. Commun.* **112:**778–786.

Yamauchi, T., Nakata, H., and Fujisawa, H., 1981, A new activator protein that activtes tryptophan 5-monooxygenase and tyrosine 3-monooxygenase in the presence of Ca^{2+}-, calmodulin-dependent protein kinase, *J. Biol. Chem.* **256:**5404–5409.

Yano, K., Nakashima, S., and Nozawa, Y., 1983, Coupling of polyphosphoinositide breakdown with calcium efflux in formyl–methionyl–leucyl–phenlalanine-stimulated rabbit neutrophils, *FEBS Lett.* **161:**296–300.

Yin, H. L., and Stossel, T. P., 1982, Calcium control of actin network structure by gelsolin, in: *Calcium and Cell Function,* Vol. II, pp. 325–337.

Zawalich, W., Brown, C., and Rasmussen, H., 1983, Insulin secretion: Combined effects of phorbol ester and A23187, *Biochem. Biophys. Res. Commun.* **117:**448–455.

Zurini, M., Krebs, J., Penniston, J. T., and Carafoli, E., 1984, Controlled proteolysis of the purified Ca^{2+}-ATPase o the erythrocyte membrane, *J. Biol. Chem.* **259:**618–627.

5

REGULATION OF PROTEIN KINASE C BY LIPID COFACTORS

Barry R. Ganong, Carson R. Loomis, Yusuf A. Hannun, and Robert M. Bell

1. INTRODUCTION

Hormone-stimulated biological responses in a number of tissues and cell types are associated with turnover of inositol phospholipids. This phenomenon was first observed over 30 years ago by Hokin and Hokin (1953) and, since then, has been demonstrated in a wide variety of systems (reviewed by Michell, 1975; Berridge, 1984; Hirasawa and Nishizuka, 1985; Hokin, 1985). The significance of phosphatidylinositol turnover remained a mystery until two relatively recent discoveries helped clarify the role of inositol phosphatides in transmembrane signaling.

The first stemmed from observations that phosphatidylinositol turnover was accompanied by calcium mobilization, though the relationship between the two was not understood. Michell (1975) suggested that phosphatidylinositol turnover itself might precede and initiate mobilization of calcium. Following observations that polyphosphoinositide degradation in liver occurs very rapidly following vasopressin treatment, Michell and co-workers (1981) proposed that polyphosphoinositide degradation is the primary response to agonist treatment. Berridge (1983) found a rapid increase in inositol-1,4,5-trisphosphate followed immediately by inositol-1,4-bisphosphate upon stimulation of insect salivary glands, over a time course during which neither inositol phosphate nor inositol increased. He suggested that inositol-1,4,5-trisphosphate or inositol-1,4-bisphosphate may function as a second messenger to control calcium release. Since then, this

Barry R. Ganong, Carson R. Loomis, and Robert M. Bell • Department of Biochemistry, Duke University Medical Center, Durham, North Carolina 27710. Yusuf A. Hannun • Departments of Biochemistry and Medicine, Duke University Medical Center, Durham, North Carolina 27710.

hypothesis has been strongly supported experimentally (Berridge, 1984; Hira-sawa and Nishizuka, 1985; Hokin, 1985). Inositol-1,4,5-trisphosphate was found to stimulate calcium release from nonmitochondrial stores in permeabilized pancreatic cells (Streb *et al.*, 1983) and hepatocytes (Burgess *et al.*, 1984; Joseph *et al.*, 1984; Thomas *et al.*, 1984).

The second discovery concerned a cyclic nucleotide–independent protein kinase that required phospholipids and calcium for activity (Nishizuka, 1984). This enzyme, known as protein kinase C, was found to be markedly stimulated by diacylglycerol (Takai *et al.*, 1979c; Kishimoto *et al.*, 1980), which also increased the affinity of the enzyme for calcium, resulting in greater activity at much lower concentrations of calcium. Since diacylglycerol is also released by phosphatidylinositol turnover, it was proposed that diacylglycerol serves as a second messenger for protein kinase C activation.

Interest in protein kinase C was also stimulated when it was found that protein kinase C is the high-affinity receptor for phorbol diesters (reviewed by Ashendel, 1985). Phorbol esters comprise a class of compounds isolated from plants that exhibit a diversity of biological responses in animal tissues and cells, including proliferation, differentiation, and tumor promotion (Blumberg, 1980, 1981). Phorbol esters were found to activate protein kinase C directly *in vitro* in the same manner as does diacylglycerol and to stimulate protein kinase C in platelets (Castagna *et al.*, 1982).

Protein kinase C is thus representative of a small class of enzymes whose activity is known to be modulated *in vivo* by lipids and unique in its regulation by diacylglycerols. It is the goal of this review to describe recent experimental investigations of lipid regulation of protein kinase C *in vivo* and *in vitro* and, from these data, to propose a molecular model of protein kinase C activation by diacylglycerol, phospholipids, and calcium, which may provide a framework to guide subsequent inquiry into the regulation of this important enzyme.

2. ACTIVATION OF PROTEIN KINASE C *IN VIVO* BY DIACYLGLYCEROLS

Studies of protein kinase C *in vitro* indicated that the enzyme's activity is regulated by diacylglycerol, suggesting a role for diacylglycerol as a second messenger in hormone-mediated responses. Diacylglycerol is a common metabolic intermediate in lipid metabolism (Bell and Coleman, 1980), which, *a priori*, makes it seem an unlikely candidate for a second messenger. Confirmation of such a role would require demonstration of protein kinase C activation and biological responses following treatment of intact cells with exogenous diacylglycerols.

There are two requirements for a diacylglycerol to exhibit biological activity in intact cells. First, the diacylglycerol must partition from the aqueous medium, presumably as dissolved monomers, into the outer leaflet of the cell's plasma membrane. Once inserted into the membrane, it must be free to flip across the membrane to become accessible to the intracellular compartment. Flip-flop of diacylglycerol across membranes is rapid (Allan *et al.*, 1978; Ganong and Bell, 1984), so the only hindrance to biological responses by diacylglycerols is diffusion of monomers to the cell surface.

Studies with typical diacylglycerols, such as dioleoylglycerol, generally failed to indicate any biological activity (Nishizuka, 1984). However, biologically relevant diacylglycerols, possessing two long fatty acyl groups, are minimally soluble in water. These compounds would phase separate rapidly upon addition to aqueous media and, therefore, would not be taken up efficiently by cells. To overcome this difficulty, a diacylglycerol was synthesized with oleate in position 1 and acetate in position 2 of the glycerol backbone. This compound activated protein kinase C *in vitro,* as well as dioleoylglycerol (Mori *et al.,* 1982), but the presence of a short chain in position 2 would be expected to increase its solubility, and, indeed, *sn*-1-oleoyl-2-acetylglycerol (OAG) has been found to have numerous biological activities.

In addition, diacylglycerols with symmetric, short acyl chains have been prepared and tested in similar systems because of three problems in interpreting some of OAG's responses. First, OAG is similar to the most potent of the phorbol diesters, PMA,* in possessing a long-chain fatty acyl ester adjacent to an acetate residue (Nishizuka, 1984). OAG may be recognized more as an analogue of PMA than as an analogue of endogenous diacylglycerol. Second, due to the great discrepancy in acyl chain lengths, OAG would probably have a different conformation in a bilayer than a diacylglycerol with acyl chains of more symmetric length, resulting in a conformation of polar groups at the membrane surface different from that of the endogenous molecular species. Third, the presence of a single long acyl chain coupled to a fairly compact, moderately polar head group may give OAG some properties of nonionic detergents. Use of diacylglycercols with two short fatty acyl moieties of the same length would overcome the solubility limitation posed by long-chain diacylglycerols without invoking these potential problems.

Reports of biological effects of phorbol esters are too numerous to cite (see Blumberg, 1980, 1981; Ashendel, 1985). This review will concentrate on those responses mimicked by exogenous diacylglycerols. These data have been interpreted as confirmations that phorbol diester effects *in vivo* are mediated through protein kinase C activation and that diacylglycerols are indeed bioregulatory second messengers.

---

* PMA, phorbol-12-myristate-13-acetate (has also been abbreviated as TPA, 12-tetradecanoylphorbol-13-acetate).

2.1. Specific Phosphorylations and Inhibition of Phorbol Ester Binding

The most straightforward ways of demonstrating association of exogenous diacylglycerols with protein kinase C in intact cells are to show that diacylglycerols mimic phorbol ester–stimulated phosphorylations and inhibit phorbol ester binding. Specific phorbol ester–induced phosphorylations were mimicked by exogenous diacylglycerols in human platelets (Sano *et al.*, 1983; Kaibuchi *et al.*, 1983), lymphocytes (Kaibuchi *et al.*, 1985), neutrophils (Fujita *et al.*, 1984), HL-60 human leukemia cells (Ebeling *et al.*, 1985; Kreutter *et al.*, 1985), human and sheep erythrocytes (Raval and Allan, 1985), rat peritoneal mast cells (Katakami *et al.*, 1984), GH$_3$ pituitary cells (Drust and Martin, 1984), and 3T3-L1 fibroblasts (Blackshear *et al.*, 1985). Protein kinase C phosphorylates serine and threonine residues in target proteins (Nishizuka, 1984). Interestingly, both OAG and PMA were found to stimulate tyrosine-specific phosphorylation of a M_r 42,000 protein in chick embryo fibroblasts (Gilmore and Martin, 1983).

Exogenous diacylglycerols were found to compete for phorbol diester binding to human lymphocytes (Guy *et al.*, 1985), neutrophils (Nishihira and O'Flaherty, 1985; O'Flaherty *et al.*, 1985b; Cox *et al.*, 1986), HL-60 cells (Ebeling *et al.*, 1985), A431 human epidermoid carcinoma cells (Davis *et al.*, 1985a,b), and 3T3 fibroblasts (Rozengurt *et al.*, 1984).

2.2. Platelets

Platelet activation by agonists, such as thrombin or platelet-activating factor, is initiated by phosphoinositide turnover. Two specific phosphorylation events signal involvement of two second messenger pathways, each stimulating a distinct protein kinase. Myosin light chain, of M_r 20,000, is phosphorylated by calmodulin-dependent myosin light-chain kinase (Hathaway and Adelstein, 1979), indicating an increase in free cytoplasmic calcium. This response, mediated physiologically by inositol-1,4,5-trisphosphate, may be mimicked by low levels of calcium ionophore A23187 (Kaibuchi *et al.*, 1983).

The other phosphorylated protein is a polypeptide of M_r 40,000 (47,000).* Its phosphorylation is not induced by low levels of A23187 but is strongly stimulated by PMA, at concentrations where phosphatidylinositol turnover does not occur and myosin light-chain phosphorylation is minimal (Castagna *et al.*, 1982; Yamanishi *et al.*, 1983). This suggests that the 40K protein is a substrate of protein kinase C, a conclusion supported by similar tryptic phosphopeptide patterns from 40K protein phosphorylated either in a purified cell-free system

---

* Recent evidence indicates that the 40K (47K) protein phosphorylated by protein kinase C in platelets is an inositol trisphosphate 5′-phosphomonoesterase (Connolly *et al.*, 1986).

with homogeneous protein kinase C or in intact platelets in response to thrombin, platelet-activating factor, or PMA (Ieyasu *et al.*, 1982; Sano *et al.*, 1983; Yamanishi *et al.*, 1983).

Neither A23187 nor PMA alone is capable of fully activating platelets at moderate concentrations that induce significant phosphorylation of the 20K or 40K protein, respectively. Stimulation of both protein kinase C and calcium-dependent events, by agonists that evoke phosphatidylinositol turnover or by synergistic treatment with low levels of PMA and A23187, appears necessary for platelet activation. Responses indicative of full activation include shape change, aggregation, and release of serotonin and ATP from intracellular stores. PMA and A23187 together stimulated release of over seven times the amount of serotonin released by either alone (Yamanishi *et al.*, 1983). However, higher concentrations of PMA or A23187 alone also caused significant release. The current hypothesis that a full platelet response requires simultaneous elevation of diacylglycerol and calcium may undergo revision as new data are accumulated.

When OAG was tested for its ability to mimic platelet responses, its effects were essentially the same as those of PMA. OAG stimulated 40K phosphorylation without phosphatidylinositol turnover and induced release of serotonin and lysosomal *N*-acetylglucosaminidase only in synergism with A23187 (Kaibuchi *et al.*, 1983; Kajikawa *et al.*, 1983; Friesen and Gerrard, 1985). However, at high concentrations, OAG unexpectedly caused secretion of a significant amount of serotonin and *N*-acetylglucosaminidase. A high concentration of OAG also caused aggregation and release of ATP without raising free calcium concentrations, as measured by quin2 (2-[(2-bis-[Carboxymethyl]amino-5-methyl-phenoxy)methyl]-6-methoxy-8-bis[carboxymethyl]amino-quinoline) fluorescence (Rink *et al.*, 1983).

In another study, OAG was compared with symmetric short-chain diacylglycerols in their ability to stimulate platelets. diC_6, diC_8, and diC_{10}* stimulated 40K phosphorylation as efficiently as OAG, without stimulating phosphatidylinositol turnover or arachidonate release (Watson *et al.*, 1984; Lapetina *et al.*, 1985). At diacylglycerol concentrations promoting maximal 40K phosphorylation, neither aggregation nor release of ATP or serotonin occurred unless calcium or suboptimal A23187 was included in the assay buffer. Monoacylglycerols and analogues were unable to promote 40K phosphorylation, in agreement with their inability to stimulate protein kinase C *in vitro* (Takai *et al.*, 1979c; Kishimoto *et al.*, 1980; Mori *et al.*, 1982). In addition, 1,2- and 1,3-dioleoylglycerol, and 1,2-dioleoylglyceryl ether were inactive, though conclusions regarding specificity could not be made due to their limited solubility. Finally, structural analogues of diC_8, with -H, -Cl, or -SH in place of the -OH, did not induce 40K phosphorylation, suggesting a stringent requirement for the free three-hydroxyl group (Lapetina *et al.*, 1985).

---

*diC_n, *sn*-1,2-diacylglycerol with saturated acyl moieties *n* carbons in length.

These results support a model for platelet activation requiring simultaneous calcium mobilization and protein kinase C activation. However, the roles for diacylglycerol and calcium appear to be more complex than originally proposed. Pretreatment of platelets with OAG or phorbol esters was shown to inhibit agonist-induced phosphoinositide turnover (Watson and Lapetina, 1985), suggesting a mechanism for negative feedback control of the activation pathway. Data have been presented that some of OAG's effects on platelets are secondary responses to ADP release and that OAG may, in fact, weakly stimulate cAMP production (Ashby *et al.*, 1985).

Hallam and co-workers (1985) reported that 20K phosphorylation was induced by thrombin or platelet-activating factor in the absence of external calcium, at intracellular calcium levels well below those required for ionophore-induced 20K phosphorylation, suggesting an additional mechanism for myosin light-chain phosphorylation. Phorbol esters and OAG were shown to induce phosphorylation of myosin light chain in platelets, though at different sites than are modified by calcium-dependent myosin light-chain kinase (Naka *et al.*, 1983; Friesen and Gerrard, 1985). The physiological significance of this phosphorylation is not known. Further, the calcium independence of PMA- and OAG-mediated aggregation and secretion, indicated by experiments with quin2, was contradicted by demonstration of a transient but significant increase in cytosolic calcium that was not detected by quin2 but was indicated by the calcium-sensitive photoprotein, aequorin (Ware *et al.*, 1985).

Arachidonic acid and its metabolites also play a role in platelet activation. Thrombin and collagen stimulate release of arachidonic acid from phospholipids, with subsequent formation of prostaglandins and thromboxane A_2 (Samuelsson, 1977). Arachidonic acid may be released from diacylglycerol via glyceride lipases or from phospholipids via phospholipase A_2. The latter pathway may be activated solely by increased intracellular calcium without inositol phosphatide turnover (Rittenhouse and Horne, 1984). PMA and OAG have been found to stimulate this reaction by an unknown mechanism (Halenda *et al.*, 1985), indicating an additional potential role of diacylglycerol in platelet activation.

2.3. Neutrophils

Activated neutrophils exhibit a variety of responses, including superoxide production, degranulation, and aggregation. Among the most potent activators are chemotactic peptides, such as C5a of the complement system and the *N*-formylated peptide fMet-Leu-Phe. Exposure of neutrophils to fMet-Leu-Phe induced rapid phosphoinositide turnover (Dougherty *et al.*, 1984) and increased intracellular calcium (Pozzan *et al.*, 1983; Korchak *et al.*, 1984), suggesting roles for protein kinase C and calcium in mediating neutrophil function. In support of this are demonstrations of synergism between PMA and calcium ionophores in stimulating superoxide production, release of granular constituents, and aggregation (di Virgilio et al., 1984; Robinson et al., 1984; White et al., 1984). Nevertheless, as in the case of platelets, these responses could be induced by high concentrations of either PMA (Korchak et al., 1984; di Virgilio et al., 1984; White et al., 1984) or A23187 (Goldstein et al., 1974; O'Flaherty et al., 1978).

Similar biological responses were elicited by OAG, its palmitoyl (PAG) and myristoyl (MAG) analogues, and short-chain symmetric diacylglycerols. OAG was synergistic with A23187 in releasing N-acetylglucosaminidase (Kajikawa et al., 1983), and with both A23187 and fMet-Leu-Phe in stimulating superoxide production (Penfield and Dale, 1984; Dewald et al., 1984). However, OAG alone stimulated oxygen consumption (Cooke and Hallett, 1985) and superoxide production (Fujita et al., 1984), with no apparent rise in cytosolic calcium. In all respects tested, OAG, MAG, and PAG resembled PMA and mezerein, a tumor promoter that also interacts with protein kinase C (O'Flaherty et al., 1985b; Nishihira and O'Flaherty, 1985).

Short-chain diacylglycerols stimulated lysozyme (but not β-glucuronidase) release and superoxide production in the absence of ionophores and inhibited binding of phorbol-12,13-dibutyrate (PDBu) to its receptor (Cox et al., 1986). In every case, diC_8 was most potent, followed by diC_7 and diC_9. Interestingly, diC_{10} effectively stimulated lysozyme release but not superoxide production, suggesting the possibility that two distinct pools of protein kinase C may mediate these responses. This was supported by the observation that diC_{10} displaced only 45% of bound PDBu from neutrophils, whereas diC_6–diC_9 displaced over 90%. None of the diacylglycerols tested was chemotactic.

Other compounds capable of inducing or potentiating neutrophil activation include platelet-activating factor and the arachidonate metabolites leukotriene B_4 and 5-HETE. To varying degrees, OAG, PAG, and MAG were synergistic with 5-HETE, leukotriene B_4, A23187, and fMet-Leu-Phe in releasing β-glucuronidase and lysozyme from primary and secondary granules, respectively, whereas a diether analogue of PAG was inactive (O'Flaherty et al., 1984; O'Flaherty et al., 1985a). MAG and 5-HETE also synergistically stimulated superoxide production (O'Flaherty et al., 1985a). Both indomethacin (Dale and Penfield, 1985) and prostaglandins of the E series (Penfield and Dale, 1985) were shown to enhance OAG-induced superoxide generation, and it was suggested that this enhancement may result from inhibition of metabolic attenuation of the diacylglycerol signal.

Superoxide production is catalyzed by oxidation of NADPH, generated by glucose oxidation via the pentose phosphate pathway. Normally dormant, NADPH oxidase is stimulated in response to neutrophil activators. Because PMA and OAG induce the oxidative burst in neutrophils, protein kinase C has been suggested to mediate activation of NADPH oxidase. Recently, protein kinases C has been shown to phosphorylate and activate NADPH oxidase from neutrophils *in vitro* (Cox *et al.*, 1985; Papini *et al.*, 1985).

McCall and co-workers (1985) have demonstrated that OAG, PMA, and mezerein stimulated hexose transport in neutrophils. Both in phorbol ester–treated cells and *in vitro* with purified enzyme, the glucose transporter of erythrocytes was phosphorylated by protein kinase C (Witters *et al.*, 1985). Kitagawa and co-workers (1985) have presented evidence that PMA may increase hexose transport in 3T3 cells by stimulating translocation of the transporter to the plasma membrane.

These studies clearly indicate a role for protein kinase C in neutrophil function. A number of phosphorylations induced by PMA (Helfman *et al.*, 1983; White *et al.*, 1984; Fujita *et al.*, 1984) or OAG (Fujita *et al.*, 1984) have been described, though none of their functions is known.

2.4. Other Release Reactions

Treatment of rat peritoneal mast cells with concanavalin A stimulated phosphatidylinositol turnover, histamine release, and phosphorylation of several cytosolic proteins. In these cells, histamine release and phosphorylations could be induced by PMA or OAG, and these responses were enhanced by low levels of A23187 (Katakami *et al.*, 1984). Purified mast-cell protein kinase C catalyzed the same phosphorylations *in vitro*.

Luteinizing hormone (LH) release from the pituitary is triggered by gonadotropin-releasing hormone (GnRH) by a mechanism involving calcium and phosphoinositide turnover. Calmodulin has been implicated in GnRH action. PMA stimulated LH release, a response mimicked by diC_8, diC_6, and OAG, with diC_8 being most effective (Conn *et al.*, 1985; Naor and Eli, 1985). Release by PMA and diacylglycerol was enhanced by A23187 (Harris *et al.*, 1985; Naor and Eli, 1985). However, although GnRH action is absolutely dependent on extracellular calcium, LH release by PMA or diacylglycerols is not (Conn *et al.*, 1985; Naor and Eli, 1985). Furthermore, unlike GnRh, neither PMA- nor diC_8-stimulated release was inhibited by antagonists of calmodulin, calcium channels, or GnRH (Harris *et al.*, 1985).

Clonal cell lines derived from rat pituitary tumors (GH cells) have provided a system for studying release mechanisms for prolactin. PMA-stimulated prolactin release from GH_4C_1 cells was enhanced by A23187 (Delbeke *et al.*, 1984). In permeabilized (Ronning and Martin, 1985) or intact (Martin and Kowalchyk, 1984a,b) GH_3 cells, roles were indicated for calcium and diacylglycerol in thyrotropin-releasing hormone-induced prolactin release. OAG, 1,2- and 1,3-dioleoylglycerol, and natural, mixed species diacylglycerols, but not 1,2-distearoyl- or 1,2-dipalmitoylglycerol, stimulated release. A similar observation was made using primary pituitary cultures, where A23187 or a calcium-channel activator enhanced PMA- and OAG-induced prolactin release (Koike *et al.*, 1985). Interestingly, PMA and diC$_8$ inhibited prolactin synthesis and release from human decidual tissue (Harman *et al.*, 1986a), whereas they stimulated placental lactogen synthesis and release from trophoblast cells (Harman *et al.*, 1986b).

Pancreatic islet cells respond to glucose and carbamylcholine by degrading polyphosphoinositides (Best and Malaisse, 1984) concomitant with insulin release. OAG was found to induce insulin release in the presence of extracellular calcium and suboptimal concentrations of glucose (Malaisse *et al.*, 1985). Protein kinase C has been implicated in release reactions in the exocrine pancreas as well. Merritt and Rubin (1985) reported synergism between ionomycin and either PDB, 1,2-, or 1,3-dioleoylglycerol in stimulating amylase secretion by pancreatic acini.

Secretion of pulmonary surfactant, a complex of predominantly disaturated phospholipid and protein, by alveolar type II cells is critical in preventing alveolar collapse. Surfactant secretion is stimulated both by agents that increase cAMP and by PMA. Sano and co-workers (1985) demonstrated that OAG stimulated phosphatidylcholine secretion from cultured type II cells alone or in synergism with terbutaline, a β$_2$-adrenergic agonist.

Gastric acid secretion may be monitored by measuring aminopyrine accumulation in acidic intracellular spaces. In rat parietal cells, PMA and OAG inhibited secretagogue-induced aminopyrine accumulation in a similar manner (Anderson and Hanson, 1985).

2.5. Receptor Regulation

Binding of epidermal growth factor (EGF) to its receptor initiates a number of events, including activation of an endogenous tyrosine kinase, tyrosine-specific autophosphorylation (reviewed by Bertics *et al.*, 1985), and phosphoinositide degradation. The human epidermoid carcinoma line A431 overproduces EGF receptor, providing and ideal system for studying the biochemistry of EGF/EGF receptor interactions. Phorbol esters and other tumor promoters have profound effects on the EGF receptor. In A431 cells exposed to phorbol esters, phosphorylation of the EGF receptor was enhanced primarily at serine residues (Davis and Czech, 1984). Phosphorylation of the EGF receptor at serine and threonine residues by purified protein kinase C was also demonstrated *in vitro*, with either A431 membranes or purified receptor as substrate (Cochet *et al.*, 1984; Downward *et al.*, 1985). A specific threonine residue, at position 654, was identified as a critical substrate site (Hunter *et al.*, 1984; Davis and Czech, 1985). Phosphorylation of the EGF receptor by protein kinase C, either *in vitro* or *in vivo* in response to tumor promoters, decreases both its endogenous tyrosine kinase activity and its affinity for EGF, without affecting receptor number (Cochet *et al.*, 1984; Davis and Czech, 1984; Friedman *et al.*, 1984; Iwashita and Fox, 1984; Bertics *et al.*, 1985; Downward *et al.*, 1985).

These effects were mimicked by exogenous diacylglycerols. Treatment of Swiss 3T3 cells or rat tracheal epithelial 2C5 cells with OAG resulted in decreased affinity of the receptor for EGF (Sinnett-Smith and Rozengurt, 1985; Jetten *et al.*, 1985). diC_5–diC_{10} had the same effect on 2C5 and A431 cell (Davis *et al.*, 1985a,b; Jetten *et al.*, 1985). In A431 cells, diC_8 was the most potent diacylglycerol in promoting EGF receptor phosphorylation at serine and threonine, and inhibition of EGF binding and EGF-dependent tyrosine-specific autophosphorylation (Davis *et al.*, 1985b). Finally, 3-OH analogues of diC_8 were unable to promote any of these responses.

Binding of transferrin to its receptor results in rapid internalization of the receptor/transferrin complex. PMA induced internalization of unoccupied transferrin receptors by a mechanism that was kinetically indistinguishable from normal internalization of occupied receptors (Klausner *et al.*, 1984). This phorbol ester–induced internalization was reversible and was associated with hyperphosphorylation of the receptor (May *et al.*, 1984). Affinity-immobilized receptor was phosphorylated *in vitro* by highly purified protein kinase C activated by either phorbol esters or diacylglycerols (May *et al.*, 1985). diC_8 stimulated transferrin receptor internalization in the same manner as PMA (May *et al.*, 1986). Internalization, but not phosphorylation, was dependent on intact microtubules, as judged by studies with specific inhibitors (May *et al.*, 1985).

A decrease in both the number and affinity of β-adrenergic receptors on cardiac myocytes was induced by phorbol esters and OAG (Limas and Limas, 1985). As with the transferrin receptor, the decrease in β-receptor number appeared to result from internalization and required intact microtubules. Phosphorylation and down-regulation of β-adrenergic receptors in erythrocytes was induced by diC_8 (B. Strulovici and R. Lefkowitz, unpublished data). OAG and 1,2-dioleoylglycerol inhibited phenylephrine-stimulated elevations of cytoplasmic calcium in hepatocytes (Cooper *et al.*, 1985) and acetylcholine-dependent depolarization in cultured myotubules (Eusebi *et al.*, 1985), respectively, suggesting that protein kinase C also down-regulates α_1-adrenergic and nicotinic cholinergic receptors.

Besides the α_1-adrenergic receptor, OAG exerts negative control over other receptors that are normally coupled to phosphoinositide degradation. Phorbol diesters and OAG inhibited thrombin-stimulated formation of inositol phosphates in platelets (Watson and Lapetina, 1985), and blocked polyphosphoinositide breakdown and cytoplasmic calcium elevation in aortic smooth muscle cells stimulated by angiotensin II (Brock *et al.*, 1985) and 70Z/3 pre-B lymphocytes stimulated with lipopolysaccharide (Rosoff and Cantley, 1985).

A similar mechanism may regulate membrane proteins in T lymphocytes. Considerable evidence indicates that antigen recognition by T lymphocytes requires noncovalent association between two cell-surface proteins, the antigen receptor (Ti) and the T3 antigen. Antigen-dependent down-regulation of the Ti/T3 complex is important in regulating T-cell immune functions. PDBu was found to induce phosphorylation of the T3 antigen, and both PDBu and OAG caused a 50% decrease in cell-surface Ti/T3 complex (Cantrell et al., 1985).

When T lymphocytes are exposed to an antigen or mitogen, receptors for the lymphokine interleukin 2 (IL-2) are induced. PMA treatment of resting T lymphocytes or T leukemic cell lines resulted in rapid induction of IL-2 receptors (Depper et al., 1984; Shackelford and Trowbridge, 1984), indicating involvement of protein kinase C in induction. diC_8 also induced IL-2 receptors on resting T lymphocytes (Depper et al., 1985).

Phagocytic cells such as neutrophils, as well as other cells, possess cell-surface receptors for C3b of the complement system (Fearon, 1980). Phagocytosis requires internalization of C3b receptors occupied by C3b-coated targets. Ligand-independent internalization of the C3b receptor, which depended on an intact cytoskeleton (O'Shea et al., 1985a), was stimulated by phorbol esters (Changelian et al., 1985), OAG, and diC_8 and enhanced by A23187 (O'Shea et al., 1985b).

A recent report suggests that protein kinase C may regulate insulin receptor affinity. In human lymphocyte and monocyte cell lines, diacylglycerols caused a decrease in insulin receptor affinity in both intact cells and isolated membranes (Grunberger et al., 1985).

2.6. Proliferation, Differentiation, and Differentiated Functions

Among the biological responses of tumor promoters linking protein kinase C to malignant transformation are their effects on proliferation and differentiation. In a number of cases, these responses have been mimicked by exogenous diacylglycerols. OAG, diC_8, and PMA stimulated [^{3}H]thymidine incorporation in quiescent Swiss 3T3 fibroblasts in synergism with low concentrations of EGF, insulin, and other growth promoters (Rozengurt et al., 1984; Davis et al., 1985a).

Human peripheral lymphocytes are stimulated to proliferate by plant lectins such as phytohemagglutinin, unless macrophages are selectively removed, in which case an additional stimulus is necessary. Kaibuchi and co-workers (1985) found that both OAG and PMA stimulated a mitogenic response in macrophage-depleted human lymphocytes in the presence of A23187 and phytohemagglutinin. Similarly, both ionomycin and either OAG or diC_8 dramatically stimulated [^{3}H]thymidine incorporation into human tonsillar B lymphocytes (Guy et al., 1985).

Although increased c-*myc* gene expression is associated with the mitogenic activity of platelet-derived growth factor, c-*myc* expression does not appear to be sufficient for mitogenic response (Coughlin *et al.*, 1985). PMA, PDBu, and OAG stimulated c-*myc* expression to the same level as platelet-derived growth factor, with less then one-tenth the mitogenic response.

The human promyelocytic leukemia line, HL-60, has been a model system for studying differentiation in response to chemical inducers. Agents such as dimethylsulfoxide (DMSO), dibutyryl-cAMP, and retinoic acid stimulate HL-60 to differentiate to a neutrophillike cell type. On the other hand, 1,25-dihydroxy vitamin D_3 and tumor promoters, such as phorbol esters and teleocidin, cause HL-60 to differentiate into a cell type with characteristics of macrophages. Although OAG was ineffective at stimulating macrophage differentiation (McNamara *et al.*, 1984; Kreutter *et al.*, 1985; Yamamoto *et al.*, 1985), treatment of HL-60 with a 1-ether analog, 1-hexadecyl-2-acetylglycerol (HEG), induced nonspecific esterase activity and some morphologic changes similar to those seen in PMA-treated HL-60, though cell adherence was not observed (McNamara *et al.*, 1984). This result is unexpected in light of Cabot and Jaken's report (1984) that HEG does not activate protein kinase C. Perhaps HEG is slowly metabolized to its phosphatidylcholine product, platelet-activating factor, which elicits responses in several cells, and, at least in platelets, induces phosphatidylinositol turnover. In contrast to OAG, diC_8 stimulated macrophage differentiation of HL-60, as judged by morphological criteria (Ebeling *et al.*, 1985).

A murine pre-B lymphocyte cell line, 70Z/3, was stimulated to differentiate by either lipopolysaccharide, PMA, or OAG (Rosoff and Cantley, 1985). Differentiation was indicated by expression of cell-surface immunoglobulin M. Murine B cells, which already express membrane immunoglobulin, are further activated by exposure to specific antigens or to heterologous antisera directed against membrane immunoglobulin. This activation, involving membrane depolarization and expression of I-A antigens, was mimicked by PMA and OAG (Coggeshall and Cambier, 1985).

To probe the mechanism of hormone-induced maturation of rat oocytes, ovarian follicles from sexually immature female rats were exposed to either PMA or OAG. Initiation of maturation was indicated by germinal vesicle breakdown, and formation of the first polar body signaled complete maturation. PMA and OAG elicited both of these responses alone, as well as enhancing the same responses to a gonadotropin-releasing hormone agonist (Aberdam and Dekel, 1985).

Tumor promoters are also known to inhibit differentiation of some cell types. In ovarian granulosa cells, follicle-stimulating hormone (FSH)-stimulated steroidogenesis was inhibited by PMA (Welsh *et al.*, 1984; Shinohara *et al.*, 1985). OAG and diC_8 were found to dramatically inhibit FSH-induced production of progesterone, 20-α-OH-progesterone, and estrogen, and induction of LH/human chorionic gonadotropin receptors (Kasson *et al.*, 1985). OAG partially blocked aggregation and aromatase induction (Shinohara *et al.*, 1985). A 3-SH analogue of diC$_8$ had no effect. Since FSH provokes elevation of cAMP levels, the ability of diacylglycerols to inhibit cAMP production was assessed. diC$_8$ and OAG inhibited FSH-stimulated increases in both intracellular and extracellular cAMP levels. In addition, they inhibited steroid production induced by both forskolin and dibutyryl-cAMP. Thus, protein kinase C blocks differentiated functions of granulosa cells in at least two ways: by inhibiting both hormone-induced production of the second messenger cAMP and responses induced by cAMP itself.

Friend erythroleukemic cells, like HL-60, are stimulated by DMSO to differentiate. In several respects, including hemoglobin accumulation, the terminal cell type resembles orthochromatic normoblasts, nucleated precursors to erythrocytes. This differentiation is inhibited by PMA. OAG was found to inhibit DMSO-induced generation of heme-containing cells (Pincus *et al.*, 1984). Dioctanoylglycerol and dilauroylglycerol had similar, though less pronounced, effects.

In liver, glycogen phosphorylase may be activated by agonists that increase cAMP or calcium. Phosphorylase activation by a calcium-sensitive phosphorylase kinase may be mimicked by calcium ionophores. Van de Werve and coworkers (1985) have found that glycogen phosphorylase activation by ionophores is enhanced by phorbol esters, 1,2-dioleoylglycerol, and OAG. They suggest that involvement of protein kinase C in phosphorylase activation may be due to inhibition of phosphorylase phosphatase.

PMA stimulated thymidine incorporation in cultured thyroid cells. Both PMA and, to a lesser extent, diC$_8$ inhibited iodine uptake and incorporation into protein (Bachrach *et al.*, 1985).

2.7. Interactions with Other Second Messenger Systems

The relationship between the protein kinase C/calcium system and the cAMP system appears to vary between different tissue types. In ovarian granulosa cells, phorbol esters or diacylglycerols antagonize FSH and other agents that increase cAMP levels (Kasson *et al.*, 1985). Unlike its effects in liver, vasopressin regulates epithelial transport by increasing cAMP levels. In toad bladder, the ability of vasopressin to stimulate water flow was inhibited by PMA and diC$_8$ (Schlondorff *et al.*, 1985). diC$_8$ also inhibited the hydroosmotic effects of cAMP and 8-bromo-cAMP, and PMA decreased cellular cAMP levels, suggesting that protein kinase C activation inhibits vasopressin-mediated responses both proximal and distal to cAMP formation.

In contrast, surfactant release from alveolar type II cells induced by β-adrenergic agonists is enhanced by protein kinase C activators (Sano *et al.*, 1985). Growth hormone–releasing factor causes an increase in pituitary cAMP levels, a response that is enhanced by the tumor promoters PMA and teleocidin, as well as by OAG (Cronin and Canonico, 1985). Similarly, in rat pinealocytes, phorbol esters, OAG, and α_1-adrenergic agonists, which stimulate phosphoinositide degradation, enhanced cAMP production stimulated by the β agonist, isoproterenol (Sugden *et al.*, 1985).

Activation of platelets and neutrophils not only involves turnover of phosphoinositides but also arachidonic acid release from phospholipids, resulting in production of active eicosanoids. Arachidonic acid is thought to arise both by calcium-dependent phospholipase A_2 action on membrane phospholipids and by glyceride lipase-dependent degradation of diacylglycerol. Protein kinase C itself may affect arachidonic acid metabolism. Recent observations indicate that protein kinase C activation by PMA and OAG stimulates arachidonate release and prostaglandin production in synergism with interleukin-1 or transforming growth factors in several cell types (Levine and Xiao, 1985; Levine *et al.*, 1985).

2.8. Effects on Ion Conductance

A plasma membrane Na^+/K^+ antiporter is found in most mammalian cells whose activation results in cytoplasmic alkalinization, which is thought to be an important step in proliferative responses. A role for protein kinase C in regulating this exchanger was indicated by the demonstration that its activity was stimulated by phorbol diesters (see Grinstein *et al.*, 1985). Stimulation of Na^+/H^+ antiport activity by OAG was observed in HeLa and murine neuroblastoma cells (Moolenaar *et al.*, 1984), rat thymic lymphocytes (Grinstein *et al.*, 1985), and Swiss 3T3 cells (Vara *et al.*, 1985).

It has been postulated that neuronal voltage-sensitive calcium channels are regulated by phosphorylation. In chick dorsal root ganglia neurons, both OAG and a phorbol ester rapidly attenuated voltage-induced calcium currents (Rane and Dunlap, 1986). Recovery was also rapid. Maximal calcium currents were restored within 2 min of OAG removal.

2.9. Other Responses Elicited by Diacylglycerols

Among the responses that have been associated with, and indicative of, tumor promoter activity is induction of ornithine decarboxylase activity. In mouse epidermal cells (Sasakawa *et al.*, 1985) and guinea pig lymphocytes (Otani *et al.*, 1985b), ornithine decarboxylase activity was induced by PMA and OAG. However, in the lymphocytes, unlike the epidermal cells, OAG did not induce activity by itself but served to enhance induction mediated by calcium ionophores and agents that increase cAMP. In rat tracheal epithelial cells, ornithine decarboxylase was induced by diC_{10} and, to a lesser extent, diC_8 (Jetten *et al.*, 1985).

Intercellular communication via small molecule exchange through gap junctional contacts between adjacent cells can be demonstrated in confluent monolayers of cultured cells. Gainer and Murray (1985) and Enomoto and Yamasaki (1985) have employed a fluorescent dye transfer assay, and Davidson and coworkers (1985) used radiolabeled citrulline incorporation in cocultures of two cell lines, defective in either argininosuccinate synthetase or argininosuccinate lyase to demonstrate intercellular communication. In both assays, PMA and OAG inhibited communication rapidly.

Injection of adrenocorticotropic hormone (ACTH) into the third cerebral ventricle induces excessive grooming behavior in rats. To test their hypothesis that this response is due to inhibition of protein kinase C, Gispen and co-workers (1985) determined the effect of subsequent intraventricular injection of diC_8 and found a significant reduction in ACTH-stimulated grooming.

2.10. Effects of Phospholipase C Treatment

Treatment of cells with exogenous bacterial phospholipase C results in increased membrane diacylglycerol levels due to nonspecific hydrolysis of membrane phospholipids. Phospholipase C treatment has been shown to mimic many responses that are induced by phorbol esters and exogenous diacylglycerols. Table I presents a list of some of these responses.

3. ACTIVATION OF PROTEIN KINASE C *IN VITRO*

3.1. Activation by Lipids

3.1.1. Phospholipid Dependence

Protein kinase C is defined as a calcium- and phospholipid-dependent, diacylglycerol-activated protein kinase. Its activity was found to depend on a membrane component that could be extracted by Triton X-100 and chloroform–methanol, and phospholipid was identified as the required activator (Takai *et al.*, 1979a,b). In studies of phospholipid dependence, anionic phospholipids were the best activators, including phosphatidylinositol, cardiolipin, phosphatidic acid, and especially phosphatidylserine (Takai *et al.*, 1979a,b). Phosphatidyl-

TABLE I. Biological Responses Induced by Exogenous Phospholipase C

Cell type	Response	References
Human platelets	40K phosphorylation	Kawahara *et al.* (1980)
Human neutrophils	Superoxide production, degranulation	Grzeskowiak *et al.* (1985)
Human T lymphocytes	IL-2 receptor induction	Depper *et al.* (1985)
Friend erythroleukemic cells	Inhibited differentiation	Pincus *et al.* (1984)
Human fibroblasts/Chinese hamster V-79 cells	Reduced junctional communication	Davidson *et al.* (1985)
Human and sheep erythrocytes	Phosphoprotein changes	Raval and Allan (1985)
Guinea pig lymphocytes	Ornithine decarboxylase induction	Otani *et al.* (1985a)
Rat tracheal epithelial cells	Reduced EGF binding, ornithine decarboxylase induction	Jetten *et al.* (1985)
Rat anterior pituitary cells	Prolactin release	Koike *et al.* (1985)
GH$_3$ pituitary cells	Prolactin release	Martin and Kowalchyk (1984a)
	Phosphoprotein changes	Drust and Martin (1984)
GH$_4$C$_1$ pituitary cells	Prolactin release, reduced binding of phorbol esters and EGF	Jaken (1985)
Mouse B lymphocytes	Depolarization, I-A expression	Coggeshall and Cambier (1985)
Mouse 3T3 fibroblasts	80K phosphorylation	Rozengurt *et al.* (1983)
Mouse mammary gland	Ornithine decarboxylase induction	Rillema *et al.* (1983)

ethanolamine was a weak activator, and the zwitterionic phospholipids, phosphatidylcholine and sphingomyelin, were essentially unable to support activity.

Although phosphatidylethanolamine alone was a poor activator, it cooperatively stimulated phosphatidylserine-dependent protein kinase C activity. On the other hand, phosphatidylcholine and sphingomyelin inhibited phosphatidylserine-stimulated activity, and phosphatidylinositol and phosphatidic acid showed no cooperative effects (Kaibuchi *et al.*, 1981).

In these experiments, phospholipid dispersions were prepared by sonication. Because mixed sonic dispersions of phosphatidylserine with other lipids are heterogeneous in size and, possibly, in composition and, furthermore, would be expected to fuse in the presence of calcium, they do not provide an ideal system for investigation of stoichiometry or specificity of lipid activators at the molecular level. To provide a physically defined vesicle population, Boni and Rando (1985) prepared unilamellar vesicles from phosphatidylcholine and phosphatidylserine with a mole ratio of 4 : 1. Compared with pure phosphatidylserine vesicles, the mixed lipid vesicles gave lower protein kinase C activities but were unaffected by calcium and showed stronger diacylglycerol dependence (see Section 3.1.2).

Despite its strengths, the phosphatidylcholine/phosphatidylserine vesicle system is not ideally suited for investigations of lipid stoichiometry. For this purpose, a detergent/lipid mixed micellar assay was developed in which lipid activators are dispersed in detergent micelles (Hannun *et al.*,1985). This assay has the advantages of homogeneity in size and composition of the micelle population. Using this assay, it was shown that protein kinase C is active as a monomer and that activation by phosphatidylserine is highly cooperative, with a minimum of four molecules required for activation.

3.1.2. Diacylglycerol Dependence

In initial studies of the membrane lipid dependence of protein kinase C activation, total membrane lipids gave much higher levels of activity at lower calcium concentrations than purified phospholipid. Diacylglycerol, in the neutral lipid fraction, was identified as the activating factor. Diacylglycerol enhanced activity in the presence of phospholipids but was inactive alone. Synthetic dioleoylglycerol stimulated activity of the kinase sixfold and decreased the calcium requirement from the millimolar to the micromolar range (Takai *et al.*, 1979c; Kishimoto *et al.*, 1980).

Dioleoylglycerol is most commonly employed as the diacylglycerol activator *in vitro*. In a study of the dependence of protein kinase C on the acyl moieties of diacylglycerol, it was found that species possessing two long-chain saturated moieties, such as dipalmitoyl-glycerol or distearolyl-glycerol, were poor activators, but as long as one of the acyl groups was a long-chain unsaturated species, the length or nature of the other moiety was irrelevant (Mori et al., 1982). Thus, OAG activated protein kinase C strongly and, subsequently, has been used widely as a model diacylglycerol for in vivo studies. From these data it was concluded that protein kinase C is specific for diacylglycerols with at least one unsaturated fatty acid. More recent studies have shown that diacylglycerols with short, saturated fatty acids activate protein kinase C in vitro, as well as dioleoylglycerol, both in the standard assay with sonicated lipid dispersions (Conn et al., 1985; Davis et al., 1985b; Ebeling et al., 1985; Jetten et al., 1985) and in the mixed micellar assay (Ganong et al., 1986). Lack of activation by long-chain saturated diacylglycerols was probably a result of physical factors, since dipalmitoylgly-cerol and distearoylglycerol activated protein kinase C efficiently when the lipid environment was controlled, as in phosphatidylcholine/phosphatidylserine ves-icles (Boni and Rando, 1985) or mixed micelles (Hannun et al., 1986).

Protein kinase C activation by diacylglycerols is highly specific. Monoa-cylglycerols, triacylglycerols, cholesterol, and free unsaturated fatty acids were inactive (Takai et al., 1979c; Kishimoto et al., 1980; Mori et al., 1982). Studies with a series of diacylglycerol structural analogues indicated a requirement for both carbonyl oxygens of vicinal esters and an adjacent free hydroxyl group (Cabot and Jaken, 1984; Davis et al., 1985b; Ebeling et al., 1985; Jetten et al., 1985; Ganong et al.,1986). Activation was specific for sn-1,2-diacylglycerols, with 1,3 and sn-2,3 isomers showing little or no activity (Couturier et al., 1984; Rando and Young, 1984; Boni and Rando, 185; Ganong et al., 1986). Reports of activity with 1,3-diacylglycerols in vitro (Wise et al., 1982) and in vivo (Martin and Kowalchyk, 1984a; Merritt and Rubin, 1985) probably result from small amounts of contaminating sn-1,2 isomer. Since diacylglycerols isomerize rapidly in aqueous media (Serdarevich, 1967), it is difficult to interpret such results. In vitro, 1,3-dioctanoylglycerol, contaminated with less than 1% of the 1,2/2,3 isomers, activated protein kinase C significantly (Ganong et al., 1986).

In studies of phorbol ester displacement from protein kinase C using sonic dispersions of phosphatidylserine, diacylglycerol was found to interact with the enzyme/receptor with a 1 : 1 stoichiometry (Konig et al., 1985b). Similarly, in kinetic studies using a mixed micellar assay, a single molecule of diacylglycerol was found to activate monomeric protein kinase C (Hannun et al., 1985; Ganong et al., 1986).

3.1.3. Activation by Other Lipids

Recently, it has been reported that unsaturated fatty acids (McPhail et al., 1984; Murakami and Routtenberg, 1985), as well as retinoic acid (Ohkubo et al., 1984), activate protein kinase C independently of phospholipid in a calcium-dependent manner. Retinol enhanced the retinoic-acid-dependent activation, whereas retinal inhibited it, consistent with a previous report of protein kinase C inhibition by retinal (Taffet *et al.*, 1983).

Protein kinase C activation by long-chain fatty acyl-CoAs (Shoyab, 1985) and certain lipopolysaccharide precursors (Wightman and Raetz, 1984) has also been described. In these cases, the investigators suggested that the lipids may mimic the function of anionic phospholipids in providing an acidic/hydrophobic lipid matrix. A similar explanation may account for the observed activations by unsaturated fatty acids, although these are contradicted by other reports that fatty acids do not activate the kinase (Kishimoto *et al.*, 1980; Mori *et al.*, 1982; Shoyab, 1985).

3.2. Activation by Phorbol Diesters

Phorbol diesters are a class of potent tumor promoters isolated from plants that elicit a wide range of biological responses in animal cells (Blumberg, 1980, 1981). A high-affinity receptor for PDBu has been identified by several groups (reviewed in Niedel *et al.*, 1983; Leach *et al.*, 1983; and Ashendel, 1985). Structure-activity relationships for binding and biological potency coincided, indicating a role for this receptor in phorbol-ester-mediated responses. Protein kinase C was subsequently found to be activated in similar fashion by phorbol esters and diacylglycerols (Castagna *et al.*, 1982). Several groups have demonstrated that protein kinase C and the phorbol ester receptor copurify and share many properties, including a requirement for phospholipid and calcium for activity and phorbol ester binding (Ashendel *et al.*, 1983; Leach *et al.*, 1983; Niedel *et al.*, 1983; Parker *et al.*,1984; Uchida and Filburn, 1984; Konig *et al.*, 1985a). Demonstrations that diacylglycerols competitively inhibit PDBu binding to its receptor *in vitro* confirmed that these compounds occupy the same site in the protein kinase C/phospholipid/calcium complex (Sharkey *et al.*, 1984; Ebeling *et al.*, 1985; Hannun and Bell, 1986). Like diacylglycerols, phorbol esters interact with protein kinase C with a 1 : 1 stoichiometry (Kikkawa *et al.*,1983; Parker *et al.*, 1984).

4. CONCLUSIONS

The discovery of a diacylglycerol-activated protein kinase provided an elegant molecular explanation for the phenomenon of hormone-stimulated phosphatidylinositol turnover, first observed over 30 years ago. As early as 1975,

Michell suggested that the diacylglycerol released in this process may be recognized by specific enzymes. With his observation that diacylglycerol activated protein kinase C *in vitro,* Nishizuka (1984) proposed that the significance of inositol phospholipid turnover lay in the formation of a transient diacylglycerol pool, which would activate protein kinase C. It remained for a number of groups to examine the possibility that exogenous diacylglycerols may stimulate biological responses. When synthetic water-soluble diacylglycerols were employed to overcome solubility limitations, a wide range of effects were found, confirming the physiological relevance of diacylglycerol activation of protein kinase C.

Identification of a protein kinase as the phorbol diester receptor also explained the mode of action of this class of potent tumor promoters isolated from croton oil over 40 years ago, as suggested by Blumberg in 1981. Direct activation of protein kinase C by phorbol esters, by a phospholipid- and calcium-dependent mechanism, and copurification of phorbol ester receptor and protein kinase C activity confirmed that these tumor promoters act on the same target as diacylglycerols and, indeed, compete for the same site in the protein kinase C/phosphatidylserine/calcium complex. Thus, the cellular responses to exogenous diacylglycerols mimic both those induced by phorbol diesters and those associated with signal-induced phosphoinositide turnover.

Inability of exogenous diacylglycerols to mimic exactly all effects of phorbol esters on intact cells has led some workers to suggest that phorbol esters have additional sites of action than protein kinase C. For example, PMA is a potent tumor promoter in tests of two-stage carcinogenesis on mouse skin, whereas diC_8 is inactive (R. Stein, unpublished data). This is a reasonable speculation but may be difficult to prove. Since phorbol esters are metabolized much more slowly than diacylglycerols, and the affinity of protein kinase C for phorbol diesters is one or two orders of magnitude higher than for diacylglycerols, the ability of phorbol esters to induce stronger and, perhaps, more responses should not be surprising.

Demonstrations of *in vitro* activation of protein kinase C and stimulation of cellular responses by diacylglycerols raises a fundamental problem. How can a compound that is a common metabolic intermediate in several synthetic and degradative pathways have such a crucial and specialized second messenger function? The metabolic pool of diacylglycerol in mammalian cells is not enormous but contributes a substantial fraction of the total diacylglycerol pool in stimulated cells. The answer must lie in control of compartmentation of protein kinase C. Pools of diacylglycerol involved in glycerolipid metabolism must be localized in the endoplasmic reticulum, where glycerolipid synthetic enzymes are found. On the other hand, signal-induced diacylglycerol pools reside in the plasma membrane. Protein kinase C association with the plasma membrane has been observed in response to treatment of intact cells with exogenous diacylglycerols, phorbol esters, and hormones, and this translocation has been proposed to be an important step in protein kinase C activation. This phenomenon needs to be explored further. What factors are important in determining the subcellular localization of protein kinase C? Membrane association appears to be reversible *in vitro* by treatment with calcium chelators, and association of the enzyme with detergent/lipid mixed micelles was dependent only on phosphatidylserine and calcium and independent of diacylglycerol.

The molecular mechanism of protein kinase C activation by lipids and calcium is another problem of fundamental enzymological significance, not only from the perspective of control of a critical regulatory protein kinase but also from the perspective of how lipid/protein interactions may modulate enzyme activity. The 4 : 1 stoichiometry of phosphatidylserine and calcium for binding and activation of protein kinase C suggests that these components may form a complex at the membrane surface in which the calcium is chelated by the four carboxyl groups of the serine head groups, in much the same way that EDTA or EGTA chelates calcium. Under these conditions, protein kinase C is associated with the surface but in an inactive conformation (Figure 1A).

Activation of surface-bound protein kinase C requires association of one

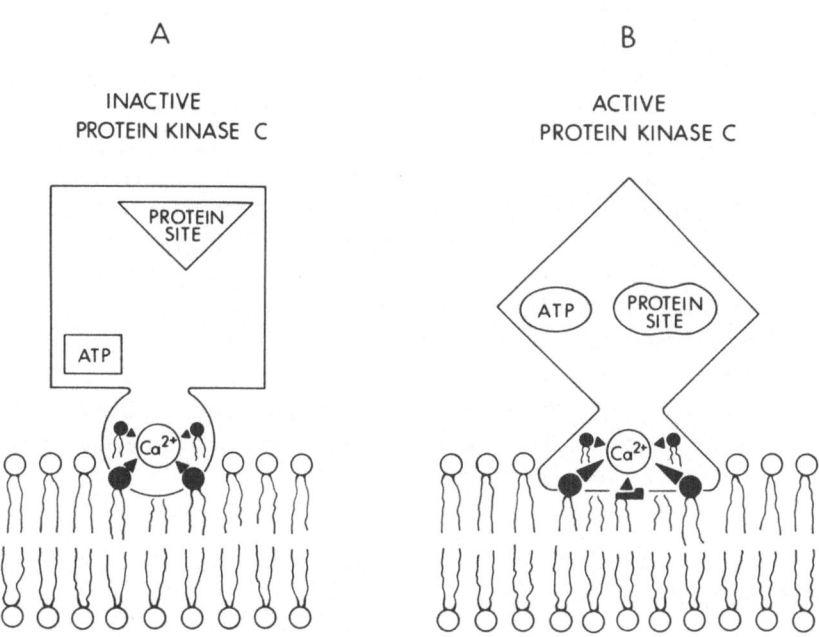

FIGURE 1. (A) Protein kinase C associated with a membrane surface complex of phosphatidylserine and calcium in an inactive form. (B) Membrane-associated protein kinase C activated by diacylglycerol.

molecule of diacylglycerol (or phorbol diester) with this complex. Three essential moieties of the diacylglycerol molecule are the free hydroxyl and the two ester carbonyl oxygens. Any of these would be capable of forming a coordinate linkage with the bound calcium, and it is proposed that the key molecular event for protein kinase C activation is formation of a direct bond between diacylglycerol and calcium (Ganong *et al.*, 1986). This ligation is envisioned to induce a conformational change in the lipid/calcium/enzyme complex resulting in kinase activation (Figure 1B).

Although this model for the molecular mechanism of protein kinase C activation by lipids and calcium is speculative, it is consistent with all available data concerning protein kinase C activation. Phospholipids that supported protein kinase C activation contain functional groups (hydroxyl and carboxyl), capable of ligating to calcium (phosphatidylinositol, cardiolipin, phosphatidylserine), whereas phospholipids lacking such functional groups (phosphatidylcholine, sphingomyelin, phosphatidylethanolamine) were essentially unable to support activation (Takai *et al.*, 1979a,b). Proteolysis of protein kinase C by an endogenous calcium-dependent protease released a fragment of M_r 51,000 whose kinase activity was simultaneously independent of calcium and lipids (Kikkawa *et al.*, 1982; Kishimoto *et al.*, 1983), indicating that the roles of these activators are related. Protein kinase C was more susceptible to calcium-dependent proteolysis when activated by diacylglycerol than when in the presence of phosphatidylserine alone (Kishimoto *et al.*, 1983), suggesting a diacylglycerol-induced conformational change. Phorbol esters with photoreactive acyl substituents labeled only lipid (Delclos *et al.*, 1983), suggesting that protein kinase C does not interact with the hydrophobic moieties of its lipid activators. Finally, this model is intriguing because of its analogy to calmodulin, in which the calcium-binding sites comprise four carboxyl groups and a carbonyl of the peptide backbone, and protein kinase C is inhibited by calmodulin antagonists (Mori *et al.*, 1980; Schatzman *et al.*, 1981; Wise *et al.*, 1982).

Despite the relative scarcity of information at present about endogenous substrates of protein kinase C and the functions of their phoshorylations, the breadth of biological responses induced by protein kinase C activators indicates the great importance of this multifunctional protein kinase in regulating the life processes of eukaryotic organisms.

REFERENCES

Aberdam, E., and Dekel, N., 1985, Activators of protein kinase C stimulate meiotic maturation of rat oocytes, *Biochem. Biophys. Res. Commun.* **132:**570–574.

Allan, D., Thomas, P., and Michell, R. H., 1978, Rapid transbilayer diffusion of 1,2-diacylglycerol and its relevance to control of membrane curvature, *Nature* **276:**289–290.

Anderson, N. G., and Hanson, P. J., 1985, Involvement of calcium-sensitive phospholipid-dependent protein kinase in control of acid secretion by isolated rat parietal cells, *Biochem. J.* **232**:609–611.

Ashby, B., Kowalska, M. A., Wernick, E., Rigmaiden, M., Daniel, J. L., and Smith, J. B., 1985, Differences in the mode of action of 1-oleoyl-2-acetyl-glycerol and phorbol ester in platelet activation, *J. Cyclic Nucl. Prot. Phos. Res.* **10**:473–483.

Ashendel, C. L., 1985, The phorbol ester receptor: A phospholipid-regulated protein kinase, *Biochim. Biophys. Acta* **822**:219–242.

Ashendel, C. L., Staller, J. M., and Boutwell, R. K., 1983, Solubilization, purification, and reconstitution of a phorbol ester receptor from the particulate protein fraction of mouse brain, *Cancer Res.* **43**:4327–4332.

Bachrach, L. K., Eggo, M. C., Mak, W. W., and Burrow, G. N., 1985, Phorbol esters stimulate growth and inhibit differentiation in cultured thyroid cells, *Endocrinology* **116**:1603–1609.

Bell, R. M., and Coleman, R. A., 1980, Enzymes of glycerolipid synthesis in eukaryotes, *Ann. Rev. Biochem.* **49**:459–487.

Berridge, M. J., 1983, Rapid accumulation of inositol trisphosphate reveals that agonists hydrolyze polyphosphoinositides instead of phosphatidylinositol, *Biochem. J.* **212**:849–858.

Berridge, M. J., 1984, Inositol trisphosphate and diacylglycerol as second messengers, *Biochem. J.* **220**:345–360.

Bertics, P. J., Weber, W., Cochet, C., and Gill, G. N., 1985, Regulation of the epidermal growth factor receptor by phosphorylation, *J. Cell. Biochem.* **29**:195–208.

Best, L., and Malaisse, W. J., 1984, Nutrient and hormone-neurotransmitter stimuli induce hydrolysis of polyphosphoinositides in rat pancreatic islets, *Endocrinology* **115**:1814–1820.

Blackshear, P. J., Witters, L. A., Girard, P. R., Kuo, J. F., and Quamo, S. N., 1985, Growth factor-stimulated protein phosphorylation in 3T3-L1 cells. Evidence for protein kinase C-dependent and -independent pathways, *J. Biol. Chem.* **260**:13304–13315.

Blumberg, P. M., 1980, *In vitro* studies on the mode of action of the phorbol esters, potent tumor promoters: Part 1, *CRC Crit. Rev. Toxicol.* **8**:153–197.

Blumberg, P. M., 1981, *In vitro* studies on the mode of action of the phorbol esters, potent tumor promoters: Part 2, *CRC Crit. Rev. Toxicol.* **8**:199–234.

Boni, L. T., and Rando, R. R., 1985, The nature of protein kinase C activation by physically defined phospholipid vesicles and diacylglycerols, *J. Biol. Chem.* **260**:10819–10825.

Brock, T. A., Rittenhouse, S. E., Powers, C. W., Ekstein, L. S., Gimbrone, M. A., Jr., and Alexander, R. W., 1985, Phorbol ester and 1-oleoyl-2-acetylglycerol inhibit angiotensin activation of phospholipase C in cultured vascular smooth muscle cells, *J. Biol. Chem.* **260**:14158–14162.

Burgess, G. M., Godfrey, P. P., McKinney, J. S., Berridge, M. J., Irvine, R. F., and Putney, J. W., Jr., 1984, The second messenger linking receptor activation to internal Ca release in liver, *Nature* **309**:63–66.

Cabot, M. C., and Jaken, S., 1984, Structural and chemical specificity of diradylglycerols for protein kinase C activation, *Biochem. Biophys. Res. Commun.* **125**:163–169.

Cantrell, D. A., Davies, A. A., and Crumpton, M. J., 1985, Activators of protein kinase C down-regulate and phosphorylate the T3/T-cell antigen receptor complex of human T lymphocytes, *Proc. Natl. Acad. Sci. USA* **82**:8158–8162.

Castagna, M., Takai, Y., Kaibuchi, K., Sano, K., Kikkawa, U., and Nishizuka, Y., 1982, Direct activation of calcium-activated, phospholipid-dependent protein kinase by tumor-promoting phorbol esters, *J. Biol. Chem.* **257**:7847–7851.

Changelian, P. S., Jack, R. M., Collins, L. A., and Fearon, D. T., 1985, PMA induced the ligand-independent internalization of CR1 on human neutrophils, *J. Immunol.* **134**:1851–1858.

Cochet, C., Gill, G. N., Meisenhelder, J., Cooper, J. A., and Hunter, T., 1984, C-kinase phosphorylates the epidermal growth factor receptor and reduces its epidermal growth factor–stimulated tyrosine protein kinase activity, *J. Biol. Chem.* **259**:2553–2558.

Coggeshall, K. M., and Cambier, J. C., 1985, B cell activation. VI. Effects of exogenous diglyceride and modulators of phospholipid metabolism suggest a central role for diacylglycerol generation in transmembrane signaling by mIg, *J. Immunol.* **134**:101–107.

Conn, P. M., Ganong, B. R., Ebeling, J., Staley, D., Niedel, J. E., and Bell, R. M., 1985, Diacylglycerols release LH: Structure-activity relations reveal a role for protein kinase C, *Biochem. Biophys. Res. Commun.* **126**:532–539.

Connolly, T. M., Lawing, W. J., and Majerus, P. W., 1986, Protein kinase C phosphorylates human platelet inositol trisphosphate 5'-phosphomonoesterase, increasing the phosphatase activity, *Cell* **46**:951–958.

Cooke, E., and Hallett, M. B., 1985, The role of C-kinase in the physiological activation of the neutrophil oxidase. Evidence from using pharmacological manipulation of C-kinase activity in intact cells, *Biochem. J.* **232**:323–327.

Cooper, R. H., Coll, K. E., and Williamson, J. R., 1985, Differential effects of a phorbol ester on phenylephrine and vasopressin-induced Ca^{2+} mobilization in isolated hepatocytes, *J. Biol. Chem.* **260**:3281–3288.

Coughlin, S. R., Lee, W. M. F., Williams, P. W., Giels, G. M., and Williams, L. T., 1985, c-*myc* gene expression is stimulated by agents that activate protein kinase C and does not account for the mitogenic effect of PDGF, *Cell* **43**:243–251.

Couturier, A., Bazgar, S., and Castagna, M., 1984, Further characterization of tumor-promoter-mediated activation of protein kinase C, *Biochem. Biophys. Res. Commun.* **121**:448–455.

Cox, C. C., Dougherty, R. W., Ganong, B. R., Bell, R. M., Niedel, J. E., and Snyderman, R., 1986, Differential stimulation of the respiratory burst and lysosomal enzyme secretion in human polymorphonuclear leukocytes by synthetic diacylglycerols, *J. Immunol.* **136**:4611–4616.

Cox, J. A., Jeng, A. Y., Sharkey, N. A., Blumberg, P. M., and Tauber, A. I., 1985, Activation of the human neutrophil nicotinamide adenine diphosphate (NADPH)-oxidase by protein kinase-C, *J. Clin. Invest.* **76**:1932–1938.

Cronin, M. J., and Canonico, P. L., 1985, Tumor promoters enhance basal and growth hormone releasing factor stimulated cyclic AMP levels in anterior pituitary cells, *Biochem. Biophys. Res. Commun.* **129**:404–410.

Dale, M. M., and Penfield, A., 1985, Superoxide generation by either 1-oleoyl-2-acetylglycerol or A23187 in human neutrophils is enhanced by indomethacin, *FEBS Lett.* **185**:213–217.

Davidson, J. S., Baumgarten, I. M., and Harley, E. H., 1985, Studies on the mechanism of phorbol ester-induced inhibition of intercellular junctional communication, *Carcinogenesis* **6**:1353–1358.

Davis, R. J., and Czech, M. P., 1984, Tumor-promoting phorbol diesters mediate phosphorylation of the epidermal growth factor receptor, *J. Biol. Chem.* **259**:8545–8549.

Davis, R. J., and Czech, M. P., 1985, Tumor-promoting phorbol diesters cause the phosphorylation of epidermal growth factor receptors in normal human fibroblasts at threonine-654, *Proc. Natl. Acad. Sci. USA* **82**:1974–1978.

Davis, R. J., Ganong, B. R., Bell, R. M., and Czech, M. P., 1985a, *sn*-1,2-dioctanoylglycerol. A cell-permeable diacylglycerol that mimics phorbol diester action on the epidermal growth factor receptor and mitogenesis, *J. Biol. Chem.* **260**:1562–1566.

Davis, R. J., Ganong, B. R., Bell, R. M., and Czech, M. P., 1985b, Structural requirements for diacylglycerols to mimic tumor-promoting phorbol diester action on the epidermal growth factor receptor, *J. Biol. Chem.* **260**:5315–5322.

Delbeke, D., Kojima, I., Dannies, P. S., and Rasmussen, H., 1984, Synergistic stimulation of prolactin release by phorbol ester, A23187 and forskolin, *Biochem. Biophys. Res. Commun.* **123**:735–741.

Delclos, K. B., Yeh, E., and Blumberg, P. M., 1983, Specific labeling of mouse brain membrane phospholipids with [20-^{3}H]phorbol 12-*p*-azidobenzoate 13-benzoate, a photolabile phorbol ester, *Proc. Natl. Acad. Sci. USA* **80**:3054–3058.

Depper, J. M., Leonard, W. J., Kronke, M., Noguchi, P. D., Cunningham, R. E., Waldmann, T. A., and Greene, W. C., 1984, Regulation of interleukin 2 receptor expression: Effects of phorbol diester, phospholipase C, and reexposure to lectin or antigen, *J. Immunol.* **133:**3054–3061.

Depper, J. M., Leonard, W. J., Drogula, C. L., Kronke, M., Waldmann, T. A., and Greene, W. C., 1985, Activators of protein kinase C and 5-azacytidine induce IL-2 receptor expression on human T lymphocytes, *J. Cell. Biochem.* **27:**267–276.

Dewald, B., Payne, T. G., and Baggiolini, M., 1984, Activation of NADPH oxidase of human neutrophils. Potentiation of chemotactic peptide by a diacylglycerol, *Biochem. Biophys. Res. Commun.* **125:**367–373.

di Virgilio, F., Lew, D. P., and Pozzan, T., 1984, Protein kinase C activation of physiological processes in human neutrophils at vanishingly small cytosolic Ca^{2+} levels, *Nature* **310:**691–693.

Dougherty, R. W., Godfrey, P. P., Hoyle, P. C., Putney, J. W., Jr., and Freer, R. J., 1984, Secretagogue-induced phosphoinositide metabolism in human leucocytes, *Biochem. J.* **222:**307–314.

Downward, J., Waterfield, M. D., and Parker, P. J., 1985, Autophosphorylation and protein kinase C phosphorylation of the epidermal growth factor receptor, *J. Biol. Chem.* **260:**14538–14546.

Drust, D. S., and Martin, T. J. F., 1984, Thyrotropin-releasing hormone rapidly activates protein phosphorylation in GH_3 pituitary cells by a lipid-linked, protein kinase C-mediated pathway, *J. Biol. Chem.* **259:**14520–14530.

Ebeling, J. G., Vandenbark, G. R., Kuhn, L. J., Ganong, B. R., Bell, R. M., and Niedel, J. E., 1985, Diacylglycerols mimic phorbol diester induction of leukemic cell differentiation, *Proc. Natl. Acad. Sci. USA* **82:**815–819.

Enomoto, T., and Yamasaki, H., 1985, Rapid inhibition of intercellular communication between BALB/c 3T3 cells by diacylglycerol, a possible endogenous functional analogue of phorbol esters, *Cancer Res.* **45:**3706–3710.

Eusebi, F., Molinaro, M., and Zani, B. M., 1985, Agents that activate protein kinase C reduce acetylcholine sensitivity in cultured myotubules, *J. Cell Biol.* **100:**1339–1342.

Fearon, D. T., 1980, Identification of the membrane glycoprotein that is the C3b receptor of the human erythrocyte, polymorphonuclear leukocyte, B lymphocyte, and monocyte, *J. Exp. Med.* **152:**20–30.

Friedman, B., Frackelton, A. R., Jr., Ross, A. H., Connors, J. M., Fujiki, H., Sugimura, T., and Rosner, M. R., 1984, Tumor promoters block tyrosine-specific phosphorylation of the epidermal growth factor receptor, *Proc. Natl. Acad. Sci. USA* **81:**3034–3038.

Friesen, L. L., and Gerrard, J. M., 1985, The effects of 1-oleoyl-2-acetylglycerol on platelet protein phosphorylation and platelet ultrastructure, *Amer. J. Pathol.* **121:**79–87.

Fujita, I., Irita, K., Takeshige, K., and Minakami, S., 1984, Diacylglycerol, 1-oleoyl-2-acetyl-glycerol, stimulates superoxidegeneration from human neutrophils, *Biochem. Biophys. Res. Commun.* **120:**318–324.

Gainer, H. St. C., and Murray, A. W., 1985, Diacylglycerol inhibits gap junctional communication in cultured epidermal cells: Evidence for a role of protein kinase C, *Biochem. Biophys. Res. Commun.* **126:**1109–1113.

Ganong, B. R., and Bell, R. M., 1984, Transmembrane movement of phosphatidylglycerol and diacylglycerol sulfhydryl analogs, *Biochemistry* **23:**4977–4983.

Ganong, B. R., Loomis, C. R., Hannun, Y. A., and Bell, R. M., 1986, Specificity and mechanism of protein kinase C activation by *sn*-1,2-diacylglycerols, *Proc. Natl. Acad. Sci. USA* **83:**1184–1188.

Gilmore, T., and Martin, G. S., 1983, Phorbol ester and diacylglycerol induce protein phosphorylation at tyrosine, *Nature* **306:**487–490.

Gispen, W. H., Schrama, L. H., and Eichberg, J., 1985, Stimulation of protein kinase C reduces ACTH-induced excessive grooming, *Eur. J. Pharmacol.* **114:**399–400.

Goldstein, I. M., Horn, J. K., Kaplan, H. B., and Weissman, G., 1974, Calcium-induced lysozyme secretion from human polymorphonuclear leukocytes, *Biochem. Biophys. Res. Commun.* **60**:807–812.

Grinstein, S., Cohen, S., Goetz, J. D., and Rothstein, A., 1985, Osmotic and phorbol ester-induced activation of Na$^+$/H$^+$ exchange: Possible role of protein phosphorylation in lymphocyte volume regulation, *J. Cell Biol.* **101**:269–276.

Grunberger, G., Rogers, J. L., and Gorden, P., 1985, Diacylglycerols: Effect on insulin receptor and tyrosine phosphorylation, *Fed. Proc.* **44**:1423.

Grzeskowiak, M., Della Bianca, V., De Togni, P., Papini, E., and Rossi, F., 1985, Independence with respect to Ca^{2+} changes of the neutrophil respiratory and secretory response to exogenous phospholipase C and possible involvement of diacylglycerol and protein kinase C, *Biochim. Biophys. Acta* **844**:81–90.

Guy, G. R., Gordon, J., Michell, R. H., and Brown, G., 1985, Synergism between diacylglycerols and calcium ionophore in the induction of human B cell proliferation mimics the inositol lipid polyphosphate breakdown signals induced by crosslinking surface immunoglobulin, *Biochem. Biophys. Res. Commun.* **131**:484–491.

Halenda, S. P., Zavoico, G. B., and Feinstein, M. B., 1985, Phorbol esters and oleoyl acetoyl glycerol enhance release of arachidonic acid in platelets stimulated by Ca^{2+} ionophore A23187, *J. Biol. Chem.* **260**:12484–12491.

Hallam, T. J., Daniel, J. L., Kendrick-Jones, J., and Rink, T. J., 1985, Relationship between cytoplasmic free calcium and myosin light chain phosphorylation in intact platelets, *Biochem. J.* **232**:373–377.

Hannun, Y. A., and Bell, R. M., 1986, Phorbol ester binding and activation of protein kinase C on Triton X-100 mixed micelles containing phosphatidylserine, *J. Biol. Chem.* **261**:9341–9347.

Hannun, Y. A., Loomis, C. R., and Bell, R. M., 1985, Activation of protein kinase C by Triton X-100 mixed micelles containing diacylglycerol and phosphatidylserine, *J. Biol. Chem.* **260**:10039–10043.

Hannun, Y. A., Loomis, C. R., and Bell, R. M., 1986, Protein kinase C activation in mixed micelles: Mechanistic implications of phospholipid, diacylglycerol, and calcium interdependencies, *J. Biol. Chem.* **261**:4184–7190.

Harman, I., Costello, A., Ganong, B., Bell, R. M., and Handwerger, S., 1986a, Activation of protein kinase C inhibits synthesis and release of decidual prolactin, *Amer. J. Physiol.* **251**:E172–E177.

Harman, I., Zeitler, P., Ganong, B., Bell, R. M., and Handwerger, S., 1986b, *sn*-1,2-diacylglycerols and phorbol esters stimulate the synthesis and release of human placental lactogen from placental cells: A role for protein kinase C, *Endocrinology* **119**:1239–1244.

Harris, C. E., Staley, D., and Conn, P. M., 1985, Diacylglycerols and protein kinase C. Potential amplifying mechanism for Ca^{2+}-mediated gonadotropin-releasing hormone-stimulated luteinizing hormone release, *Mol. Pharmacol.* **27**:532–536.

Hathaway, D. R., and Adelstein, R. S., 1979, Human platelet myosin light chain kinase requires the calcium-binding protein calmodulin for activity, *Proc. Natl. Acad. Sci. USA* **76**:1653–1657.

Helfman, D. M., Appelbaum, B. D., Vogler, W. R., and Kuo, J. F., 1983, Phospholipid-sensitive Ca^{2+}-dependent protein kinase and its substrates in human neutrophils, *Biochem. Biophys. Res. Commun.* **111**:847–853.

Hirasawa, K., and Nishizuka, Y., 1985, Phosphatidylinositol turnover in receptor mechanism and signal transduction, *Annu. Rev. Pharmacol. Toxicol.* **25**:147–170.

Hokin, L. E., 1985, Receptors and phosphoinositide-generated second messengers, *Annu. Rev. Biochem.* **54**:205–235.

Hokin, M. R., and Hokin, L. E., 1953, Enzyme secretion and the incorporation of ^{32}P into phospholipids of pancreas slices, *J. Biol. Chem.* **203**:967–977.

Hunter, T., Ling, N., and Cooper, J. A., 1984, Protein kinase C phosphorylation of the EGF receptor at a threonine residue close to the cytoplasmic face of the plasma membrane, *Nature* **311**:480–483.

Ieyasu, H., Takai, Y., Kaibuchi, K., Sawamura, M., and Nishizuka, Y., 1982, A role of calcium-activated, phospholipid-dependent protein kinase in platelet-activating factor-induced serotonin release from rabbit platelets, *Biochem. Biophys. Res. Commun.* **116**:743–750.

Iwashita, S., and Fox, C. F., 1984, Epidermal growth factor and potent phorbol tumor promoters induce epidermal growth factor receptor phosphorylation in a similar but distinctively different manner in human epidermoid carcinoma A431 cells, *J. Biol. Chem.* **259**:2559–2567.

Jaken, S., 1985, Increased diacylglycerol content with phospholipase C or hormone treatment: Inhibition of phorbol ester binding and induction of phorbol ester-like biological responses, *Endocrinology* **117**:2301.

Jetten, A. M., Ganong, B. R., Vandenbark, G. R., Shirley, J. R., and Bell, R. M., 1985, Role of protein kinase C in diacylglycerol-mediated induction of ornithine decarboxylase and reduction of epidermal growth factor binding, *Proc. Natl. Acad. Sci. USA* **82**:1941–1945.

Joseph, S. K., Thomas, A. P., Williams, R. J., and Williamson, J. R., 1984, *myo*-inositol 1,4,5-trisphosphate. A second messenger for the hormonal mobilization of intracellular Ca^{2+} in liver, *J. Biol. Chem.* **259**:3077–3081.

Kaibuchi, K., Takai, Y., and Nishizuka, Y., 1981, Cooperative roles of various membrane phospholipids in the activation of calcium-activated, phospholipid-dependent protein kinase, *J. Biol. Chem.* **256**:7146–7149.

Kaibuchi, K., Takai, Y., Sawamura, M., Hoshijima, M., Fujikura, T., and Nishizuka, Y., 1983, Synergistic functions of protein phosphorylation and calcium mobilization in platelet activation, *J. Biol. Chem.* **258**:2010–2013.

Kaibuchi, K., Takai, Y., and Nishizuka, Y., 1985, Protein kinase C and calcium ion in mitogenic response of macrophage-depleted human peripheral lymphocytes, *J. Biol Chem.* **260**:1366–1369.

Kajikawa, N., Kaibuchi, K., Matsubara, T., Kikkawa, U., Takai, Y., and Nishizuka, Y., 1983, A possible role of protein kinase C in signal-induced lysosomal enzyme release, *Biochem. Biophys. Res. Commun.* **116**:743–750.

Kasson, B. G., Conn, P. M., and Hsueh, A. J. W., 1985, Inhibition of granulosa cell differentiation by dioctanoylglycerol—A novel activator of protein kinase C, *Mol. Cell. Endocrinol.* **42**:29–37.

Katakami, Y., Kaibuchi, K., Sawamura, M., Takai, Y., and Nishizuka, Y., 1984, Synergistic action of protein kinase C and calcium for histamine release from rat peritoneal mast cells, *Biochem. Biophys. Res. Commun.* **121**:573–578.

Kawahara, Y., Takai, Y., Minakuchi, R., Sano, K., and Nishizuka, Y., 1980, Phospholipid turnover as a possible transmembrane signal for protein phosphorylation during human platelet activation by thrombin, *Biochem. Biophys. Res. Commun.* **97**:309–317.

Kikkawa, U., Takai, Y., Minakuchi, R., Inohara, S., and Nishizuka, Y., 1982, Calcium-activated, phospholipid-dependent protein kinase from rat brain: Subcellular distribution, purification, and properties, *J. Biol. Chem.* **257**:13341–13348.

Kikkawa, U., Takai, Y., Tanaka, Y., Miyake, R., and Nishizuka, Y., 1983, Protein kinase C as a possible receptor protein of tumor-promoting phorbol esters, *J. Biol. Chem.* **258**:11442–11445.

Kishimoto, A., Takai, Y., Mori, T., Kikkawa, U., and Nishizuka, Y., 1980, Activation of calcium and phospholipid-dependent protein kinase by diacylglycerol; its possible relation to phosphatidylinositol turnover, *J. Biol. Chem.* **255**:2273–2276.

Kishimoto, A., Kajikawa, N., Shiota, M., and Nishizuka, Y., 1983, Proteolytic activation of calcium-activated, phospholipid-dependent protein kinase by calcium-dependent neutral protease, *J. Biol. Chem.* **258**:1156–1164.

Kitagawa, K., Nishino, H., and Iwashima, A., 1985, Tumor promoter-stimulated translocation of glucose transport system in mouse embryo fibroblast Swiss 3T3 cell, *Biochem. Biophys. Res. Commun.* **128**:1303–1309.

Klausner, R. D., Harford, J., and van Renswoude, J., 1984, Rapid internalization of the transferrin receptor in K562 cells is triggered by ligand binding or treatment with a phorbol ester, *Proc. Natl. Acad. Sci. USA* **81**:3005–3009.

Koike, K., Judd, A. M., Yasumoto, T., and MacLeod, R. M., 1985, Calcium mobilization potentiates prolactin release induced by protein kinase C activators, *Mol. Cell. Endocrinol.* **40**:137–143.

Konig, B., DiNitto, P. A., and Blumberg, P. M., 1985a, Phospholipid and Ca^{2+} dependency of phorbol ester receptors, *J. Cell. Biochem.* **27**:255–265.

Konig, B., DiNitto, P. A., and Blumberg, P. M., 1985b, Stoichiometric binding of diacylglycerol to the phorbol ester receptor, *J. Cell. Biochem.* **29**:37–44.

Korchak, H. M., Vienne, K., Rutherford, L. E., Wilkenfeld, C., Finkelstein, M. C., and Weissman, G., 1984, Stimulus response coupling in the human neutrophil. II. Temporal analysis of changes in cytosolic calcium and calcium efflux, *J. Biol. Chem.* **259**:4076–4082.

Kreutter, D., Caldwell, A. B., and Morin, M. J., 1985, Dissociation of protein kinase C activation from phorbol ester–induced maturation of HL-60 leukemia cells, *J. Biol. Chem.* **260**: 5979–5984.

Lapetina, E. G., Reep, B., Ganong, B. R., and Bell, R. M., 1985, Exogenous *sn*-1,2-diacylglycerols containing saturated fatty acids function as bioregulators of protein kinase C in human platelets, *J. Biol. Chem.* **260**:1358–1361.

Leach, K. L., James, M. L., and Blumberg, P. M., 1983, Characterization of a specific phorbol ester aporeceptor in mouse brain cytosol, *Proc. Natl. Acad. Sci. USA* **80**:4208–4212.

Levine, L., and Xiao, D.-M., 1985, The stimulations of arachidonic acid metabolism by recombinant murine interleukin 1 and tumor promoters or 1-oleoyl-2-acetylglycerol are synergistic, *J. Immunol.* **135**:3430–3433.

Levine, L., Xiao, D.-M., Worth, N., and Lilley, W. E., 1985, The stimulation of prostaglandin production by transforming growth factor-α and 12-*O*-tetradecanoyl-phorbol-13-acetate or 1-oleoyl-2-acetyl-glycerol is synergistic, *Biochem. Biophys. Res. Commun.* **130**:110–117.

Limas, C. J., and Limas, C., 1985, Phorbol ester- and diacylglycerol-mediated desensitization of cardiac β-adrenergic receptors, *Circ. Res.* **57**:443–449.

Malaisse, W. J., Dunlop, M. E., Mathias, P. C. F., Malaisse-Lagae, F., and Sener, A., 1985, Stimulation of protein kinase C and insulin release by 1-oleoyl-2-acetyl-glycerol, *Eur. J. Biochem.* **149**:23–27.

Martin, T. F. J., and Kowalchyk, J. A., 1984a, Evidence for the role of calcium and diacylglycerol as dual second messengers in thyrotropin-releasing hormone action: Involvement of diacylglycerol, *Endocrinology* **115**:1517–1526.

Martin, T. J. F., and Kowalchyk, J. A., 1984b, Evidence for the role of calcium and diacylglycerol as dual second messengers in thyrotropin-releasing hormone action: Involvement of Ca^{2+}, *Endocrinology* **115**:1527–1536.

May, W. S., Jacobs, S., and Cuatrecasas, P., 1984, Association of phorbol ester–induced hyperphosphorylation and reversible regulation of transferrin membrane receptors in HL-60 cells, *Proc. Natl. Acad. Sci. USA* **81**:2016–2020.

May, W. S., Lapetina, E. G., and Cuatrecasas, P., 1986, Intracellular activation of protein kinase C and regulation of the surface transferrin receptor by diacylglycerol is a spontaneously reversible process that is associated with rapid formation of phosphatidic acid, *Proc. Natl. Acad. Sci. USA* **83**:1281–1284.

May, W. S., Sahyoun, N., Jacobs, S., Wolf, M., and Cuatrecasas, P., 1985, Mechanism of phorbol diester–induced regulation of surface transferrin receptor involves the action of activated protein kinase C and an intact cytoskeleton, *J. Biol. Chem.* **260**:9419–9426.

McCall, C., Schmitt, J., Cousart, S., O'Flaherty, J., Bass, D., and Wykle, R., 1985, Stimulation of hexose transport by human polymorphonuclear leukocytes: A possible role for protein kinase C, *Biochem. Biophys. Res. Commun.* **126**:450–456.

McNamara, M. J. C., Schmitt, J. D., Wykle, R. L., and Daniel, L. W., 1984, 1-*O*-hexadecyl-2-acetyl-*sn*-glycerol stimulates differentiation of HL-60 human promyelocytic leukemia cells to macrophage-like cells, *Biochem. Biophys. Res. Commun.* **122:**824–830.

McPhail, L. C., Clayton, C. C., and Snyderman, R., 1984, A potential second messenger role for unsaturated fatty acids: Activation of Ca^{2+}-dependent protein kinase, *Science* **224:**622–625.

Merritt, J. E., and Rubin, R. P., 1985, Pancreatic amylase secretion and cytoplasmic free calcium. Effects of ionomycin, phorbol dibutyrate, and diacylglycerols alone and in combination, *Biochem. J.* **230:**151–159.

Michell, R. H., 1975, Inositol phospholipids and cell surface receptor function, *Biochim. Biophys. Acta* **415:**81–147.

Michell, R. H., Kirk, C. J., Jones, L. M., Downes, C. P., and Creba, J. A., 1981, The stimulation of inositol phospholipid metabolism that accompanies calcium mobilization in stimulated cells: Defined characteristics and unanswered questions, *Philos. Trans. R. Soc. London Ser. B* **296:**123–137.

Moolenaar, W. H., Tertoolen, L. G. J., and de Laat, S. W., 1984, Phorbol ester and diacylglycerol mimic growth factors in raising cytoplasmic pH, *Nature* **312:**371–374.

Mori, T., Takai, Y., Minakuchi, R., Yu, B., and Nishizuka, Y., 1980, Inhibitory action of chlorpromazine, dibucaine, and other phospholipid-interacting drugs on calcium-activated, phospholipid-dependent protein kinase, *J. Biol. Chem.* **255:**8378–8380.

Mori, T., Takai, Y., Yu, B., Takahashi, J., Nishizuka, Y., and Fujikura, T., 1982, Specificity of the fatty acyl moieties of diacylglycerol for the activation of calcium-activated, phospholipid-dependent protein kinase, *J. Biochem. (Tokyo)* **91:**427–431.

Murakami, K., and Routtenberg, A., 1985, Direct activation of purified protein kinase C by unsaturated fatty acids (oleate and arachidonate) in the absence of phospholipids and Ca^{2+}, *FEBS Lett.* **192:**189–193.

Naka, M., Nishikawa, M., Adelstein, R. S., and Hidaka, H., 1983, Phorbol ester–induced activation of human platelets is associated with protein kinase C phosphorylation of myosin light chains, *Nature* **306:**490–492.

Naor, Z., and Eli, Y., 1985, Synergistic stimulation of luteinizing hormone (LH) release by protein kinase C activators and Ca^{2+}-ionophore, *Biochem. Biophys. Res. Commun.* **130:**848–853.

Niedel, J. E., Kuhn, L. J., and Vandenbark, G. R., 1983, Phorbol diester receptor copurifies with protein kinase C, *Proc. Natl. Acad. Sci. USA* **80:**36–40.

Nishihira, J., and O'Flaherty, J. T., 1985, Phorbol myristate acetate receptors in human polymorphonuclear neutrophils, *J. Immunol.* **135:**3439–3447.

Nishizuka, Y., 1984, The role of protein kinase C in cell surface signal transduction and tumour promotion, *Nature* **308:**693–698.

O'Flaherty, J. T., Showell, H. J., Becker, E. L., and Ward, P. A., 1978, Substances which aggregate neutrophils. Mechanisms of action, *Am. J. Pathol.* **92:**155–166.

O'Flaherty, J. T., Schmitt, J. D., McCall, C. E., and Wykle, R. L., 1984, Diacylglycerols enhance human neutrophil degranulation responses: Relevancy to a multiple mediator hypothesis of cell function, *Biochem. Biophys. Res. Commun.* **123:**64–70.

O'Flaherty, J. T., Schmitt, J. D., and Wykle, R. L., 1985a, Interactions of arachidonate metabolism and protein kinase C in mediating neutrophil function, *Biochem. Biophys. Res. Commun.* **127:**916–923.

O'Flaherty, J. T., Schmitt, J. D., Wykle, R. L., Redman, J. F., Jr., and McCall, C. E., 1985b, Diacylglycerols and mezerein activate neutrophils by a phorbol myristate acetate-like mechanism, *J. Cell. Physiol.* **125:**192–199.

Ohkubo, S., Yamada, E., Endo, T., Itoh, H., and Hidaka, H., 1984, Vitamin A acid-induced activation of Ca^{2+}-activated, phospholipid-dependent protein kinase from rabbit retina, *Biochem. Biophys. Res. Commun.* **118:**460–466.

O'Shea, J. J., Brown, E. J., Gaither, T. A., Takahashi, T., and Frank, M. M., 1985a, Tumor-promoting phorbol esters induce rapid internalization of the C3b receptor via a cytoskeleton-dependent mechanism, *J. Immunol.* **135**:1325–1330.

O'Shea, J. J., Siwik, S. A., Gaither, T. A., and Frank, M. M., 1985b, Activation of the C3b receptor: Effect of diacylglycerols and calcium mobilization, *J. Immunol.* **135**:3381–3387.

Otani, S., Matsui, I., Kuramoto, A., and Morisawa, S., 1985a, Induction of ornithine decarboxylase in guinea pig lymphocytes. Synergistic effect of diacylglycerol and calcium, *Eur. J. Biochem* **147**:27–31.

Otani, S., Matsui-Yuasa, I., Hashikawa, K., Kasai, S., Matsui, K., and Morisawa, S., 1985b, Synergistic induction of ornithine decarboxylase by diacylglycerol, A23187, and cholera toxin in guinea pig lymphocytes, *Biochem. Biophys. Res. Commun.* **130**:389–395.

Papini, E., Grzeskowiak, M., Bellavite, P., and Rossi, F., 1985, Protein kinase C phosphorylates a component of NADPH oxidase of neutrophils, *FEBS Lett.* **190**:204–208.

Parker, P. J., Stabel, S., and Waterfield, M. D., 1984, Purification to homogeneity of protein kinase C from bovine brain—Identity with the phorbol ester receptor, *EMBO J.* **3**:953–959.

Penfield, A., and Dale, M. M., 1984, Synergism between A23187 and 1-oleoyl-2-acetyl-glycerol in superoxide production by human neutrophils, *Biochem. Biophys. Res. Commun.* **125**:332–336.

Penfield, A., and Dale, M. M., 1985, Prostaglandins E_1 and E_2 enhance the stimulation of superoxide release by 1-oleoyl-2-acetylglycerol from human neutrophils, *FEBS Lett.* **181**:335–338.

Pincus, S. M., Beckman, B. S., and George, W. J., 1984, Inhibition of dimethylsulfoxide-induced differentiation in Friend erythroleukemic cells by diacylglycerols and phospholipase C, *Biochem. Biophys. Res. Commun.* **125**:491–499.

Pozzan, T., Lew, D. P., Wollheim, C. B., and Tsien, R. Y., 1983, Is cytosolic ionized calcium regulating neutrophil activation? *Science* **221**:1413–1415.

Rando, R. R., and Young, N., 1984, The stereospecific activation of protein kinase C, *Biochem. Biophys. Res. Commun.* **122**:818–823.

Rane, S. G., and Dunlap, K., 1986, Kinase C activator 1,2-oleoylacetylglycerol attenuates voltage-dependent calcium current in sensory neurons, *Proc. Natl. Acad. Sci. USA* **83**:184–188.

Raval, P. J., and Allan, D., 1985, The effects of phorbol ester, diacylglycerol, phospholipase C, and Ca^{2+} ionophore on protein phosphorylation in human and sheep erythrocytes, *Biochem. J.* **232**:43–47.

Rillema, J. A., Wing, L.-Y. C., and Foley, K., 1983, Effects of phospholipases on ornithine decarboxylase activity in mammary gland explants from midpregnant mice, *Endocrinology* **113**:2024–2028.

Rink, T. J., Sanchez, A., and Hallam, T. J., 1983, Diacylglycerol and phorbol ester stimulate secretion without raising cytoplasmic free calcium in human platelets, *Nature* **305**:317–319.

Rittenhouse, S. E., and Horne, W. C., 1984, Ionomycin can elevate intraplatelet Ca^{2+} and activate phospholipase A without activating phospholipase C, *Biochem. Biophys. Res. Commun.* **123**:393–397.

Robinson, J. M., Badwey, J. A., Karnovsky, M. L., and Karnovsky, M. J., 1984, Superoxide release by neutrophils: Synergistic effects of a phorbol ester and a calcium ionophore, *Biochem. Biophys. Res. Commun.* **122**:734–739.

Ronning, S. A., and Martin, T. J. F., 1985, Prolactin secretion in permeable GH_3 pituitary cells is stimulated by Ca^{2+} and protein kinase C activators, *Biochem. Biophys. Res. Commun.* **130**:524–532.

Rosoff, P. M., and Cantley, L. C., 1985, Lipopolysaccharide and phorbol esters induce differentiation but have opposite effects on phosphatidylinositol turnover and Ca^{2+} mobilization in 70Z/3 pre-B lymphocytes, *J. Biol. Chem.* **260**:9209–9215.

Rozengurt, E., Rodriguez-Pena, M., and Smith, K. A., 1983, Phorbol esters, phospholipase C, and growth factors rapidly stimulate the phosphorylation of a M_r 80,000 protein in intact quiescent 3T3 cells, *Proc. Natl. Acad. Sci. USA* **80**:7244–7248.

Rozengurt, E., Rodriguez-Pena, A., Coombs, M., and Sinnett-Smith, J., 1984, Diacylglycerol stimulates DNA synthesis and cell division in mouse 3T3 cells: Role of Ca^{2+}-sensitive phospholipid-dependent protein kinase, *Proc. Natl. Acad. Sci. USA* **81**:5748–5752.

Samuelsson, B., 1977, The role of prostaglandin endoperoxides and thromboxanes in human platelets, in: *Prostaglandins in Hematology* (M. J. Silver, J. B. Smith, and J. J. Kocsis, eds.), Spectrum Publications, New York, pp. 1–10.

Sano, K. Takai, Y., Yamanishi, J., and Nishizuka, Y., 1983, A role of calcium-activated phospholipid-dependent protein kinase in human platelet activation. Comparison of thrombin and collagen actions, *J. Biol. Chem.* **258**:2010–2013.

Sano, K., Voelker, D. R., and Mason, R. J., 1985, Involvement of protein kinase C in pulmonary surfactant secretion from alveolar type II cells, *J. Biol. Chem.* **260**:12725–12729.

Sasakawa, N., Ishii, K., Yamamoto, S., and Kato, R., 1985, Induction of ornithine decarboxylase by 1-oleoyl-2-acetyl-glycerol in isolated mouse epidermal cells, *Biochem. Biophys. Res. Commun.* **128**:913–920.

Schatzman, R. C., Wise, B. C., and Kuo, J. F., 1981, Phospholipid-sensitive calcium-dependent protein kinase: Inhibition by antipsychotic drugs, *Biochem. Biophys. Res. Commun.* **98**:669–676.

Schlondorff, D., Levine, S. D., Satriano, J., and Jacoby, M., 1985, Inhibition of vasopressin-stimulated water flow in toad bladder by phorbol myristate acetate, dioctanoylglycerol, and RHC-80267, *J. Clin. Invest.* **76**:1071–1078.

Serdarevich, B., 1967, Glyceride isomerizations in lipid chemistry, *J. Am. Oil Chem. Soc.* **44**:381–393.

Shackelford, D. A., and Trowbridge, I. S., 1984, Induction of expression and phosphorylation of the human interleukin 2 receptor by a phorbol diester, *J. Biol. Chem.* **259**:11706–11712.

Sharkey, N. A., Leach, K. L., and Blumberg, P. M., 1984, Competitive inhibition by diacylglycerol of specific phorbol ester binding, *Proc. Natl. Acad. Sci. USA* **81**:607–610.

Shinohara, O., Knecht, M., and Catt, K. J., 1985, Differential actions of phorbol ester and diacylglycerol on inhibition of granulosa cell maturation, *Biochem. Biophys. Res. Commun.* **133**:468–474.

Shoyab, M., 1985, Long-chain fatty acyl-coenzyme A's activate both the ligand-binding and protein kinase activities of phorboid and ingenoid receptor, *Arch. Biochem. Biophys.* **236**:435–440.

Sinnett-Smith, J. W., and Rozengurt, E., 1985, Diacylglycerol treatment rapidly decreases the affinity of the epidermal growth factor receptors of Swiss 3T3 cells, *J. Cell. Physiol.* **124**:81–86.

Streb, H., Irvine, R. F., Berridge, M. J., and Schultz, I., 1983, Release of Ca^{2+} from a nonmitochondrial intracellular store in pancreatic acinar cells by inositol-1,4,5-trisphosphate, *Nature* **306**:67–69.

Sugden, D., Vanecek, J., Klein, D. C., Thomas, T. P., and Anderson, W. B., 1985, Activation of protein kinase C potentiates isoprenaline-induced cyclic AMP accumulation in rat pinealocytes, *Nature* **314**:359–361.

Taffet, S. M., Greenfield, A. R. L., and Haddox, M. K., 1983, Retinal inhibits TPA-activated, calcium-dependent, phospholipid-dependent protein kinase ("C" kinase), *Biochem. Biophys. Res. Commun.* **114**:1194–1199.

Takai, Y., Kishimoto, A., Iwasa, Y., Kawahara, Y., Mori, T., and Nishizuka, Y., 1979a, Calcium-dependent activation of a multifunctional protein kinase by membrane phospholipids, *J. Biol. Chem.* **254**:3692–3695.

Takai, Y., Kishimoto, A., Iwasa, Y., Kawahara, Y., Mori, T., Nishizuka, Y., Tamura, A., and Fujii, T., 1979b, A role of membranes in the activation of a new multifunctional protein kinase system, *J. Biochem. (Tokyo)* **86**:575–578.

Takai, Y., Kishimoto, A., Kikkawa, U., Mori, T., and Nishizuka, Y., 1979c, Unsaturated diacylglycerol as a possible second messenger for the activation of calcium-activated, phospholipid-dependent protein kinase system, *Biochem. Biophys. Res. Commun.* **91**:1218–1224.

Thomas, A. P., Alexander, J., and Williamson, J. R., 1984, Relationship between inositol poly-phosphate production and the increase of cytosolic free Ca^{2+} induced by vasopressin in isolated hepatocytes, *J. Biol. Chem.* **259**:5574–5584.

Uchida, T., and Filburn, C. R., 1984, Affinity chromatography of protein kinase C–phorbol ester receptor on polyacrylamide-immobilized phosphatidylserine, *J. Biol. Chem.* **259**:12311–12314.

Van de Werve, G., Proietto, J., and Jeanrenaud B., 1985, Control of glycogen phosphorylase interconversion by phorbol esters, diacylglycerols, Ca^{2+}, and hormones in isolated rat hepa-tocytes, *Biochem. J.* **231**:511–516.

Vara, F., Schneider, J. A., and Rozengurt, E., 1985, Ionic responses rapidly elicited by activation of protein kinase C in quiescent Swiss 3T3 cells, *Proc. Natl. Acad. Sci. USA* **82**:2384–2388.

Ware, J. A., Johnson, P. C., Smith, M., and Salzman, E. M., 1985, Aequorin detects increased cytoplasmic calcium in platelets stimulated with phorbol ester or diacylglycerol, *Biochem. Biophys. Res. Commun.* **133**:98–104.

Watson, S. P., and Lapetina, E. G., 1985, 1,2-Diacylglycerol and phorbol ester inhibit agonist-induced formation of inositol phosphates in human platelets: Possible implications for negative feedback regulation of inositol phospholipid hydrolysis, *Proc. Natl. Acad. Sci. USA* **82**:2623–2626.

Watson, S. P., Ganong, B. R., Bell, R. M., and Lapetina, E. G., 1984, 1,2-Diacylglycerols do not potentiate the action of phospholipase A_2 and C in human platelets, *Biochem. Biophys. Res. Commun.* **121**:386–391.

Welsh, T. H., Jr., Jones, P. B. C., and Hsueh, A. J. W., 1984, Phorbol ester inhibition of ovarian and testicular steroidogenesis *in vitro*, *Cancer Res.* **44**:885–892.

White, J. R., Huang, C.-K., Mill, J. M., Jr., Naccache, P. H., Becker, E. L., and Sha'afi, R. I, 1984, Effect of phorbol 12-myristate 13-acetate and its analogue 4α-phorbol 12,13-didecanoate on protein phosphorylation and lysosomal enzyme release in rabbit neutrophils, *J. Biol. Chem.* **259**:8605–8611.

Wightman, P. D., and Raetz, C. R. H., 1984, The activation of protein kinase C by biologically active lipid moieties of lipopolysaccharide, *J. Biol. Chem.* **259**:10048–10052.

Wise, B. C., Glass, D. B., Jen Chou, C.-H., Raynor, R. L., Katoh, N., Schatzman, R. C., Turner, R. S., Kibler, R. F., and Kuo, J. F., 1982, Phospholipid-sensitive Ca^{2+}-dependent protein kinase from heart. II. Substrate specificity and inhibition by various agents, *J. Biol. Chem.* **257**:8489–8495.

Witters, L. A., Vater, C. A., and Lienhard, G. E., 1985, Phosphorylation of the glucose transporter *in vitro* and *in vivo* by protein kinase C, *Nature* **315**:777–778.

Yamamoto, S., Gotoh, H., Aizu, E., and Kato, R., 1985, Failure of 1-oleoyl-2-acetylglycerol to mimic the cell-differentiating action of 12-*O*-tetradecanoylphorbol 13-acetate in HL-60 cells, *J. Biol. Chem.* **260**:14230–14234.

Yamanishi, J., Takai, Y., Kaibuchi, K., Sano, K., Castagna, M., and Nishizuka, Y., 1983, Syn-ergistic functions of phorbol ester and calcium in serotonin release from human platelets, *Biochem. Biophys. Res. Commun.* **112**:778–786.

6

PROTEIN KINASE C AND ITS ROLE IN CELL GROWTH

James R. Woodgett, Tony Hunter, and Kathleen L. Gould

1. INTRODUCTION

Protein phosphorylation is recognized as the primary mechanism for the transduction of extracellular stimuli into intracellular events (Cohen, 1982; Nestler *et al.*, 1984). In eukaryotes, for example, all of the biochemical actions of cAMP have been attributed to activation of cAMP-dependent protein kinase (Krebs and Beavo, 1979). Ca^{2+} has also been implicated as a second messenger by its ability to activate several calmodulin-dependent protein kinases and at least one phosphoprotein phosphatase (Cohen, 1985; Nairn *et al.*, 1985). Another signal transduction pathway, the ubiquity of which has only recently been appreciated, is that of phosphatidylinositol (PI) turnover (Hokin and Hokin, 1953; Michell, 1975; for reviews, see Berridge, 1984; Hokin, 1985). Activation of this system causes hydrolysis of polyphosphoinositol lipids to form polyphosphoinositides, the most thoroughly studied of which is inositol-1,4,5-trisphosphate (hereafter termed IP_3) and diacyglycerols (DAG). The primary role of IP_3 is to stimulate the release of Ca^{2+} from intracellular stores, most probably the endoplasmic reticulum (Berridge, 1983; Streb *et al.*, 1983), thus causing activation of Ca^{2+}-dependent processes. The major function of DAGs is to activate a serine/threonine-specific protein kinase, termed protein kinase C (Takai *et al.*, 1979a). The simultaneous formation of two second messengers by agonists of PI turnover endows this pathway with enhanced flexibility of response, greater possibilities of synergistic effects, and complex interactions with other signaling pathways, some aspects of which will be described below.

James R. Woodgett, Tony Hunter, and Kathleen L. Gould • Molecular Biology and Virology Laboratory, The Salk Institute, San Diego, California 92138.

Protein kinase C was first described by Nishizuka and colleagues as a protein kinase activatable *in vitro* by partial proteolysis (Inoue *et al.*, 1977; Takai *et al.*, 1977). Subsequently, this group demonstrated that the activity of this enzyme could be stimulated by Ca^{2+} and acidic phospholipids in place of proteolysis (Takai *et al.*, 1979b). A physiological role for protein kinase C became apparent upon the discovery that DAG stimulated protein kinase activity by reducing dependence on Ca^{2+} and phospholipids (Takai *et al.*, 1979a; Kishimoto *et al.*, 1980).

In addition to a role in normal cellular regulation, protein kinase C was implicated in tumorigenesis by the observation that this protein represents the major cellular receptor for tumor-promoting phorbol esters (Castagna *et al.*, 1982; Ashendel *et al.*, 1983a, 1983b; Niedel *et al.*, 1983; Leach *et al.*, 1983; Sando and Young, 1983; Parker *et al.*, 1984a; for review, see Nishizuka, 1984; Ashendel, 1985). This one discovery revolutionized the field of tumor promotion and allowed the biochemical effects underlying tumor formation to be probed with far greater insight than before.

Here, we review the role of protein kinase C in the regulation of cell growth and discuss the mechanisms through which this control is achieved.

2. PROTEIN KINASE C

2.1. The Enzyme

Protein kinase C is distributed broadly among the tissues of vertebrates and throughout the phyla of the animal kingdom, being detectable in simpler organisms such as lobster and cockroach (Kuo *et al.*, 1980; Minakuchi *et al.*, 1981). The enzyme was initially purified to near homogeneity from bovine heart (Wise *et al.*, 1982a) and pig spleen (Schatzman *et al.*, 1983b) but is more abundant in brain (Kuo *et al.*, 1980) from where it is now more commonly isolated (Kikkawa *et al.*, 1982, 1983a, 1986; Le Peuch *et al.*, 1983; Parker *et al.*, 1984a; Walsh *et al.*, 1984; Wolf *et al.*, 1984; Jeng *et al.*, 1986). The enzyme from these preparations exhibits a single protein staining band of 77–83 kD upon sodium dodecyl sulfate (SDS)–polyacrylamide gel electrophoresis, behaves as a monomer upon gel permeation chromatography, and has an isoelectric point of approximately pH 5.6 (Kikkawa *et al.*, 1982; Wise *et al.*, 1982a). Preparations of varying purity have also been described from renal cortex (Uchida and Filburn, 1984), pituitary glands (Turgeon *et al.*, 1984), chick oviduct (Horn *et al.*, 1985), and leukocytes (Christiansen and Juhl, 1986). Although various methods of purification have been described, most require several days for completion, resulting in low yields, or use Ca^{2+} during the initial stages, which can lead to proteolysis. Furthermore, the extreme instability of the purified enzyme, which, in many cases, exhibited a half-life of only days, made study of protein kinase C laborious.

Recently, we developed a purification scheme from rat brain that can be completed in less than 2 days and results in the isolation of milligram quantities of homogeneous enzyme (Woodgett and Hunter, 1987a). Moreover, this preparation retains over 50% of initial activity after storage for 1 year on ice, enabling the accumulation of large quantities of active enzyme. Unlike the preparations described above, the final step of our purification scheme yields a 78/80-kD doublet, both components of which represent protein kinase C. A similar doublet has been observed recently by two other groups (Kikkawa et al., 1986; Wooten et al., 1987). The physiological relevance of these two proteins is discussed in Section 2.10.

In vitro, enzyme activity is absolutely dependent on Ca^{2+} (K_a of 5×10^{-6} M) and phospholipids, of which phosphatidylserine is the most effective (K_a of 1 $\mu g/ml$); although PI, cardiolipin, and phosphatidic acid can substitute, albeit less potently (Takai et al., 1979b; Wise et al., 1982b; Konig et al., 1985a). Some phospholipids, such as phosphatidylethanolamine, potentiate binding of phosphatidylserine at low Ca^{2+} concentrations, whereas others, such as phosphatidylcholine or sphingomyelin, appear inhibitory (Kaibuchi et al., 1981). Hence, in vivo, the phospholipid composition of membranes may have profound effects on the requirements for activation of protein kinase C in any particular cell type. Indeed, the lipid composition may determine which cellular membranes can support binding of protein kinase C. Since the inner leaflet of the plasma membrane is particularly rich in acidic phospholipids, and other membranes contain greater proportions of phosphatidylcholine, one might predict that the enzyme will exhibit a higher affinity for the former, which is also the site of agonist-induced DAG production.

In vitro, 1,2-sn DAGs stimulate protein kinase C by reducing the dependency of the enzyme for Ca^{2+} and phospholipids (Takai et al., 1979a; Kishimoto et al., 1980; Parker et al., 1984a). The enzyme exhibits specificity for the 1,2-configuration of DAGs and is not significantly activated by either 1,3- or 2,3-DAGs. DAGs promote a shift in the apparent subcellular localization of the enzyme from the cytosol to the membranes in vivo (see Section 2.7). Complete activation of the enzyme in vitro is accomplished by 6–8 mole % of phosphatidylserine at saturating Ca^{2+} concentrations (Hannun et al., 1985; see Chapter 5, this volume). In contrast, only one molecule of DAG is required to activate each molecule of protein kinase C (Konig et al., 1985b). Certain unsaturated fatty acids, such as oleic and arachidonic acids, have been reported to activate protein kinase C in the absence of Ca^{2+} and phospholipid, although the physiological relevance of these observations is uncertain due to the high concentrations of lipids (>100 μM) required for the Ca^{2+}-independent effect (Murakami *et al.*, 1986). The enzyme binds up to 4 moles of Ca^{2+} per mole, and this binding appears to be independent of phospholipid, suggesting that there are separate binding sites for Ca^{2+} and phospholipid (J. R. Glenney, Jr., and J. R. Woodgett, unpublished observations). It is not known whether DAG binds to an autonomous site on protein kinase C or whether it interacts with one of the phospholipid binding sites, although the latter is more likely since DAG normally does not bind to protein kinase C in the absence of phospholipid. DAG probably alters the microenvironment of the phospholipids in the bilayer in such a way as to generate a productive complex, since these compounds are known to perturb membrane bilayer structure *in vitro* (Das and Rand, 1986). Active protein kinase C therefore appears to exist as a ternary complex, comprising the enzyme, phospholipid, DAG, and Ca^{2+}. The physiological role of Ca^{2+} in the activation of protein kinase C is discussed further in Section 5.2. *In vivo,* dissolution of the complex and inactivation would occur primarily via phosphorylation of the DAG to phosphatidic acid by DAG kinases, but degradation through the action of DAG lipases may also be important.

Protein kinase C was isolated initially as a protease-activated protein kinase (Takai *et al.*, 1977). Incubation of the protein with low concentrations of Ca^{2+}-dependent neutral protease (calpain I), or trypsin, generated a catalytic fragment of 50 kD that was no longer dependent upon Ca^{2+} or phospholipid, termed protein kinase M (Kishimoto *et al.*, 1983). Sustained proteolysis destroyed activity. Proteolytic cleavage by calpain I was enhanced in the presence of phospholipids and DAG, which suggests that the activated conformation of the enzyme is more susceptible to cleavage (Kishimoto *et al.*, 1983). The physiological significance of this alternative activation pathway is discussed in Section 2.8. The phospholipid and Ca^{2+} independence of the 50-kD fragment implies that a 20- to 30-kD domain of the enzyme is responsible for binding these compounds and is inhibitory in their absence. Indeed, protein kinase M fails to bind to phosphatidylserine affinity columns in the presence of Ca^{2+}, unlike the native protein, and a 35-kD phorbol ester–binding fragment can be generated upon proteolysis of protein kinase C (Hoshijima *et al.*, 1986; Lee and Bell, 1986; J. R. Woodgett; unpublished observations).

Like other protein kinases, protein kinase C undergoes an autophosphorylation reaction *in vitro* in the presence of activators. Phosphorylation is on serine and threonine residues and occurs at multiple sites (Huang *et al.*, 1986a; Woodgett and Hunter, 1987a). These sites appear to be contained within a 15- to 18-kD amino (N) or carboxy (C)-terminal domain, since the intact 80-kD protein can be cleaved by an endogenous brain protease during purification to generate a 62-kD fragment, which retains Ca^{2+} and phospholipid and ATP binding properties, but is essentially inactive as a protein kinase and cannot be phosphorylated by intact protein kinase C (Woodgett and Hunter, 1987a). The effect of autophosphorylation on enzyme activity is unclear as yet, although effects of self-phosphorylation on other protein kinases have been documented (for example, see Lai *et al.*, 1986; Miller and Kennedy, 1986).

Protein kinase C in brain is developmentally regulated with the highest levels of expression of activity occurring about 30 days after birth (Nagle *et al.*, 1981; Turner *et al.*, 1984b). Embryonic (day 16) neuronal cultures have been reported to undergo a twenty-fold induction in protein kinase C levels during the first 7 days of culture relative to other cellular proteins, suggesting that the cellular concentration of this protein kinase is under specific regulation (Burgess *et al.*, 1986). The recent isolation of cDNA clones to protein kinase C will allow the mechanism of this induction to be determined.

2.2. Inhibitors

A variety of compounds have been described that inhibit the activity of protein kinase C, which can be subdivided into several classes (Table I). The first inhibitors to be reported were members of the phenothiazine antipsychotic drug family (Mori *et al.*, 1980; Schatzman *et al.*, 1981). However, these drugs are also potent inhibitors of calmodulin-dependent processes, which include calmodulin-dependent protein kinases and, thus, are not suitable as specific inhibitors of any one enzyme. These compounds have a high affinity for hydrophobic regions of proteins and appear to inhibit protein kinase C by interfering with phospholipid binding, since the inhibition can be overcome by adding phospholipids *in vitro* (Uratsuji *et al.*, 1985). It should be noted that protein kinase C does not bind (or phosphorylate) calmodulin in the presence of Ca^{2+}, although phosphorylation of certain proteins *in vitro* has been reported to be inhibited partially by addition of calmodulin (Albert *et al.*, 1984a).

Another class of inhibitors that interact with phospholipids include surface-active polypeptides, such as marine worm cytotoxin AIV and cobra cytotoxin 1, the antibiotic polymyxin B, the spermatogenesis inhibitor gossypol, and the antiestrogen tamoxifen and its derivatives (for appropriate references, see Table I). Of these, cytotoxins 1 and AIV appear most selective with respect to inhibition of other protein kinases. Tamoxifen, for example, is an inhibitor of cAMP phosphodiesterase (Lam, 1984). Unfortunately, the toxicity of these compounds restricts their usefulness *in vivo*. Recently, sphingosine, a component of sphingomyelin phospholipids, has been shown to inhibit purified protein kinase C competitively with respect to Ca^{2+}, DAG, and phosphatidylserine (Hannun *et al.*, 1986). Furthermore, this compound inhibits thrombin-stimulated phosphorylation of p47 (see Section 3.2.2) and reduces phorbol dibutyrate binding upon addition to platelets (Hannun *et al.*, 1985). Similar effects were observed with a related compound, sphinganine, which is a long-chain base like sphingosine.

TABLE I. Inhibitors of Protein Kinase C

Inhibitor	50% inhibition (μM)	Other actions	References
Lipophilic compounds			
Phenothiazines	10–50	Anticalmodulin	Mori et al. (1980)
W7	60	Anticalmodulin	Tanaka et al. (1982)
R-24571	5	Anticalmodulin	Mazzei et al. (1984)
Marine worm cytotoxin AIV	1.3	Cell toxin	Kuo et al. (1983)
Cobra cytotoxin I	3.7	Cell toxin	Kuo et al. (1983)
Gossypol	31	Adenylate cyclase inhibitor	Kimura et al. (1985)
Polymyxin B	2.0	Antibiotic	Mazzei et al. (1982)
Tamoxifen	100	cAMP phosphodiesterase inhibitor	O'Brian et al. (1985)
			Su et al. (1985)
CP-46,665-1	10	Anticalmodulin	Shoji et al. (1985)
Retinal	10		Taffet et al. (1983)
Palmitoylcarnitine	27	Myosin light-chain kinase inhibitor	Katoh et al. (1981)
Alkyllysophospholipid	5	Inhibitor of phospholipid metabolism, lysogen	Helfman et al. (1983b)
Sphingosine	100	Metabolite of phospholipids	Hannun et al. (1985)
Nucleotide analogues			
Isoquinolinesulfonamides (H-7)	6	General inhibitors of protein kinases	Hidaka et al. (1984)
Quercetin	10	Enzymes and protein kinase inhibitor	Gschwendt et al. (1983)
Ap₄A	100	Inhibition of some protein kinases	Shoyab (1985)
Substrate analogues			
Tosyl lysine chloromethyl ketone	1000	Protease inhibitor	Solomon et al. (1985)
Tosyl phenylalanine chloromethyl ketone	8000	Protease inhibitor	Solomon et al. (1985)
Synthetic peptides (substrate analogues)	28–145		Su et al. (1986a)
Inhibitor proteins	?		Schwantke and Le Peuch (1984)
Ca²⁺-binding inhibitor proteins	~1		McDonald and Walsh (1986)
Others			
Staurosporine	0.0027	cAMP-dependent protein kinase inhibitor	Tamaoki et al. (1986)
Polyamines	~500	Activators of phosphatases, etc.	Qi et al. (1983)
Selenium compounds	~60	Myosin light-chain kinase inhibitor	Su et al. (1986b)
Calmodulin	?	Cellular Ca²⁺ mediator	Albert et al. (1984a)

These reagents have been used to test for involvement of protein kinase C in processes such as phorbol ester–dependent differentiation of HL-60 cells (see Section 4.2) (Merrill et al., 1986) and the oxidative burst of human neutrophils (Wilson et al., 1984). Indeed, these long-chain bases have been postulated as natural antagonists of protein kinase C although, currently, there is no direct evidence to support this idea. It is also not known whether these compounds have effects on cellular processes independent of protein kinase C.

All protein kinases require nucleoside triphosphates as cofactors, usually ATP. Various other nucleoside phosphates and their nonhydrolyzable analogues are competitive inhibitors with respect to ATP, with K_i values in the 100-μM range. Recently, Hidaka and colleagues reported the use of a series of isoquinolinesulphonamide compounds that inhibited several protein kinases with respect to ATP but did so with different selectivities. For example, one compound, termed H-8, inhibited cyclic nucleotide–dependent protein kinases fifteen- to thirty-fold more potently than protein kinase C, whereas another, termed H-7, inhibited protein kinase C with a potency similar to the other protein kinases (Hidaka et al., 1984). Thus, judicious, comparative use of these compounds may allow the selective effects of inhibition of protein kinase C to be evaluated (Matsui et al., 1986; Tohmatsu et al., 1986; Wright and Hoffman, 1986). However, a major caveat is that the inhibition can only be monitored in conjunction with inhibition of cyclic nucleotide–dependent protein kinases, since none of the compounds inhibits protein kinase C selectively.

The most potent inhibitor described thus far is staurosporine, an antifungal microbial alkaloid, which exhibits half-maximal inhibition of protein kinase C at 27 nM (Tamaoki et al., 1986). Unfortunately, this compound also inhibits cAMP-dependent protein kinase potently and is toxic to growing cells.

Protein kinase C phosphorylates a variety of synthetic peptides, corresponding to both physiological and in vitro phosphorylation site sequences (see Section 3.3) (O'Brian et al., 1984; Ferrari et al., 1985; Kishimoto et al., 1985; Turner et al., 1985; Woodgett et al., 1986). Substitution of the phosphorylated residue in such peptides with a nonphosphorylatable residue such as alanine results in the formation of selective but rather weak inhibitors of the enzyme that are competitive with respect to substrates (O'Brian et al., 1984; Su et al., 1986a).

Although the number of inhibitory compounds described so far is impressive, their potency and selectivity do not compare favorably with the inhibitory protein of cAMP-dependent protein kinase, which has a K_i for the catalytic subunit of this protein kinase in the nanomolar range and has little, if any, effect on any other protein kinases (Ashby and Walsh, 1972; Scott et al., 1985). Recently, Schwantke and Le Peuch (1984) reported the detection of a proteinaceous inhibitory factor specific for protein kinase C. McDonald and Walsh (1985, 1986) reported identification of similar factors that were also Ca^{2+}-binding proteins and inhibited protein kinase C with micromolar potency. The physiological relevance of these factors awaits further characterization.

In summary, although there is a multitude of protein kinase C inhibitors, with the possible exception of the sphingosine compounds, their usefulness is restricted. If employed to determine whether protein kinase C is involved in a particular response, corroborative data from other sources should be sought to be sure that the inhibitors are not acting through other mechanisms.

2.3. Antibodies

Until recently, there has been a dearth of immunological reagents directed against protein kinase C, partly due to difficulties in obtaining sufficient quantities of homogeneous protein for immunization. Several groups have now generated polyclonal antisera to the rat brain and human spleen protein kinase C (Ballester and Rosen, 1985; Fry *et al.*, 1985; Girard *et al.*, 1985; Huang and Huang, 1986; Woodgett and Hunter, 1987b; R. Erikson, personal communication; and G. Gill, personal communication). Additionally, monoclonal antibodies have been isolated in the laboratories of Y. Nishizuka and D. Koshland (Nishizuka, 1986; Mochly-Rosen and Koshland, 1987). Kuo's group has used antisera to detect protein kinase C by immunoblotting and observes a minor immunoreactive band of 80 kD and a major band of 67 kD, which is postulated to represent a partially degraded form of protein kinase C (Girard *et al.*, 1985; 1986). Ballester and Rosen (1985) raised antisera to SDS–polyacrylamide gel–purified protein kinase C and detect a single bank of 80 kD upon immunoprecipitation of the rat pituitary cell line, GH$_3$. The antiserum used by Fry *et al.*, (1985) immunoprecipitates an 80-kD band from Abelson murine leukemia virus (Ab-MuLV)-transformed mouse NIH 3T3 cells labeled with [^{32}P]orthophosphate. Antipeptide antiserum have also been raised to various regions of the protein (Young *et al.*, 1987).

The preparation of protein kinase C that we used to raise antibodies comprised a 78/80-kD doublet, as described in Section 2.1. Antiserum from one rabbit recognized only the 80-kD component, as determined by immunoblotting of purified protein kinase C and immunoprecipitation from biosynthetically labeled cell lines (Woodgett and Hunter, 1987b). This band was shown to be protein kinase C by several criteria, including correspondence of relative abundance in cell lines with the number of 12-*O*-tetradecanoylphorbol-13-acetate (TPA)-binding sites, similar *Staphylococcus aureus* V8 protease cleavage map to purified protein and inhibition of 80-kD protein immunoprecipitation by preabsorption of antiserum with purified protein kinase C. In contrast, antiserum from a second rabbit detected both polypeptides of the purified enzyme and immunoprecipitated a 78/80-kD doublet from a variety of cell lines. The 78-kD polypeptide had a structure related to, but distinct from, that of the 80kD protein, with the differences being similar to those between the 78- and 80-kD components of purified protein kinase C. The possibilities that the doublet was generated by either proteolysis of the 80-kD protein or differences in phosphorylation state were ruled out by several lines of evidence, including the use of SDS-boiled lysates for immunoprecipitation and treatment of the enzyme or immunoprecipitates with purified phosphoprotein phosphatases. This demonstration of the physiological existence of more than one form of protein kinase C is consistent with the recent identification of multiple genes for this protein kinase (see Section 2.10).

Currently, the function of the two forms is unknown, although by analogy to the multiple forms of both the regulatory and catalytic subunits of cAMP-dependent protein kinase, the two proteins may be regulated differentially by agonists or other transduction systems (see Section 5). The two forms might also be independently distributed, as in the case of the two brain forms of the multifunctional calmodulin-dependent protein kinase (Miller and Kennedy, 1985; McGuinness *et al.*, 1985).

2.4. Natural Agonists

The concentration of DAG in the plasma membrane of resting cells is extremely low but is rapidly elevated upon agonist-induced breakdown of inositol-containing phospholipids. There are multiple agonists of this breakdown pathway which, upon binding to their receptors, induce the hydrolysis of polyphosphorylated phosphatidylinositols by phospholipase C to form IP_3 and DAG (Table II) (for review, see Berridge and Irvine, 1984). Addition of phospholipase C to cells predictably has similar effects to tumor promoters (Jeng *et al.*, 1985) (see Section 2.5). In certain cell types, the DAG is metabolized to arachidonic acid, as well as phosphatidic acid, thus activating icosanoid synthesis (Bell *et al.*, 1979; Lapetina, 1982). IP_3 is water soluble and provokes Ca^{2+} release from intracellular stores by binding to saturable receptors (Baukal *et al.*, 1985; Hirata *et al.*, 1985; Spat *et al.*, 1986). The major function of the DAG in the plasma membrane appears to be the binding and activation of protein kinase C. Indeed, addition of cell-permeant synthetic DAGs to cells elicits transient activation of protein kinase C *in vivo* (Davis *et al.*, 1985a,b; see Section 4.5; see Chapter 5). Hence, at least part of the physiological response to agonists of PI turnover is directly attributable to protein kinase C activation. Many of the agonists, such as PDGF, operate through other signal transduction pathways, in addition to PI turnover, and so their effects are not exclusively directed through protein kinase C (Coughlin *et al.*, 1985).

The role of Ca^{2+} in the activation mechanism is understood less clearly since resting Ca^{2+} levels ($\sim 5 \times 10^{-7}$ M) appear sufficient to allow at least partial activation of the enzyme by DAG. Although synthetic DAGs apparently activate protein kinase C without the need for Ca^{2+} ionophores, there is evidence that

TABLE II. Natural Agonists of Protein Kinase C[a]

Agonist	Cell or tissue type	References
Acetylcholine	Smooth muscle	Abdel-Latif et al. (1977)
		Akhtar and Abdel-Latif (1980)
		Brock et al. (1985)
	Parotid	Berridge et al. (1983)
	Avian salt gland	Fisher et al. (1983)
ACTH	Adrenal cortex	Vilgrain et al. (1984a)
Adrenaline	Liver	Creba et al. (1983)
ADP	Platelets	Leung et al. (1983)
Angiotensin	Liver	Creba et al. (1983)
		Garrison et al. (1984)
	Smooth muscle	Smith et al. (1984)
	Adrenal cortex	Farese et al. (1984)
Antigen	Basophils	Beaven et al. (1984)
	Mast cells	White et al. (1985)
Bombesin	Swiss 3T3 fibroblasts	Brown et al. (1984)
Bradykinin	Fibroblasts	Vicentini and Villereal (1984)
	Neurobastoma/glioma hybrid	Yano et al. (1984)
Carbachol	Smooth muscle	Park and Rasmussen (1985)
	Astrocytoma cells	Masters et al. (1984)
EGF	A431 epidermoid carcinoma	Sawyer and Cohen (1981)
FGF	Swiss 3T3 fibroblasts	Tsuda et al. (1985)
Glucose	Pancreatic islets	Laychock (1983)
Gonadotropin-releasing hormone (GRH)	Pituitary	Hirota et al. (1985)
		Naor et al. (1985)

Agonist	Cell type	Reference
Heat shock	CHO cells	Stevenson et al. (1986)
5-hydroxytryptamine	Blowfly salivary gland	Litosch et al. (1984)
Interleukin-2 (IL-2)	T lymphocytes	Farrar and Anderson (1985)
Interleukin-3 (IL-3)	FDC-P1 hematopoetic cells	Farrar et al. (1985)
Lipopolysaccharide	B lymphocytes	Wightman and Raetz (1984)
Formyl–methionyl–leucyl–phenylalanine (FMLP)	Neutrophils	Volpi et al. (1983)
		Yano et al. (1983)
	HL-60 leukemic cells	Dougherty et al. (1984)
Light	Photoreceptors	Vandenburg and Montal (1984)
Phytohemagglutinin (PHA)	Lymphoblastoid T cells	Hasegawa-Sasaki and Sasaki (1983)
Platelet-activating factor	Platelets	Shukla and Hanahan (1983)
Platelet-derived growth factor (PDGF)	Swiss 3T3 fibroblasts	Habenicht et al. (1981)
		Berridge et al. (1984)
Prostaglandin $F_{2\alpha}$	Swiss 3T3 fibroblasts	MacPhee et al. (1984)
Serotonin	Blowfly salivary gland	Berridge et al. (1983)
		Litosch et al. (1984)
Serum	Swiss 3T3 fibroblasts	Rodriguez-Pena and Rozengurt (1985)
Spermatozoa	Sea urchin eggs	Turner et al. (1984)
Thrombin	Platelets	Billah and Lapetina (1982)
Thyrotropin-releasing hormone (TRH)	Pituitary (GH_3 cells)	Rebecchi and Gershengorn (1983)
		Drust and Martin (1985)
Vasopressin	Liver	Rhodes et al. (1983)
		Garrison et al. (1984)
	Swiss 3T3 fibroblasts	Brown et al. (1984)
	Hepatocytes	Seyfred and Wells (1984)

[a] This does not account for all of the natural agonists of protein kinase C; rather, it is intended to be a representative selection. The criteria for identification of agonists include stimulation of PI turnover, translocation of protein kinase C activity from the soluble to particulate fraction, and phosphorylation of known substrates.

addition of these compounds in platelets elicits a detectable, transient elevation of Ca^{2+} levels (Ware et al., 1985). Ca^{2+} ionophores have been used to determine whether certain processes that are activated by agonists of protein kinase C, such as c-myc gene expression (see Section 4.4), are also stimulated by Ca^{2+}-dependent mechanisms (Kaibuchi et al., 1986). However, such ionophores also cause limited but detectable DAG production, making studies of the actions of Ca^{2+} that are totally independent of DAG difficult (Sano et al., 1985b). Ca^{2+} has been shown to synergize with activators of protein kinase C in certain processes, such as platelet activation, smooth muscle contraction, and lymphocyte mitogenesis, although these complex responses probably involve additional mechanisms (Kaibuchi et al., 1983, 1985; Park and Rasmussen, 1985; Truneh et al., 1985; Takai et al., 1985; see Section 4). The role of Ca^{2+} in protein kinase C activation is discussed further in Section 2.7.

It should be noted that all of the natural agonists of pI turnover generate DAGs at the plasma membrane since this is the location of the agonist–receptor molecules. Furthermore, DAGs are confined to this locale since their short half-lives preclude lateral diffusion or vesicle-mediated transfer to other membrane compartments. Thus, naturally activated protein kinase C is restricted to this environment, and it is here that one might expect to find substrates of physiological relevance (see Section 3). It should be noted that endogenous DAGs are being formed continuously in cells via pathways of triglyceride and phospholipid metabolism. The inability of these DAG molecules to promote activation of protein kinase C is due to their being predominantly of the 2,3-conformation, which is incapable of activating protein kinase C. They are also primarily localized in the smooth endoplasmic reticulum, the phospholipid composition of which is inhibitory for binding of protein kinase C (see Section 2.1).

2.5. Activation by Tumor Promoters

Studies of experimental multistage carcinogenesis systems identified several compounds with the capacity to stimulate progression of tumor growth following an initiation event (for review, see Boutwell, 1974). The mouse skin system revealed a series of compounds isolated from croton oil, the phorbol esters, as potent tumor-promoting agents (for review, see Slaga, 1984). Although these compounds were shown to exert pleiotropic effects on a variety of cells (see Section 4), it was not until a specific receptor protein was identified for phorbol esters that their mechanism of action began to be understood. Phorbol esters are lipophilic molecules that intercalate into phospholipid bilayers. Using [^{3}H]phorbol dibutyrate, a less lipid-soluble analogue of the most potent phorbol ester, TPA, saturable, high-affinity binding sites (K_d = 8–30 nM) were revealed in cultured cells (Driedger and Blumberg, 1980) and mouse epidermis (Delclos et al., 1980).

The number of receptors was estimated at 10^5–10^6/cell (Shoyab and Todaro, 1980; Blumberg et al., 1984). By monitoring displacement of [^{3}H]phorbol dibutyrate, several other tumor promoters such as mezerein, aplysiatoxin, and teleocidin were demonstrated to bind to the same putative receptor (Horowitz et al., 1983; Blumberg et al., 1984).

Binding of phorbol esters to the receptor was dependent on the presence of acidic phospholipids and Ca^{2+} (Sando and Young, 1983), although others have found the binding to be independent of Ca^{2+} (Arcoleo and Weinstein, 1985). Highest levels were found in brain tissue, but specific binding was detected in all mammalian tissues examined (Shoyab et al., 1981). Indeed, binding sites have been identified in all vertebrate species and in *Drosophila melanogaster*, and the sea urchin *Lytechinus pictus* (Blumberg et al., 1983).

2.6. Read Protein Kinase C for Phorbol Ester Receptor

In 1982, Castagna et al. reported that phorbol esters could substitute for DAG in activating partially purified protein kinase C. Moreover, the amounts required for activation of the enzyme were similar to the doses necessary for tumor promotion in intact cells. Several subsequent studies demonstrated that the phorbol ester receptor copurified with protein kinase C through various separation procedures (Ashendel et al., 1983a,b; Kikkawa et al., 1983b; Leach et al., 1983; Niedel et al., 1983; Arcoleo and Weinstein, 1985) and, indeed, through to homogeneous preparations of the protein kinase (Parker et al., 1984a). Other nonphorbol tumor promoters such as teleocidin, debromoaplysiatoxin (Fujiki et al., 1984), and mezerein (Miyake et al., 1984) subsequently were shown to activate protein kinase C directly. Furthermore, the biological potency of a variety of tumor promoters correlated with their relative ability to activate protein kinase C *in vitro*. Although there is now considerable experimental evidence supporting the notion that protein kinase C represents the major cellular receptor for tumor-promoting phorbol esters, the possibility that some of the biological effects of these compounds are directly mediated through other interactions cannot be excluded yet. For example, reports of nonlinear Scatchard binding plots have raised the possibility of multiple binding sites (Blumberg et al., 1984). There is also growing evidence for the existence of more than one form of protein kinase C (see Sections 2.1 and 2.10). In addition, the lipophilic nature of these molecules causes perturbations of phospholipid bilayers at relatively low concentrations (Tran et al., 1983). Such changes in fluidity may affect a variety of proteins that are associated with membranes, including ion pumps, receptors, and cytoskeletal structures, independently of activation of protein kinase C.

Indeed, some biological effects of phorbol esters cannot be elicited by other activators of protein kinase C. For example, DAG and bryostatin, a macrocyclic lactone that binds and activates protein kinase C competitively with respect to TPA, are unable to mimic the property of TPA to induce differentiation of the human promyelocytic cell line, HL-60 (Yamamoto et al., 1985; Kraft et al., 1986). In fact, bryostatin actually antagonizes TPA-induced differentiation (Kraft et al., 1986). These observations suggest that activation of protein kinase C is not sufficient itself to cause differentiation of these cells and that phorbol esters have additional effects. One might argue in this particular case, however, that bryostatin competes out binding of TPA to protein kinase C and prevents translocation of the active protein kinase from the plasma membrane to a critical subcellular compartment inaccessible to bryostatin, which is required for TPA to induce differentiation. This might result from a lower rate of movement of bryostatin to other compartments compared with TPA or by interaction with a plasma membrane constituent that constrains it to the plasma membrane, like DAG (see Section 2.7). Nonetheless, that phorbol esters operate exclusively through the activation of protein kinase C still remains an assumption. Therefore, evidence for the involvement of protein kinase C in a pathway relying solely on the use of phorbol esters should be viewed with caution.

Although of diverse overall structure, tumor promoters share several backbone features with 1,2-DAGs, the natural activators of protein kinase C, which presumably account for their affinity for this protein (Brasseur et al., 1985; Jeffrey and Liskamp, 1986; Wender et al., 1986). However, they exhibit dissociation constants up to one thousandfold lower than their natural counterparts. Furthermore, tumor promoters are far more stable in cells than DAGs, which are rapidly metabolized by DAG kinases and lipases. These differences in binding affinity and in vivo longevity underlie the physiological potency of tumor promoters in activating protein kinase C. Their stability may enable these activators to distribute, by vesicle-mediated processes from the plasma membrane to membrane compartments normally beyond the range of short-lived DAGs, thus introducing active protein kinase C to new environments and potentially causing ectopic phosphorylation of proteins. Indeed, in HL-60 and liver cells, DAGs induce the phosphorylation of only a subset of proteins phosphorylated in response to TPA (Kreutter et al., 1985; Kiss and Luo, 1986) and cannot mimic all of the effects of phorbol esters on monocyte maturation (Shinohara et al., 1985a). However, in leukocytes and platelets, the two agonists appear to induce patterns of phosphorylation that are indistinguishable (Kawahara et al., 1980; Fujiki et al., 1984).

2.7. Subcellular Distribution of Protein Kinase C

As mentioned in Section 2.1, brain is the richest source of protein kinase C. Efficient extraction of activity from this tissue requires the presence of Ca^{2+}

chelators such as ethyleneglycoltetraacetic acid (EGTA). This property has been exploited to purify the protein kinase by homogenization of brain tissue in the presence of Ca^{2+}, followed by extraction of the insoluble fraction with EGTA, yielding a twenty-fold enrichment in protein kinase C activity (Wolf et al., 1984). The basis for this phenomenon has been demonstrated recently using inverted erythrocyte ghost membranes and purified protein kinase C assayed by [^{3}H]phorbol dibutyrate binding (May et al., 1985a; Wolf et al., 1985b). At Ca^{2+} levels below 100 nM, the protein kinase shows little affinity for the membranes. Elevation of Ca^{2+} levels promotes association of the protein kinase with the membranes, which is reversible upon addition of EGTA; half-maximal binding occurs at 0.5 μM Ca^{2+}. However, the membrane-associated enzyme is catalytically inactive. Ca^{2+}-induced binding of protein kinase C to membranes has also been measured in cell lysates using immunological detection methods, although half-maximal association occurred at 4 μM Ca^{2+} by this criterion (Woodgett and Hunter, 1987b). This apparent discrepancy may be explained by the fact that the effective phospholipid concentration was very low in the latter procedure, which would result in an increase in the requirement for Ca^{2+} (Woodgett and Hunter, 1987b). Recently, Dougherty and Niedel (1986) reported that Ca^{2+} increased the affinity of protein kinase C for TPA, thus providing a mechanism for Ca^{2+}-induced sensitization of the enzyme to activation.

In 1982, Kraft et al., reported that treatment of EL4 thymoma cells with TPA caused the amount of protein kinase C activity in the cytosolic fraction to decrease about five-fold with respect to levels in control cells. Kraft and Anderson (1983) then reported that TPA caused a specific increase in the amount of protein kinase C associated with the plasma membrane. The interpretation of these data was that TPA caused protein kinase C to redistribute from a predominantly cytosolic location to the particulate (membrane) fraction. Subsequently, it was shown that the decrease in cytosolic enzyme activity was accompanied by a corresponding increase in activity in the particulate fraction (for example, see Anderson et al., 1985; Palfrey and Waseem, 1985; Avissar et al., 1986). Translocation has since been used to determine whether protein kinase C is activated by various physiological stimuli such as IL-2 (Farrar and Anderson, 1985) and IL-3 stimulation of lymphocytes (Farrar et al., 1985), TRH activation of GH$_3$ cells (Drust and Martin, 1985), GRH activation of primary pituitary cells (Naor et al., 1985), cross-linking of immunoglobulins in B lymphocytes (Nel et al., 1986), hormonal activation of pituitary gonadotrophs (Hirota et al., 1985), and long-term synaptic plasticity (Akers et al., 1986). Usually the translocation is documented to occur from the cytoplasm to the membrane, but there are reports of different patterns of redistribution caused by plant lectins and other stimuli (Costa-Casnellie et al., 1985; Mire et al., 1986; Nishihara et al., 1986). The significance of these latter, variable translocation events is presently unclear and requires further study, not only of activity but also of distribution of the enzyme protein, since changes in specific activity of the membrane-bound and soluble forms of protein kinase C cannot be ruled out. The subcellular localization of the enzyme is also sensitive to the growth state of the cell; in proliferating cells there is a greater proportion of membrane-bound enzyme than in resting cells, which presumably reflects activation of protein kinase C preceding and/or during mitosis (Adamo *et al.*, 1986).

Recently, direct evidence for altered partitioning of the enzyme following TPA treatment between soluble to particulate fractions has been provided using antibodies toward protein kinase C (Ballester and Rosen, 1985; Woodgett and Hunter, 1987b), and the phenomenon has been visualized by immunofluorescence of intact HL-60 cells (Shoji *et al.*, 1986). In contrast to the Ca^{2+}-induced redistribution, the TPA-induced membrane-associated fraction exhibited protein kinase activity and required detergent for extraction, suggesting the binding was hydrophobic in nature (Wolf *et al.*, 1985a,b). It should be noted that since most of the aforementioned translocation studies were performed in the presence of EGTA, the contribution of Ca^{2+} to membrane binding was overlooked. Indeed, the "translocation" event simply may reflect a change in affinity of protein kinase C for membranes rather than a physical redistribution between subcellular compartments (see Section 2.7.2).

2.7.1. Down-Regulation

One of the first properties of the phorbol ester receptor to be documented was its down-regulation in response to chronic treatment of certain intact cells with phorbol ester (Shoyab and Todaro, 1980; Jaken *et al.*, 1981; Solanki *et al.*, 1981; Collins and Rozengurt, 1982; Yamasaki *et al.*, 1982). Recently, down-regulation of protein kinase C has been shown to occur in rat brain slices (Shenolikar *et al.*, 1986). Cells in which the receptor had been down-regulated were refractory to further stimulation by phorbol esters or other agonists of protein kinase C. This property has been exploited to determine involvement of protein kinase C in cellular processes (see, for example, Rozengurt *et al.*, 1983; Coughlin *et al.*, 1985). The basis for desensitization recently has been elucidated using antibodies to protein kinase C (Ballester and Rosen, 1985; Woodgett and Hunter, 1987b; Young *et al.*, 1987). Upon exposure of cells to TPA, protein kinase C undergoes translocation to the membrane as in the normal activation pathway. However, the activated form has a half-life of only 50–300 min compared with a stability of greater than 24 hr in resting cells (Woodgett and Hunter, 1987b; Young *et al.*, 1987). Although this reduction in half-life acts as a limit on the length of time that molecules of protein kinase C can remain active, it probably would not be a determining factor in naturally activated cells due to the rapid metabolism of DAGs and feedback inhibition of the signal that generates new

DAGs (see Sections 3.2.2 and 5.2). Persistent stimulation of cells with TPA causes newly synthesized protein kinase C molecules to bind to the membrane and thus immediately become subject to the enhanced degradation rate. Pulse-label experiments have shown that continuous exposure to TPA does not inhibit synthesis of protein kinase C; indeed, the rate may be somewhat increased (Woodgett and Hunter, 1987b). Similarly, there is no reduction in protein kinase C mRNA levels in cells treated with TPA (Young *et al.*, 1987). Therefore, very low but, nonetheless, finite levels of active protein kinase C are present in cells that have been chronically treated with TPA, raising the possibility that certain critical substrates may still be phosphorylated to low levels even after "down-regulation." It appears likely that prolonged activation of protein kinase C is deleterious to cells and that enhanced proteolysis represents a last-ditch mechanism for removal of the activated protein.

The most likely candidate protease responsible for cleavage of protein kinase C *in vivo* is calpain I, a Ca^{2+}-dependent enzyme (Kishimoto *et al.*, 1983; Melloni *et al.*, 1986). This protein binds to membranes in the presence of Ca^{2+}, which places it in a similar subcellular location to that of protein kinase C. Treatment of cells with leupeptin inhibits TPA-induced down-regulation of protein kinase C (Chida *et al.*, 1986b). Although calpain I is inhibited by leupeptin, so are some other protease activities. Therefore, identification of the proteins involved in the process of degrading protein kinase C *in vivo* and, indeed, the actual pathway of destruction, still await determination.

2.7.2. A Model for Protein Kinase C Activation

Assimilation of the detailed observations given above generates a model in which protein kinase C in resting cells is normally present in an inactive state (Figure 1). The affinity of the enzyme for phospholipids, membrane-bound substrate proteins, and the presence of low but finite concentrations of Ca^{2+} probably maintains the protein in close proximity to the plasma membrane. Activation of PI turnover generates DAGs within the plasma membrane, which enter a high-affinity ternary complex with phospholipids and molecules of protein kinase C that previously were loosely associated with the membrane. This complex formation activates protein kinase C, and by removing protein kinase C molecules from the loosely associated membrane population also promotes, by mass action, the binding of more molecules of the enzyme to the membrane where they too enter the high-affinity ternary complex. This produces a net "translocation" of protein kinase C molecules from the cytosol to the membranous fraction. The simultaneous formation of IP_3 elicits an elevation in intracellular Ca^{2+} and potentiates DAG binding and, thus, activation of protein kinase C. DAG transiently released by the ternary complex due to mass action is either reincorporated

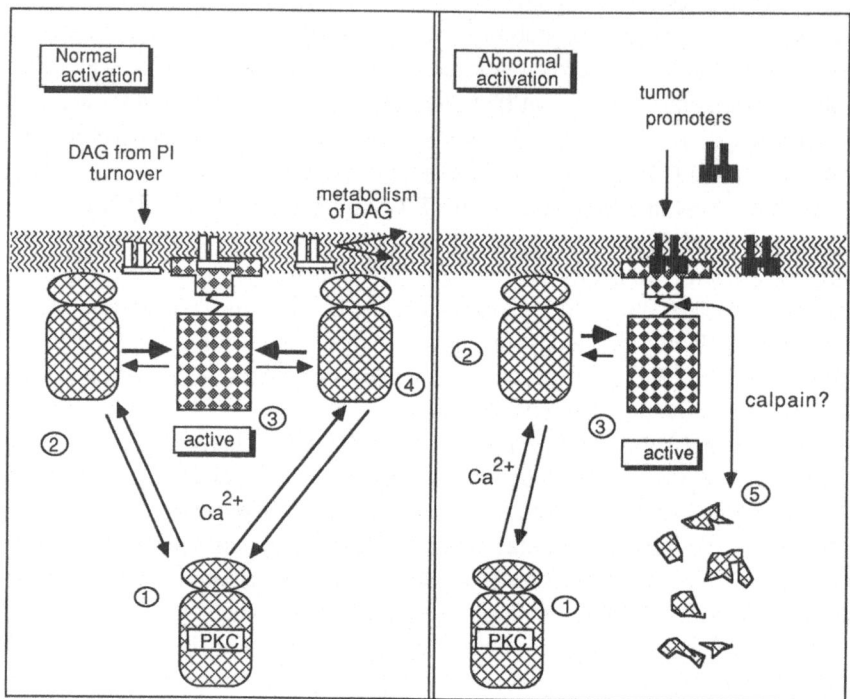

FIGURE 1. Model of protein kinase C regulation. In the resting state, protein kinase C fluxes between a loose association with the plasma membrane (2) and the cytosol (1), the former process being promoted by Ca^{2+}. Upon binding the activator (DAG formed by agonist-induced PI turnover or tumor promoter), the membrane-associated enzyme (2) undergoes a conformational change and is activated (3). Inactivation of the protein kinase occurs reversibly by metabolism of the DAG predominantly via DAG kinase (4) or irreversibly by degradation of the tumor promoter–activated protein kinase by a calpainlike protease (5). (For further details, see Section 2.5.)

into the ternary complex or metabolized. The latter outcome causes dissociation of the ternary complex, inactivating protein kinase C and reducing its affinity for the membrane. TPA usurps this pathway by substituting for DAG and causing protracted stimulation of activity. This activation can only be reversed by complete proteolysis of protein kinase C due to the low rate of metabolism of the phorbol esters by hydrolases (see Figure 1).

2.8. Is M Kinase Involved in Signal Transduction?

As mentioned in Section 2.1, protein kinase C can be cleaved proteolytically to a form that no longer requires phospholipids or Ca^{2+} for activity (termed M kinase). The physiological relevance of this fragment recently has been championed by several groups who suggest that since M kinase is soluble, it may phosphorylate substrates in the cytosol and nucleus, thus transducing the effects of agonists from the membrane to other compartments (Tapley and Murray, 1984a,b, 1985; Melloni et al., 1985; Mizuta et al., 1985). These groups have detected the formation of M kinase in a variety of cells in response to agonists of protein kinase C. Furthermore, formation of M kinase could be inhibited by treatment of the cells with leupeptin. TPA is a weak mitogen (see Section 4.3) and stimulates transcription of a variety of genes that would require transmission of membrane signals into the nucleus (Rabin et al., 1986) (see Section 4.4). Moreover, substrate proteins for protein kinase C in vitro have been documented that are not membrane-associated or are of nuclear origin and, thus, would not be accessible to intact, activated protein kinase C (see Table 5). Since TPA decreases the half-life of protein kinase C (Woodgett and Hunter, 1987b), some active fragments of the enzyme presumably are formed in an agonist-dependent manner.

Although proteolytically activated products of protein kinase C undoubtedly can be formed in cells, we have failed to observe an accumulation of these fragments in stimulated cells by immunological methods. The fragments that are generated are of extremely transient nature (J. R. Woodgett and T. Hunter, unpublished observations). The M kinase activity detected by Tapley and Murray (1985) and Mizuta et al. (1985) might be generated during isolation by TPA-enhanced, limited proteolysis of the activated enzyme. These fragments might be stabilized by the extraction conditions. By pretreatment of neutrophils with protease inhibitors, Pontremoli et al. (1986a) have determined which responses to TPA are dependent on intracellular proteolysis. The neutrophils produced oxygen radicals in response to TPA regardless of pretreatment, indicating that M kinase was not involved in this pathway. However, another response to TPA, exocytosis of serine proteinase, only occurred in the absence of the inhibitors. It is not known whether this latter process is dependent on cellular proteolytic activity per se or if there is a specific requirement for proteolysis of protein kinase C. Vilgrain et al. (1984a) detected no increase in M kinase activity following adrenocorticotropic hormone treatment of adrenal cortex suspension cells, despite activation of protein kinase C. Similarly, TPA treatment of MCF-7 cells activates protein kinase C but does not stimulate generation of M kinase (Darbon et al., 1986a).

It is worth noting that none of the nuclear or cytoplasmic in vitro substrates has yet been shown to be phosphorylated directly by protein kinase C in vivo (see Section 3.2). In fact, virtually all of the proteins that have been demonstrated as physiological substrates are membrane associated. Indeed, in the case of one, the transforming protein of Rous sarcoma virus (RSV), $pp60^{v-src}$, membrane association is an absolute requisite for phosphorylation by protein kinase C in

TPA-treated cells (Buss *et al.,* 1986; see Section 3.2.1). In our opinion, protein kinase C activity is confined to the environment of the cell membrane, and soluble degradation products are inactivated too swiftly to be of physiological relevance. The mechanisms by which nuclear events are stimulated by phorbol esters thus remain to be elucidated.

2.9. Regulation of Protein Kinase C by Phosphorylation

As mentioned in Section 2.1, purified protein kinase C incorporates phosphate into itself in the absence of substrates. This autophosphorylation reaction is common to many protein kinases (see, for instance, Roach, 1981; Miller and Kennedy, 1986). Additionally, several protein kinases are regulated by phosphorylation by other serine/threonine-specific protein kinases; indeed, myosin light-chain kinase is phosphorylated by protein kinase C *in vitro* (see Section 3.7.2). Recent discoveries that many growth factor receptors and the transforming proteins of certain retroviruses have protein–tyrosine kinase activity have focused attention on the identification of the proteins phosphorylated on tyrosine residues (for review, see Hunter and Cooper, 1985). Since protein kinase C is implicated in growth control (Sections 4.2 and 4.3) and PI turnover is reported to be elevated in some retrovirally transformed cells (Diringer and Friis, 1977; Macara *et al.,* 1984; Fry *et al.,* 1985), the enzyme itself may be regulated by phosphorylation under various physiological conditions. The phosphorylation state of protein kinase C may thus reflect control by other signal transduction and regulatory mechanisms.

Two studies have determined the phosphorylation state of protein kinase C *in vivo.* Fry *et al.* (1985) immunoprecipitated protein kinase C from ^{32}P-labeled NIH 3T3 fibroblasts and fibroblasts transformed by Ab-MuLV, a retrovirus encoding a protein–tyrosine kinase, $P160^{gag-abl}$, as its transforming protein (Witte *et al.,* 1980). Phosphorylated protein kinase C was detected in the fibroblasts and predominantly contained phosphoserine with some phosphothreonine. The phosphoamino acid content of protein kinase C from Ab-MuLV-transformed cells was similar to that of the enzyme from their untransformed counterparts, with no phosphotyrosine being detected. We have immunoprecipitated protein kinase C from ^{32}P-labeled mouse and human cell lines with antiserum that recognizes two forms of the enzyme (see Section 2.3) and find that in resting cells, both forms predominantly contain phosphoserine with some phosphothreonine (Woodgett and Hunter, 1987b). Although the enzyme contains stoichiometric amounts of phosphate *in vivo,* the majority appears to be present in a site(s) that turns over rather slowly. Treatment of these cells with TPA causes up to a three-fold increase in ^{32}P incorporation without altering the relative amounts of phosphoserine and phosphothreonine. Preliminary phosphopeptide mapping experiments indicate that several peptides are labeled, some of which correspond to sites phosphorylated on the purified enzyme after *in vitro* auto-phosphorylation, suggesting that this reaction occurs *in vivo* (J. R. Woodgett, unpublished observations). In agreement with the results of Fry *et al.* (1985), we detect no phosphotyrosine in protein kinase C immunoprecipitated from human A431 cells infected with either RSV or Sneider Theilen-Feline sarcoma virus (ST-FeSV), which encode the protein–tyrosine kinases pp60^{v-src} and P85$^{gag-fes}$, respectively (Collett and Erikson, 1978; Barbacid and Lauver, 1981). Protein kinase C therefore does not appear to be a direct substrate of protein–tyrosine kinases but could be phosphorylated potentially by other serine/threonine–specific protein kinases.

2.10. Multiple Forms of Protein Kinase C

Until recently, only one form of protein kinase C was detected in various tissues from several species, supporting the idea that all of the effects of TPA and DAG were directed through one protein. In Sections 2.1 and 2.3 we stated that, in fact, two forms of protein kinase C could be purified from rat brain, the physiological relevance of which was demonstrated by the use of antiserum to this preparation. The two proteins appear to be the products of distinct RNA molecules and, thus, represent true isozymes of protein kinase C (Woodgett and Hunter, 1987b).

Recently, several groups have isolated cDNA clones of protein kinase C. Up to four distinct cDNAs have been observed that are highly related but predict distinct amino acid sequences (Coussens *et al.*, 1986a; Knopf *et al.*, 1986; Ono *et al.*, 1986a,b; Parker *et al.*, 1986; Makowske *et al.*, 1986; Ohno *et al.*, 1987; Housey *et al.*, 1987). The predicted protein sequences of the clones reveal a region homologous to known protein kinase domains towards the C terminus, a centrally located acidic region that possibly represents the Ca^{2+}-binding sites, and a somewhat hydrophobic N-terminal region that contains two clusters of four cysteine residues that are candidates for metal ion–binding sites. At least three of these clones hybridize to genes located on distinct human chromosomes (Coussens *et al.*, 1986a). Hence, a family of protein kinase C-like proteins appears to exist in mammalian cells, the functions of which are unknown as yet.

Clearly, the assumption that one gene product mediates all of the actions of tumor-promoting phorbol esters is wrong, and the mechanisms through which agonists activate protein kinase C are likely to be far more complex than believed previously. Analysis of the expression of these genes indicates that their mRNAs are regulated independently and that their relative abundance varies widely in different cell lines and tissues (Coussens *et al.*, 1986a; Knopf *et al.*, 1986; Ohno *et al.*, 1987). In rat and rabbit brain, there is evidence of differential RNA splicing to generate two forms of protein kinase C that differ only in their C-terminal 50 amino acids (Ono *et al.*, 1986a; Ohno *et al.*, 1987).

Three apparently distinct forms of protein kinase C have been resolved from rat and rabbit brain by hydroxylapatite chromatography (Huang *et al.*, 1986b; Jaken and Kiley, 1987). The three forms all exhibit molecular masses of 80 kD but show antigenic differences. Preliminary results indicate varying dependencies upon Ca^{2+} and phorbol esters. Thus, differential activation of the protein kinase C isozymes by differing agonist conditions may occur *in vivo*. The relationships between these enzymes and the two forms of protein kinase C discussed in Section 2.3 to the cDNA clones remain to be determined.

The presence of multiple protein kinase C enzymes may also explain the failure to isolate mutants defective in protein kinase C since multiple lesions would be required to generate a null phenotype. Several mutants of mouse and human promyelocytic cells and fibroblasts have been isolated, however, which are defective in their response to phorbol esters (Butler-Gralla and Herschman, 1981; Huberman *et al.*, 1981, 1982; Murao *et al.*, 1983; Lotem and Sachs, 1979) or dihydroteleocidin B (Shimizu *et al.*, 1986). These mutants were either selected by their ability to grow in increasing concentrations of phorbol ester or their inability to respond mitogenically to the activators. The mutant cells tend to be defective in their response to activator, failing to differentiate in the case of HL-60 variants or to undergo mitosis in the case of the Swiss 3T3 or 3T3-L1 cells (Huberman *et al.*, 1982; Herschman, 1985; Shimuzu *et al.*, 1986). The molecular basis for the defects in these mutants is still uncertain but is not due to the absence of protein kinase C since the mutant cell lines contain "normal" levels of the enzyme. One mutant (clone R-35 of HL-60 cells), however, may have defective regulation of protein kinase C, since phorbol ester binding apparently is not down-regulated in these cells (Solanki *et al.*, 1981). Similarly, three other clonal lines of HL-60 cells resistant to TPA-induced differentiation were found to be defective in TPA-induced translocation of protein kinase C activity from the cytosol to the membrane (Homma *et al.*, 1986). These latter results are difficult to reconcile with current models of protein kinase C activation since phorbol ester binding was normal in these cells. The defect might possibly lie in a membrane-associated protein component that stabilizes binding of protein kinase C. In this regard, it was reported recently that the enzyme does associate with 110/115-kD polypeptides that are enriched in cytoskeletal preparations (Wolf and Sahyoun, 1986). It is clear, however, that the mutational events leading to cellular resistance to the effects of phorbol esters are complex and probably distal to the involvement of protein kinase C.

3. IDENTIFICATION OF SUBSTRATES FOR PROTEIN KINASE C

In order to understand the mechanisms of action of agonists of protein kinase C, the proteins that are phosphorylated by this enzyme must be identified. This has not been an easy task since the purified enzyme exhibits unbounded promiscuity in phosphorylating proteins offered to it *in vitro*. Hence, a rather daunting list of substrates has been created that includes proteins of seemingly every function, subcellular location, and structure (for graphic representation of the growth curve of putative substrates, see Figure 2). The very length of this list has complicated attempts to rationalize observations of effects of agonists *in vivo* with changes in the phosphorylation state of proteins phosphorylated by protein kinase C *in vitro*. A hint of the exaggerated nature of the *in vitro* substrate list is to be found in two-dimensional gel analyses of ^{32}P-labeled lysates from cells treated with or without TPA (for example, see Kiss and Steinberg, 1985). Surprisingly, few changes in phosphorylation occur, with the exception of the dramatic appearance of a cluster of spots in the 80-kD range (see Section 3.2.3). The sensitivity of this technique is not very high and would not detect proteins of low abundance or proteins that do not focus in the first dimension.

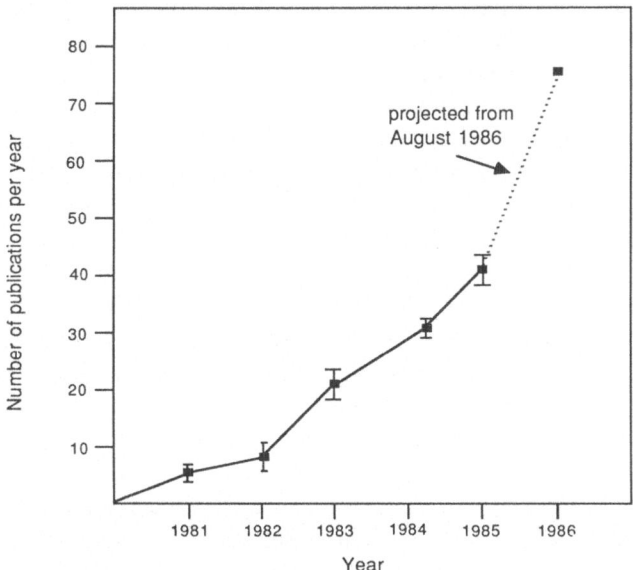

FIGURE 2. Number of publications reporting substrates for protein kinase C since 1981. Error bars indicate estimated errors in our interpretation of the data and papers that were overlooked accidentally.

In view of the difficulties in evaluating whether a protein is a true substrate of protein kinase C, we propose the following criteria as a guideline for investigation of putative targets:

1. Since many proteins are phosphorylated at multiple sites by several protein kinases, the specific site(s) phosphorylated by protein kinase C must be identified. The site(s) should also be phosphorylated stoichiometrically *in vitro* by protein kinase C with reasonable kinetics with respect to other substrate proteins, i.e., have K_m values near the predicted physiological concentration for that protein. If the amino acid sequence of the putative substrate is known, the exact residue(s) should be determined. Use of synthetic peptides that contain the site often can be helpful (see Section 3.3). If protein sequence is not available, peptides containing the site should be generated by proteolytic or chemical cleavage and characterized, for example, with respect to high-performance liquid chromatography retention times or behavior upon two-dimensional thin-layer analysis at different pHs, to enable identification of the site in a mixture of other phosphorylated peptides. Since some protein kinases have overlapping specificities, other enzymes should be tested for their ability to phosphorylate the identified residue.

2. The protein should be shown to be phosphorylated rapidly at the same site(s) identified in 1 (above) *in vivo* in response to agonists of protein kinase C. If specific antisera are available, the protein of interest may be immunoprecipitated from ^{32}P-labeled cultured cell lines and analyzed by SDS–gel electrophoresis. If such reagents are not available, the protein should be purified from ^{32}P-labeled cells. If sufficient amounts of protein can be isolated (milligram quantities), unlabeled tissue may be used, followed by chemical analysis of phosphate content (requires > 0.5 nmole phosphate per sample) or fast atom bombardment mass spectrometry (requires > 100 pmole and exact peptide sequence must be known). Suitable agonists include TPA, phorbol dibutyrate, teleocidin, synthetic DAGs and phospholipase C, which should be used at concentrations that previously have been shown to be biologically effective (for example, 1–50 nM for TPA). Controls should also be performed using compounds that do not activate protein kinase C, such as 4-β-phorbol. It should be noted that none of these "pure" agonists can be administered effectively to whole animals due to their lipid solubility. Hence, isolated cells are required. Other agonists have mixed effects and should be avoided in initial studies. It is important to maintain the phosphorylation state of the proteins at cell lysis during subsequent isolation by inclusion of ethylenediaminetetraacetic acid and NaF to inhibit protein kinases and phosphoprotein phosphatases, respectively. Once the protein has been purified sufficiently, the phosphorylation state of the putative site within the protein should be determined and shown to increase specifically after agonist treatment.

Although fulfillment of these two criteria provides substantial evidence in favor of the protein of interest being a physiological substrate, additional corroborative data are useful. For example, DAGs have a much shorter biological half-life than TPA and, thus, the time course of phosphorylation should be different for the two agonists. Treatment of certain cells with high levels of TPA for 1–2 days causes a dramatic decrease in protein kinase C levels (see Section 2.7.1). Subsequent challenge of the cells with agonists should not stimulate phosphorylation of the putative substrate, although this does depend on the degree of down-regulation achieved (which is cell type specific) and affinity for substrate. If the phosphorylation of the protein *in vitro* has some measurable effect, such as a change in catalytic activity, this parameter may be monitored in lysates before and after stimulation. However, this does not rule out the possibility of an indirect effect of protein kinase C on the parameter in question.

In Section 2.7.2, we proposed that activated protein kinase C is confined to the membrane environment. This restriction of locale also constrains the number of potential substrates. For this reason, we suggest that an additional criteria be applied to putative substrates, namely that at least a fraction of the protein of interest should be either loosely or intrinsically associated with the cytoplasmic face of the membrane. Indeed, all of the proteins identified to date as physiological substrates appear to fulfill this last requirement.

3.1. Receptor Substrates

In contrast to the multitude of proteins demonstrated as targets *in vitro*, only a handful of these proteins satisfy most of the criteria detailed above. Of these, the substrates that are ligand receptors will be discussed first.

3.1.1. Epidermal Growth Factor Receptor

The receptor for epidermal growth factor (EGF) is a 170-kD glycoprotein, which has an N-terminal, extracellular EGF-binding domain and a C-terminal, intracellular protein–tyrosine kinase domain (Ushiro and Cohen, 1980; Ullrich et al., 1984; Xu et al., 1984; Hunter, 1984b). Addition of EGF stimulates phosphorylation of the purified receptor at tyrosine (residues 1068, 1148, and 1173) (Downward et al., 1984a). Tyrosine 1173 corresponds to the major site of tyrosine phosphorylation *in vivo* upon addition of EGF, and occupancy of this site correlates with increased protein kinase activity of the receptor toward exogenous substrates.

Addition of tumor promoters to cell lines that are responsive to EGF causes a rapid decrease in high-affinity binding of ^{125}I-EGF (Lee and Weinstein, 1978; Brown et al., 1979; Shoyab et al., 1979), blocks EGF-induced tyrosine phosphorylation of the EGF receptor in the intact cells and activity toward exogenous substrates (Friedman et al., 1984; Cochet et al., 1984; Decker, 1984; Downward et al., 1985), and can induce transient, reversible internalization of the receptor in KB cells (Beguinot et al., 1985). The effect of TPA clearly depends on the cell type, to a large extent. TPA was shown to increase phosphate incorporation into the EGF receptor at serine and threonine residues (Cochet et al., 1984; Davis and Czech, 1984; Iwashita and Fox, 1984). The molecular basis of this effect was explained partially by the demonstration that protein kinase C phosphorylates the EGF receptor directly. This phosphorylation correlated with inhibition of high-affinity EGF binding and reduction of the EGF-stimulated protein–tyrosine kinase activity of the receptor (Cochet et al., 1984; Downward et al., 1985; Fearn and King, 1985). The site of phosphorylation by protein kinase C was identified as threonine 654, a residue located nine amino acids from the cytoplasmic boundary of the receptor's transmembrane domain (Hunter et al., 1984; Davis and Czech, 1985). Threonine 654 is flanked by basic residues: phosphorylation introduces a negatively charged phosphate group into this region that might be expected to perturb the secondary structure. Signal transduction is postulated to occur via receptor dimerization since the single passage of the receptor polypeptide chain through the plasma membrane would appear insufficient to transmit a structural change from the extracellular EGF-binding domain. Perhaps, the region of the protein containing threonine 654 is involved in the dimerization process. Experiments in which TPA was added to cells after EGF, followed by immunoprecipitation of the receptor with antiphosphotyrosine antibodies, have demonstrated that phosphorylation of threonine 654 can occur on molecules that have already been activated by EGF, indicating that the tyrosine and threonine phosphorylation events are not mutually exclusive (King and Cooper, 1986).

That phosphorylation of threonine 654 is responsible for the documented effects of tumor promoters on the EGF receptor was recently confirmed by the generation of mutant receptor molecules in which threonine 654 was replaced by a nonphosphorylatable alanine residue. Cells transfected with this mutant DNA no longer lost their high-affinity EGF-binding component upon treatment with TPA nor were the receptors internalized in response to the tumor promoter (Lin et al., 1986). In contrast, EGF-induced internalization of the mutant receptors was unaffected.

The function of the serine phosphorylation sites on the EGF receptor is unknown, as are the protein kinases that are responsible for their occupancy. The EGF receptors from hepatocytes (Rackoff et al., 1984) and A431 cells (Ghosh-Dastidar and Fox, 1984), however, are substrates for cAMP-dependent protein kinase in vitro, although this phosphorylation has no apparent effect on the protein–tyrosine kinase activity of the receptor.

The effects of tumor promoters on EGF affinity and phosphorylation of the receptor are mimicked by synthetic DAGs (McCaffrey *et al.*, 1984; Davis *et al.*, 1985b; Sinnett-Smith and Rozengurt, 1985), PDGF, and bombesin (Brown *et al.*, 1984), substantiating the role of protein kinase C. One of the effects of EGF in A431 epidermoid carcinoma cells is to enhance PI turnover (Sawyer and Cohen, 1981). Therefore, in this cell type, EGF promotes the phosphorylation (Iwashita and Fox, 1984) and desensitization (Friedman *et al.*, 1984) of its own receptor. However, it should be noted that the response of A431 cells to EGF is rather unusual with respect to activation of PI turnover. In other cell types, phosphorylation of threonine 654 is likely to represent heterologous receptor transmodulation, being promoted by the more consistent agonists of protein kinase C, such as PDGF.

A cellular gene that is highly related to the EGF receptor, termed c-*erb*-B2 or *neu* (Stern *et al.*, 1986), encodes a putative receptor that has a threonine placed similarly to threonine 654 (T. Yamamoto, personal communication). This threonine apparently is phosphorylated in response to TPA treatment, although definitive identification of the phosphorylated residue and effect of phosphorylation on protein–tyrosine kinase activity have yet to be determined.

Recently, the transforming protein gp75^{verb-B} of avian erythroblastosis virus (AEV) has been shown to represent a truncated form of avian EGF receptor (Downward *et al.*, 1984b). This protein has lost most of the extracellular domain, which contains the binding site for EGF, but has retained the transmembrane and cytoplasmic domains and, hence, a threonine residue equivalent to threonine 654 (threonine 105). Treatment of one clonal line of AEV-infected chicken fibroblasts with TPA causes increased phosphorylation of this threonine residue (Decker, 1985). TPA treatment inhibits growth of this fibroblast line but has no effect on AEV-transformed chick erythroblasts, the natural target cells of this virus (Decker, 1985; J. Meisenhelder, personal communication).

3.1.2. IL-2 Receptor

The human IL-2 receptor is a glycosylated protein of ~50 kD that comprises 251 amino acids (Cosman *et al.*, 1984; Leonard *et al.*, 1984; Nikaido *et al.*, 1984). Approximately 90% of the protein is extracellular and contains the binding domain for IL-2. A putative transmembrane domain is followed by 11 C-terminal amino acids. It is appropriate to note here that the signal transduction mechanism of the IL-2 receptor is not understood, since the protein has no enzymatic activity and only a very short cytoplasmic tail.

Treatment of human T cells with TPA induces rapid phosphorylation of the IL-2 receptor, predominantly on a serine residue(s) (Shackelford and Trowbridge, 1984; Leonard *et al.*, 1985). Protein kinase C directly phosphorylates immunoprecipitated IL-2 receptor at the same tryptic peptides that are labeled following

TPA treatment of cells (Shackelford and Trowbridge, 1986; Gallis et al., 1986; Taguchi et al., 1986). The major site of phosphorylation was identified as serine 247, with minor phosphorylation of threonine 250. Serine 247 is located nine residues C-terminal to the transmembrane domain (Shackelford and Trowbridge, 1986; Gallis et al., 1986). When mutant receptor molecules, in which serine 247 had been substituted by an alanine residue, were expressed in transfected cells, they no longer became phosphorylated in response to TPA (Gallis et al., 1986). The function of this phosphorylation is not fully understood. Unlike TPA-induced phosphorylation of the EGF receptor, phosphorylation of the IL-2 receptor by protein kinase C does not appear to alter ligand binding affinity (Robb and Rusk, 1986). In fact, preliminary evidence indicates that long-term TPA treatment, like IL-2 itself, actually causes an increase in the total number of IL-2 receptors, as determined by binding of a monoclonal antibody, termed anti-Tac, to the receptor (Depper et al., 1984). The mechanism of this induction appears to be through an increase in mRNA level (D. A. Shackelford and I. S. Trowbridge, personal communication). The relationship, if any, between the rapid phosphorylation of serine 247 and the slower induction of IL-2 receptors by TPA is presently unclear. Although IL-2 induces translocation of protein kinase C from soluble to particulate fractions (Farrar and Anderson, 1985; see Section 2.7), IL-2 apparently does not stimulate phosphorylation of its receptor at serine 247 (Leonard et al., 1985). The reason for this discrepancy is unknown but may reflect the transient nature of the activation of protein kinase C by DAG, as opposed to the persistent activation by TPA (see Section 2.6).

3.1.3. Transferrin Receptor

Treatment of human HL-60 and K562 cells with phorbol esters causes increased internalization of transferrin receptors and an inhibition of transferrin binding (Rovera et al., 1982; Klausner et al., 1984; May et al., 1984). In macrophages, however, TPA causes an increase in cell-surface expression of the receptor, indicating that the effects are probably more complex than thought at first (Buys et al., 1984). An increase in receptor phosphorylation at a serine residue(s) was observed following stimulation with TPA (May et al., 1984) or synthetic DAGs (May et al., 1986). Incubation of transferrin receptor with purified protein kinase C yielded the same phosphotryptic peptides that were observed after TPA treatment of intact cells (May et al., 1985b). Unlike the EGF and IL-2 receptors, which are located with their C termini inside the cell, the transferrin receptor has the opposite orientation with the N terminus being cytoplasmic. The major phosphorylated residue has been identified as serine 24, which is present in the N-terminal portion of the receptor and is located 41 residues from the membrane (Davis et al., 1986). The mechanism by which the receptor is internalized in response to phosphorylation is unclear but requires an intact cytoskeleton (May *et al.*, 1985b). Teleologically, it is unclear why a cell might want to reduce iron transport transiently in response to activation of PI turnover. Perhaps, the internalized receptor itself or transferrin has a function in addition to iron transport. It is worth noting that transferrin receptor recycling is inhibited during mitosis, a desirable effect, since intracellular vesicle fusion processes are disrupted during this phase of the cell cycle (Warren *et al.*, 1984). The role of phosphorylation of the transferrin receptor at serine 24 in response to mitogens and TPA will clearly need to be addressed by site-directed mutagenesis.

3.1.4. Insulin, Insulinlike Growth Factor I, and Insulinlike Growth Factor II Receptors

The receptors for insulin and a related, but distinct, polypeptide termed insulinlike growth factor 1 (IGF-I) have a similar structure and consist of a disulfide-linked heterodimer ($\alpha_2\beta_2$) of ~135 kD (α) and 95 kD (β) (Pilch and Czech, 1979; Kasuga *et al.*, 1982; Massague and Czech, 1982). Like the EGF receptor, the insulin and IGF-I receptors share intrinsic ligand-stimulated protein–tyrosine kinase activity that is a property of the β subunits (Avruch *et al.*, 1982; Kasuga *et al.*, 1982; Petruzelli *et al.*, 1982; Jacobs *et al.*, 1983; Rubin *et al.*, 1983; Van Obberghen *et al.*, 1983; Le Bon *et al.*, 1986).

Treatment of IM-9, Fao, and Hep G2 cells with phorbol esters causes a 1.3- to 4-fold increase in the phosphate content of these two receptors on serine and threonine residues (Jacobs *et al.*, 1983a; Takayama *et al.*, 1984; Jacobs and Cuatrecasas, 1986). Insulin receptors isolated from certain cells treated with tumor promoter exhibit lower insulin-stimulatable protein–tyrosine kinase activity than receptors from control cells, which may be explained partly by a 4- to 5-fold increase in K_m of the receptor for ATP (Haring *et al.*, 1986). In addition, phorbol esters reduce insulin receptor affinity in U-937 and HL-60 cells (Thomopoulos *et al.*, 1982), increase internalization of [^{125}I]insulin in Hep G2 cells (Grunberger and Gordon, 1982; Blake and Strader, 1986), and decrease glucose transport in adipocytes (Haring *et al.*, 1986).

The insulin and IGF-I receptors are basally phosphorylated in resting cells, predominantly on serine residues. TPA treatment increases the phosphorylation of tryptic peptides containing these sites and induces phosphorylation of additional tryptic peptides (Jacobs and Cuatrecasas, 1986). The direct involvement of protein kinase C, however, remains to be established. Although protein kinase C recently has been shown to be capable of phosphorylating purified insulin receptor *in vitro* (Bollag *et al.*, 1986), the sites of phosphorylation have not been identified and correlated with sites phosphorylated in intact cells exposed to tumor promoters. Furthermore, analysis of the amino acid sequence of the cytoplasmic part of the β-subunit of the insulin receptor, which contains the protein–tyrosine kinase domain, does not reveal the presence of serine or threonine residues in a primary sequence environment favored by protein kinase C (be it proximal or not to the transmembrane region), suggesting that the effect of tumor promoters on these receptors is indirect (Ullrich et al., 1985; Ebina et al., 1985; see Section 3.3).

IGF-II cross-links specifically to a 250-kD membrane protein (Kasuga et al., 1981; Massague and Czech, 1982) which, unlike the insulin and IGF-I receptors, has no intrinsic protein–tyrosine kinase activity (Corvera et al., 1986b). Insulin increases surface expression of IGF-II receptors, and this effect is blocked by prior administration of TPA, suggesting a role for protein kinase C in the recycling of this receptor (M. P. Czech, personal communication). However, as for the insulin and IGF-I receptors, direct phosphorylation of the IGF-II receptor by protein kinase C also remains to be demonstrated.

3.1.5. Adrenergic Receptors

Involvement of protein kinase C in the regulation of the β-adrenergic receptor was implied by the observations that phorbol esters desensitize isoproterenol-stimulated adenylate cyclase activity in astrocytoma cells (Mallorga et al., 1980), mouse epidermis (Mufson et al., 1977; Garte and Belman, 1980), and rat reticulocytes (Yamashita et al., 1986). It was demonstrated later that phorbol esters induce a threefold increase in the phosphate content of turkey erythrocyte β-adrenergic receptors, which correlated with uncoupling of the receptor from the stimulatory guanine nucleotide–binding protein G_s (Kelleher et al., 1984; Sibley et al., 1984). Although the sequence of mammalian β-adrenergic receptor is now known (Dixon et al., 1986), the site at which it is phosphorylated in response to TPA has yet to be determined, partly due to the very small amounts of the receptor that can be isolated. Recently, the mechanism of homologous (agonist-induced) desensitization of the β-adrenergic receptor has been elucidated in S49 lymphoma cells. This system also involves phosphorylation of the receptor, but the protein kinase responsible for this event has been shown to be distinct from enzymes characterized previously, including protein kinase C (Benovic et al., 1986). This "desensitizing" protein kinase has a preference for phosphorylating the agonist-occupied receptor. It is possible, therefore, that instead of protein kinase C directly phosphorylating the receptor, it activates this other protein kinase. Analysis of the phosphorylation sites induced on the receptor by TPA and β agonists may resolve this question.

Phorbol esters also desensitize gonadotropin-responsive adenylate cyclase in a murine Leydig tumor cell line (Rebois and Patel, 1985). In S49 cells, however, phorbol esters have been reported to enhance the β-adrenergic response by facilitating productive interaction with G_s (Bell *et al.*, 1985). TPA-induced augmentation of cAMP production has also been reported in pinealocytes (Sugden *et al.*, 1985), pituitary cells (Summers and Cronin, 1986), and pheochromocytoma cells (Hollingsworth *et al.*, 1986). The contradictory effects of phorbol esters on the β-adrenergic receptor in S49 cells, i.e., desensitization or augmentation, appear to depend on the time of treatment and amount of activation. Low concentrations of phorbol esters act to desensitize the receptor, whereas higher amounts cause a time-dependent stimulation of agonist-induced cAMP production (Johnson *et al.*, 1986; Yamashita *et al.*, 1986).

Phorbol esters appear to regulate adenylate cyclase through yet another mechanism in human platelets. In this system, adenylate cyclase is activated by prostaglandin E_1 and inhibited by epinephrine (Akatories and Jakobs, 1984). Phorbol esters interfere with the inhibitory component of this complex, namely the inhibitory guanine nucleotide–binding protein G_i (Watanabe *et al.*, 1985; Jakobs *et al.*, 1985). Protein kinase C has been shown to phosphorylate the α subunit of G_i *in vitro* (Katada *et al.*, 1985). However, this phosphorylation reaction only occurred with free α subunit and was potently inhibited by the addition of free βγ subunits. The possibility that α-subunit phosphorylation by protein kinase C is responsible for uncoupling G_i in platelets must await demonstration that such an event occurs *in vivo*. Although the effects of phorbol esters on the β-adrenergic systems of hepatocytes and platelets appear contradictory, i.e., inhibiting the former and relieving inhibitory constraints for the latter, TPA acts to block the effect of epinephrine in both cases. It is clear, however, that the effects of phorbol esters on the β-adrenergic system are complex and cannot be explained by a single paradigm (see Section 5.1).

α_1-adrenergic receptors also appear to be modulated by tumor promoters. In this case, phorbol esters attenuate the α_1-adrenergic response in hepatocytes (Cooper *et al.*, 1985b; Garcia-Sainz *et al.*, 1985; Lynch *et al.*, 1985), adipocytes (Corvera *et al.*, 1986a; Schimmel *et al.*, 1986), and smooth muscle (Danthuluri and Deth, 1984; Leeb-Lundberg *et al.*, 1985; McMillan *et al.*, 1986). Phorbol esters induce a rapid increase in phosphate content of vas deferens smooth muscle cell–purified receptor (80 kD) from a basal level of 1.2 mole phosphate per mole receptor to 3.6 mole per mole (Leeb-Lundberg *et al.*, 1985). Since α_1-adrenergic receptors operate through activation of PI turnover, the uncoupling effect of phorbol esters may reflect a feedback inhibition mechanism that prevents protracted stimulation of this pathway (see Section 5.3). It remains to be determined whether protein kinase C phosphorylates the α_1-adrenergic receptor directly.

3.1.6. Other Potential Receptor Substrates

There are several other receptor molecules for which there are preliminary data suggesting regulation by protein kinase C, including the PDGF receptor (Whiteley *et al.*, 1985), the colony-stimulating factor (CSF-1) receptor (Chen and Wilkins, 1985), the vitronectin and fibronectin receptors (E. K. Freed and T. Hunter, personal communication), the leukotriene B_4 receptor (O'Flaherty *et al.*, 1986), the tumor necrosis factor receptor (Scheurich *et al.*, 1986), the T3/T-cell antigen receptor (Cantrell *et al.*, 1985), the T4 antigen (Acres *et al.*, 1986), and the muscarinic and nicotinic ACh receptors (Liles *et al.*, 1986; Huganir *et al.*, 1984).

In the case of the PDGF receptor, TPA does not down-regulate PDGF binding (Coughlin *et al.*, 1985; Eide *et al.*, 1986; Sturani *et al.*, 1986), and there does not appear to be a suitable site for phosphorylation near the transmembrane domain (Yarden *et al.*, 1986; see Section 3.3).

The CSF-1 receptor recently has been the subject of intense interest, owing to its close relationship with the c-*fms* protein, the cellular homologue of the v-*fms* protein of the McDonough strain of FeSV which, like the receptors for EGF and insulin, has protein–tyrosine kinase activity (Sherr *et al.*, 1985). Treatment of peritoneal exudate macrophages with TPA causes a rapid transient decrease in CSF-1 binding, which returns to normal levels after 6 hr, even in the continued presence of TPA (Chen and Wilkins, 1985). Down-regulation of protein kinase C by phorbol ester treatment prevented subsequent modulation of CSF-1 binding. However, there is no current evidence that the CSF-1 receptor is a direct substrate for protein kinase C, and there does not appear to be a suitable phosphorylation site for this protein kinase within the sequence of the c-*fms* protein (Coussens *et al.*, 1986b; see Section 3.3).

The receptors for tumor necrosis factor (Scheurich *et al.*, 1986) and leukotriene B_4 (O'Flaherty *et al.*, 1986) recently have been found to be down-regulated in response to tumor promoter treatments. Although there is no current evidence that these proteins are phosphorylated by protein kinase C, one might speculate that they are regulated by phosphorylation in analogy with other receptor molecules.

The cellular receptors for vitronectin and fibronectin belong to a family of receptors involved in cell adhesion that recognizes an arginine–glycine–aspartic acid motif present in their ligand proteins, such as vitronectin, fibronectin, and von Willebrand factor (Pierschbacher and Ruoslahti, 1984). The vitronectin receptor is composed of a 150-kD α subunit and a 115-kD β subunit, the former of which contains two disulfide-linked polypeptides of 125 kD and 25 kD (Pytela *et al.*, 1985). Treatment of human MG63 osteosarcoma cells with TPA induces phosphorylation of the β subunit on a serine residue(s) (E. K. Freed and T.

Hunter, unpublished observations). Similarly, phosphorylation of the related small subunit (gpIIIa) of the platelet fibrinogen receptor is stimulated by TPA, although threonine phosphorylation is increased in this case (E. K. Freed and T. Hunter, unpublished observations). From recent cDNA cloning studies, both subunits of all these receptors appear to have short cytoplasmic tails, which must contain these phosphorylation sites. Direct evidence that protein kinase C mediates these phosphorylations, however, is not available, and the sites have not been mapped yet. The functional effect of these phosphorylations has yet to be determined, but since these receptors appear to be linked internally to the cytoskeleton and externally to the cell substratum, one might speculate that these phosphorylations are involved in morphological changes induced by TPA (see Section 4.5).

Antigen recognition by T lymphocytes is apparently mediated by the T3/T-cell antigen complex in human and murine cells, and down-regulation of cell-surface expression of this complex correlates with inhibition of antigen recognition (Reinherz et al., 1982; Zanders et al., 1983; Oettgen et al., 1986). Treatment of human T lymphoblasts with TPA causes a 50% decrease in surface expression of the T3 antigen (Cantrell et al., 1985). In HPB-ALL cells, TPA causes a rapid decrease in surface expression of the T3 antigen and the T-cell receptor (Cantrell et al., 1985). The 26-kD γ chain and, to a lesser extent, the 21-kD δ chains of T3 antigen were phosphorylated specifically upon TPA treatment. The T4 antigen, which is expressed primarily on helper T cells, is believed to function in helper T-cell binding or recognition, although its exact function in the immune response is unknown (Acuto and Reinherz, 1985). Besides its normal biological role, it also appears to serve as the receptor for human immunodeficiency virus (Dalgleish et al., 1984; Klatzmann et al., 1984). TPA treatment of helper T cells also induces the rapid phosphorylation and loss of surface expression of T4 antigen (Acres et al., 1986). Although direct involvement of protein kinase C remains to be determined in both cases, this may represent a mechanism whereby agonists of PI turnover can desensitize T cells toward antigen recognition. It should be noted that activation of T cells causes an elevation in PI turnover, providing a potential autogenous down-regulation of further antigen recognition upon initial antigen binding.

Pretreatment of cells with TPA reduces the ability of the muscarinic acetylcholine (mACh) receptor to activate PI turnover in response to agonist (Labarca et al., 1984; Orellana et al., 1985; Vincentini et al., 1985). Concomitant with TPA treatment is a rapid internalization of the mACh receptor with subsequent degradation, with similar kinetics to carbachol-induced removal of the receptor (Liles et al., 1986). These data suggest that protein kinase C is involved in the agonist-induced desensitization of the mACh receptor. The mechanism of this feedback inhibition in vivo is unknown so far. It is of interest to note that the α and δ subunits of the nicotinic ACh receptor are phosphorylated by protein kinase C *in vitro* (Huganir *et al.*, 1984), and preliminary results suggest that phosphorylation of this receptor is stimulated by phorbol esters *in vivo* (R. Huganir and P. Greengard, personal communication).

There are several receptors that have not been investigated with respect to regulation by phorbol esters, namely the receptors for fibroblast growth factor (FGF), bombesin, nerve groth factor (NGF), transforming growth factor-β (TGF-β), interleukin-1, interferon, etc. It will not be surprising at all if some of these receptors join the list of protein kinase C substrates in Section 3.1 within the next few years.

Review of the effects of protein kinase C on the ligand receptors described in Section 3.1 uncovers a common trend. Protein kinase C appears to have a generally negative effect on the function of these proteins. This inhibitory role is discussed further in Sections 4, 5.1, and 5.2.

3.2. Nonreceptor Substrates

There is a plethora of other proteins whose phosphorylation has been linked to protein kinase C, and we have organized these remaining proteins into three categories. The first division consists of proteins whose phosphorylation is enhanced *in vivo* by tumor promoters and are targets of protein kinase C *in vitro*. These proteins are listed in Table III. Other than the receptors that were reviewed in Section 3.1, above, we will discuss only proteins that are likely to be physiological substrates of protein kinase C and proteins whose phosphorylation by various protein kinases are the subject of general interest.

3.2.1. pp60$^{c\text{-src}}$

The transforming protein of RSV, pp60$^{v\text{-}src}$, and its normal cellular homologue, pp60$^{c\text{-}src}$, are membrane-bound phosphoproteins that exhibit protein–tyrosine kinase activity (reviewed in Krueger *et al.*, 1983; Hunter and Cooper, 1985). Although these two enzymes have very similar molecular weights and amino acid sequences, they differ in specific activity markedly. *In vitro*, pp60$^{v\text{-}src}$ catalyzes the phosphorylation of endogenous substrates about ten times more effectively than pp60$^{c\text{-}src}$ (Iba *et al.*, 1985; Cooper *et al.*, 1985; Coussens *et al.*, 1985). Presumably, this underlies the inability of pp60$^{c\text{-}src}$ to transform tissue culture cells or cause tumors in animals even when it is expressed at higher levels than its transforming counterpart, pp60$^{v\text{-}src}$ (Iba *et al.*, 1984; Parker *et al.*, 1984b; Shalloway *et al.*, 1984; Cooper *et al.*, 1985; Johnson *et al.*, 1985). Although there are scattered point mutations, pp60$^{v\text{-}src}$ varies structurally from pp60$^{c\text{-}src}$, primarily at its C terminus (Czernilovsky *et al.*, 1980; Takeya *et al.*,

TABLE III. Proteins Phosphorylated by Protein Kinase C *in Vitro* and Whose Phosphorylation is Stimulated by TPA *in Vivo*

Protein	Section	References
EGF receptor	3.1.1	Cochet *et al.* (1984)
IL-2 receptor	3.1.2	Shackelford and Trowbridge (1984)
Transferrin receptor	3.1.3	Davis *et al.* (1986)
pp60^{src}	3.2.1	Tamura *et al.* (1984); Gould *et al.* (1985); Purchio *et al.* (1985, 1986); Gentry *et al.* (1986)
47 kD (p47, IP$_3$ phosphatase)	3.2.2	Castagna *et al.* (1982); Imaoka *et al.* (1983); Kaibuchi *et al.* (1983); Sano *et al.* (1983); Yamanishi *et al.* (1983); Connolly *et al.* (1986)
Myosin light chain (MLC)	3.2.2	Castagna *et al.* (1982); Endo *et al.* (1982); Naka *et al.* (1983); Yamanishi *et al.* (1983); Nishikawa *et al.* (1984)
80 kD (p80)	3.2.3	Albert *et al.* (1984a, 1986); Rozengurt *et al.* (1984); Blackshear *et al.* (1985, 1986); Coughlin *et al.* (1985); Tsuda *et al.* (1985); Isacke *et al.* (1986b); Rodriguez-Pena and Rozengurt (1986)
p36/p35	3.2.4	Summers and Creutz (1985); Gould *et al.* (1986); Khanna *et al.* (1986a); Michener *et al.* (1986); Khanna *et al.* (1986b); Johnsson *et al.* (1986)
Vinculin	3.2.5	Werth *et al.* (1983); Werth and Pastan, (1984); Kawamoto and Hidaka (1984)
Ribosomal protein S6	3.2.6	Le Peuch *et al.* (1983); Blenis *et al.* (1984); Trevillyan *et al.* (1984); Padel and Soling (1985); Parker *et al.* (1985); Trevillyan *et al.* (1985)
Glycogen synthase	3.2.7	Roach and Goldman (1983); Ahmad *et al.* (1984); Ciudad *et al.* (1984); Imazu *et al.* (1984); Arino and Guinovart (1986); Blackmore *et al.* (1986); Wang *et al.* (1986)
Tyrosine hydroxylase (TH)	3.2.8	Albert *et al.* (1984b); Pocotte *et al.* (1985); McTigue *et al.* (1985); Pocotte and Holz (1986)
Actin-binding protein (ABP) (filamin)	3.2.5	Carroll *et al.* (1982); Kawamoto and Hidaka (1984)
CaM-BP$_{103/97}$	3.2.5	Ling *et al.* (1986)
Glucose transporter	4.5	Witters *et al.* (1985); Allard *et al.* (1986); Gibbs *et al.* (1986); Kitagawa *et al.* (1986)
Class I HLA antigens	3.2.9	Feuerstein *et al.* (1985); Rouis *et al.* (1986); Shackelford and Trowbridge (1986)
NADPH oxidase (31.5-kD component)	3.2.9	Papini *et al.* (1985)
Myelin basic protein	3.2.9	Vartanian *et al.* (1986); Kishimoto *et al.* (1985); Turner *et al.* (1984a); Wise *et al.* (1982b)

1982; Takeya and Hanafusa, 1982, 1983; Schwartz et al., 1983). The last 19 amino acids in pp60$^{c\text{-}src}$ have been deleted during the genesis of pp60$^{v\text{-}src}$ and have been replaced with 12 unrelated amino acids derived from a downstream region in the chicken genome. This mechanism of generating a transforming protein by deletion of the C terminus of pp60$^{c\text{-}src}$ is not unique to RSV, since it has also been reported for two other avian sarcoma viruses, S1 and S2, which have acquired the c-src gene (Hagino-Yamagishi et al., 1984).

Under most conditions, serine 17 is the major site of serine phosphorylation in both pp60$^{c\text{-}src}$ and pp60$^{v\text{-}src}$ (Patschinsky et al., 1986). The major site of tyrosine phosphorylation in pp60$^{v\text{-}src}$ is tyrosine 416, the primary site of auto-phosphorylation in vitro (Smart et al., 1981; Patschinsky et al., 1982). In pp60$^{c\text{-}src}$, the major site of tyrosine phosphorylation is not tyrosine 416 but, rather, lies in the C-terminal tryptic peptide (Cooper et al., 1986; Laudano and Buchanan, 1986) and has been identified as tyrosine 527 (Cooper et al., 1986). If dephosphorylation at tyrosine 527 occurs during the isolation of pp60$^{c\text{-}src}$ (Courtneidge, 1985) or the phosphate content of the protein is reduced deliberately with acid–phosphatase (Cooper and King, 1986), the specific activity of pp60$^{c\text{-}src}$ is augmented about fivefold. Phosphorylation of tyrosine 527, therefore, functions to suppress pp60$^{c\text{-}src}$ enzymatic activity. Further support for this model comes from studies of pp60$^{c\text{-}src}$ complexed to middle-T antigen of polyoma virus and from site-directed mutagenesis experiments. The middle-T pp60$^{c\text{-}src}$ complex exhibits increased immune complex kinase activity compared with free pp60$^{c\text{-}src}$ (Bolen et al., 1984; Cartwright et al., 1985; Courtneidge, 1985; Yonemoto et al., 1985). Analysis of phosphorylation sites has shown that in contrast to free pp60$^{c\text{-}src}$, pp60$^{c\text{-}src}$ associated with middle-T antigen is phosphorylated at tyrosine 416 and not at tyrosine 527 (Cartwright et al., 1986). Replacement of tyrosine at residue 527 with phenylalanine creates a mutant pp60$^{c\text{-}src}$ protein with augmented protein kinase activity (Cartwright et al., 1987a; Kmiecik et al., 1987, Piwnica-Worms et al., 1987). Furthermore, when this mutant pp60$^{c\text{-}src}$ is expressed in tissue culture cells or in animals, it exhibits transforming and tumorigenic properties, respectively. These data argue that phosphorylation/dephosphorylation of tyrosine 527 is one mechanism whereby pp60$^{c\text{-}src}$ function is controlled. Deletion of the region containing residue 527 in RSV has allowed pp60$^{v\text{-}src}$ to escape regulation.

pp60$^{c\text{-}src}$ can also be phosphorylated at an additional serine(s) by protein kinase C. Purified protein kinase C phosphorylates pp60$^{c\text{-}src}$ and pp60$^{v\text{-}src}$ in vitro at serine (Gould et al., 1985; Purchio et al., 1986). In contrast, neither protein is phosphorylated readily by purified preparations of the multifunctional calmodulin-dependent protein kinase, casein kinase I, or casein kinase II (Gould et al., 1985). cAMP-dependent protein kinase also phosphorylates both pp60$^{c\text{-}src}$ and pp60$^{v\text{-}src}$ but at a site (serine 17) distinct from the targets of protein kinase C (Gould et al., 1985). The site of in vitro protein kinase C phosphorylation in mammalian pp60$^{c\text{-}src}$ is serine 12 (Gould *et al.*, 1985). pp60$^{v\text{-}src}$ and chicken pp60$^{c\text{-}src}$ are phosphorylated additionally at serine 48 (Gould *et al.*, 1985).

In intact cells, TPA treatment leads to a marked increase in the amount of phosphate in pp60$^{v\text{-}src}$ (Gould *et al.*, 1985; Purchio *et al.*, 1985) and pp60$^{c\text{-}src}$ (Tamura *et al.*, 1984; Gould *et al.*, 1985; Gentry *et al.*, 1986) at the same site(s) observed *in vitro* (Gould *et al.*, 1985). Nearly all of the pp60$^{c\text{-}src}$ molecules are phosphorylated within 15 min of TPA treatment (Gould *et al.*, 1985). Other activators of protein kinase C, including 1-oleoyl-2-acetylglycerol (OAG), serum, PDGF, vasopressin, prostaglandin F$_{2\alpha}$, bradykinin, and EGF in A431 cells also stimulate phosphorylation of serine 12 in pp60$^{c\text{-}src}$ (K. L. Gould and T. Hunter, unpublished observations). In contrast to TPA, these agents result in transient and substoichiometric phosphorylation of pp60$^{c\text{-}src}$ at serine 12, which is maximal between 2 and 5 min after treatment (K. L. Gould and T. Hunter, unpublished results). This difference in kinetics can be explained by the transient, rather than prolonged, activation of protein kinase C by these agents and the restriction of protein kinase C activated by DAG to the plasma membrane. Insulin and EGF, which do not activate protein kinase C in fibroblasts (Habenicht *et al.*, 1981; Coughlin *et al.*, 1985), do not enhance serine 12 phosphorylation (K. L. Gould and T. Hunter, unpublished data). These data are consistent with the hypothesis that protein kinase C mediates phosphorylation of pp60$^{c\text{-}src}$ and pp60$^{v\text{-}src}$ directly. Additional evidence that protein kinase C itself phosphorylates pp60^{src} *in vivo* is provided by the properties of mutant pp60$^{v\text{-}src}$ molecules, in which the N-terminal glycine has been changed to an alanine or a glutamic acid (Kamps *et al.*, 1985). These mutant proteins are not myristylated and are unable to associate with membranes where the active form of protein kinase C is located (Buss *et al.*, 1986). pp60$^{v\text{-}src}$ phosphorylation is not induced by TPA treatment of these mutant-infected cells (Buss *et al.*, 1986), supporting the idea that protein kinase C activity is restricted to the plasma membrane.

Several investigators have concluded that tumor promoter treatment of cells does not result in an altered immune complex kinase activity of pp60$^{c\text{-}src}$ (Goldberg *et al.*, 1980; Tamura *et al.*, 1984; Gould *et al.*, 1985; Gentry *et al.*, 1986). This might indicate that the protein kinase activity of pp60$^{c\text{-}src}$ *in vivo* is also unchanged. However, this type of assay for pp60$^{c\text{-}src}$ protein kinase activity relies on the maintenance of its functional state following disruption of its normal environment. This assumption might be erroneous. For instance, the effect of protein kinase C phosphorylation could depend on the interations of pp60$^{c\text{-}src}$ with a lipid bilayer and/or other proteins.

Treatment of quiescent fibroblast cells with PDGF not only leads to serine 12 phosphorylation but to a novel N-terminal tyrosine phosphorylation event in 5–10% of pp60$^{c\text{-}src}$ molecules (Ralston and Bishop, 1985; K. L. Gould and T. Hunter, unpublished results). This is accompanied by a two- to threefold increase in immune complex kinase activity (Ralston and Bishop, 1985; K. L. Gould and

T. Hunter, unpublished results). No detectable decrease in tyrosine 527 phosphorylation occurs (K. L. Gould and T. Hunter, unpublished results). pp60$^{c\text{-}src}$ from certain neuroblastomas also exhibits increased immune complex kinase activity and contains phosphotyrosine in its N-terminal region (Bolen *et al.* 1985). It is not clear yet whether N-terminal tyrosine phosphorylation in either situation is a cause or a result of augmented protein kinase activity. It suggests, however, that there might be mechanisms other than tyrosine 527 phosphorylation whereby the activity of pp60$^{c\text{-}src}$ is controlled. Further evidence for this notion comes from the study of pp60$^{c\text{-}src}$ in neural tissues. pp60$^{c\text{-}src}$ is abundant in both glial cells and neurons (Brugge *et al.*, 1985; Cotton and Brugge, 1983). In neurons, the enzyme exhibits enhanced immune complex kinase activity and differs in its N-terminal half (Brugge *et al.*, 1985; Cartwright *et al.*, 1987b; G. Walter; personal communication). The nature of this difference is unclear but represents yet another possible mechanism of pp60$^{c\text{-}src}$ regulation. A point worth noting here is that the level of tyrosine 527 phosphorylation in pp60$^{c\text{-}src}$ from normal cells does not appear to change with any physiological stimuli. Thus, the role of this residue in normal regulation of pp60$^{c\text{-}src}$ remains to be clarified.

pp60$^{c\text{-}src}$ is present apparently in all cell types but is particularly prominent in platelets (Golden *et al.*, 1986), peripheral blood lymphocytes (Golden *et al.*, 1986), neural tissue (Cotton and Brugge, 1983; Sorge *et al.*, 1984; Brugge *et al.*, 1985), and in adrenal medullary chromaffin cells (Parsons and Creutz, 1986). One possible role for pp60$^{c\text{-}src}$ suggested by its abundance in these cell types is in exocytosis, which is a major activity of all these cells. Certainly, the high level of pp60$^{c\text{-}src}$ expression in fully differentiated tissues makes a pivotal function in mitogenesis unlikely. A high percentage of pp60$^{c\text{-}src}$ molecules are phosphorylated whenever protein kinase C is activated. Although there is no evidence at the present time that this phosphorylation event has biological consequence, it might indicate a role for pp60$^{c\text{-}src}$ in the PI turnover signal transduction pathway (see Section 5.3).

3.2.2. p47 and Myosin Light Chain

Two of the substrates of protein kinase C studied most extensively are the 40- to 47-kD (p47) and 20-kD proteins found in platelets. The 20-kD protein has been identified as MLC (Daniel *et al.*, 1981). Recently, p47 has been shown to have IP$_3$-5' phosphatase activity (Connolly *et al.*, 1986). Phosphorylation of these proteins was first described in ^{32}P-labeled platelets stimulated with thrombin or collagen (Lyons *et al.*, 1975). Inhibitors of thrombin-induced activation, such as dibutyryl-cAMP (db-cAMP) (Lyons *et al.*, 1975; Ieyasu *et al.*, 1982; Sano *et al.*, 1983; Yamanishi *et al.*, 1983), prostaglandin E$_1$ (Lyons *et al.*, 1975; Kawahara *et al.*, 1980; Ieyasu *et al.*, 1982; Sano *et al.*, 1983; Yamanishi *et al.*,

1983), aspirin (Chiang *et al.*, 1981), sodium nitroprusside (Takai *et al.*, 1981; Sano *et al.*, 1983; Yamanishi *et al.*, 1983), and 8-bromo-cGMP (Takai *et al.*, 1981; Sano *et al.*, 1983; Yamanishi *et al.*, 1983) prevent phosphorylation of the two proteins. Both p47 and p20 are also phosphorylated when platelets are treated with platelet-activating factor (Ieyasu *et al.*, 1982), collagen (Haslam and Lynham, 1977), OAG (Kaibuchi *et al.*, 1983; Lapetina *et al.*, 1985), the synthetic DAGs, dihexanoylglycerol, dioctanoylglycerol, didecanoylglycerol, and dibutyrylglycerol (Lapetina *et al.*, 1985), phospholipase C (Kawahara *et al.*, 1980), TPA (Castagna *et al.*, 1982; Yamanishi *et al.*, 1983), Ca^{2+} ionophores (Haslam and Lynham, 1977; Kaibuchi *et al.*, 1983; Yamanishi *et al.*, 1983; Sano *et al.*, 1985b), or concanavalin A (Con A) (Bennett *et al.*, 1979). However, the rate and extent to which each is phosphorylated depend on the inducing agent.

Physiological activators of platelets stimulate the hydrolysis of inositol phospholipids and, thus, activate protein kinase C and raise the concentration of intracellular Ca^{2+} concurrently. TPA activates the protein kinase C–mediated arm of this bifurcating pathway, whereas the Ca^{2+} ionophore, A23187, acts to raise intracellular Ca^{2+} levels without stimulating protein kinase C significantly (see Section 2.4). p47 phosphorylation is stimulated to the same level and at the same rate by physiological effectors or TPA (Carroll *et al.*, 1982; Yamanishi *et al.*, 1983) or synthetic DAGs (Kaibuchi *et al.*, 1983), but only a high concentration of the Ca^{2+} ionophore A23187 is able to induce p47 phosphorylation (Kaibuchi *et al.*, 1983; Yamanishi *et al.*, 1983; Sano *et al.*, 1985b). In contrast, A23187 stimulates MLC phosphorylation to the same extent and at the same rate as thrombin does (Kaibuchi *et al.*, 1983; Yamanishi *et al.*, 1983; Sano *et al.*, 1985b), whereas MLC phosphorylation is slower and reduced in the case of TPA (Yamanishi *et al.*, 1983) or OAG treatments (Kaibuchi *et al.*, 1983). Since p47 phosphorylation correlates with activation of protein kinase C, rather than the rise in Ca^{2+} concentration, it is an excellent candidate substrate. Indeed, p47 exhibits many of the characteristics expected of a physiological substrate of protein kinase C. Purified (Imaoka *et al.*, 1983) and partially purified p47 (Kaibuchi *et al.*, 1983; Sano *et al.*, 1983) can be phosphorylated by homogeneous preparations of protein kinase C *in vitro* on a serine residue(s) (Sano *et al.*, 1983). The sites phosphorylated *in vivo* in response to thrombin have been compared by tryptic peptide mapping with those phosphorylated in response to TPA (Yamanishi *et al.*, 1983), collagen (Sano *et al.*, 1983), OAG (Kaibuchi *et al.*, 1983), platelet-activating factor (Ieyasu *et al.*, 1982), or high levels of A23187, (Sano *et al.*, 1985b) and found to be identical. Furthermore, the sites phosphorylated *in vitro* by purified protein kinase C were shown to be the same as those phosphorylated *in vivo* (Kaibuchi *et al.*, 1983; Sano *et al.*, 1983). Thus, the accumulated evidence suggests that protein kinase C is directly responsible for p47 phosphorylation *in vivo*.

This is particularly exciting in light of the recent identification of p47 as an IP_3-5' phosphatase (Connolly et al., 1986). Phosphorylation of p47 by protein kinase C in vitro stimulates IP_3 phosphatase activity (Connolly et al., 1986). Since IP_3 phosphatase is partially phosphorylated as isolated, it is possible that the completely dephosphorylated enzyme is inactive. If this were the case, activation of protein kinase C would desensitize release of intracellular Ca^{2+} by promoting dephosphorylation of IP_3, since the product inositol 1,4-biphosphate is inactive (Streb et al., 1983) (see Section 2.4). Phosphorylation of p47 thus represents a feedback inhibitory mechanism.

A number of questions remain regarding p47. First, p47, purified from thrombin-treated platelets, consists of several isoelectric forms, which contain phosphothreonine, as well as phosphoserine (Imaoka et al., 1983). Since protein kinase C phosphorylates only serine residues in p47 (Sano et al., 1983), p47 might be subject to phosphorylation and regulation by another protein kinase(s) as well. Second, phosphorylated p47 is a cytosolic protein (Lyons and Atherton, 1979; Imaoka et al., 1983), as would be predicted for a phosphatase acting on soluble IP_3. How does it become phosphorylated within seconds by a membrane-bound protein kinase? Perhaps p47 is membrane-associated in its basal state and, upon phosphorylation, is activated and translocates from the plasma membrane to the cytosol. Translocation of protein kinase C substrates is a potential mechanism whereby a signal generated at the plasma membrane could be transduced to other regions of the cell, since active protein kinase C itself does not appear to leave the membrane (see Section 2.7). Now that the function of p47 has been identified, these possibilities can be addressed.

As mentioned above, MLC phosphorylation parallels Ca^{2+} mobilization more closely than protein kinase C activation during the platelet response. In fact, MLC is a substrate of a Ca^{2+}/calmodulin-dependent protein kinase, MLC kinase (MLCK) (Naka et al., 1983; Nishikawa et al., 1983; 1984). MLC, however, can also be phosphorylated by protein kinase C either as the isolated subunit (Endo et al., 1982; Nishikawa et al., 1984) or in association with myosin heavy chain in the form of heavy meromysin (Nishikawa et al., 1983; 1984) or intact myosin (Endo et al., 1982; Naka et al., 1983; Nishikawa et al., 1984). When MLC is associated with the heavy chain of myosin, either kinase will incorporate 1 mole of phosphate per mole of MLC in vitro (Nishikawa et al., 1984) but into different sites (Naka et al., 1983; Nishikawa et al., 1983). Protein kinase C phosphorylates a threonine residue, whereas the target of MLCK is a serine (Nishikawa et al., 1984). Thus, phosphorylation of MLC by these two protein kinases can be distinguished in vivo. In thrombin-treated platelets, MLC is phosphorylated predominantly at the site specific to MLCK (Naka et al., 1983). In contrast, MLC from TPA-stimulated platelets contains two sites of phosphorylation (Naka et al., 1983). The major site corresponds to that phosphorylated in vitro by protein kinase C, whereas the minor site specific to MLCK (Naka et al., 1983). Phosphorylation of myosin by MLCK stimulates its actin-activated Mg-ATPase activity (Nishikawa *et al.*, 1984). When MLC is phosphorylated sequentially by MLCK and protein kinase C, there is a decrease in the actin-activated Mg-ATPase activity (Nishikawa *et al.*, 1983, 1984) due to a decrease in the affinity for actin (Nishikawa *et al.*, 1983). Furthermore, prior phosphorylation by protein kinase C decreases the rate of subsequent phosphorylation by MLCK *in vitro*. Thus, *in vivo*, protein kinase C phosphorylation of MLC might decrease the rate of phosphorylation by MLCK and contribute to the disruption of cytoskeletal organization in TPA-stimulated platelets. In support of this idea are the findings of Inagaki *et al.*, (1984), who reported that addition of the protein kinase inhibitor, H-7 (see Section 2.2), prior to TPA treatment increased the relative proportion of MLCK-specific phosphorylation in MLC and concomitantly enhanced serotonin secretion. Thus, protein kinase C phosphorylation of MLC might influence the activation process resulting from TPA treatment. However, during platelet activation stimulated by physiological agents such as thrombin, it is not clear yet whether protein kinase C phosphorylation of MLC occurs or what role it might play. The studies of MLC phosphorylation cited here illustrate the importance of identifying the sites phosphorylated in the substrate protein both *in vitro* and *in vivo* and examining the substrate's ability to be phosphorylated by other protein kinases when assessing the significance of a given phosphorylation event.

Recently, it has been shown that MLCK, as well as MLC, is a target for protein kinase C *in vitro* (Ikebe *et al.*, 1985; Nishikawa *et al.*, 1985). MLCK is also a substrate for cAMP-dependent protein kinase (Ikebe *et al.*, 1985; Nishikawa *et al.*, 1985). The two groups that examined these phosphorylation events concluded that phosphorylation by protein kinase C diminished the affinity of MLCK for calmodulin (Ikebe *et al.*, 1985; Nishikawa *et al.*, 1985). However, there were significant discrepancies between the findings of the two groups. Nishikawa *et al.* (1985) reported that each kinase phosphorylated two sites in MLCK, one of which was shared. In contrast, Ikebe *et al.* (1985) concluded that cAMP-dependent protein kinase phosphorylated two sites, and protein kinase C a different pair of sites. In any case, if MLCK were inhibited by protein kinase C activation *in vivo*, this could explain the reduced extent of MLC phosphorylation in TPA-activated platelets, compared with thrombin activation. Investigation of the phosphorylation state of MLCK *in vivo* awaits further study.

Stimulation of p47 and MLC phosphorylation to variable levels always accompanies platelet activation (see Section 3.2.2). Several studies have been aimed at dissecting the activation process and have examined these phosphorylation events in parallel with morphological changes and release of serotonin. For example, platelet shape change and release reactions can be separated by treating cells with Con A in the absence of extracellular Ca^{2+} (Bennett *et al.*, 1979). Under these conditions, shape change and phosphorylation of p47 and MLC occur at the normal rate, but serotonin is not released (Bennett *et al.*,

1979). Pretreatment of platelets with papaverine inhibits Con A stimulation of p47 and MLC phosphorylation but inhibits shape change to a lesser extent (Bennett *et al.*, 1979). Pretreatment of platelets with cytochalasin B does not alter the rate or level of thrombin-induced phosphorylation but does inhibit the extension of pseudopodia (Carroll *et al.*, 1982). The interpretation of these studies is that phosphorylation of p47 and MLC is not necessary for shape change and, although it might be required, is not sufficient to induce the release reaction.

3.2.3. p80

The induction of phosphorylation of an 80-kD protein in parallel with protein kinase C activation is well documented. This prominent phosphorylation event has been studied primarily in 3T3-L1 cells by the group of Blackshear, in human fibroblasts by Coughlin *et al.* (1985), and in Swiss 3T3 cells by Rozengurt's group, Takai and his colleagues, and ourselves. When cells are stimulated with TPA (Rozengurt *et al.*, 1984; Coughlin *et al.*, 1985; Blackshear *et al.*, 1985; Tsuda *et al.*, 1985; Isacke *et al.*, 1986b), phorbol dibutyrate (PDBU) (Rozengurt *et al.*, 1984; Coughlin *et al.*, 1985), exogenous phospholipase C (Rozengurt *et al.*, 1983), PDGF (Rozengurt *et al.*, 1983; Coughlin *et al.*, 1985; Blackshear *et al.*, 1985; Tsuda *et al.*, 1985; Isacke *et al.*, 1986b), serum (Rodriguez-Pena and Rozengurt, 1985; Coughlin *et al.*, 1985), FGF (Blackshear *et al.*, 1985; Tsuda *et al.*, 1985), bombesin and gastrin-releasing peptide (Isacke *et al.*, 1986b; Zachary *et al.*, 1986), bradykinin (Coughlin *et al.*, 1985), synthetic DAGs diC6–diC8 (Blackshear *et al.*, 1985), or OAG (Rozengurt *et al.*, 1984), p80 phosphorylation is rapidly enhanced. Agents unable to stimulate PI turnover (Habenicht *et al.*, 1981; Coughlin *et al.*, 1985), such as insulin (Rozengurt *et al.*, 1983; Blackshear *et al.*, 1985; Isacke *et al.*, 1986b) or EGF (Rozengurt *et al.*, 1983; Isacke *et al.*, 1986b), do not induce p80 phosphorylation. Furthermore, the phosphorylation state of p80 is unaffected by agents that increase cAMP levels, such as prostaglandin E_1 (Rozengurt *et al.*, 1983), db-cAMP (Blackshear *et al.*, 1985), adenosine agonists (Rozengurt *et al.*, 1983), or cholera toxin (Rozengurt *et al.*, 1983). Finally, if protein kinase C is down-regulated by prolonged treatments with tumor promoters (see Section 2.7.1), stimulation of p80 phosphorylation is refractory to subsequent treatment with PDBU (Rozengurt *et al.*, 1983), PDGF (Blackshear *et al.*, 1985; Rozengurt *et al.*, 1983), TPA (Blackshear *et al.*, 1985), bombesin (Zachary *et al.*, 1986), serum (Rodriguez-Pena and Rozengurt, 1985), phospholipase C (Rozengurt *et al.*, 1983), or OAG (Rozengurt *et al.*, 1984). The site(s) of *in vivo* phosphorylation have been examined by phosphoamino acid analysis (Blackshear *et al.*, 1985; Isacke *et al.*, 1986b), one-dimensional V8 protease mapping (Rodriguez-Pena and Rozengurt, 1985; Rozengurt *et al.*, 1983, 1984; Blackshear *et al.*, 1986; Zachary *et al.*,

1986), and two-dimensional tryptic peptide analysis (Isacke *et al.*, 1986b). These studies have indicated that p80 is basally phosphorylated at several sites on serine, with some phosphorylation on threonine, and that stimulation of phosphorylation occurs primarily at serine residues. Furthermore, the sites of enhanced phosphorylation are the same irrespective of the inducing agent. p80 phosphorylation can be detected easily on one-dimensional SDS–polyacrylamide gels and unambiguously identified on two-dimensional gels. The protein is also heat stable, allowing significant enrichment prior to analysis (Blackshear *et al.*, 1986). These properties make p80 a convenient indicator for cellular activation of protein kinase C. For example, in a mutant NIH 3T3 cell line resistant to TPA-stimulated mitogenesis, normal TPA-enhanced p80 phosphorylation has been used to show that the mutation in these cells does not lie in protein kinase C (Bishop *et al.*, 1985).

p80 phosphorylation also can be stimulated *in vitro* by protein kinase C (Albert *et al.*, 1986), in cell-free extracts treated with phorbol esters and phosphatidylserine (Rodriguez-Pena and Rozengurt, 1986), and in heat-stable extracts upon the addition of partially purified protein kinase C (Blackshear *et al.*, 1986). The site(s) of *in vitro* phosphorylation have been compared to those phosphorylated *in vivo* by one-dimensional V8 protease mapping and found to be identical (Blackshear *et al.*, 1985; Rodriguez-Pena and Rozengurt, 1986). These results support the hypothesis that p80 is a direct substrate of protein kinase C. However, a point to bear in mind is that p80 is phosphorylated constitutively at a site(s) that is phosphorylated more heavily when protein kinase C is activated (Isacke *et al.*, 1986b). This suggests that p80 also serves as a substrate for another protein kinase(s) with site specificity similar to that of protein kinase C.

Unfortunately, the function of p80 is unknown. Although it has a similar molecular weight, it is not protein kinase C (Blackshear *et al.*, 1986; Isacke *et al.*, 1986b). The tissue distribution of p80 has been examined using its heat-resistant properties (Blackshear *et al.*, 1986), and antisera have been raised against partially purified (Blackshear *et al.*, 1986) or purified (Albert *et al.*, 1986) p80. It is most abundant in brain but detected in all tissues (Albert *et al.*, 1986; Blackshear *et al.*, 1986). Albert *et al.* (1986) have demonstrated that the protein is tightly associated with membranes. It was described earlier by Wu *et al.* (1982) as a constituent of synaptosomal membranes (Blackshear *et al.*, 1986) and is probably the same as the 80-kD phosphoprotein described in lymphocyte membranes by Denis *et al.* (1986) and in rat embryo fibroblasts by Sato *et al.* (1985). A complete understanding of the role p80 plays in protein kinase C–mediated events awaits its further characterization. An intriguing observation, especially in light of the identification of p47 as an IP_3 phosphatase (Section 3.2.2), is that pig DAG kinase exhibits a molecular weight of 80 kD (Kanoh and Ono, 1986). As discussed in Section 5.2, there is evidence that protein kinase C regulates DAG kinase. Another candidate to be p80 is one of the membrane-associated

PI kinases that has a molecular weight of 85 kD (M. Whitman, D. Kaplan, T. M. Roberts, and L. Cantley, personal communication).

3.2.4. Protein–Tyrosine Kinase Substrates p36 and p35

p36 was first recognized by virtue of its phosphotyrosine content in RSV-transformed chicken cells (Radke et al., 1980). Variously termed p34, p36, and p39, it has been detected readily in two-dimensional gel analyses of phospho-proteins from cells transformed by retroviruses encoding protein–tyrosine kinases and from certain growth-factor-treated cells (for review, see Cooper and Hunter, 1983). Because of its possible role in transformation and growth control, much effort has been directed toward understanding its characteristics and function. It is an abundant, basic protein, comprising between 0.1 and 0.4% of total cell protein (Cooper and Hunter, 1983; Isacke et al., 1986a). p36 is located intra-cellularly at the cytoplasmic surface of the plasma membrane where it appears to form a part of the cortical cytoskeleton (Cooper and Hunter, 1983). It is found in a variety of tissues and cell types (Gould et al., 1984; Greenberg et al., 1984). The protein exists as a monomer or as a heterotetramer in conjunction with an 11-kD protein, p11, forming a $(p36)_2(p11)_2$ 85-kD complex (Erikson et al., 1984; Gerke and Weber, 1984; Gerke and Weber, 1985; Glenney and Tack, 1985). p36 binds to F-actin and fodrin in a Ca^{2+}-dependent manner (Gerke and Weber, 1984; Glenney, 1986b) and recently has been shown to interact with anionic phospholipids in a Ca^{2+}-dependent manner with high affinity (Glenney, 1985, 1986a,b). p36 shares functional homology with a 35-kD protein (Glenney, 1986b), initially described as a substrate for the EGF receptor protein–tyrosine kinase (Fava and Cohen, 1984; Guigni et al., 1985; Sawyer and Cohen, 1985).

The sequence of p36 from several species has been deduced from analysis of cDNA clones (Huang et al., 1986c; Kristensen et al., 1986; Saris et al., 1986), and the purified bovine protein has been partially sequenced (Glenney and Tack, 1985). The salient feature of this sequence is a fourfold internal repeat (Saris et al., 1986). The sequences also reveal very high conservation of the protein through evolution. With these reports, a new complication in nomen-clature arose. The protein is referred to variously as p36, calpactin I, and li-pocortin II. We will refer to it here as p36. Comparison of the p36 sequence with that of the phospholipase A_2 inhibitor, lipocortin I (Wallner et al., 1986), reveals a 50% identity over the C-terminal 300 amino acids. Lipocortin I appears to be identical to p35 or calpactin II (Glenney, 1986b; Pepinsky and Sinclair, 1986). Thus, p35 and p36 are closely related structurally and functionally.

p36 is basally phosphorylated on serine residues (Cooper and Hunter, 1983). Treatment of human fibroblasts with TPA causes a rapid eightfold increase in phosphoserine content (Gould et al., 1986). p36 can be phosphorylated by protein kinase C *in vitro* at serine (Gould *et al.*, 1986; Khanna *et al.*, 1986a,b; Johnsson *et al.*, 1986). The K_m for the reaction is 1 μM (Gould *et al.*, 1986), and protein kinase C can catalyze the incorporation of 0.5–0.8 mole of phosphate per mole of p36 (Gould *et al.*, 1986; Khanna *et al.*, 1986a,b). p36 is not phosphorylated to a significant extent by cAMP-dependent protein kinase, casein kinase I, casein kinase II, phosphorylase kinase, cGMP-dependent protein kinase, or multifunctional calmodulin-dependent protein kinase (Gould *et al.*, 1986; Khanna *et al.*, 1986a,b). There is a single site of phosphorylation *in vitro*, which is also the only major site of phosphorylation observed *in vivo* in response to TPA (Gould *et al.*, 1986). The target residue is serine 25 (Gould *et al.*, 1986; Johnsson *et al.*, 1986), which lies just two amino acids C-terminal to the major site of *in vitro* tyrosine phosphorylation, tyrosine 23, and in the region of the protein that binds p11 (Glenney and Tack, 1985). Phosphorylation of serine 25 and tyrosine 23 appears to be mutually exclusive, since molecules containing phosphate at both sites have yet to be observed. These data, coupled with the subcellular localization of p36, indicate that p36 serves as a direct target for protein kinase C *in vivo*. The stoichiometry of phosphorylation *in vivo* in response to TPA varies with cell type and, at best, reaches only 0.1 mole phosphate per mole of p36, whereas incorporation can exceed 0.5 mole per mole *in vitro*. Since the stoichiometry of phosphorylation at tyrosine 23 *in vivo* is also low (<0.1 mole per mole, even in transformed cells), only a critical subpopulation of the p36 molecules may be accessible to these membrane-associated protein kinases.

A further twist to the terminology of these proteins comes from the finding that two proteins from bovine adrenal chromaffin cells are homologous to p35 and p36 (Creutz *et al.*, 1987). These proteins are among several partially purified on the basis of their Ca^{2+}-dependent interaction with chromaffin granule membranes (Creutz *et al.*, 1983). Chromobindin 9 is homologous to p35 and chromobindin 8 is homologous to p36 (Creutz *et al.*, 1987). Chromobindin 9 (Summers and Creutz, 1985) and p35 (Khanna *et al.*, 1986a; K. L. Gould, J. R. Woodgett, and J. R. Glenney, Jr., unpublished observations) also serve as targets for protein kinase C *in vitro*. The phosphorylation of chromobindin 9 *in vivo* is stimulated in response to TPA treatment of chromaffin cells (Michener *et al.*, 1986). Although further study of these phosphorylation events is desirable, it is probable that p35, like p36, will prove to be a physiological substrate for protein kinase C.

Despite the accumulation of data, the functions of p35 and p36 remain enigmatic. Both p36 and p35 have been reported to act as phospholipase A_2 inhibitors *in vitro* (Huang *et al.*, 1986c). However, this activity appears to be a consequence of the phospholipid-binding properties of p36 and p35 (Davidson *et al.*, 1987) and probably does not reflect the physiological function of the two proteins. How protein kinase C or protein-tyrosine phosphorylation might affect these proteins is currently a subject for speculation. Potential effects would be changes in the known properties of these proteins, such as their binding to Ca^{2+} and phospholipid, F-actin, or fodrin. In the case of p36, phosphorylation might affect its interaction with the 11-kD subunit (Johnsson *et al.*, 1986). Definitive answers to these points need further study, such as site-specific mutation of serine 25 and tyrosine 23 of p36 to nonphosphorylatable residues.

3.2.5. Vinculin and Associated Cytoskeletal Proteins

Vinculin is a 130-kD cytoskeletal protein that is localized in cultured cells predominantly at the termini of microfilament bundles in adhesion plaques (Geiger, 1979; Burridge and Feramisco, 1980), in some areas of cell–cell contact (Geiger, 1979), and underlying fibronectin (Burridge and Feramisco, 1980). By immunoelectron microscopy, vinculin is found at the zonula adherens in the junctional complex of intestinal epithelial brush border cells, at dense plaques in smooth muscle cells, and at the fascia adherens of the intercalated disk membranes of cardiac muscle cells (Geiger *et al.*, 1980). This localization suggests that vinculin might be involved in mediating the attachment of actin to membranes. Although vinculin cannot bind directly to actin (Evans and Robson, 1983) and is not an integral membrane protein (Geiger, 1979), there is evidence that vinculin might be a component of a bridge linking actin to the extracellular matrix. Vinculin binds a 215-kD protein termed talin (Burridge and Mangeat, 1984). This protein also localizes to adhesion plaques and regions underneath cell-surface fibronectin (Burridge and Connell, 1983). Furthermore, talin recently has been shown to bind the fibronectin receptor (Horwitz *et al.*, 1986). How this tripartite complex is attached to actin remains to be determined. PDGF treatment causes vinculin, but not talin, to disappear rapidly from adhesion plaques prior to the disruption of stress fibers (Herman and Pledger, 1985). This indicates that vinculin cannot be the sole mediator of an interaction between stress fiber components and adhesion plaques.

Vinculin phosphorylation was first examined in chicken cells and their RSV-transformed counterparts during a survey of cytoskeletal proteins aimed at detecting phosphotyrosine-containing proteins (Sefton *et al.*, 1981). Vinculin contained elevated levels of phosphotyrosine in RSV-transformed cells but was phosphorylated primarily at phosphoserine and phosphothreonine in both normal and transformed cells (Sefton *et al.*, 1981). Vinculin could also be phosphorylated *in vitro* by pp60[v-src], and this reaction was stimulated by the presence of anionic phospholipids (Ito *et al.*, 1982). Vinculin, along with α-actinin, exhibits an altered distribution in RSV-transformed cells that parallels the dissolution of adhesion plaques and general cytoskeletal disorganization consequent to transformation (David-Pfeuty and Singer, 1980; Shriver and Rohrschneider, 1981). For this reason, several studies have addressed the possible correlation between morphological changes in avian sarcoma virus–transformed cells and phosphorylation of vinculin on tyrosine. The consensus of these investigations is that the extent of vinculin phosphorylation on tyrosine is not correlated with transformed cell morphology but, rather, reflects the nature of the transforming protein (Iwashita *et al.*, 1983; Rohrschneider and Rosok, 1983; Antler *et al.*, 1985; Nigg *et al.*, 1986). In addition, vinculin phosphorylation on tyrosine is not prerequisite for stress fiber dissolution prior to mitosis in normal cells (Rosok and Rohrschneider, 1983).

More recently, it was found that protein kinase C also phosphorylates vinculin *in vitro* (Werth *et al.*, 1983; Kawamoto and Hidaka, 1984), whereas other serine/threonine–specific protein kinases (cAMP-dependent, cGMP-dependent, and calmodulin-dependent) do not catalyze vinculin phosphorylation to a significant extent (Werth *et al.*, 1983). *In vivo*, vinculin is phosphorylated at multiple sites in its basal state, although only a small percentage of vinculin molecules contain phosphate (Sefton *et al.*, 1981; Werth and Pastan, 1984). Phosphorylation of vinculin at serine and threonine is enhanced within minutes of TPA treatment but is not affected by the addition of db-cAMP or db-cGMP (Werth and Pastan, 1984). TPA treatment results in increased phosphorylation of two tryptic phosphopeptides and the appearance of a novel tryptic phosphopeptide (Werth and Pastan, 1984). All three of these are among those phosphorylated *in vitro* by protein kinase C (Werth and Pastan, 1984). Vinculin is located primarily at the cytoplasmic surface of the plasma membrane. Furthermore, it has the potential to bind directly to anionic phospholipids (Ito *et al.*, 1983). Thus, it appears to be located correctly to serve as a direct target of protein kinase C. A point worth noting here is that talin, which shares a common intracellular localization, can also be phosphorylated *in vitro* by protein kinase C (Litchfield and Ball, 1986), although it is not clear whether this event occurs *in vivo*. The multiple phosphorylation sites in vinculin suggest that other protein kinases, in addition to protein kinase C, contribute to its complicated phosphorylation state, raising the possibility that one or more of these is activated in response to TPA as well.

Since TPA treatment leads to many of the same morphological changes as transformation (see Section 4.6), there is a possibility that protein kinase C phosphorylation of vinculin mediates this process. However, as in the case of tyrosine phosphorylation, protein kinase C phosphorylation of vinculin does not correlate with morphological alterations. First, vinculin phosphorylation has returned to basal levels by the time these changes occur (Werth and Pastan, 1984). Second, stress fiber dissolution appears to precede vinculin reorganization, indicating that vinculin is not a primary mediator of the cytoskeletal alterations (Schliwa *et al.*, 1984). Therefore, the role of protein kinase C–mediated phosphorylation of vinculin remains elusive.

Filamin (Kawamoto and Hidaka, 1984) or platelet ABP (Carroll and Gerrard, 1982; Carroll *et al.*, 1982), erythrocyte protein CaM-BP$_{103/97}$ (Ling *et al.*,

1986), and erythrocyte bands 4.1 and 4.9 (Ling and Sapirstein, 1984; Horne *et al.*, 1985; Cohen and Foley, 1986; Faquin *et al.*, 1986) are also cytoskeletal proteins that have been reported to be targets of protein kinase C *in vitro* (Kawamoto and Hidaka, 1984; Cohen and Foley, 1986) and whose phosphorylation is enhanced in response to TPA treatment (Carroll and Gerrard, 1982; Carroll *et al.*, 1982; Ling and Sapirstein, 1984; Horne *et al.*, 1985; Cohen and Foley, 1986; Ling *et al.*, 1986). The phosphorylation sites in these proteins have not been characterized nor have the kinetics of *in vitro* phosphorylation. However, phosphorylation of these proteins might contribute to the morphological changes observed following TPA treatment.

3.2.6. Ribosomal Protein S6

The phosphorylation of ribosomal protein S6 can be catalyzed *in vitro* by a multitude of purified protein kinases. These include cAMP-dependent (Del Grande and Traugh, 1982; Wettenhall and Cohen, 1982), cGMP-dependent (Del Grande and Traugh, 1982), H4P (Donahue and Masaracchia, 1984), calmodulin-dependent (Gorelick *et al.*, 1983), casein kinase I (Cobb and Rosen, 1983), protease-activated kinase II (Perisic and Traugh, 1983), and Ca^{2+}/phospholipid-dependent (Le Peuch *et al.*, 1983; Padel and Soling, 1985; Parker *et al.*, 1985) protein kinases. *In vitro*, protein kinase C can incorporate phosphate into at least three sites in S6 (Padel and Soling, 1985; Parker *et al.*, 1985), which are also phosphorylated *in vivo* (Padel and Soling, 1985). Peptide mapping studies have shown that these sites are largely distinct from those phosphorylated *in vitro* by cAMP-dependent protein kinase (Le Peuch *et al.*, 1983; Padel and Soling, 1985; Parker *et al.*, 1985). In combination, these two protein kinases can generate highly phosphorylated forms of S6 (Parker *et al.*, 1985) but, separately, are unable to drive S6 phosphorylation to the pentaphosphorylated forms (Parker *et al.*, 1985; Martin-Perez *et al.*, 1984) that have been detected *in vivo* following mitogen treatment (Thomas *et al.*, 1982).

In intact cells, S6 phosphorylation is stimulated to various extents by treatment with phorbol esters (Trevillyan *et al.*, 1984; Padel and Soling, 1985), prostaglandin $F_{2\alpha}$ (Thomas *et al.*, 1982), insulinlike growth factor (Haselbacher *et al.*, 1979), OAG (Blenis *et al.*, 1984), or serum (Thomas *et al.*, 1979; Blenis *et al.*, 1984). It is also phosphorylated in oocytes treated with progesterone or maturation-promoting factor (Nielsen *et al.*, 1982). Despite the evidence suggesting a role for protein kinase C in the phosphorylation of S6, this ribosomal protein is unlikely to be a physiological substrate. First, S6 phosphorylation is stimulated independently of protein kinase C activation by insulin (Lastick and McConkey, 1980; Thomas *et al.*, 1982; Trevillyan *et al.*, 1985), EGF (Lastick and McConkey, 1980; Thomas *et al.*, 1982), and agents that increase cAMP

(Lastick and McConkey, 1980). Second, maximum phosphorylation of S6 is observed only after about 1 hr of serum stimulation (Novak-Hofer and Thomas, 1985), whereas p80 (Rodriguez-Pena and Rozengurt, 1985) and p47 (Lyons *et al.*, 1975; Haslam and Lynham, 1977) are maximally phosphorylated in 1–2 min following serum or thrombin treatments, respectively (see Sections 3.2.2 and 3.2.3). Third, there is little evidence to date indicating that protein kinase C is constitutively activated in RSV- (Decker, 1981; Blenis *et al.*, 1984), polyoma virus– (Kennedy and Leader, 1981), SV40- (Kennedy and Leader, 1981; Rance *et al.*, 1985), or spontaneously transformed (Rance *et al.*, 1985) cell lines where S6 phosphorylation is chronically enhanced (but see Section 5.3). Fourth, there is also no evidence that protein kinase C is stimulated in parallel with S6 phosphorylation in oocytes injected with an active fragment of Ab-MuLV-encoded protein–tyrosine kinase (Maller *et al.*, 1985) or purified pp60^{v-src} (Spivack *et al.*, 1984). Fifth, and most convincing, is the identification of novel S6-specific kinase activities from stimulated *Xenopus* oocytes (Erikson and Maller, 1985, 1986) and fibroblasts (Novak-Hofer and Thomas, 1984, 1985; Blenis and Erikson, 1985; Lawen and Martini, 1985; Tabarini *et al.*, 1985) whose activities are increased under the same conditions that enhance S6 phosphorylation. Examples include the activation of S6 kinase activity in oocytes by insulin (Stefanovic *et al.*, 1986) and in fibroblasts by phorbol esters (Blenis and Erikson, 1985; Tabarini *et al.*, 1985), serum (Novak-Hofer and Thomas, 1984; Blenis and Erikson, 1985; Lawen and Martini, 1985), EGF (Novak-Hofer and Thomas, 1984, 1985), insulin (Tabarini *et al.*, 1985), sodium orthovanadate (Novak-Hofer and Thomas, 1985), and transformation by RSV (Blenis and Erikson, 1985). Unlike protein kinase C (Padel and Soling, 1985; Parker *et al.*, 1985) and cAMP-dependent protein kinase (Martin-Perez *et al.*, 1984), the S6-specific kinase activity is able to produce more highly phosphorylated forms of S6 observed *in vivo* (Novak-Hofer and Thomas, 1984). Furthermore, the kinetics of its activation in intact cells parallels that of S6 phosphorylation (Novak-Hofer and Thomas, 1985). Interestingly, the presence of phosphatase inhibitors during isolation stabilizes S6 kinase activity (Novak-Hofer and Thomas, 1984; Tabarini *et al.*, 1985), indicating that it might be regulated by phosphorylation/dephosphorylation. Thus, it might be an intermediate in a protein kinase cascade initiated by protein kinase C and/or protein–tyrosine kinase activation. Although S6-specific protein kinase activity will probably be a major, if not *the* major, protein kinase responsible for S6 phosphorylation *in vivo*, it must be remembered that many protein kinases are capable of phosphorylating S6. The activation of different protein kinases (perhaps different S6-specific kinases) by different physiological stimuli may explain the variable degree of S6 phosphorylation observed *in vivo* (Lastick and McConkey, 1980; Thomas *et al.*, 1982).

A discussion of S6 phosphorylation is not complete without a discussion of its biological consequence. There is a strong correlation between the extent of S6 phosphorylation and activation of protein synthesis (Thomas et al., 1982). However, a definitive relationship between these two events has proved difficult to establish. There is evidence that phosphorylated small ribosomal subunits are incorporated preferentially into initiation complexes (Duncan and McConkey, 1982a; Thomas et al., 1982) and that S6 phosphorylation influences specific mRNA recruitment (Duncan and McConkey, 1982b). It has also been demonstrated that phosphorylation of 40S ribosomal subunits increases the binding of AUG and poly(A,U,G) (Burkhard and Traugh, 1983). However, the function of S6 in initiation and translation of protein synthesis is not understood fully. Therefore, the effect rendered by phosphorylation of S6 on this process remains to be clarified.

3.2.7. Glycogen Synthase

The activity of glycogen synthase, the rate-limiting enzyme in glycogen synthesis, is regulated by multisite phosphorylation (Roach, 1981; Cohen, 1982). To date, nine serine/threonine–specific protein kinases (cAMP-dependent and cGMP-dependent protein kinases, phosphorylase kinase, glycogen synthase kinase 4, multifunctional calmodulin-dependent protein kinase, casein kinase II, glycogen synthase kinase 3, casein kinase I, and protein kinase C) are known to be capable of phosphorylating glycogen synthase in vitro, and their specificity for at least seven different phosphorylation sites has been determined in vitro (reviewed in Cohen, 1982; Kuret et al., 1985; Wang et al., 1986). Glycogenolytic hormones increase the phosphorylation state and, consequently, inactivate glycogen synthase in vivo (Roach, 1981; Cohen, 1982). Given the protein kinases capable of phosphorylating this enzyme, assigning responsibility for phosphorylation events occurring in vivo presents a daunting task.

One of the effectors of glycogen synthase inactivation in intact hepatocytes is TPA (Roach and Goldman, 1983; Arino and Guinovart, 1986; Blackmore et al., 1986). TPA treatment results in the phosphorylation of glycogen synthase within a cyanogen bromide fragment (termed CB-1) (Arino and Guinovart, 1986) that contains a tryptic peptide shown to be phosphorylated in vitro by protein kinase C (Ciudad et al., 1984; Imazu et al., 1984). Based on this evidence, it is tempting to speculate that protein kinase C mediates the phosphorylation of glycogen synthase directly. However, certain issues must be clarified before this conclusion can be reached. First, there is disagreement regarding the effect protein kinase C phosphorylation has on glycogen synthase activity. It is reported to inactive the muscle enzyme (Kishimoto et al., 1978; Ahmad et al., 1984) and to be without effect on the liver isozyme (Imazu et al., 1984) in vitro. Second, protein kinase C is not unique in its ability to phosphorylate CB-1 since this peptide contains phosphorylation sites for seven protein kinases (Cohen,

1982; Kuret *et al.*, 1985). A more detailed analysis of the *in vivo* phosphorylation sites would aid determination of which protein kinases might be involved. It is also entirely possible that TPA does not affect the phosphorylation state of glycogen synthase by activating a protein kinase but, rather, inactivates a phosphoprotein phosphatase activity, Finally, glycogen synthase is intimately associated with glycogen particles in liver and skeletal muscle. The subcellular localization of the structures is not consistent with access to a membrane-bound protein kinase such as protein kinase C. Indeed, all of the known proteins involved in the regulation of glycogen synthase are either tightly associated with the glycogen particle (see, e.g., Stralfors *et al.*, 1985) or are soluble (such as cAMP-dependent protein kinase).

3.2.8. Tyrosine Hydroxylase

TH catalyzes the rate-limiting step in the biosynthesis of the catecholamine neurotransmitters, dopamine, norepinephrine, and epinephrine (Nagatsu *et al.*, 1964). Like glycogen synthase, its activity recently has been shown to be regulated by multisite phosphorylation. To date, four serine residues have been identified as *in vitro* sites of TH phosphorylation (Campbell *et al.*, 1986). TH enzymatic activity is enhanced *in vitro* upon phosphorylation by cAMP-dependent protein kinase (Joh *et al.*, 1978; Yamauchi and Fujisawa, 1979; Vulliet *et al.*, 1980), cGMP-dependent protein kinase (Roskoski, 1986), or protein kinase C (Albert *et al.*, 1984b). Multifunctional calmodulin-dependent protein kinase (type II) also phosphorylates TH *in vitro* (Vulliet *et al.*, 1984). cAMP-dependent protein kinase, cGMP-dependent protein kinase, and protein kinase C all phosphorylate the same residue on TH, serine 40, which is also phosphorylated slowly by multifunctional calmodulin-dependent protein kinase (Campbell *et al.*, 1986). However, this latter protein kinase phosphorylates preferentially a different site (Vulliet *et al.*, 1985), recently identified as serine 19 (Campbell *et al.*, 1986). Although phosphorylation by calmodulin-dependent protein kinase has no apparent effect on the activity of TH, the phosphorylated form of this enzyme is rather unstable, making activity assessments difficult (Vulliet *et al.*, 1984). An unidentified protein kinase found as a trace contaminant in purified TH can phosphorylate serine 8 in TH, and serine 153 can be phosphorylated under some conditions by cAMP-dependent protein kinase *in vitro* (Campbell *et al.*, 1986).

TH phosphorylation can be augmented *in vivo* by agents that increase the activity of cAMP-dependent protein kinase, such as cAMP (Halegoua and Patrick, 1980; Meligeni *et al.*, 1982), cholera toxin (Halegoua and Patrick, 1980; McTigue *et al.*, 1985). db-cAMP (Meligeni *et al.*, 1982; McTigue *et al.*, 1985), and 8-bromo-cAMP (Haycock *et al.*, 1982a; Meligeni *et al.*, 1982) with concomitant TH activation (Meligeni *et al.*, 1982; McTigue *et al.*, 1985; Roskoski and Roskoski, 1987). Treatment of PC12 pheochromocytoma cells or adrenal chromaffin cells with other stimulators of catecholamine biosynthesis, such as EGF (Halegoua and Patrick, 1980; McTigue et al., 1985), Ca^{2+}-dependent potassium depolarization (Haycock et al., 1982b; McTigue et al., 1985; Nose et al., 1985), activators of guanylate cyclase (Roskoski and Roskoski, 1987), ACh (Haycock et al., 1982a,b; Nose et al., 1985) and ACh analogues (Haycock et al., 1982b; Nose et al., 1985), A23187 (Haycock et al., 1982b; Nose et al., 1985), NGF (Letendre et al., 1977; Halegoua and Patrick, 1980; McTigue et al., 1985), or TPA (McTigue et al., 1985; Pocotte et al., 1985; Pocotte and Holz, 1986), also results in enhanced TH phosphorylation. In vivo, there are several sites of phosphorylation in TH (Haycock et al., 1982a; McTigue et al., 1985), probably corresponding to four serine residues (McTigue et al., 1985). Various agonists of catecholamine biosynthesis induce phosphorylation of different subsets of these sites (Haycock et al., 1982a; McTigue et al., 1985; Cremins et al., 1986). How the four sites identified in vivo might interact, their relationship to the four identified in vitro phosphorylation sites, and the role of protein kinase C in this scheme are subjects for future study. With regard to the latter question, TH appears to be a soluble protein. Hence, like glycogen synthase and S6, this protein is unlikely to be a direct substrate for protein kinase C in vivo, with the effect of TPA being indirect, perhaps through protein kinase C–mediated activation of other protein kinases.

3.2.9. Likely Candidates

The class I HLA antigens have been demonstrated to be phosphorylated in vivo upon addition of TPA to a variety of cells (Feuerstein et al., 1985; Rouis et al., 1986) and in membrane preparations by purified protein kinase C (Shackelford and Trowbridge, 1986), suggesting that they serve as a physiological substrate for protein kinase C. Questions remain regarding the stoichiometry, kinetics, and site(s) of this phosphorylation event. Also, it is not clear whether stimuli other than TPA induce this phosphorylation or what parameter of cell function it could affect.

NADPH oxidase in phagocytes is activated by a variety of stimuli, including TPA (see Section 4.7). It has been suggested that the 31.5-kD component of NADPH oxidase is a physiological substrate of protein kinase C (Papini et al., 1985). This hypothesis was drawn from in vitro studies using partially purified NADPH oxidase and in vivo studies in which, due to a lack of antiserum, a TPA-induced phosphoprotein could be identified only tentatively as NADPH oxidase (Papini et al., 1985). Nevertheless, it is a provocative observation that raises the possibility that protein kinase C may regulate the activity of this enzyme directly.

Myelin basic protein was first recognized as an *in vitro* substrate of protein kinase C by Kuo's group (Turner *et al.*, 1982). The site(s) of phosphorylation were determined later by Turner *et al.*, (1984a, 1985) and Kishimoto *et al.* (1985), although some discrepancies were apparent; the former group favoring a single site of phosphorylation (serine 115), with the latter detecting several sites. Recently, the phosphorylation of myelin basic protein has been observed in response to phorbol ester treatment *in vivo* (Vartanian *et al.*, 1986). It remains to be determined whether the sites that are stimulated *in vivo* correspond to sites phosphorylated by purified protein kinase C.

3.2.10. Other Possible Substrates

A multitude of other proteins have been postulated to be substrates for protein kinase C. Some of these have known identities, but most are described only by molecular weights on SDS–polyacrylamide gels. We have grouped proteins whose phosphorylation is augmented when cells are treated with tumor promoters into our second category and listed them in Table IV. Some of these proteins might prove to be physiological targets of protein kinase C once they have been characterized further. Table V lists proteins in the last category. These proteins simply have been reported to be *in vitro* substrates of protein kinase C.

It is clear from Tables III–V that multiple proteins are phosphorylated either by protein kinase C *in vitro* or in response to protein kinase C activation *in vivo*, or both. The cumulative increase in substrates reported since the identification of protein kinase C is illustrated graphically in Figure 2. This graph is somewhat misleading due to the promiscuity of protein kinase C *in vitro*. In light of this, demonstrating that a protein is phosphorylated by protein kinase C *in vitro* is hardly noteworthy unless this observation is substantiated by *in vivo* studies coupled with kinetic and/or site specificity analyses (see Section 3). Similarly, *in vivo* studies alone are less informative without accompanying *in vitro* data. Bearing in mind that protein kinase C represents the major cellular receptor for TPA, it is predictable that novel phosphorylation events will occur when cells are treated with TPA. However, although the phosphorylation of a protein in response to tumor promoter treatment might indicate that the protein plays a role in the cellular response, it does not prove that it is a direct substrate of protein kinase C. A clear example of this is a 42-kD protein whose phosphorylation on tyrosine is augmented upon stimulation of protein kinase C (see Section 5.3). Obviously, protein kinase C is not directly responsible for this event. Rather, protein kinase C activates a protein–tyrosine kinase or, alternatively, inhibits a phosphotyrosine–protein phosphatase.

In summary, despite the large amount of interest in identifying potential

TABLE IV. Proteins Whose Phosphorylation is Stimulated by TPA Treatment *in Vivo*

Protein	Cell type	Reference
Identified proteins		
α-Adrenergic receptor	DDT₁, MF-2	Leeb-Lundberg et al. (1985)
β-Adrenergic receptor	Erythrocytes	Kelleher et al. (1984)
	Erythrocytes	Sibley et al. (1984)
Insulin receptor	Hep G2, IM-9	Jacobs and Cuatrecasas (1986)
IGF-I receptor	Hep G2, IM-9	Jacobs and Cuatrecasas (1986)
Vitronectin receptor	MG 63	E.K. Freed and T. Hunter (unpublished observations)
v-erb-B	Fibroblasts	Decker (1985)
p56 (protein–tyrosine kinase)	LSTRA, T lymphocytes	Casnellie and Lamberts (1986)
Polyoma mT antigen	Fibroblasts	Matthews and Benjamin (1986)
Heat shock protein (27–28 kD)	HL-60	Feuerstein and Cooper (1983)
	HL-60, U937, normal monocytes	Feuerstein and Cooper (1984)
	Rat embryo fibroblasts	Welch (1985)
T3 γ and δ chains	T lymphocytes	Cantrell et al. (1985)
T4 antigen	T helper cells	Acres et al. (1986)
Lipomodulin	Thymocytes	Hirata et al. (1984)
Band 4.1	Erythrocytes	Horne et al. (1985)
		Ling and Sapirstein (1984)
		Cohen and Foley (1986)
		Faquin et al. (1986)
Band 4.9	Erythrocytes	Horne et al. (1985)
		Cohen and Foley (1986)
		Faquin et al. (1986)
Histone H2B and H4	Splenocytes	Patskan and Baxter (1985)
Histone H1⁰	Reuber H35 hepatoma	Butler et al. (1986)
	Reuber H35 hepatoma	Butler et al. (1986)

Unidentified proteins (kD)

59, 47, 45, 27	Neutrophils	Schneider et al. (1981)
80, 69, 55, 48, 22, 13	Neutrophils	Andrews and Babior (1983)
98, 62, 20, 15, 13	Neutrophils	Fujita et al. (1984)
62, 46, 43, 18.5	Neutrophils	Genmaro et al. (1985)
90, 70, 64, 55, 50, 40	Neutrophils	White et al. (1985)
88, 64, 47	Neutrophils	Barrowman et al. (1986)
60, 34, 31, 22	Mast	Katakami et al. (1984)
17	HL-60	Feuerstein and Cooper (1983)
	HL-60, U937, normal monocytes	Feuerstein and Cooper (1984)

Ten proteins ranging from 12–120	HL-60, R-94	Anderson et al. (1985)
80, 33	HL-60	Macfarlane (1986)
~70, 68, 68, 62, 55, 47	Murine B lymphocytes	Hornbeck and Paul (1986)
130, 90, 72	Platelets	Chiang et al. (1981)
180	Platelets	Bourguignon et al. (1985a)
30	Platelets	Bourguignon et al. (1985b)
80, 34, 27, 17	A431	Sahai et al. (1986)
83, 69, 62, 35–40	Pancreatic acinar	Burnham et al. (1986)
82, 80, ~59, 59, 65	GH$_3$	Drust and Martin (1983)
Unspecified proteins	S49	Kiss and Steinberg (1985)
28	MCF-7	Issandou et al. (1986)
27	Vascular endothelial	Darbon et al. (1986b)
80, 40, 29	Pancreatic islets	Dunlop and Larkins (1986)
70, 56, 35	Hepatocytes	Garrison et al. (1984)
83, 75, 47, 43, 21	Primary rat neurons	Burgess et al. (1986)
37	Chicken embryo	Laszlo et al. (1981)
90	Balb/3T3, C3H/10T1/2	Chida et al. (1986a)
350, 80	Rat embryo fibroblasts	Sato et al. (1985)
22	3T3L1 fibroblasts, adipocytes	Blackshear et al. (1985)
42	Chicken embryo fibroblasts	Bishop et al. (1983)
	Chicken embryo fibroblasts	Gilmore and Martin (1983)
	Chicken embryo fibroblasts	Cooper et al. (1984)
	Human skin fibroblasts	Kohno (1985)

TABLE V. Substrates of Protein Kinase C *in Vitro*

Proteins	Phosphorylation[a]	Phosphate incorporated[b]	Phosphorylation affect	References
Enzymes				
ATP-citrate lyase	ND[d]	0.5	Unknown	Hardie et al. (1986)
Acetyl-CoA carboxylase	ND	0.9	Inactivation	Hardie et al. (1986)
Ca^{2+}-activated neutral protease	ND	ND	Unknown	Hincke and Tolnai (1986)
Cytochrome P-450	0.14	4	Unknown	Vilgrain et al. (1984b)
DNA methyltransferase	ND	1	Activation	De Paoli-Roach et al. (1986)
Guanylate cyclase	ND	1	Activation	Zwiller et al. (1985)
High mobility group (HMG)-CoA reductase	80	1	Inactivation	Bet et al. (1985)
Lactate dehydrogenase	10	ND	Unknown	Woodgett et al. (1986)
MLCK	ND	2	Increase in Ka for calmodulin	Ikebe et al. (1985)
	4.2	2	Increase in Ka for calmodulin	Nishikawa et al. (1985)
Phosphofructokinase	ND	ND	Activation	Hofer et al. (1985)
	0.7	1.5	Unknown	Nettelblad et al. (1986)
	ND	0.95	No change	Rider and Hue (1986)
Phosphorylase kinase	ND	ND	Activation	Kishimoto et al. (1977, 1978)
Pyruvate kinase type M_2	ND	ND	Unknown	Noda et al. (1986)
Topoisomerase II	0.1	0.85	Activation	Sahyoun et al. (1986)
Cytoskeletal elements				
C protein	ND	1.6	Unknown	Lim et al. (1985)
Caldesmon	9.7	8	Inhibition of MLCK	Umekawa and Hidaka et al. (1985)
Fibrinogen	300–600	5	Unknown	Humble et al. (1984)
	ND	ND	Unknown	Papanikolaou et al. (1982)

Protein				
Microtubule-associated protein (MAP-2)	ND	25–31	Unknown	Tsuyama et al. (1986)
	0.9	10	Decreases actin interaction	Akiyama et al. (1986)
Talin	ND	0.8	Unknown	Litchfield and Ball (1986)
Troponin I	3.4	1.7	Unknown	Katoh et al. (1983)
	6.66	0.9	Unknown	Mazzei and Kuo (1984)
Troponin T	0.3	2	Unknown	Katoh et al. (1983)
	0.13	2	Unknown	Mazzei and Kuo (1984)
Other proteins				
B-50	ND	ND	Unknown	Aloyo et al. (1983)
	ND	ND	Unknown	Eichberg et al. (1986)
eIF-2 β subunit	0.13	1.6	Unknown	Schatzman et al. (1983a)
G_i regulatory component of adenylate cyclase	ND	ND	Inhibition	Katada et al. (1985)
α subunit of transducin	~1	ND	Unknown	Zick et al. (1986)
Histone H1	ND	2	Unknown	Iwasa et al. (1980)
	0.6	2	Unknown	Wise et al. (1982b)
HMG 1 and HMG 2	ND	1	Unknown	Ramachandran et al. (1984)
HMG 14	5.1	1	Unknown	Ramachandran et al. (1984)
HMG 17	4.6	1	Unknown	Ramachandran et al. (1984)
Gap junction protein, MP26	ND	ND	Unknown	Lampe et al. (1986)
	ND	0.33	Unknown	Takeda et al. (1986)
Phospholamban	ND	ND	Unknown	Iwasa and Hosey (1984)
	ND	ND	Unknown	Movsesian et al. (1984)
Protamine	ND	ND	Unknown	Takai et al. (1977)
Protein F1	ND	ND	Unknown	Akers and Routtenberg (1985)
Retinol-binding protein	6.2	ND	Unknown	Cope et al. (1984)
Retinoic acid–binding protein	5.1	ND	Unknown	Cope et al. (1984)
Rhodopsin	ND	ND	Decreased ability to activate [35]-GTPγS binding to transducin	Kelleher and Johnson (1986)
T200	ND	ND	Unknown	Shackelford and Trowbridge (1986)
Vitamin D–binding protein	ND	ND	Unknown	Wooten et al. (1985)

TABLE V. (*Continued*)

Proteins	Source	References
Unidentified proteins (kD)		
88, 51, 42	Sarcolemma	Iwasa and Hosey (1984)
15	Sarcolemma	Presti *et al.* (1985)
94, 87, 78, 51, 46, 11.5, 10	Sarcolemma	Yuan and Sen (1986)
36	Pituitary secretory granules	Turgeon and Cooper (1986)
87, 47	Synaptosomes	Wu *et al.* (1982)
69, 37, 17	Mitochondria	Backer *et al.* (1986)
89, 38, 34, 17, 15, 14	Leukemic cells	Helfman *et al.* (1983a)
89, 73, 51, 17, 15, 15, 14	Neutrophils	Helfman *et al.* (1983b)
130, 43, 41, and 34	Neutrophils	Huang *et al.* (1983)
46 and 46	Neutrophils	Ohtsuka *et al.* (1986)
245, 200, 170, 120, 69, 68, 58, 52, 49, 29	HL-60	Durham *et al.* (1985)
92, 84, 70, 67, 53, 45, 40, 36	EL4 cytosol	Kramer and Sando (1986)
36	Lung	Malkinson *et al.* (1985)

[a] K_m for phosphorylation by protein kinase C.
[b] Moles of phosphate incorporated per mole of substrate.
[c] Effect of protein kinase C phosphorylation on the function of the substrate.
[d] ND, not done.

targets of protein kinase C, as evidenced by the volume of literature, there are but a dozen proteins that fit the criteria of physiological substrates.

3.3. Predicting Phosphorylation Sites for Protein Kinase C

Amino acid sequencing of peptides containing sites of phosphorylation for cAMP-dependent protein kinase revealed the presence of one or, more commonly, two basic amino acids N-terminal to the target residue. The importance of these residues was confirmed by the use of synthetic peptides (Kemp *et al.*, 1976; Zetterqvist *et al.*, 1976). A similar but distinct requirement for basic amino acids N-terminal to the target residue has since been determined for MLCK (Kemp *et al.*, 1983) and the multifunctional calmodulin-dependent protein kinase (Pearson *et al.*, 1985), whereas casein kinase II appears to favor a string of acidic residues C-terminal to a phosphorylation site (Meggio *et al.*, 1984). In addition to the N-terminal basic residue requirement, MLCK has been shown to be sensitive to C-terminal noncharged residues (Pearson *et al.*, 1986).

Initial experiments designed to evaluate the structural requirements of protein kinase C were performed using peptides derived from sites phosphorylated by the enzyme on myelin basic protein, which is a substrate for protein kinase C *in vitro* (Turner *et al.*, 1982, 1984a, 1985; Kishimoto *et al.*, 1985). These studies arrived at conflicting conclusions that protein kinase C preferred N-terminal basic residues in one case or C-terminal residues in the other. One study using model peptides indicated that C-terminal basic residues were sufficient for phosphorylation (Ferrari *et al.*, 1985), whereas another argued that such residues should be N-terminal (O'Brian *et al.*, 1984).

In an attempt to resolve these anomalies, we synthesized synthetic peptides that contained the phosphorylation sites for protein kinase C from proteins we and others had identified as physiological substrates for this enzyme (Woodgett *et al.*, 1986). The best substrates had basic residues flanking the target residue on both sides. However, one substrate, p36, contained only one basic residue, and this was located C-terminal to the target serine (Gould *et al.*, 1986). The peptide derived from pp60^{src} contained the phosphorylation sites for both protein kinase C (serine 12) and cAMP-dependent protein kinase (serine 17). Each protein kinase exhibited complete specificity for its own site, demonstrating that although the two enzymes share a preference for basic residues near to their targets, their actual specificities do not overlap. We concluded from these studies that the minimal physiological requirement for primary sequence of protein kinase C is Ser/Thr-X-Arg/Lys, where X is an uncharged residue. However, the presence of additional basic residues on each side of the target enhances phosphorylation significantly as in the sequence, B-B-X-Ser/Thr-X-B-B, where B is an arginine or lysine residue.

Similar conclusions recently have been drawn by House *et al.*, (1987). The sequences around the phosphorylation sites for protein kinase C in all of the physiological substrates identified to date conform to this paradigm (Figure 3). It should be emphasized, however, that fulfillment of this requirement is only one aspect in determining the site of phosphorylation on a substrate for protein kinase C (see Section 3).

We have noticed that protein kinase C phosphorylation sites lie in close proximity to the plasma membrane. For example, the major site of phosphorylation on pp60^{src} lies ten amino acids from the lipid bilayer if one assumes that the N-terminal myristylated glycine residue is engulfed by the lipid head groups (Section 3.2.1). Threonine 654 of the EGF receptor is again ten amino acids C-terminal to the transmembrane domain. Serine 247 of the IL-2 receptor is nine residues from the membrane. There are some potential exceptions. The phosphorylation site on the transferrin receptor is 41 residues distal to the transmembrane domain (Section 3.2.3) although, in this case, the inverse orientation of the receptor in the membrane may mean that the protein chain has to fold around, placing the peptide backbone in the vicinity of serine 24 close to the membrane in the same orientation as substrates like the EGF receptor. In the case of substrates like p36, which are peripherally associated with the membrane, it is difficult to predict where the target residue lies with respect to the lipid bilayer, but the acceptor amino acid could readily be brought into an appropriate location

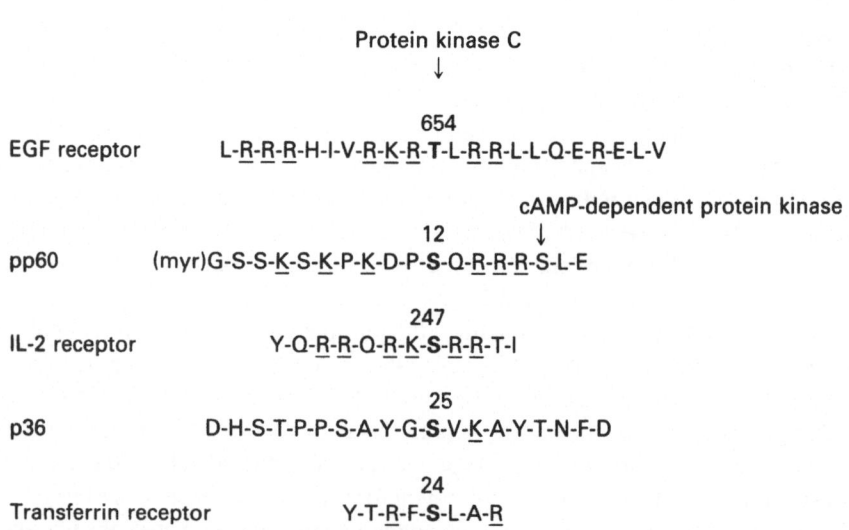

FIGURE 3. Sequences around physiological phosphorylation sites of protein kinase C (from Woodgett *et al.*, 1986; Davis *et al.*, 1986). (myr) Myristic acid. Phosphorylated residues are in bold type; basic residues are underlined.

and orientation by protein folding. It is likely that a limitation is placed on protein kinase C by its phospholipid binding property, such that it can only phosphorylate residues in membrane-associated proteins within a small distance of the bilayer. Therefore, estimation of the physical distance of a putative site of phosphorylation from a membrane-binding or membrane-traversing region of the substrate protein may prove to be valuable in predicting protein kinase C phosphorylation sites.

4. CELLULAR EFFECTS OF TUMOR PROMOTERS

Since the discovery of their potent tumor-promoting activity in the two-stage carcinogenesis model, phorbol esters have been used widely to probe biological mechanisms (see Section 2.5). They have been found to elicit profound and diverse effects on numerous physiological processes. So much effort has been devoted to TPA-induced activities that each aspect deserves its own review. Thus, we will discuss only briefly the better-documented effects on events related to cell growth and the role protein kinase C might play in them.

4.1. Carcinogenic Effects

Following initial application of a low dose of carcinogen to mouse skin, repeated additions of a suitable promoter led to the formation of papillomas and occasional carcinomas (for review, see Slaga, 1984). The current interpretation of this process is that a genetic alteration modifying the growth potential of target cells serves as the initiating event (Hennings and Yuspa, 1985). Promoters modulate the dynamics of growth and differentiation of the epidermal cell population such that initiated cells escape regulation selectively (Hennings and Yuspa, 1985). Studies in both cultured epidermal cells and mouse skin have substantiated this view. The activated Ha-*ras* gene can function alone as an initiator of mouse skin carcinogenesis (Brown *et al.*, 1986) and, further, the c-Ha-*ras* gene has been found to be activated in certain dimethylbenz[*a*]anthracene-induced carcinomas (Quintanilla *et al.*, 1986). In cultured mouse epidermal basal cells, TPA induces half the population to differentiate while it stimulates DNA synthesis in the other half (Yuspa *et al.*, 1982). *In vivo*, TPA treatment results in the commitment of a subpopulation of basal cells to differentiate and accelerates terminal differentiation of committed cells (Reiners and Slaga, 1983). TPA promotes growth and transformation of C3H 10T1/2 cells (Hsiao *et al.*, 1984) and rat fibroblasts or embryo cells (Dotto *et al.*, 1985; Hsiao *et al.*, 1986) expressing the v-Ha-*ras* oncogene. Moreover, it is unable to induce the differentiation of keratinocytes that are infected with the v-Ha-*ras* oncogene and are blocked at an early reversible stage of maturation (Yuspa et al., 1985). Thus, TPA could expand a population of initiated cells selectively by increasing its relative number and/or growth rate.

TPA also might select for proliferation of initiated cells by inducing the amplification of a gene(s) that provides a growth advantage (Varshavsky, 1981). Evidence in favor of this possibility is that addition of TPA increases the incidence of drug-resistant colonies significantly in mouse 3T3 and 3T6 cells (Varshavsky, 1981; Barsoum and Varshavsky, 1983; Bojan et al., 1983) and Chinese hamster lung cells (Hayashi et al., 1983) but only modestly in hamster Chinese ovary and V79 cells (Bojan et al., 1983). Resistance to cadmium, methotrexate, or (phosphonacetyl)-L-aspartate (PALA) is accompanied by amplification of the metallothionein 1, dihydrofolate reductase, or aspartate transcarbamylase genes, respectively (Varshavsky, 1981; Barsoum and Varshavsky, 1983; Bojan et al., 1983; Hayashi et al., 1983). Furthermore, a TPA-nonproliferative variant of 3T3 cells resists TPA-augmented amplification of the metallothionein 1 gene (Herschman, 1985).

Presumably, protein kinase C mediates the effects of tumor promoters. It will be interesting to determine whether synthetic DAGs can duplicate the effects of TPA in mouse skin carcinogenesis through transient activation of protein kinase C or whether tumor promoters have other critical effects in this process (Section 2.7.1).

4.2. Effects on Cell Differentiation

Tumor promoters exhibit diverse and unpredictable effects on cell differentiation. TPA inhibits the differentiation of a large number of cell types, including Friend virus–transformed murine erythroleukemia cells (Miao et al., 1978; Rovera et al., 1977; Yamasaki et al., 1977), cultured embryonic chick ganglia (Ishii, 1978), chick myogenic cells (Cohen et al., 1977; Dlugosz et al., 1983), rat mammary cell lines (Sonnenberg et al., 1983), cultured hamster epidermal cells (Sisskin and Barrett, 1981), normal avian melanoblasts and melanocytes (Payette et al., 1980; Sieber-Blum and Sieber, 1981), human HO (Huberman et al., 1979) and mouse B-16 (Mufson et al., 1979a) melanoma cells, mouse neuroblastoma lines (Ishii et al., 1978) and Balb/3T3 T-proadipocytes (Diamond et al., 1977; Yun and Scott, 1983). In marked contrast, TPA promotes the differentiation of the human malignant T-lymphoblast cell line, Molt-3 (Nagasawa and Mak, 1980), the neuroblastoma line, SH-SY5Y (Pahlman et al., 1983), human chronic lymphocytic leukemic cells (Totterman et al., 1980), medullary thyroid carcinoma cells (de Bustros et al., 1985), human erythroleukemia cells (Papayannopoulou et al., 1983), human erythroleukemia virus–transformed murine erythroleukemic cells (Miao *et al.*, 1978), the pre-lymphocyte cell line 70Z/3 (Rosoff *et al.*, 1984), human myeloid leukemic cells (Pegoraro *et al.*, 1980), human T-lymphoid leukemia cells (Ryffel *et al.*, 1982), chick chondroblasts to fibroblasts (Lowe *et al.*, 1978), and the human promy-elocytic leukemia cell lines, HL-60 (Huberman and Callaham, 1979; Rovera *et al.*, 1979a,b; Lotem and Sachs, 1979; Koeffler *et al.*, 1981) and KG-1 and ML-3 (Koeffler *et al.*, 1981) to macrophagelike cells. The differentiation of certain cells is unaffected by TPA. For example, TPA does not induce PC12 cells to differentiate nor does it interfere with NGF-stimulated differentiation (Burstein *et al.*, 1982). The maturation stage of K562 and KG-1a similarly is unaffected by TPA addition (Koeffler *et al.*, 1981).

The mechanism whereby TPA elicits these varied responses is not known. It is not clear yet whether activation of protein kinase C alone is responsible for these effects or whether other actions of tumor promoters play a role (see Section 2.7.1). This point can be addressed by distinguishing the effects of synthetic DAGs and tumor promoters on these cells.

4.3. Mitogenic Effects

4.3.1. Lymphocytes

Tumor promoters are, at best, weakly mitogenic for lymphocytes (Truneh *et al.*, 1985). The inability of tumor promoters to mobilize Ca^{2+} (Tsien *et al.*, 1982; Gelfand *et al.*, 1985) appears to be one reason for this, since a significant mitogenic response is obtained with a combination of Ca^{2+} ionophore and TPA in lymphocytes (Wang *et al.*, 1975; Mastro and Smith, 1983), mouse T lymphocytes (Truneh *et al.*, 1985), resting human B cells (Guy *et al.*, 1985), mouse thymocytes (Moore *et al.*, 1986), and human peripheral blood lymphocytes, splenic lymphocytes, and thymocytes (Delia *et al.*, 1984). However, this syn-ergistic effect on T-cell mitogenesis depends on the production of IL-2 by the target cells (Truneh *et al.*, 1985). Thus, even the combination of TPA and Ca^{2+} mobilization is insufficient to elicit a maximal mitogenic response in lympho-cytes. Further examples of this are resting B cells, where a combination of A23187 and TPA has been demonstrated to drive resting B lymphocytes into the cell cycle (G_0 to G_1), but not into S phase (Monroe and Kass, 1985) and macrophage-depleted peripheral blood lymphocytes, where a low level of PHA in addition to TPA and A23187, is required for a full mitogenic response (Kai-buchi *et al.*, 1985).

TPA also potentiates markedly the proliferative activity of mitogenic lectins, such as PHA and Con A (Mastro and Mueller, 1974; Wang et al., 1975; Rosenstreich and Mizel, 1979), which alone submaximally raise intracellular Ca^{2+} (Tsien et al., 1982) and induce IL-2 production by T lymphocytes (Truneh et al., 1985). It also has been reported that TPA enhances the mitogenic activity of PHA and S. aureus in a Ca^{2+}-independent manner (Gelfand et al., 1985). When used in combination with lectins, IL-1 has a similar effect, quantitatively, on mitogenesis (Rosenstreich and Mizel, 1979) and IL-2 production (Farrar et al., 1980) as TPA. Despite this functional similarity, IL-1 and TPA possess distinct comitogenic capabilities, since IL-1 is able to synergize with the combination of TPA and PHA in stimulating T-lymphocyte proliferation (Krakauer et al., 1982). Together, these data indicate that other factors complementing the activation of protein kinase C and the mobilization of Ca^{2+} are critical for lymphocyte proliferation. Furthermore, because of the possibility that TPA induces the production of lymphocyte growth factors, such as IL-1 from macrophages (Mizel et al., 1978), identifying the necessary components and their functions in proliferative responses is more straightforward when defined lymphocyte populations are utilized.

4.3.2 Other Cells

TPA alone is mitogenic for quiescent Swiss 3T3 fibroblasts (Driedger and Blumber, 1977; Dicker and Rozengurt, 1979), certain other fibroblasts (O'Brien et al., 1979a), and vascular smooth muscle cells (Owen, 1985). It stimulates proliferation of myeloid and erythroid precursor cells in culture (Fibach et al., 1980) and formation of bone marrow hematopoietic colonies in agar (Stuart and Hamilton, 1980). These proliferative responses to TPA are cell dependent (O'-Brien et al., 1979a) and suboptimal (Dicker and Rozengurt, 1979). Maximal mitogenic effects can be obtained with TPA in combination with other growth factors (Dicker and Rozengurt, 1979; Stuart and Hamilton, 1980) which, in contrast to TPA, mobilize Ca^{2+} substantially in fibroblasts (Moolenaar et al., 1984a; Morris et al., 1984). Recently, the role of protein kinase C in mitogenesis has been called into question by studies utilizing the ability of phorbol esters to down-regulate the enzyme (see Section 2.7.1). For example, PDGF maintained the ability to induce a full mitogenic response in fibroblasts chronically stimulated with TPA (Coughlin et al., 1985). This result suggests that PDGF-induced mitogenesis occurs independently of protein kinase C. However, as mentioned in Section 2.7.1, a small level of active protein kinase C molecules that might still be playing a role in the mitogenic process remains in down-regulated cells.

In some cell types that respond mitogenically to EGF, such as human breast cancer cells (Osborne et al., 1981), tumor promoters are growth inhibitory. Presumably, this is due to EGF receptor down-regulation (see Section 2.1.1).

4.4. Effects on Gene Expression and Enzymatic Activities

Correlative alterations in mRNA and protein synthesis accompany the changes in cell growth and differentiation elicited by TPA. Indeed, TPA induces the expression of a whole spectrum of genes, several of which have been identified. Among the known effects of TPA is the induction of the c-*fos* gene in HL-60 cells (Mitchell *et al.*, 1985; Muller *et al.*, 1984), fibroblasts (Greenberg *et al.*, 1984; Kruijer *et al.*, 1984), A431 cells (Bravo *et al.*, 1985), and murine thymocytes (Moore *et al.*, 1986). The c-*myc* gene is also activated in response to TPA in murine thymocytes (Moore *et al.*, 1986), human lymphocytes (Reed *et al.*, 1985), and fibroblasts (Kelly *et al.*, 1983; Greenberg and Ziff, 1984; Coughlin *et al.*, 1985; Kaibuchi *et al.*, 1986). The β-actin gene in fibroblasts (Greenberg and Ziff, 1984), the prolactin gene in GH_4 cells (Osborne and Tashjian, 1981; Murdoch *et al.*, 1985), the ornithine decarboxylase gene in mouse skin (Verma *et al.*, 1986), the IL-2 gene in sensitive EL4 cells (Harrison *et al.*, 1987), the metallothionein gene (Imbra and Karin, 1987), and the c-*sis* gene in K562 cells (Colamonici *et al.*, 1986) are further examples of TPA-stimulated genes. Other induced sequences currently are being identified from cDNA libraries based on their preferential expression in TPA-stimulated cells. Among such sequences are the *pro-1* and *pro-2* genes, which appear to confer sensitivity to growth in the presence of TPA (Colburn *et al.*, 1986).

TPA also inhibits the expression of several genes, which are (as above) too numerous to cover comprehensively. Examples include the phosphoenolpyruvate carboxykinase gene (Chu and Granner, 1986), the glycophorin A and B genes (Siebert and Fukuda, 1986), and a collagen gene (Delclos and Blumber, 1979; Sobel *et al.*, 1983). This latter effect might contribute to TPA-induced morphological changes (Section 4.6).

A variety of proteins are expressed selectively in response to TPA, including a 32-kD protein in Balb/c 3T3 cells (Hiwasa *et al.*, 1982), a 54-kD protein in Friend cells (Mitsuse and Oishi, 1985), and several unidentified proteins in mouse epidermal cell cultures (Cabral *et al.*, 1981), mouse fibroblasts (Haarr *et al.*, 1986), and chicken embryo cells (Laszlo *et al.*, 1981). Significantly, addition of TPA augments the synthesis and secretion and/or release of several proteases, including plasminogen activator (Wigler and Weinstein, 1976; Ojakian, 1981; Degen *et al.*, 1985; Stoppelli *et al.*, 1986), acid-activated protease from mouse epidermal cells (Gottesman and Yuspa, 1981) and fibroblasts (Gottesman and Sobel, 1980; Rabin *et al.*, 1986), collagenase (Hersh *et al.*, 1986), and a membrane-bound neutral proteinase (Pontremoli *et al.*, 1986b). TPA-induced release of proteases permits angiogenesis to occur *in vitro* (Montesano and Orci, 1985) and is responsible for at least some of the morphological changes accompanying TPA treatment (Quigley, 1979).

TPA stimulates the activity of several enzymes that are indicative of growth or functional phenotype. Examples are ornithine decarboxylase in certain fibro-blasts (O'Brien *et al.*, 1979a; Gilmour *et al.*, 1985), tracheal epithelial cells (Jetten *et al.*, 1985) and mouse basal epidermal cell cultures (Lichti *et al.*, 1981), TH in chromaffin cells (Pocotte and Holz, 1986), acid phosphatase in neutrophils (Vorbrodt *et al.*, 1979), NADPH oxidase in neutrophils (Suzuki and Lehrer, 1980), and transglutaminase in epidermal cells (Yuspa *et al.*, 1982). Conversely, activities of other enzymes, such as histidase in mouse epidermis (Colburn *et al.*, 1975), are suppressed by TPA.

The molecular mechanisms whereby activation of protein kinase C leads to marked changes in gene and protein expression are not well understood. Recently, however, short DNA sequences have been identified in the 5'-flanking region of the prolactin gene and in the Moloney murine leukemia virus enhancer, which confer phorbol ester inducibility in *cis* to heterologous transcription units (Elsholtz *et al.*, 1986). By using synthetic oligonucleotides, other workers have narrowed down the requirement for phorbol ester inducibility to an 8-base-pair consensus sequence (TGAGTCAG) derived from the 5'-flanking region of the fibroblast collagenase and metallothionein genes (Angel *et al.*, 1987). Since the active form of protein kinase C apparently is restricted to cytoplasmic membranes (Section 2.5), protein kinase C presumably alters gene expression indirectly. For instance, protein kinase C might indirectly affect the binding of a specific transcription factor to such an enhancerlike sequence or modify its activity.

4.5. Effects on Transport and Ion Fluxes

TPA induces many biochemical changes in common with other growth factors. TPA stimulates 2-deoxyglucose uptake in chicken embryo cells (Driedger and Blumberg, 1977), Balb/c 3T3 preadipocytes (O'Brien, 1982), Swiss 3T3 cells (Dicker and Rozengurt, 1979), rat adipocytes (Martz *et al.*, 1986), and a variety of other cell types (Lee and Weinstein, 1979). This stimulation has been reported to be dependent (O'Brien, 1982) or independent (Dicker and Rozengurt, 1979; Lee and Weinstein, 1979) of mRNA and protein synthesis. The activity of the glucose transporter might be affected directly by protein kinase C, since it can be phosphorylated *in vivo* upon addition of TPA (Witters *et al.*, 1985; Gibbs *et al.*, 1986; Allard *et al.*, 1986) and *in vitro* by protein kinase C (Witters *et al.*, 1985). However, recent studies have dissociated the phosphorylation of the glucose transporter from increases in its activity, suggesting that protein kinase C may have no direct effect on the activity of the transporter or the transporter might be subject to more than one mode of regulation (Allard *et al.*,

1986; Gibbs *et al.*, 1986; Kitagawa *et al.*, 1986). Glycolytic flux is also enhanced in the presence of TPA (O'Brien *et al.*, 1979b), and purine base phosphoribosylation is increased (Dicker and Rozengurt, 1979; Becker *et al.*, 1983).

Ion fluxes are also affected by the addition of TPA to cells. TPA has been reported to induce the influx of Mg^{2+} (Grubbs and Maguire, 1986). TPA also stimulates uptake of Na^+ ions, as measured by atomic absorption or $^{22}Na^+$ accumulation (Rosoff *et al.*, 1984; Grinstein *et al.*, 1985; Owen, 1985). Because this process depends on extracellular Na^+ and is blocked by amiloride, these results indicate that TPA treatment activates the Na^+/H^+ exchanger by altering its pH dependence (Rosoff *et al.*, 1984; Grinstein *et al.*, 1985; Owen, 1985; Swann and Whitaker, 1985; Lau *et al.*, 1986). However, TPA appears to stimulate Na^+/H^+ exchange to various degrees in different cell types. In certain cells, the antiporter is only weakly stimulated (Whiteley *et al.*, 1984) or is unaffected (Vicentini and Villereal, 1985; Vigne *et al.*, 1985) by TPA treatment. Vara and Rozengurt (1985) noted that the Na^+/H^+ exchanger can be partly activated by EGF and insulin in a protein kinase C–independent manner. They and, subsequently, others have proposed that Na^+ influx can be regulated by diverse mechanisms (Vara and Rozengurt, 1985; Vicentini and Villereal, 1985; Wiener *et al.*, 1986) or that the Na^+/H^+ exchanger is active in certain cells constitutively (Vigne *et al.*, 1985). In any case, the Na^+/H^+ exchanger should be considered a potential protein kinase C substrate.

Activation of the Na^+/H^+ exchanger leads to an increase of between 0.05 and 0.15 units in intracellular pH (Burns and Rozengurt, 1983; Besterman and Cuatrecasas, 1984; Moolenaar *et al.*, 1984b; Hesketh *et al.*, 1985; Wiener *et al.*, 1986). In addition, elevation of intracellular levels of Na^+ ions can lead to activation of the Na^+/K^+ ATPase and an influx of K^+, measured by $^{86}Rb^+$ influx (Moroney *et al.*, 1978; Dicker and Rozengurt, 1981). Furthermore, an increase in internal Na^+ concentration could activate a Na^+/Ca^{2+} exchanger and promote an increase in the intracellular Ca^{2+} content (Balk *et al.*, 1984). This mechanism could explain how agents, such as EGF, that do not stimulate PI turnover, induce an increase in intracellular Ca^{2+} levels.

Other ion channels appear to be regulated either directly or indirectly by TPA treatment. There is evidence that Ca^{2+} efflux is stimulated by TPA, probably via a Ca^{2+} ATPase activity (Lagast *et al.*, 1984; Mottola and Romeo, 1982; Rickard and Sheterline, 1985; Pollack *et al.*, 1987). Furthermore, TPA induced increases in voltage-dependent Ca^{2+} currents (De Reimer *et al.*, 1985; Wakade *et al.*, 1986), blocked Ca^{2+}-dependent K^+ conductance in hippocampal pyramidal neurons (Baraban *et al.*, 1985; Malenka *et al.*, 1986), and blocked a voltage-sensitive Cl^- current (Madison *et al.*, 1986). Perhaps, protein kinase C will be found to phosphorylate and, thus, regulate the activity of several ion channel proteins.

4.6. Effects on Cell Morphology

TPA has a profound influence on cultured cell morphologies. First, cells reach higher saturation densities in the presence of TPA and often cross over one another (Driedger and Blumberg, 1977; Boreiko et al., 1980). Within hours following addition of TPA, fibroblasts appear thinner, more enlongated or rounded, and acquire prominent processes (Driedger and Blumberg, 1977; Rifkin et al., 1979; Boreiko et al., 1980). RSV-transformed fibroblasts cluster and round up (Quigley, 1979). Surface-associated fibronectin is reduced (Blumberg et al., 1976). The increased production of serine proteases may be responsible for many of these alterations (Quigley, 1979).

Other morphological changes occur within minutes and are independent of mRNA and protein synthesis (Ojakian, 1981; Schliwa et al., 1984; Kellie et al., 1985). TPA treatment causes a redistribution of F-actin, vinculin, and α-actinin and a reduction in the number of adhesion plaques in epithelial cells (Schliwa et al., 1984; Kellie et al., 1985). TPA also causes a reorganization of actin in fibroblasts (Rifkin et al., 1979) and keratinocytes (Kitano et al., 1986). In myotubes, it induces the disassembly of myofibrils but not subsarcolemmal microfilaments (Croop et al., 1980). TPA stimulates reorganization of the cytokeratin network (Fey and Penman, 1984; Ben-Ze'ev, 1986) and nuclear matrix–intermediate filament scaffold (Fey and Penman, 1984). Junctional complexes also are disrupted by TPA addition (Ojakian, 1981; Yancey et al., 1982; Fitzgerald et al., 1983; Ben-Ze'ev, 1986), with coordinated reduction in cytokeratin and high-molecular-weight desmoplakin synthesis (Ben-Ze'ev, 1986). Consequently, intercellular communication, as measured by inhibition of transepithelial electrical resistance (Ojakian, 1981), dye (Enomoto et al., 1984) or uridine (Murray and Fitzgerald, 1979) transfer, or metabolic cooperation (Yotti et al., 1979), is ablated. The interruption of intercellular communication might be of critical importance in allowing genetically mutated cells to escape negative growth regulation.

In many ways, the morphological changes wrought by TPA have been likened to those occurring upon cellular transformation. One important difference is that the alterations induced by TPA almost always are transient. Cells revert to normal morphologies in roughly 24–72 hr after addition of TPA and, when replated in the absence of TPA, behave normally. There are exceptions to this. For example, TPA imparts irreversibly anchorage independence and tumorigenicity to epidermal cells isolated from sensitized mice pretreated with a carcinogen (Colburn et al., 1979).

Since TPA-induced morphological changes are rapid and energy dependent (Schliwa et al., 1984), it is logical to predict that protein kinase C mediates the initial events directly by phosphorylating one or more critical substrates. However, evidence in support of this hypothesis is still lacking. Phosphorylation of vinculin by protein kinase C and its redistribution does not appear to be related causally to altered morphology (Schliwa *et al.*, 1984). The phosphorylation of other cytoskeletal proteins, such as talin (Litchfield and Ball, 1986), the gap junction protein, MP 26 (Lampe *et al.*, 1986; Takeda *et al.*, 1986), and filamin (Carroll *et al.*, 1982; Kawamoto and Hidaka, 1984), by protein kinase C requires further characterization. In light of the rapid changes in morphology elicited by TPA, other cytoskeletal elements should be considered potential targets for protein kinase C.

4.7. Effects on Secretion

A great amount of information on the role of protein kinase C in secretion elicited by extracellular signals has been obtained in platelet studies. Stimulation of platelets with tumor promoters mimics many of the responses associated with activated platelets. TPA will cause extension of pseudopodia (Carroll *et al.*, 1982), aggregation (Chiang *et al.*, 1981; White *et al.*, 1974; Zucker *et al.*, 1974), secretion of serotonin (White *et al.*, 1974; Zucker *et al.*, 1974; Chiang *et al.*, 1981) and adenine nucleotides (White *et al.*, 1974; Zucker *et al.*, 1974), a rise in cGMP levels (Chiang *et al.*, 1981), a decrease in the level of cAMP (Chiang *et al.*, 1981), and alterations in protein phosphorylation (Chiang *et al.*, 1981; Carroll *et al.*, 1982). However, platelets respond to TPA in a manner fundamentally different than they do to physiological activators (White *et al.*, 1974; Zucker *et al.*, 1974; Carroll *et al.*, 1982). Thrombin treatment, for instance, induces platelets to lose their discoid shape, extend pseudopodia, centralize their granules, and form a dense mass of microfilaments, termed the contractile gel (White *et al.*, 1974; Zucker *et al.*, 1974; Carroll *et al.*, 1982). During this process of internal transformation, storage granules are squeezed and release their products into channels of the open canalicular system (White *et al.*, 1974; Carroll *et al.*, 1982). The presence of extracellular Ca^{2+} is critical to this activation process (White *et al.*, 1974; Zucker *et al.*, 1974). In marked contrast, TPA induces storage organelles to swell rather than to contract (White *et al.*, 1974; Carroll *et al.*, 1982). A contractile gel is not formed and, hence, the release reaction stimulated by TPA develops slowly and is submaximal (White *et al.*, 1974; Carroll *et al.*, 1982). Furthermore, TPA does not induce thromboxane synthesis (Mufson *et al.*, 1979b; Chiang *et al.*, 1981; Kaibuchi *et al.*, 1983) or arachidonic acid release (Kaibuchi *et al.*, 1983). Inhibitors of thrombin-induced aggregation and release, such as prostaglandin E_1 (White *et al.*, 1974; Zucker *et al.*, 1974), adenosine (White *et al.*, 1974; Zucker *et al.*, 1974), aspirin (White *et al.*, 1974; Zucker *et al.*, 1974; Chiang *et al.*, 1981), or the cyclooxygenase inhibitor, indomethacin (Mufson *et al.*, 1979b), fail to inhibit TPA-induced reaction.

A likely explanation for these differences is that unlike physiological activators (Hallam et al., 1984), TPA does not cause a measurable rise in intracellular Ca^{2+} levels (monitored using the intracellularly trapped fluorescent dye, quin-2) in platelets (Rink et al., 1983). An increase in intracellular Ca^{2+} levels alone, produced by application of a Ca^{2+} ionophore, also proves insufficient to induce maximal responses (Rittenhouse-Simmons, 1981; Rink et al., 1982; Kaibuchi et al., 1983; Kajikawa et al., 1983; Yamanishi et al., 1983; Sano et al., 1985b; Pollack and Rink, 1986). However, simultaneous treatment with a Ca^{2+} ionophore (A23187 or ionomycin) and TPA (Kajikawa et al., 1983; Kaibuchi et al., 1983; Yamanishi et al., 1983) elicits a response quantitatively and qualitatively similar to that induced by physiological effectors.

Similar findings have since been made in many other systems, especially in secreting cells. For example, TPA stimulates a submaximal and slowly developing release response in neutrophils (Estensen et al., 1974; Goldstein et al., 1975; Wright et al., 1977; Whitin et al., 1980), without raising intracellular Ca^{2+} levels measurably (Smith and Iden, 1979; Sha'afi et al., 1983; Korchak et al., 1984). In contrast, A23187 induces a rapid and transient release reaction (Pozzan et al., 1983; Cockcroft et al., 1981). Stimulation with a combination of tumor promoter and Ca^{2+} ionophore mimics the chemotactic factor FMLP in eliciting a full physiological response (Kajikawa et al., 1983; Di Virgilio et al., 1984; Robinson et al., 1984; White et al., 1984). Synergistic effects of Ca^{2+} and protein kinase C activation have also been noted in tracheal smooth muscle contraction (Park and Rasmussen, 1985), mitogenesis of chicken heart mesenchymal cells (Balk et al., 1984), induction of ornithine decarboxylase (Otani et al., 1985), ACh release from guinea pig ileum (Tanaka et al., 1984), neurotransmitter release from PC12 cells (Pozzan et al., 1984) and rat neurons (Sakamoto et al., 1985), cortisol production from adrenocortical cells (Culty et al., 1984), histamine release from mast cells (Katakami et al., 1984) and basophils (Schliemer et al., 1982), insulin release from pancreatic islet cells (Zawalich et al., 1983), phosphate accumulation in cultured renal kidney cells (Kinoshita et al., 1986), Na^+/H^+ exchange in human WI-38 cells (Vicentini and Villereal, 1985), prolactin secretion from GH_4 and GH_3 cells (Delbeke et al., 1984; Ronning and Martin, 1986), catecholamine release from bovine adrenal chromaffin cells (Knight and Baker, 1983; Brocklehurst and Pollard, 1985), aldosterone secretion from adrenal glomerulosa cells (Kojima et al., 1983, 1984), and amylase secretion from parotid cells (Takuma and Ichida, 1986) and pancreatic acini (de Pont and Fleuren-Jakobs, 1984; Merritt and Rubin, 1985). It should be borne in mind that quin-2 fluorescence might not be sensitive enough to detect protein kinase C–induced changes in intracellular Ca^{2+} (Ware et al., 1985) and that elevated Ca^{2+} levels might cause a low level of PI breakdown (see Section 5.2). Nevertheless, it seems apparent from the studies listed above that protein kinase C and elevated intracellular Ca^{2+} levels cooperate to bring about many signal-induced responses.

4.8. Effects of Phorbol Esters in the Presence of Agonists

Initially, unexpected results were obtained when TPA and agonist were added together to cells. Rather than potentiating the effects of the agonist in many systems, TPA inhibited them. For example, prior treatment of platelets with TPA inhibited thrombin-induced formation of inositol phosphates (MacIntyre et al., 1985; Rittenhouse and Sasson, 1985; Watson and Lapetina, 1985; Zavoico et al., 1985), Ca^{2+} mobilization (MacIntyre et al., 1985; Zavoico et al., 1985; Cox and Carroll, 1986; Poll and Westwick, 1986: Yoshida et al., 1986), shape change (Yoshida et al., 1986), and secretion (Poll and Westwick, 1986). TPA pretreatment blocked gonadotropin-induced granulosa cell differentiation (Shinohara et al., 1985b), angiotensin-induced secretion from adrenal glomerulosa cells (Kojima et al., 1986), anti-IgM-induced proliferation of resting B cells (Hawrylowicz and Klaus, 1984; Mizuguchi et al., 1986), and stimulation of neutrophil secretion by FMLP (Naccache et al., 1985; Barrowman et al., 1986). Prior treatment with TPA also prevented the induction of Ca^{2+} mobilization by antigen in basophils (Sagi-Eisenberg et al., 1985), carbachol in cultured astrocytoma cells (Orellana et al., 1985), growth factors in Swiss 3T3 cells (Hesketh et al., 1985; McNeil et al., 1985), FMLP in neutrophils (Naccache et al., 1985; Barrowman et al., 1986), angiotensin II in adrenal glomerulosa cells (Kojima et al., 1986), ACh in bovine adrenal medullary cells (Misbahuddin et al., 1985), anti-IgM in resting B cells (Mizuguchi et al., 1986), TRH in GH_3 cells (Drummond, 1985), and carbachol in PC12 cells (Vicentini et al., 1985). Hydrolysis of phosphoinositides in response to angiotensin II in adrenal glomerulosa cells (Kojima et al., 1986), carbachol in astrocytes (Orellana et al., 1985) and PC12 cells (Vicentini et al., 1985), FMLP in differentiated HL-60 cells (Kikuchi et al., 1986), and anti-Ig in resting B cells (Mizuguchi et al., 1986) was also blocked by pretreatment with TPA. In the pre-B lymphocyte cell line, 70Z/3, TPA treatment reduced the levels of intracellular Ca^{2+} and PI hydrolysis (Rosoff and Cantley, 1985). Similarly, TPA reduced PI breakdown in Friend cells (Faletto et al., 1985). These data argue that protein kinase C plays a critical role in negative feedback control of PI hydrolysis (see Section 5.2). Indeed, Pandol and Schoeffield (1986) suggest that the primary function of protein kinase C might be to inhibit PI turnover induced by agonist, since they detected 1,2-DAG formation only at supramaximal levels of agonist for secretion of amylase from pancreatic acini.

Other data have also implied that protein kinase C functions to attenuate agonist-induced processes in addition to PI hydrolysis. For example, prior treatment with TPA blocked growth factor–stimulated Na^+/H^+ exchange (Whiteley et al., 1984; Owen, 1985; Vicentini and Villereal, 1985; Mendoza et al., 1986b), Ca^{2+} efflux (Mendoza et al., 1986a), and EGF-stimulated DNA synthesis (Owen, 1985). Furthermore, TPA pretreatment inhibits voltage-gated Ca^{2+} channel functions in PC12 and RINm5F cells (Di Virgilio et al., 1986).

4.9. Can the Effects of Tumor Promoters on Cells Be Mimicked by Synthetic Diacylglycerols?

The question of whether transient protein kinase C activation can account for the multitude of cellular responses elicited by tumor promoters is currently being addressed through the use of synthetic DAGs. Considerable evidence gathered to date suggests that it can. As examples, OAG can replace TPA as a comitogen with A23187 and PHA in macrophage-depleted lymphocytes (Kaibuchi et al., 1985) or with insulin in Swiss 3T3 fibroblasts (Rozengurt et al., 1984). Similarly OAG can stimulate ornithine decarboxylase activity (Jetten et al., 1985), c-myc gene expression (Kaibuchi et al., 1986), and hexose uptake (Farese et al., 1985). Like TPA, DAG can induce secretion of pulmonary surfactant in alveolar type II cells (Sano et al., 1985a) and several hormones from anterior pituitary cells (Conn et al., 1985; Negro-Vilar and Lapetina, 1985). OAG can stimulate Na^+ uptake and a rapid increase in intracellular pH by activating Na^+/H^+ exchange (Moolenaar et al., 1984a; Vara et al., 1985; Siffert and Scheid, 1986). OAG attenuates voltage-dependent Ca^{2+} currents in sensory neurons (Rane and Dunlap, 1986) and inhibits gap-junctional communication, as measured by dye transfer (Gainer and Murray, 1985). DAG can synergize with Ca^{2+} to elicit maximal release reactions from platelets (Kaibuchi et al., 1983; Kajikawa et al., 1983; Lapetina et al., 1985), mast cells (Katakami et al., 1984), rMTC 6-23 cells (Hishikawa et al., 1985), pituitary cells (Naor and Eli, 1985), and neutrophils (Penfield and Dale, 1984; Fujita et al., 1984) and can interact with Ca^{2+} in prolactin release from GH_3 cells (Ronning and Martin, 1986). Synthetic DAGs are capable of preventing gonadotropin-induced granulosa cell differentiation (Shinohara et al., 1985b), Friend cell differentiation (Faletto et al., 1985), histamine-induced acid secretion from parietal cells (Anderson and Hanson, 1985), follicle-stimulating hormone-induced progesterone production from granulosa cells (Veldhuis and Demers, 1986), and angiotensin-induced activation of phospholipase C (Brock et al., 1985). These results suggest that protein kinase C mediates many of the pleiotropic effects elicited by TPA.

However, there might be some exceptions. As discussed in Section 2.7.1, phorbol esters, in contrast to physiological stimuli, down-regulate protein kinase C. Thus, certain effects of phorbol esters could be due to the depletion of protein kinase C in addition to, or rather than, its activation. This explanation has been suggested for the antagonistic effects of phorbol esters in lymphocyte activation (Russell et al., 1986). The phorbol ester–induced differentiation of HL-60 cells might serve as another example of this phenomenon, although the use of synthetic DAGs to probe this question has led to conflicting results. On the one hand, OAG treatment does not induce the differentiation process even when applied numerous times at 30-min intervals (Kreutter et al., 1985; Yamamoto et al., 1985), whereas other synthetic 1,2-DAGs do (Ebeling et al., 1985). Because

OAG is metabolized so rapidly, data derived from the use of more stable synthetic DAGs are more influential. On the other hand, mutants of HL-60 cells provide additional evidence that transient activation of protein kinase C is insufficient to induce HL-60 differentiation. These mutant cells resist TPA-induced differentiation, apparently through a defect in the down-regulation of protein kinase C (Solanki *et al.*, 1981) (see Section 2.10). This suggests that it might be the removal of protein kinase C rather than, or in addition to, its activation, which is obligatory for HL-60 differentiation.

5. INTERACTION OF PROTEIN KINASE C WITH OTHER SIGNALING SYSTEMS

Thus far, we have discussed the properties of protein kinase C with respect to its role in mediating many of the effects of agonists of PI turnover and have described the consequences of abnormal regulation of this enzyme by compounds such as the phorbol esters. However, an understanding of the role of protein kinase C in normal cell growth requires consideration of the cellular metabolic interactions as a whole and not in defined isolation. Obviously, no one agonist operates on any tissue in the absence of other competing or complementary effectors. Circulating levels of hormones do not change in unison, and their individual effectiveness depends on the background levels of other agonists/antagonists. It is therefore important to discern the interactions between the different signal transduction pathways in order to understand the physiological role of a particular pathway.

5.1. Interaction with the cAMP Pathway

The transduction mechanism involving the generation of the second messenger $3',5'$-cAMP was the first to be characterized extensively and remains the model system for hormonal regulation of metabolism (for reviews, see Krebs and Beavo, 1979; Smigel *et al.*, 1985). Like part of the PI turnover pathway, the cAMP system operates by activating a serine/threonine–specific protein kinase, cAMP-dependent protein kinase. However, in many cells the two pathways elicit quite different responses, presumably, in part, by altering the phosphorylation state of distinct sets of proteins. Some interactions between the cAMP pathway and PI turnover have already been alluded to, namely the inhibitory effects of TPA on β-adrenergic receptors (Section 3.1.5). The inverse interaction also occurs, i.e., agonists of cAMP production in certain cell types act to inhibit breakdown of inositol phospholipids, thus preventing activation of protein kinase

C. For example, in neutrophils, elevation of cAMP inhibits PI turnover, phosphatidic acid production, and resynthesis of inositol phospholipids (Della-Bianca et al., 1986). In other cell types such as platelets and lymphocytes, there is evidence that cGMP is elevated upon TPA treatment and has similar inhibitory effects to cAMP (Coffey and Hadden, 1983; Nishizuka, 1983). In yet other cells, such as pinealocytes (Sugden et al., 1985) and cerebral cortex tissue (Hollingsworth et al., 1986), agonists of protein kinase C stimulate the cAMP pathway. For the most part, the molecular mechanisms for these negative and positive actions between the cyclic nucleotide and phosphoinositide pathways remain to be elucidated. However, the recent discovery that hormonal activation of phospholipase C is mediated through a guanine nucleotide–binding protein, (termed G_x or G_p) as yet uncharacterized, introduces a potential target for regulation by analogy with the G_s and G_i proteins of adenylate cyclase (Cockcroft and Gomberts, 1985; Litosch et al., 1985). It is clear, however, that these two second messenger systems do interact extensively, albeit in a rather complex cell-specific manner, which also reflects the adaptability of control pathways (Nishizuka, 1986).

5.2. Interactions between the Two Limbs of the Phosphatidylinositol Turnover Pathway

The generation of two second messengers upon agonist-induced breakdown of PI, i.e., DAG and IP_3, endows this pathway with a potential for separate regulation of each product through control of their metabolizing enzymes, such as DAG kinase and IP_3-5' phosphatase (see Section 3.2.2). Since the two second messengers operate through distinct mechanisms, this allows attenuation of one limb of the signal transduction process, without necessarily affecting the other. Indeed, the production of these two second messengers does not have to be coordinated. DAG could be produced in the absence of IP_3 formation by phospholipase C hydrolysis of inositol phospholipids other than 4,5-bisphosphophosphatidylinositol (PIP_2), since inositol 1-phosphate and inositol 1,4-bisphosphate, the hydrolysis products of the other inositol phospholipids, are biologically inactive (Streb et al., 1983). Although there is still disagreement with regard to the nature of the hormone-sensitive phospholipase, activities have been purified from several sources that vary in their ability to hydrolyze PI and PIP_2, depending on the Ca^{2+} concentration (Wilson et al., 1984; Nakanishi et al., 1985). The potential for such finite regulation is exemplified by the inhibitory action of phorbol esters on PI turnover (see Section 4.8) caused by feedback inhibition of PI hydrolysis catalyzed by protein kinase C (see Section 3.2.2). Other second messengers might also interfere with this signaling pathway in a manner similar to phorbol esters, allowing cross talk and modulation between different agonists.

As discussed in Section 2.4, IP_3 causes an elevation of intracellular Ca^{2+}. Ca^{2+} is regarded by many as a "third messenger" since a wide variety of cellular processes are sensitive to levels of this metal ion. Many of the effects of Ca^{2+} are mediated through the Ca^{2+}-binding protein, calmodulin. In the presence of Ca^{2+}, this protein binds to a series of target enzymes that include at least three protein kinases and a phosphoprotein phosphatase (for review, see Nairn *et al.*, 1985). IP_3, like DAG, therefore mediates changes in protein phosphorylation, albeit of a potentially different set of substrates.

Of course, Ca^{2+} levels are not only regulated by PI breakdown and therefore linked to DAG production. Influx of extracellular Ca^{2+} plays a role in a variety of processes and tissues. For example, skeletal muscle uses Ca^{2+} fluxes to regulate contractile processes. The role of PI turnover in such systems is somewhat unclear, since elevation of intracellular Ca^{2+} might be expected to activate Ca^{2+}-stimulated phospholipase C. However, protein kinase C levels are rather low in muscle tissue, and it is generally believed that Ca^{2+} influxes generated by muscle depolarization act primarily through calmodulin-dependent processes such as the activation of phosphorylase kinase and MLCK (Cohen, 1985).

5.3. Interactions between Protein Kinase C and Protein–Tyrosine Kinases

There are a number of observations that suggest interplay between protein kinase C and protein–tyrosine kinases. (1) As mentioned in Sections 3.1.1 and 3.2.1, the EGF receptor, $pp60^{src}$ and, possibly, $pp56^{lck}$, the LSTRA T-cell lymphoma kinase, which are all protein–tyrosine kinases, are substrates for protein kinase C. (2) A major cellular substrate for protein–tyrosine kinases, p36, is also a physiological substrate for protein kinase C (Section 3.2.5). (3) Treatment of certain cell lines with phorbol esters stimulates the phosphorylation of a 42-kD protein (termed p42) on tyrosine (Bishop *et al.*, 1983; Gilmore and Martin, 1983; Cooper *et al.*, 1984; Kohno, 1985). This latter observation indicates that protein kinase C must either activate a protein–tyrosine kinase or inhibit a protein–phosphotyrosine phosphatase. (4) TPA and synthetic DAGs are partial mitogens for certain cell lines and, thus, mimic some of the effects of polypeptide mitogens that act through receptors having protein–tyrosine kinase activity, such as PDGF (Dicker and Rozengurt, 1978). Similarly, the mitogenic activity of thrombin in hamster fibroblasts can be blocked by neomycin, an inhibitor of PIP_2 hydrolysis, indicating that PI turnover is required for initiation of DNA synthesis (Carney *et al.*, 1985).

These interactions suggest a close relationship between the pathways regulated by protein–tyrosine kinases and protein kinase C. It is likely that the two systems overlap at certain points, perhaps involving common substrates or the activation of convergent pathways. Indeed, since protein–tyrosine kinases have been demonstrated as substrates for protein kinase C, and TPA causes phosphorylation of p42 on tyrosine, it is possible that the mitogenic effects of phorbol esters are mediated via the same pathway as that used by polypeptide mitogens acting through protein–tyrosine kinases. The molecular mechanisms by which phorbol esters and polypeptide mitogens stimulate DNA synthesis and mitosis are unknown, but primary events are likely to involve the activation of specific transcription factors that promote transcription of certain genes containing appropriate regulatory elements. Solving the mechanism of action of one type of mitogen could provide an understanding of the other.

The mitogenic effects of protein kinase C (Section 4.3) are somewhat paradoxical, given the apparent propensity of the enzyme for negative regulation of pathways. Perhaps this indicates a dual role for the enzyme: desensitization to external signals and preparation for cell division. These two functions may serve to protect a cell uncommitted to cell division from growth inhibitory signals. This protective or isolative effect may underlie the property of phorbol esters to synergize with other mitogens in stimulating cell division in cells such as lymphocytes (Section 4.3.1). Alternatively, the synergistic effects might be due to activation of a completely distinct mitogenic pathway for phorbol esters.

6. CONCLUSIONS AND PERSPECTIVES

Literature searches for papers pertaining to protein kinase C indicate an almost exponential rate of publication since 1981. This explosion of information has led to significant advances in discerning the functions of protein kinase C and the molecular basis for the action of agonists of PI turnover and tumor promoters. Despite this surge of research, however, there are still many basic questions that remain unanswered. At a descriptive level, the activation of the enzyme is reasonably well understood, but the molecular aspects of DAG/tumor promoter, phospholipid and Ca^{2+} binding to the enzyme, and the way in which these sites interact to activate the enzyme are largely undefined. Mechanistic details of enzyme inactivation are also scanty. For instance, the relative contributions of direct reversal of activation by dissociation of DAG and irreversible inactivation by proteolysis have not been assessed. There is no doubt that the activity of protein kinase C is regulated by DAG, but we do not know whether the activity or the localization of the enzyme is modulated by other processes such as autophosphorylation and, more importantly, phosphorylation by other protein kinases.

The realization that the enzyme is not a single entity but, instead, a family of highly related proteins raises many questions. What is the function of these isozymes? Are they regulated differentially? Do they respond to the same ago-nists? Do they phosphorylate the same proteins? To what extent are they ex-pressed differentially in different cell types? One or more of the protein kinase C genes is expressed in almost every cell type, but the overall level of expression varies. As yet, nothing is known about what regulates the expression of the protein kinase C genes. The high level of protein kinase C in the brain suggests that the enzyme plays an important role in neural function, but there are no real clues as to the nature of this role.

We have learned the identity of a number of substrates of protein kinase C, but these seem to be largely part of negative feedback loops either regulating the PI cycle or the cross talk between growth factor receptors. No substrates have been described that are good candidates for participation in the stimulatory pathways induced by protein kinase C, such as mitogenesis, secretion, and neurotransmission. The subcellular location of protein kinase C may well dictate its access to substrates, but the exact location of the resting enzyme is unknown. In addition, it is unclear whether the DAG-activated enzyme remains at the plasma membrane or moves to other membrane compartments as well. Another lacuna is any understanding of the mechanism of nuclear signaling by activated protein kinase C. Does it involve the translocation of the enzyme or a phosphor-ylated substrate, or the stimulation of a diffusible messenger system?

There are many avenues to pursue in answering these questions. The newly isolated cDNA clones will enable site-directed mutagenesis to probe struc-ture/function relationships and provide a molecular basis for studies of the purified protein kinase. In addition, cDNA clones will allow us to establish the exact structural relatedness of the protein kinase C family members and to analyze the transcriptional and translational control of expression for each protein kinase C gene separately, particularly during development and differentiation. Expression of the cDNA clones in bacteria, yeast, or insect cells possibly may lead to the production of enough native enzyme for crystallographic structural studies to be contemplated. The availability of polyvalent and monoclonal antibodies to the enzyme (and its different forms) now allows investigation of the protein *in situ* by immunoelectron microscopy, including the precise subcellular locations of the unactivated and activated forms of the enzyme. A high priority will be the discovery of new substrates. Their identification may direct interest into other areas of metabolism, as well as solve the involvement of protein kinase C in processes such as mitogenesis and secretion. Identification of the genes that are transcriptionally activated by tumor promoters recently has allowed delineation of nucleotide sequences required for induction of these genes. The identification of the DNA-binding proteins that recognize these sequences and determination of how their binding is regulated will be important steps in elucidating the nuclear signaling pathway used by protein kinase C. Further probing of the relationship between protein kinase C and protein–tyrosine kinases may provide clues to the enigmatic mechanism of action of the latter enzymes in growth and other cellular processes.

There is every indication that protein kinase C has a central role in cellular growth control. Unfortunately, the difficulty in carrying out straightforward genetic analysis in mammalian systems has been a major limitation in determining the exact function of protein kinase C in growth. Mutant cells have been selected that are nonresponsive to TPA and, therefore, defective in the protein kinase C–induced mitogenic pathway. The mutants obtained so far appear to have functional protein kinase C but are affected at some step distal to the action of protein kinase C, the nature of which remains to be defined. The failure to isolate structural mutations for protein kinase C might imply that protein kinase C is essential. On the other hand, the existence of multiple genes for protein kinase C may preclude the ready isolation of such mutants, particularly if the isozymes have equivalent functions in tissue culture cells. An alternative way of obtaining cells deficient in protein kinase C would be to express the cDNA clones in antisense constructs driven by inducible promoters in case the absence of protein kinase C is lethal.

Given the near universal distribution of protein kinase C in eukaryotic organisms, one might hope to carry out genetic analysis of its function in a simple eukaryote more amenable to genetic manipulation. *D. melanogaster* would be the organism of choice, and, based on the precedents with other highly conserved protein kinase genes, the availability of cDNA clones to the mammalian enzyme should enable the cloning of the *Drosophila* gene. Whether protein kinase C will have similar functional roles in simpler metazoans and mammals remains to be seen.

The likelihood that protein kinase C plays a central role in growth control provokes the question of whether the protein kinase C genes are protooncogenes, i.e., would expression of an altered form of protein kinase C be sufficient to usurp normal growth control and transform cells? Although the definitive answer to this question must await transfection studies with protein kinase C cDNAs, one might predict that protein kinase C will earn the dubious honor of a three-letter designation, such as c-*pkc*, based on the mitogenic activity of tumor promoters, their ability to cause morphological changes, and the fact that PI turnover appears to be elevated in certain transformed cells, suggesting a function for protein kinase C in their transformed phenotype (Diringer and Friis, 1977; Macara *et al.*, 1984; Fry *et al.*, 1985; Preiss *et al.*, 1986). Although none of the characterized oncogenes are cognate with the protein kinase C genes, there are precedents for oncogenes that encode serine/threonine–specific protein kinases, such as the v-*mil/raf* and v-*mos* genes (Hunter, 1984a).

Under what circumstances might protein kinase C be an oncoprotein? Overexpression of the cellular protein–tyrosine kinases, which have oncogenic counterparts, does not result in transformation, apparently due to the ability of cells to regulate these enzymes in a negative fashion. Likewise, overexpression of protein kinase C *per se* is unlikely to result in transformation, since the enzyme would be largely inactive unless DAG were present continuously. Mutation or loss of negative regulatory domains appears to be necessary for the oncogenic activation of protein–tyrosine kinases. In the case of protein kinase C, however, the separate requirement for activators to translocate the enzyme to the membrane and cause stimulation means that very subtle mutations will be needed to generate a protein kinase C molecule that is spontaneously active and still able to bind to membranes. If membrane substrates are not critical for mitogenesis, however, a form analogous to the proteolytically activated fragment of protein kinase C (M kinase) might be oncogenic if stabilized against rapid degradation. If an unregulated form of protein kinase C can act as an oncoprotein, one might equally expect a constitutively activated form of phospholipase C or an inhibitor of DAG kinase (de Chaffoy de Courcelles *et al.,* 1985) to act in this manner since, in both cases, continuous activation of protein kinase C would result. However, this mode of activation might not be as effective as mutations in protein kinase C itself, since the DAG-activated protein kinase C would be subject to an enhanced proteolytic rate, as observed in TPA-treated cells.

In the arena of tumor promotion, the current dogma is that tumor promoters work solely through protein kinase C. This raises the issue of how the activation of protein kinase C by tumor promoters differs from activation by DAG. Agonist-induced production of DAG is tightly regulated by feedback mechanisms which, coupled with the rapid metabolism of DAG, limits the duration of protein kinase C activation. In contrast, tumor promoters are eliminated only slowly and have a higher affinity for the enzyme than DAG. As a result, tumor promoters cause a prolonged activation of protein kinase C, which may permit transport of the activated enzyme into membrane compartments normally inaccessible to protein kinase C. This might be sufficient to explain differences in cellular response to tumor promoters and DAG, but one should not dismiss the possibility that there is another much less abundant tumor promoter receptor that has escaped detection so far.

The interrelationship of the protein kinase C signal pathway with other signal systems has become an important field of research. The stimuli to which each cell is sensitive depends on the selective expression of receptors and their regulatory state (for example, the phosphate occupancy of threonine 654 of the EGF receptor). Thus, in an organism, the binding of extracellular effectors to their receptors is interpreted and integrated by an intracellular mediator network, resulting in a composite response tailored to the needs of each cell. This allows different cell types to react to the same stimulus in different ways, depending on the precise mediator pathways that are present. As additional substrates for protein kinase C are revealed, more may be learned about the spectrum of possible responses to agonists of PI turnover and the integration of the protein kinase C pathway with other signal transduction mechanisms.

So what has this detailed knowledge of protein kinase C added to our understanding of the umbrella subject of growth control and tumorigenesis? In our view, the most exciting aspect of research on protein kinase C is that it has drawn together such disparate disciplines as enzymology, tumor promotion, and signal transduction, in much the same way as the fields of viral oncogenes and growth factor receptors became strange bedfellows through their shared interest in tyrosine as a phosphate acceptor. Such interconnections provide us with a tantalizing first glimpse of the complex regulatory networks that remain to be unraveled.

ACKNOWLEDGMENTS. We thank numerous colleagues in our lab for their encouragement. We are especially grateful to Jon Cooper, Ellen Freed, John Glenney, Clare Isacke, Jill Meisenhelder, David Shalloway, and to those who communicated their results prior to publication. We also acknowledge the inspirational support provided by Jennifer, Caroline, Steve, and the Salmon River. This work was supported by U.S. Public Health Service Grant number CA 39780. J.R.W. was supported by SERC/NATO postdoctoral fellowship.

REFERENCES

Abdel-Latif, A. A., Akhtar, R. A., and Hawthorne, J. N., 1977, Acetylcholine increases the breakdown of triphosphoinositide of rabbit iris muscle prelabelled with [^{32}P]phosphate, *Biochem. J.* **162**:61–73.

Acres, R. B., Conlon, P. J., Mochizuki, D. Y., and Gallis, B., 1986, Rapid phosphorylation and modulation of the T4 antigen on cloned helper T cells induced by phorbol myristate acetate or antigen, *J. Biol. Chem.* **261**:16210–16214.

Acuto, O., and Reinherz, E. L., 1985, The human T-cells receptor: Structure and function, *N. Engl. J. Med.* **312**:1100–1112.

Adamo, S., Caporale, C., Aguanno, S., Lazdins, J., Faggioni, A., Belli, L., Cortesi, E., Nervi, C., Gastaldi, R., and Molinaro, M., 1986, Proliferating and quiescent cells exhibit different subcellular distribution of protein kinase C activity, *FEBS Lett.* **195**:352–356.

Ahmad, Z., Lee, F.-T., DePaoli-Roach, A., and Roach, P. J., 1984, Phosphorylation of glycogen synthase by the Ca^{2+}- and phospholipid-activated protein kinase (protein kinase C), *J. Biol. Chem.* **259**:8743–8747.

Akatories, K., and Jakobs, K. H., 1984, N$_i$-mediated inhibition of human platelet adenylate cyclase, *Eur. J. Biochem.* **145**:333–338.

Akers, R. F., and Routtenberg, A., 1985, Protein kinase C phosphorylates a 47 M$_r$ protein (F1) directly related to synaptic plasticity, *Brain Res.* **334**:147–151.

Akers, R. F., Lovinger, D. M., Colley, P. A., Linden, D. J., and Routtenberg, A., 1986, Translocation of protein kinase C activity may mediate hippocampal long-term potentiation, *Science* **231**:587–589.

Akhtar, R. A., and Abdel-Latif, A. A., 1980, Requirement for calcium ions in acetylcholine-stimulated phosphodiesteratic cleavage of phosphatidyl-*myo*inositol 4,5-bisphosphate in rabbit iris smooth muscle, *Biochem. J.* **192**:783–791.

Akiyama, T., Nishida, E., Ishida, J., Saji, N., Ogawara, H., Hoshi, M., Miyata, Y., and Sakai,

H., 1986, Purified protein kinase C phosphorylates microtubule-associated protein 2, *J. Biol. Chem.* **261**:15648–15651.

Albert, K. A., Wu, W. C.-S., Nairn, A. C., and Greengard, P., 1984a, Inhibition by calmodulin of calcium/phospholipid-dependent protein phosphorylation, *Proc. Natl. Acad. Sci. USA* **81**:3622–3625.

Albert, K. A., Helmer-Matyjek, E., Nairn, A. C., Muller, T. H., Haycock, J. W., Greene, L. A., Goldstein, M., and Greengard, P., 1984b, Calcium/phospholipid-dependent protein kinase (protein kinase C) phosphorylates and activates tyrosine hydroxylase, *Proc. Natl. Acad. Sci. USA* **81**:7713–7717.

Albert, K. A., Walaas, S. I., Wang, J. K.-T., and Greengard, P., 1986, Widespread occurrence of "87 kDa," a major specific substrate for protein kinase C, *Proc. Natl. Acad. Sci. USA* **83**:2822–2826.

Allard, W. J., Gibbs, E. M., Witters, L. A., and Lienhard, G. E., 1986, Increased glucose transport in response to phorbol ester, growth factors, and insulin: Relationship to phosphorylation of the glucose transporter, *Fed. Proc.* **45**:1552.

Aloyo, V. J., Zwiers, H., and Gispen, W. H., 1983, Phosphorylation of B-50 protein by calcium-activated, phospholipid-dependent protein kinase and B-50 protein kinase, *J. Neurochem.* **41**:649–653.

Anderson, N. G., and Hanson, P. J., 1985, Involvement of calcium-sensitive phospholipid-dependent protein kinase in control of acid secretion by isolated rat parietal cells, *Biochem. J.* **232**:609–611.

Anderson, N. L., Gemmell, M. A., Coussens, P. M., Murao, S.-I., and Huberman, E., 1985, Specific protein phosphorylation in human promyelocytic HL-60 leukemia cells susceptible or resistant to induction of cell differentiation by phorbol-12-myristate-13-acetate, *Cancer Res.* **45**:4955–4962.

Anderson, W. B., Estival, A., Tapiovaara, H., and Goopalakristna, R., 1985, Altered subcellular distribution of protein kinase C (a phorbol ester receptor). Possible role in tumor promotion and the regulation of cell growth: Relationship to changes in adenylate cyclase activity, *Adv. Cyc. Nuc. Res.* **19**:287–306.

Andrews, P. C., and Babior, B. M., 1983, Endogenous protein phosphorylation by resting and activated human neutrophils, *Blood* **61**:333–340.

Angel, P., Imagawa, M., Chiu, R., Stein, B., Imbra, R. J., Rahmsdorf, H. J., Jonat, C., Herrlich, P., and Karin, M., 1987, Phorbol ester-inducible genes contain a common *cis* element recognized by a TPA-modulated *trans*-acting factor, *Cell* **49**:729–739.

Antler, A. M., Greenberg, M. E., Edelman, G. M., and Hanafusa, H., 1985, Increased phosphorylation of tyrosine in vinculin does not occur upon transformation by some avian sarcoma viruses, *Mol. Cell. Biol.* **5**:263–267.

Arcoleo, J. P., and Weinstein, I. B., 1985, Activation of protein kinase C by tumor promoting phorbol esters, teleocidin, and aplysiatoxin in the absence of added calcium, *Carcinogenesis* **6**:213–217.

Arino, J., and Guinovart, J. J., 1986, Phosphorylation and inactivation of rat hepatocyte glycogen synthase by phorbol esters and mezerein, *Biochem. Biophys. Res. Commun.* **134**:113–119.

Ashby, C. D., and Walsh, D. A., 1972, Characterization of the interaction of a protein inhibitor with adenosine 3′,5′-monophosphate–dependent protein kinases, *J. Biol. Chem.* **247**:6637–6642.

Ashendel, C. L., 1985, The phorbol ester receptor: A phospholipid-regulated protein kinase, *Biochim. Biophys. Acta* **822**:219–242.

Ashendel, C. L., Staller, J. M., and Boutwell, R. K., 1983a, Identification of a calcium- and phospholipid-dependent phorbol ester binding activity in the soluble fraction of mouse tissues, *Biochem. Biophys. Res. Commun.* **111**:340–345.

Ashendel, C. L., Staller, J. M., and Boutwell, R. K., 1983b, Protein kinase activity associated with a phorbol ester receptor purified from mouse brain, *Cancer Res.* **43**:4333–4337.

Avissar, S., Shanitzki, B., Stenzel, K. H., and Novogrodsky, A., 1986, Early effects of TPA on protein kinase activity in murine thymocytes, *Exp. Cell Res.* **165:**353–361.

Avruch, J., Nemenoff, R. A., Blackshear, P. J., Pierce, M. N., and Osathanondh, R., 1982, Insulin-stimulated tyrosine phosphorylation of the insulin receptor in detergent extracts of human placental membranes, *J. Biol. Chem.* **257:**15162–15166.

Backer, J. M., Arcoleo, J. P., and Weinstein, I. B., 1986, Protein phosphorylation in isolated mitochondria and the effects of protein kinase C, *FEBS Lett.* **200:**161–164.

Balk, S. D., Morisi, A., and Gunther, H. S., 1984, Phorbol 12-myristate 13-acetate, ionomycin, or ouabain, and raised extracellular magnesium induce proliferation of chicken heart mesen-chymal cells, *Proc. Natl. Acad. Sci. USA* **81:**6418–6421.

Ballester, R., and Rosen, O. M., 1985, Fate of immunoprecipitable protein kinase C in GH₃ cells treated with phorbol 12-myristate 13-acetate, *J. Biol. Chem.* **260:**15194–15199.

Baraban, J. M., Synder, S. H., and Alger, B. E., 1985, Protein kinase C regulates ionic conductance in hippocampal pyramidal neurons: Electrophysiological effects of phorbol esters, *Proc. Natl. Acad. Sci. USA* **82:**2538–2542.

Barbacid, M., and Lauver, A. V., 1981, Gene products of McDonough feline sarcoma virus have an *in vitro*-associated protein kinase that phosphorylates tyrosine residues: Lack of detection of this enzymatic activity *in vivo*, *J. Virol.* **40:**812–821.

Barrowman, M. M., Cockcroft, S., and Gomperts, B. D., 1986, Potentiation and inhibition of secretion from neutrophils by phorbol ester, *FEBS Lett.* **201:**137–142.

Barsoum, J., and Varshavsky, A., 1983, Mitogenic hormones and tumor promoters greatly increase the incidence of colony-forming cells bearing amplified dihydrofolate reductase genes, *Proc. Natl. Acad. Sci. USA* **80:**5330–5334.

Baukal, A. J., Guillemette, G., Rubin, R., Spat, A., and Catt, K. J., 1985, Binding sites for inositol trisphosphate in the bovine adrenal cortex, *Biochem. Biophys. Res. Commun.* **133:**532–538.

Beaven, M. A., Moore, J. P., Smith, G. A., Hesketh, T. R., and Metcalfe, J. C., 1984, The calcium signal and phosphatidylinositol breakdown in 2H3 cells, *J. Biol. Chem.* **259:**7137–7142.

Becker, M. A., Dicker, P., and Rozengurt, E., 1983, Mitogenic enhancement of purine base phosphoribosylation in Swiss mouse 3T3 cells, *Am. J. Physiol.* **244:**C288–C296.

Beg, Z. H., Stonik, J. A., and Brewer, Jr., H. B., 1985, Phosphorylation of hepatic 3-hydroxy-3-methylglutaryl coenzyme A reductase and modulation of its enzymic activity by calcium-activated and phospholipid-dependent protein kinase, *J. Biol. Chem.* **260:**1682–1687.

Beguinot, L., Hanover, J. A., Ito, S., Richert, N. D., Willingham, M. C., and Pastan, I., 1985, Phorbol esters induce transient internalization without degradation of unoccupied epidermal growth factor receptors, *Proc. Natl. Acad. Sci. USA* **82:**2774–2778.

Bell, J. D., Buxton, I. O., and Brunton, L. L., 1985, Enhancement of adenylate cyclase activity in S49 lymphoma cells by phorbol esters, *J. Biol. Chem.* **260:**2625–2628.

Bell, R. L., Kennerly, D. A., Stanford, N., and Majerus, P. W., 1979, Diglyceride lipase: A pathway for arachidonate release from human platelets, *Proc. Natl. Acad. Sci. USA* **76:**3238–3241.

Bennett, W. F., Belville, J. S., and Lynch, G., 1979, A study of protein phosphorylation in shape change and Ca^{++}-dependent serotonin release by blood platelets, *Cell* **18:**1015–1023.

Benovic, J. L., Strasser, R. H., Caron, M. G., and Lefkowitz, R. J., 1986, β-adrenergic receptor kinase: Identification of a novel protein kinase that phosphorylates the agonist-occupied form of the receptor, *Proc. Natl. Acad. Sci. USA* **83:**2797–2801.

Ben-Ze'ev, A., 1986, Tumor promoter–induced disruption of junctional complexes in cultured epithelial cells is followed by the inhibition of cytokeratin and desmoplakin synthesis, *Exp. Cell Res.* **164:**335–352.

Berridge, M. J., 1983, Rapid accumulation of inositol trisphosphate reveals that agonists hydrolyse polyphosphatidylinositol instead of phosphatidylinositol, *Biochem. J.* **212:**849–858.

Berridge, M. J., 1984, Inositol trisphosphate and diacylglycerol as second messengers, *Biochem. J.* **220**:345–360.

Berridge, M. J., and Irvine, R. F., 1984, Inositol trisphosphate, a novel second messenger in cellular signal transduction, *Nature* **312**:315–321.

Berridge, M. J., Dawson, R. M. C., Downes, C. P., Heslop, J. P., and Irvine, R. F., 1983, Changes in the levels of inositol phosphates after agonist-dependent hydrolysis of membrane phosphoinositides, *Biochem. J.* **212**:473–482.

Berridge, M. J., Heslop, J. P., Irvine, R. F., and Brown, K. D., 1984, Inositol trisphosphate formation and calcium mobilization in Swiss 3T3 cells in response to platelet-derived growth factor, *Biochem. J.* **222**:195–201.

Besterman, J. M., and Cuatrecasas, P., 1984, Phorbol esters rapidly stimulate amiloride-sensitive Na^+/H^+ exchange in a human leukemic cell line, *J. Cell Biol.* **99**:340–343.

Billah, M. M., and Lapetina, E. G., 1982, Rapid decrease of phosphatidylinositol 4,5-bisphosphate in thrombin-stimulated platelets, *J. Biol. Chem.* **257**:12705–12708.

Bishop, R., Martinez, R., Nakamura, K., and Weber, M. J., 1983, A tumor promoter stimulates phosphorylation on tyrosine, *Biochem. Biophys. Res. Commun.* **115**:536–543.

Bishop, R., Martinez, R., Weber, M. J., Blackshear, P. J., Beatty, S., Lim, R., and Herschman, H. R., 1985, Protein phosphorylation in tetradecanoyl phorbol acetate–nonproliferative variant of 3T3 cells, *Mol. Cell. Biol.* **5**:2231–2237.

Blackmore, P. F., Strickland, W. G., Bocckino, S. B., and Exton, J. H., 1986, Mechanism of hepatic glycogen synthase inactivation induced by Ca^{2+}-mobilizing hormones, *Biochem. J.* **237**:235–242.

Blackshear, P. J., Witters, L. A., Girard, P. R., Kuo, J. F., and Quamo, S. N., 1985, Growth factor–stimulated protein phosphorylation in 3T3-L1 cells: Evidence for protein kinase C–dependent and –independent pathways, *J. Biol. Chem.* **260**:13304–13315.

Blackshear, P. J., Wen, L., Glynn, B. P., and Witters, L. A., 1986, Protein kinase C–stimulated phosphorylation *in vitro* of a M_r 80,000 protein phosphorylated in response to phorbol esters and growth factors in intact fibroblasts: Distinction from protein kinase C and prominence in brain, *J. Biol. Chem.* **261**:1459–1469.

Blake, A. D., and Strader, C. D., 1986, Potentiation of specific association of insulin with HepG2 cells by phorbol esters, *Biochem. J.* **236**:227–234.

Blenis, J., and Erikson, R. L., 1985, Regulation of a ribosomal protein S6 kinase activity by the Rous sarcoma virus transforming protein, serum, or phorbol ester, *Proc. Natl. Acad. Sci. USA* **82**:7621–7625.

Blenis, J., Spivack, J. G., and Erikson, R. L., 1984, Phorbol ester, serum, and Rous sarcoma virus transforming gene product induce similar phosphorylations of ribosomal protein S6, *Proc. Natl. Acad. Sci. USA* **81**:6408–6412.

Blumberg, P. M., Driedger, P. E., and Rossow, P. W., 1976, Effect of a phorbol ester on a transformation-sensitive surface protein of chick fibroblasts, *Nature* **264**:446–447.

Blumberg, P. M., Jaken, S., Konig, B., Sharkey, N. A., Leach, K. L., Jeng, A. Y., and Yeh, E., 1984, Mechanism of action of the phorbol ester tumor promoters: Specific receptors for lipophilic ligands, *Biochem. Pharmacol.* **33**:933–940.

Bojan, F., Kinsella, A. R., and Fox, M., 1983, Effect of tumor promoter 12-O-tetradecanoylphorbol-13-acetate on recovery of methotrexate-, N-(phosphonacetyl)-L-aspartate-, and cadmium-resistant colony-forming mouse and hamster cells, *Cancer Res.* **43**:5217–5221.

Bolen, J. B., Thiele, C. J., Israel, M. A., Yonemoto, W., Lipsich, L. A., and Brugge, J. S., 1984, Enhancement of cellular *src* gene product associated tyrosyl kinase activity following polyoma virus infection and transformation, *Cell* **38**:767–777.

Bolen, J. B., Rosen, N., and Israel, M. A., 1985, Increased $pp60^{c-src}$ tyrosyl kinase activity in human neuroblastomas is associated with aminoterminal tyrosine phosphorylation of the src gene product, *Proc. Natl. Acad. Sci. USA* **82**:7275–7279.

Bollag, G. E., Roth, R. A., Beaudoin, J., Mochly-Rosen, D., and Koshland, D. E., 1986, Protein kinase C directly phosphorylates the insulin receptor *in vitro* and reduces its protein–tyrosine kinase activity, *Proc. Natl. Acad. Sci. USA* **83**:5822–5824.

Boreiko, C., Mondal, S., Narayan, K. S., and Heidelberger, C., 1980, Effect of 12-O-tetradeca-noylphorbol-13-acetate on the morphology and growth of C3H/10T1/2 mouse embryo cells, *Cancer Res.* **40**:4709–4716.

Bourguignon, L. Y. W., Walker, G., and Bourguignon, G. J., 1985a, Phorbol ester–induced phosphorylation of a transmembrane glycoprotein (GP 180) in human blood platelets, *J. Biol. Chem.* **260**:11775–11780.

Bourguignon, L. Y. W., Field, S., and Bourguignon, G. J., 1985b, Phosphorylation of a tropom-yosin-like (30 KD) protein during platelet activation, *J. Cell. Biochem.* **29**:19–30.

Boutwell, R. K., 1974, The function and mechanism of promoters of carcinogenesis, *CRC Crit. Rev. Toxicol.* **2**:419–443.

Brasseur, R., Cabiaux, V., Huart, P., Castagna, M., Baztar, S., and Ruysschaert, J. M., 1985, Structural analogies between protein kinase C activators, *Biochem. Biophys. Res. Commun.* **127**:969–976.

Bravo, R., Burckhardt, J., Curran, T., and Muller, R., 1985, Stimulation and inhibition of growth by EGF in different A431 cell clones is accompanied by the rapid induction of c-*fos* and c-*myc* proto-oncogenes, *EMBO J.* **4**:1193–1197.

Brock, T. A., Rittenhouse, S. E., Powers, C. W., Ekstein, L. S., Gimbrone, M. A., Jr., and Alexander, R. W., 1985, Phorbol ester and 1-oleoyl-2-acetylglycerol inhibit angiotensin acti-vation of phospholipase C in cultured vascular smooth muscle cells, *J. Biol. Chem.* **260**:14158–14162.

Brocklehurst, K. W., and Pollard, H. B., 1985, Enhancement of Ca^{2+}-induced catecholamine release by the phorbol ester TPA in digitonin-permeabilized cultured bovine adrenal chromaffin cells, *FEBS Lett.* **183**:107–110.

Brown, K., Quintanilla, M., Ramsden, M., Kerr, I. B., Young, S., and Balmain, A., 1986, v-*ras* genes for Harvey and BALB murine sarcoma viruses can act as initiators of two-stage mouse skin carcinogenesis, *Cell* **46**:447–456.

Brown, K. D., Dicker, P., and Rozengurt, E., 1979, Inhibition of epidermal growth factor binding to surface receptors by tumor promoters, *Biochem. Biophys. Res. Commun.* **86**:1037–1043.

Brown, K. D., Blay, J., Irvine, R. F., Heslop, J. D., and Berridge, M. J., 1984, Reduction of epidermal growth factor receptor affinity by heterologous ligands: Evidence for a mechanism involving the breakdown of phosphoinositides and the activation of protein kinase C, *Biochem. Biophys. Res. Commun.* **123**:377–384.

Brugge, J. S., Cotton, P. C., Queral, A. E., Barrett, J. N., Nonner, D., and Keane, R. W., 1985, Neurones express high levels of a structurally modified, activated form of $pp60^{c-src}$, *Nature* **316**:555–557.

Burgess, S. K., Sayhoun, N., Blanchard, S. G., LeVine III, H., Chang, K.-J., and Cuatrecasas, P., 1986, Phorbol ester receptors and protein kinase C in primary neuronal cultures: Development and stimulation of endogenous phosphorylation, *J. Cell Biol.* **102**:312–319.

Burkhard, S. J., and Traugh, J. A., 1983, Changes in ribosome function by cAMP-dependent and cAMP-independent phosphorylation of ribosomal protein S6, *J. Biol. Chem.* **258**:14003–14008.

Burnham, D. B., Munowitz, P., Hootman, S. R., and Williams, J. A., 1986, Regulation of protein phosphorylation in pancreatic acini: Distinct effects of Ca^{2+} ionophore A23187 and 12-O-tetradecanoylphorbol 13-acetate, *Biochem. J.* **235**:125–131.

Burns, C. P., and Rozengurt, E., 1983, Serum, platelet-derived growth factor, vasopressin, and phorbol esters increase intracellular pH in Swiss 3T3 cells, *Biochem. Biophys. Res. Commun.* **116**:931–938.

Burridge, K., and Connell, L., 1983, A new protein of adhesion plaques and ruffling membranes, *J. Cell Biol.* **97**:359–367.

Burridge, K., and Feramisco, J. R., 1980, Microinjection and localization of a 130K protein in living fibroblasts: A relationship to actin and fibronectin, *Cell* **19**:587–595.

Burridge, K., and Mangeat, P., 1984, An interaction between vinculin and talin, *Nature* **308**:744–746.

Burstein, D. E., Blumberg, P. M., and Greene, L. A., 1982, Nerve growth factor–induced neuronal differentiation of PC12 pheochromocytoma cells: Lack of inhibition by a tumor promoter, *Brain Res.* **247**:115–119.

Buss, J. E., Kamps, M. P., Gould, K., and Sefton, B. M., 1986, The absence of myristic acid decreases membrane binding of pp60^{src} but does not affect tyrosine protein kinase activity, *J. Virol.* **58**:468–474.

Butler, A. P., Byus, C. V., and Slaga, T. J., 1986, Phosphorylation of histones is stimulated by phorbol esters in quiescent Reuber H35 hepatoma cells, *J. Biol. Chem.* **261**:9421–9425.

Butler-Gralla, E., and Herschmann, H. R., 1981, Variants of 3T3 cells lacking mitogenic response to the tumor promoter tetradecanoyl-phorbol-acetate, *J. Cell. Physiol.* **107**:59–68.

Buys, S. S., Keogh, E. A., and Kaplan, J., 1984, Fusion of intracellular membrane pools with cell surfaces of macrophages stimulated by phorbol esters and calcium ionophores, *Cell* **38**:569–576.

Cabral, F., Gottesman, M. M., and Yuspa, S. H., 1981, Induction of specific protein synthesis by phorbol esters in mouse epidermal cell culture, *Cancer Res.* **41**:2025–2031.

Campbell, D. G., Hardie, D. G., and Vulliet, P. R., 1986, Identification of four phosphorylation sites in the N-terminal region of tyrosine hydroxylase, *J. Biol. Chem.* **261**:10489–10492.

Cantrell, D. A., Davies, A. A., and Crumpton, M. J., 1985, Activators of protein kinase C down-regulate and phosphorylate the T3/T-cell antigen receptor complex of human T lymphocytes, *Proc. Natl. Acad. Sci. USA* **82**:8158–8162.

Carney, D. H., Scott, D. L., and Gordon, E. A., and LaBelle, E. F., 1985, Phosphoinositides in mitogenesis: Neomycin inhibits thrombin-stimulated phosphoinositide turnover and initiation of cell proliferation, *Cell* **42**:479–488.

Carroll, R. C., and Gerrard, J. M., 1982, Phosphorylation of platelet actin-binding protein during platelet activation, *Blood* **59**:466–471.

Carroll, R. C., Butler, R. G., Morris, P. A., and Gerrard, J. M., 1982, Separable assembly of platelet pseudopodal and contractile cytoskeletons, *Cell* **30**:385–393.

Cartwright, C. A., Hutchinson, M. A., and Eckhart, W., 1985, Structural and functional modification of pp60^{c-src} associated with polyoma middle tumor antigen from infected or transformed cells, *Mol. Cell. Biol.* **5**:2647–2652.

Cartwright, C. A., Kaplan, P. L., Cooper, J. A., Hunter, T., and Eckhart, W., 1986, Altered sites of tyrosine phosphorylation in pp60^{c-src} associated with polyoma virus middle tumor antigen, *Mol. Cell. Biol.* **6**:1562–1570.

Cartwright, C. A., Eckhart, W., Simon, S., and Kaplan, P. L., 1987a, Cell transformation by pp60^{c-src} mutated in the carboxy-terminal regulatory domain, *Cell* **49**:83–91.

Cartwright, C. A., Simantov, R., Kaplan, P. L., Hunter, T., and Eckhart, W., 1987b, Alterations in pp60^{c-src} accompany differentiation of neurons from rat embryo striatum, *Mol. Cell. Biol.* **7**:1830–1840.

Casnellie, J. E., and Lamberts, R. J., 1986, Tumor promoters cause changes in the state of phosphorylation and apparent molecular weight of a tyrosine protein kinase in T lymphocytes, *J. Biol. Chem.* **261**:4921–4925.

Castagna, M., Takai, Y., Kaibuchi, K., Sano, K., Kikkawa, U., and Nishizuka, Y., 1982, Direct activation of calcium-activated, phospholipid-dependent protein kinase by tumor-promoting phorbol esters, *J. Biol. Chem.* **257**:7847–7851.

Chen, B. D.-M., and Wilkins, K. L., 1985, Role of phorbol ester receptors in the 12-O-tetradecanoyl-phorbol-13-acetate (TPA)-induced down-regulation of colony-stimulating factor (CSF-1) binding to murine peritoneal exudate macrophages, *J. Cell. Physiol.* **124**:305–312.

Chiang, T. M., Cagen, L. M., and Kang, A. H., 1981, Effects of 12-O-tetradecanoyl phorbol 13-acetate on platelet aggregation, *Thromb. Res.* **21**:611–622.

Chida, K., Hashiba, H., Sasaki, K., and Kuroki, T., 1986a, Activation of protein kinase C and specific phosphorylation of a M$_r$ 90,000 membrane protein of promotable BALB/3T3 and C3H/10T1/2 cells by tumor promoters, *Cancer Res.* **46**:1055–1062.

Chida, K., Kato, N., and Kuroki, T, 1986b, Down regulation of phorbol diester receptors by proteolytic degradation of protein kinase C in a cultured line of fetal rat skin keratinocytes, *J. Biol. Chem.* **261**:13013–13018.

Christiansen, N. O., and Juhl, H., 1986, Purification and properties of protein kinase C from bovine polymorphonuclear leucocytes, *Biochim. Biophys. Acta* **885**:170–175.

Chu, D. T. W., and Granner, D. K., 1986, The effect of phorbol esters and diacylglycerol on expression of the phosphoenolpyruvate carboxykinase (GTP) gene in rat hepatoma H4IIE cells, *J. Biol. Chem.* **261**:16848–16853.

Ciudad, C., Camici, M., Ahmad, Z., Wang, YL., DePaoli-Roach, A., and Roach, P. J., 1984, Control of glycogen synthase phosphorylation in isolated rat hepatocytes by epinephrine, vasopressin, and glucagon, *Eur. J. Biochem.* **142**:511–520.

Cobb, M. H., and Rosen, O. M., 1983, Description of a protein kinase derived from insulin-treated 3T3-L1 cells that catalyzes the phosphorylation of ribosomal protein S6 and casein, *J. Biol. Chem.* **258**:12472–12481.

Cochet, C., Gill, G. N., Meisenhelder, J., Cooper, J. A., and Hunter, T., 1984, C-kinase phosphorylates the epidermal growth factor receptor and reduces its epidermal growth factor–stimulated tyrosine protein kinase activity, *J. Biol. Chem.* **259**:2553–2558.

Cockroft, S., and Gomperts, B. D., 1985, Role of guanine nucleotide binding protein in the activation of polyphosphoinositide phosphodiesterase, *Nature* **314**:534–536.

Cockroft, S., Bennett, J. P., and Gomperts, B. D., 1981, The dependence of Ca^{2+} of phosphatidylinositol breakdown and enzyme secretion in rabbit neutrophils stimulated by formylmethionyl–leucylphenylalanine or ionomycin, *Biochem. J.* **200**:501–508.

Coffey, R. G., and Hadden, J. W., 1983, Phorbol myristate acetate stimulation of lymphocyte guanylate cyclase and cyclic guanosine 3′:5′-monophosphate phosphodiesterase and reduction of adenylate cyclase, *Cancer Res.* **43**:150–158.

Cohen, C. M., and Foley, S. F., 1986, Phorbol ester- and Ca^{2+}-dependent phosphorylation of human red cell membrane skeletal proteins, *J. Biol. Chem.* **261**:7701–7709.

Cohen, P., 1982, The role of protein phosphorylation in neural and hormonal control of cellular activity, *Nature* **296**:613–620.

Cohen, P., 1985, The role of protein phosphorylation in the hormonal control of enzyme activity, *Eur. J. Biochem.* **151**:439–448.

Cohen, R., Pacifici, M., Rumbinstein, N., Biehl, J., Holtzer, H., 1977, Effect of a tumour promoter on myogenesis, *Nature* **266**:538–540.

Colamonici, O. R., Trepel, J. B., Vidal, C. A., and Neckers, L. M., 1986, Phorbol ester induces c-*sis* gene transcription in stem cell line K-562, *Mol. Cell. Biol.* **6**:1847–1850.

Colburn, N. H., Lau, S., and Head, R., 1975, Decrease of epidermal histidase activity by tumor-promoting phorbol esters, *Cancer Res.* **35**:3154–3159.

Colburn, N. H., Former, B. F., Nelson, K. A., and Yuspa, S. H., 1979, Tumour promoter induces anchorage independence irreversibly, *Nature* **281**:589–591.

Colburn, N. H., Lerman, M. L., Hegamyer, G. A., Yao, K.-T., Dowjat, W. K., and Shimada, T., 1986, Genes that cooperate with tumor promoters in transformation, *J. Cell. Biochem.* **10**(suppl. C):100.

Collett, M. S., and Erikson, R. L., 1978, Protein kinase activity associated with the avian sarcoma virus *src* gene product, *Proc. Natl. Acad. Sci. USA* **75**:2021–2024.

Collins, M. K. L., and Rozengurt, E., 1982, Binding of phorbol esters to high affinity sites on murine fibroblastic cells elicits a mitogenic response, *J. Cell. Physiol.* **112**:42–50.

Conn, P. M., Ganong, B. R., Ebeling, J., Staley, D., Neidel, J. E., and Bell, R. M., 1985,

Diacylglycerols release LH: Structure–activity relations reveal a role for protein kinase C, *Biochem. Biophys. Res. Commun.* **126:**532–539.

Connolly, T. M., Lawing, W. J., and Majerus, P. W., 1986, Protein kinase C phosphorylates human platelet inositol 5′-phosphomonoesterase, increasing the phosphatase activity, *Cell* **46:**951–958.

Cooper, J. A., and Hunter, T., 1983, Regulation of cell growth and transformation by tyrosine-specific protein kinases: The search for important cellular substrate proteins, *Curr. Top. Microbiol. Immunol.* **107:**125–161.

Cooper, J. A., and King, C. S., 1986, Dephosphorylation or antibody binding to the carboxy terminus stimulates $pp60^{c-src}$, *Mol. Cell. Biol.* **6:**4467–4477.

Cooper, J. A., Sefton, B. M., and Hunter, T., 1984, Diverse mitogenic agents induce the phosphorylation of two related 42,000-dalton proteins on tyrosine in quiescent chick cells, *Mol. Cell. Biol.* **4:**30–37.

Cooper, J. A., Hunter, T., and Shalloway, D., 1985, Protein–tyrosine kinase activity of $pp60^{c-src}$ is restricted in intact cells, in: *Cancer Cells,* Vol. 3 (J. Feramisco, C. Stiles, and B. Ozanne, eds.), Cold Spring Harbor Laboratory, Cold Spring Harbor, New York, pp. 321–328.

Cooper, J. A., Gould, K. L., Cartwright, C. A., and Hunter, T., 1986, Tyr^{527} is phosphorylated in $pp60^{c-src}$: Implications for regulation, *Science* **231:**1431–1433.

Cooper, R. H., Coll, K. E., and Williamson, J. R., 1985, Differential effect of phorbol esters on phenylephrine and vasopressin-induced Ca^{2+} mobilization in isolated hepatocytes, *J. Biol. Chem.* **260:**3281–3288.

Cope, F. O., Staller, J. M., Mahsem, R. A., and Boutwell, R. K., 1984, Retinoid-binding proteins are phosphorylated *in vitro* by soluble Ca^{+2}- and phosphatidylserine-dependent protein kinase from mouse brain, *Biochem. Biophys. Res. Commun.* **120:**593–601.

Corvera, S., Schwartz, K. R., Graham, R. M., and Garcia-Sainz, J. A., 1986a, Phorbol esters inhibit α_1-adrenergic effects and decrease the affinity of liver cell α_1-adrenergic receptors for (−)-epinephrine, *J. Biol. Chem.* **261:**520–526.

Corvera, S., Whitehead, R. E., Mottola, C., and Czech, M. P., 1986b, The insulin-like growth factor II receptor is phosphorylated by a tyrosine kinase in adipocyte plasma membranes, *J. Biol. Chem.* **261:**7675–7679.

Cosman, D., Cerretti, D. P., Larsen, A., Park, L., March, C., Dower, S., Gillis, S., and Urdal, D., 1984, Cloning, sequence, and expression of human interleukin-2 receptor, *Nature* **312:** 768–771.

Costa-Casnellie, M. R., Segel, G. B., and Lichtman, M. A., 1985, Concanavalin A and phorbol ester cause opposite subcellular redistribution of protein kinase C, *Biochem. Biophys. Res. Commun.* **133:**1139–1144.

Cotton, P. C., and Brugge, J. S., 1983, Neural tissues express high levels of the cellular *src* gene product $pp60^{c-src}$, *Mol. Cell. Biol.* **3:**1157–1162.

Coughlin, S. R., Lee, W. M. F., Williams, P. W., Giels, G. M., and Williams, L. T., 1985, c-*myc* gene expression is stimulated by agents that activate protein kinase C and does not account for the mitogenic effect of PDGF, *Cell* **43:**243–251.

Courtneidge, S. A., 1985, Activation of the $pp60^{c-src}$ kinase by middle T antigen binding or by dephosphorylation, *EMBO J.* **4:**1471–1477.

Coussens, L., Parker, P. J., Rhee, L., Yang-Feng, T. L., Chen, E., Waterfield, M. D., Francke, U., and Ullrich, A., 1986a, Multiple, distinct forms of bovine and human protein kinase C suggest diversity in cellular signaling pathways, *Science* **233:**859–866.

Coussens, L., Van Beveren, C., Smith, D., Chen, E., Mitchell, R. L., Isacke, C. M., Verma, I. M., and Ullrich, A., 1986b, Structural alteration of viral homologue of receptor proto-oncogene *fms* at carboxyl terminus, *Nature* **320:**269–272.

Coussens, P. M., Cooper, J. A., Hunter, T., and Shalloway, D., 1985, Restriction of the *in vitro* and *in vivo* tyrosine protein kinase activities of pp60^{c-src} relative to pp60^{v-src}, *Mol. Cell. Biol.* **5:**2753–2763.

Cox, A. C., and Carroll, R. C., 1986, The effect of tetradecanoylphorbol acetate on calcium–ion mobilization, protein phosphorylation, and cytoskeletal assembly induced by thrombin or arachidonate, *Biochim. Biophys. Acta* **886:**390–398.

Creba, J. A., Downes, C. P., Hawkins, P. T., Brewster, G., Michell, R. H., and Kirk, C. J., 1983, Rapid breakdown of phosphatidylinositol 4-phosphate and phosphatidylinositol 4,5-bis-phosphate in rat hepatocytes stimulated by vasopressin and other Ca^{2+} mobilizing hormones, *Biochem. J.* **212:**733–747.

Cremins, J., Wagner, J. A., and Halegoua, S., 1986, Nerve growth factor action is mediated by cyclic AMP– and Ca^{2+}/phospholipid-dependent protein kinases, *J. Cell Biol.* **103:**887–893.

Creutz, C. E., Dowling, L. G., Sando, J. J., Villar-Palasi, C., Whipple, J. H., and Zaks, W. J., 1983, Characterization of the chromobindins. Soluble proteins that bind to the chromaffin granule membrane in the presence of Ca^{2+}, *J. Biol. Chem.* **258:**14664–14674.

Creutz, C. E., Zaks, W. J., Hamman, H. C., Crane, S., Martin, W. H., Gould, K. L., Oddie, K. M., and Pawson, S. J., 1987, Identification of chromaffin granule-binding proteins, *J. Biol. Chem.* **262:**1860–1868.

Croop, J., Toyama, Y., Dlugosz, A. J., and Holtzer, H., 1980, Selective effects of phorbol 12-myristate 13-acetate on myofibrils and 10-nm filaments. *Proc. Natl. Acad. Sci. USA* **77:**5273–5277.

Culty, M., Vilgrain, I., and Chambaz, E. M., 1984, Steroidogenic properties of phorbol ester and a Ca^{2+} ionophore in bovine adrenocortical cell suspensions, *Biochem. Biophys. Res. Commun.* **121:**499–506.

Czernilofsky, A. P., Levinson, A. D., Varmus, H. E., Bishop, J. M., Tischer, E., and Goodman, H. M., 1980, Nucleotide sequence of an avian sarcoma virus oncogene (src) and proposed amino acid sequence for gene product, *Nature* **287:**198–203.

Dalgleish, A. G., Beverley, P. C. L., Clapham, P. R., Crawford, D. H., Greaves, M. F., and Weiss, R. A., 1984, The CD4 (T4) antigen is an essential component of the receptor for the AIDS retrovirus, *Nature* **312:**763–767.

Daniel, J. L., Molish, I. R., and Holmsen, H., 1981, Myosin phosphorylation in intact platelets, *J. Biol. Chem.* **256:**7510–7514.

Danthuluri, N. R., and Deth, R. C., 1984, Phorbol-ester-induced contraction of arterial smooth muscle and inhibition of α-adrenergic response, *Biochem. Biophys. Res. Commun.* **125:**1103–1109.

Darbon, J.-M., Issandou, M., Delassus, F., and Bayard, F., 1986a, Phorbol esters induce both intracellular translocation and down-regulation of protein kinase C in MCF-7 cells, *Biochem. Biophys. Res. Commun.* **137:**1159–1166.

Darbon, J.-M., Tournier, J.-F., Tauber, J.-P., and Bayard, F., 1986b, Possible role of protein phosphorylation in the mitogenic effect of high density lipoproteins on cultured vascular endothelial cells, *J. Biol. Chem.* **261:**8002–8008.

Das, S., and Rand, R. P., 1986, Modification by diacylglycerol of the structure and interaction of various phospholipid bilayer membranes, *Biochemistry* **25:**2882–2889.

David-Pfeuty, T., and Singer, S. J., 1980, Altered distributions of the cytoskeletal proteins vinculin and α-actinin in cultured fibroblasts transformed by Rous sarcoma virus, *Proc. Natl. Acad. Sci. USA* **77:**6687–6691.

Davidson, F. F., Dennis, E. A., Powell, M., and Glenney, J. R., Jr., 1987, Inhibition of phospholipase A$_2$ by "lipocortins" and calpactins, *J. Biol. Chem.* **262:**1698–1705.

Davis, R. J., and Czech, M. P., 1984, Tumor-promoting phorbol diesters mediate phosphorylation of the epidermal growth factor receptor, *J. Biol. Chem.* **259:**8545–8549.

Davis, R. J., and Czech, M. P., 1985, Tumor-promoting phorbol diesters cause the phosphorylation of epidermal growth factor receptors in normal human fibroblasts at threonine-654, *Proc. Natl. Acad. Sci. USA* **82:**1974–1978.

Davis, R. J., Ganong, B. R., Bell, R. M., and Czech, M. P., 1985a, sn-1,2-dioctanoylglycerol: A cell permeable diacylglycerol that mimics phorbol diester action on the epidermal growth factor receptor and mitogenesis, *J. Biol. Chem.* **260:**1562–1566.

Davis, R. J., Ganong, B. R., Bell, R. M., and Czech, M. P., 1985b, Structural requirements for diacylglycerols to mimic tumor-promoting phorbol diester action on the epidermal growth factor receptor, *J. Biol. Chem.* **260:**5315–5322.

Davis, R. J., Johnson, G. L., Kelleher, D. J., Anderson, J. K., Mole, J. E., and Czech, M. P., 1986, Identification of serine 24 as the unique site on the transferrin receptor phosphorylated by protein kinase C, *J. Biol. Chem.* **261:**9034–9041.

De Bustros, A., Baylin, S. B., Berger, C. L., Roos, B. A., Leong, S. S., and Nelkin, B. D., 1985, Phorbol esters increase calcitonin gene transcription and decrease c-*myc* mRNA levels in cultured human medullary thyroid carcinoma, *J. Biol. Chem.* **260:**98–104.

De Chaffoy de Courcelles, D., Roevens, P., and Van Belle, H., 1985, R 59 022, a diacylglycerol kinase inhibitor, *J. Biol. Chem.* **260:**15762–15770.

Decker, S., 1981, Phosphorylation of ribosomal protein S6 in avian sarcoma virus–transformed chicken embryo fibroblasts, *Proc. Natl. Acad. Sci. USA* **78:**4112–4115.

Decker, S. J., 1984, Effects of epidermal growth factor and 12-O-tetradecanoylphorbol-13-acetate on metabolism of the epidermal growth factor receptor in normal human fibroblasts, *Mol. Cell. Biol.* **4:**1718–1724.

Decker, S. J., 1985, Phosphorylation of the erbB gene product from an avian erythroblastosis virus–transformed chick fibroblast cell line, *J. Biol. Chem.* **260:**2003–2006.

Degen, J. L., Estensen, R. D., Nagamine, Y., and Reich, E., 1985, Induction and desensitization of plasminogen activator gene expression by tumor promoters, *J. Biol. Chem.* **260:**12426–12433.

Delbeke, D., Kojima, I., Dannies, P. S., and Rasmussen, H., 1984, Synergistic stimulation of prolactin release by phorbol ester, A23187, and forskolin, *Biochem. Biophys. Res. Commun.* **123:**735–741.

Delclos, K. B., and Blumberg, P. M., 1979, Decrease in collagen production in normal and Rous sarcoma virus–transformed chick embryo fibroblasts induced by phorbol myristate acetate, *Cancer Res.* **39:**1667–1672.

Delclos, K. B., Nagle, D. S., and Blumberg, P. M., 1980, Specific binding of phorbol ester tumor promoters to mouse skin, *Cell* **19:**1025–1032.

Del Grande, R. W., and Traugh, J., 1982, Phosphorylation of 40-S ribosomal subunits by cAMP-dependent, cGMP-dependent, and protease-activated protein kinases, *Eur. J. Biochem.* **123:**421–428.

Delia, D., Greaves, M., Villa, S., and DeBraud, F., 1984, Characterization of the response of human thymocytes and blood lymphocytes to the synergistic mitogenicity of 12-O-tetradeca-noylphorbol-13-acetate (TPA)-ionomycin, *Eur. J. Immunol.* **14:**720–724.

Della Bianca, V., De Togni, P., Grzeskowiak, M., Vicentini, L. M., and Di Virgilio, 1986, Cyclic AMP inhibition of phosphoinositide turnover in human neutrophils, *Biochim. Biophys. Acta* **886:**441–447.

Denis, G. V., Toyoshima, S., and Osawa, T., 1986, Concanavalin A- and calcium-dependent phosphorylation of a protein of 80 kDa in mouse lymphocytes rendered permeable to exoge-nously added [γ-^{32}P]ATP, *Biochim. Biophys. Acta* **885:**136–145.

De Paoli-Roach, A., Roach, P. J., Zucker, K. E., and Smith, S. S., 1986, Selective phosphorylation of human DNA methyltransferase by protein kinase C, *FEBS Lett.* **197:**149–153.

De Pont, J. J. H. H. M., and Fleuren-Jackobs, A. M. M., 1984, Synergistic effect of A23187 and a phorbol ester on amylase secretion from rabbit pancreatic acini, *FEBS Lett.* **170:**64–68.

Depper, J. M., Leonard, W. J., Kronke, M., Noguchi, P. D., Cunningham, R. E., Waldmann, T. A., and Greene, W. C., 1984, Regulation of interleukin 2 receptor expression: Effects of phorbol diester, phospholipase C, and reexposure to lectin or antigen, *J. Immunol.* **133:**3054–3061.

De Riemer, S. A., Strong, J. A., Albert, K. S., Greengard, P., and Kaczmarek, L. K., 1985, Enhancement of calcium current in aplysia neurones by phorbol ester and protein kinase C, *Nature* **313**:313–315.

Diamond, L., O'Brien, T. G., and Rovera, G., 1977, Inhibition of adipose conversion of 3T3 fibroblasts by tumour promoters, *Nature* **269**:247–249.

Dicker, P., and Rozengurt, E., 1978, Stimulation of DNA synthesis by tumour promoter and pure mitogenic factors, *Nature* **276**:723–725.

Dicker, P., and Rozengurt, E., 1979, Synergistic stimulation of early events and DNA synthesis by phorbol esters, polypeptide growth factors, and retinoids in cultured fibroblasts, *J. Supramol. Struct.* **11**:79–93.

Dicker, P., and Rozengurt, E., 1981, Phorbol ester stimulation of Na influx and Na-K pump activity in Swiss 3T3 cells, *Biochem. Biophys. Res. Commun.* **100**:433–441.

Diringer, H., and Friis, R. R., 1977, Changes in phosphatidylinositol metabolism correlate to the growth state of normal and Rous sarcoma virus–transformed Japanese quail cells, *Cancer Res.* **37**:2979–2984.

Di Virgilio, F., Lew, D. P., and Pozzan, T., 1984, Protein kinase C activation of physiological processes in human neutrophils at vanishingly small cytosolic Ca^{2+} levels, *Nature* **310**:691–693.

Di Virgilio, F., Pozzan, T., Wollheim, C. B., Vicentini, L. M., and Meldolesi, J., 1986, Tumor promoter phorbol myristate acetate inhibits Ca^{2+} influx through voltage-gated Ca^{2+} channels in two secretory cell lines, PC12 and RINm5F, *J. Biol. Chem.* **261**:32–35.

Dixon, R. A., Kobilka, B. K., Strader, D. J., Benovic, J. L., Dohlman, H. G., Frielle, T., Bolanowski, M. A., Bennett, C. D., Rands, E., Diehl, R. E., Mumford, R. A., Slater, E. E., Sigal, I. S., Caron, M. G., Lefkowitz, R. J., and Strader, C. D., 1986, Cloning of the gene and cDNA for mammalian β-adrenergic receptor and homology with rhodopsin, *Nature* **321**:75–79.

Dlugosz, A. A., Tapscott, S. J., and Holtzer, H., 1983, Effects of phorbol 12-myristate 13-acetate on the differentiation program of embryonic chick skeletal myoblasts, *Cancer Res.* **43**:2780–2789.

Donahue, M. J., and Masaracchia, R. A., 1984, Phosphorylation of ribosomal protein S6 at multiple sites by a cyclic AMP–independent protein kinase from lymphoid cells, *J. Biol. Chem.* **259**:435–440.

Dotto, G. P., Parada, L. F., and Weinberg, R. A., 1985, Specific growth response of ras-transformed embryo fibroblasts to tumour promoters, *Nature* **318**:472–475.

Dougherty, R. W., and Niedel, J. E., 1986, Cytosolic calcium regulates phorbol diester binding affinity in intact phagocytes, *J. Biol. Chem.* **261**:4097–4100.

Dougherty, R. W., Godfrey, P. P., Hoyle, P. C., Putney, J. W., and Freer, R. J., 1984, Secretagogue-induced phosphoinositide metabolism in human leucocytes, *Biochem. J.* **222**:307–314.

Downward, J., Parker, P. J., and Waterfield, M. D., 1984a, Autophosphorylation sites on the epidermal growth factor receptor, *Nature* **310**:483–485.

Downward, J., Yarden, Y., Mayes, E., Scrace, G., Totty, N., Stockwell, P., Ullrich, A., Schlessinger, J., and Waterfield, M. D., 1984b, Close similarity of epidermal growth factor receptor and v-*erb*-B oncogene protein sequences, *Nature* **307**:521–527.

Downward, J., Waterfield, M. D., and Parker, P. J., 1985, Autophosphorylation and protein kinase C phosphorylation of the epidermal growth factor receptor, *J. Biol. Chem.* **260**:14538–14546.

Driedger, P. E., and Blumberg, P. M., 1977, The effect of phorbol diesters on chicken embryo fibroblasts, *Cancer Res.* **37**:3257–3265.

Driedger, P. E., and Blumberg, P. M., 1980, Specific binding of phorbol ester tumor promoters, *Proc. Natl. Acad. Sci. USA* **77**:567–571.

Drummond, A. H., 1985, Bidirectional control of cytosolic free calcium by thyrotropin-releasing hormone in pituitary cells, *Nature* **315**:752–755.

Drust, D. S., and Martin, T. F. J., 1983, Thyrotropin-releasing hormone rapidly activates protein phosphorylation in GH$_3$ pituitary cells by a lipid-linked, protein kinase C-mediated pathway, *J. Biol. Chem.* **259**:14520–14530.

Drust, D. S., and Martin, T. F. J., 1985, Protein kinase C translocates from cytosol to membrane upon hormone activation: Effects of thyrotropin-releasing hormone in GH₃ cells, *Biochem. Biophys. Res. Commun.* **128**:531–537.

Duncan, R., and McConkey, E. H., 1982a, Preferential utilization of phosphorylated 40-S ribosomal subunits during initiation complex formation, *Eur. J. Biochem.* **123**:535–538.

Duncan, R., and McConkey, E. H., 1982b, Rapid alterations in initiation rate and recruitment of inactive RNA are temporally correlated with S6 phosphorylation, *Eur. J. Biochem.* **123**:539–544.

Dunlop, M. E., and Larkins, R. G., 1986, Glucose-induced phospholipid-dependent protein phosphorylation in neonatal rat islets, *Arch. Biochem. Biophys.* **248**:562–569.

Durham, J. P., Emler, C. A., Butcher, F. R., and Fontana, J. A., 1985, Calcium-activated, phospholipid-dependent protein kinase activity and protein phosphorylation in HL60 cells induced to differentiate by retinoic acid, *FEBS Lett.* **185**:157–161.

Ebeling, J. G., Vandenbark, G. R., Kuhn, L. J., Ganong, B. R., Bell, R. M., and Niedel, J. E., 1985, Diacylglycerols mimic phorbol diester induction of leukemic cell differentiation, *Proc. Natl. Acad. Sci. USA* **82**:815–819.

Ebina, Y., Ellis, E., Jarnagin, K., Edery, M., Graf, L., Clauser, E., Ou, J., Masiarz, F., Kan, Y. W., Goldfine, I. D., Roth, R. A., and Rutter, W. J., 1985, The human insulin receptor cDNA: The structural basis for hormone-activated transmembrane signalling, *Cell* **40**:747–758.

Eichberg, J., de Graan, P. N. E., Schrama, L. H., and Gispen, W. H., 1986, Dioctanoylglycerol and phorbol diesters enhance phosphorylation of phosphoprotein B-50 in native synaptic plasma membranes, *Biochem. Biophys. Res. Commun.* **136**:1007–1012.

Eide, B. L., Krebs, E. G., Ross, R., Pike, L. J., and Bowen-Pope, D. F., 1986, Tumor promoter enhances mitogenesis by PDGF with little effect on PDGF binding, *J. Cell. Physiol.* **126**:254–258.

Elsholtz, H. P., Mangalam, H. J., Potter, E., Albert, V. R., Supowit, S., Evans, R. M., and Rosenfeld, M. G., 1986, Two different *cis*-active elements transfer the transcriptional effects of both EGF and phorbol esters, *Science* **234**:1552–1557.

Endo, T., Naka, M., and Hidaka, H., 1982, Ca²⁺-phospholipid-dependent phosphorylation of smooth muscle myosin, *Biochem. Biophys. Res. Commun.* **105**:942–948.

Enomoto, T., Martel, N., Kanno, Y., and Yamasaki, H., 1984, Inhibition of cell communication between Balb/c 3T3 cells by tumor promoters and protection by cAMP, *J. Cell. Physiol.* **121**:323–333.

Erikson, E., and Maller, J. L., 1985, A protein kinase from Xenopus eggs specific for ribosomal protein S6, *Proc. Natl. Acad. Sci. USA* **82**:742–746.

Erikson, E., and Maller, J. L., 1986, Purification and characterization of a protein kinase from Xenopus eggs highly specific for ribosomal protein S6, *J. Biol. Chem.* **261**:350–355.

Erikson, E., Tomasiewicz, H. G., and Erikson, R. L., 1984, Biochemical characterization of a 34-kilodalton normal cellular substrate of pp60*src* and an associated 6-kilodalton protein, *Mol. Cell. Biol.* **4**:77–85.

Estensen, R. D., White, J. G., and Holmes, B., 1974, Specific degranulation of human polymorphonuclear leukocytes, *Nature* **248**:347–348.

Evans, R. R., and Robson, R. M., 1983, Purified vinculin increases the viscosity of F-actin as measured by low-shear viscometry, *J. Cell Biol.* **97**:282a.

Faletto, D. L., Arrow, A. S., and Macara, I. G., 1985, An early decrease in phosphatidylinositol turnover occurs on induction of Friend cell differentiation and precedes the decrease in c-*myc* expression, *Cell* **43**:315–325.

Faquin, W. C., Chahwala, S. B., Cantley, L. C., and Branton, D., 1986, Protein kinase C of human erythrocytes phosphorylates bands 4.1 and 4.9, *Biochim. Biophys. Acta* **887**:142–149.

Farese, R. V., Larson, R. E., and Davis, J. S., 1984, Rapid effects of angiotensin II on polyphosphoinositide metabolism in the rat adrenal glomerulosa, *Endocrinology* **114**:302–304.

Farese, R. V., Standaert, M. L., Barnes, D. E., Davis, J. S., and Pollet, R. J., 1985, Phorbol ester provokes insulin-like effects on glucose transport, amino acid uptake, and pyruvate dehydrogenase activity in BC3H-1 cultured myocytes, *Endocrinology* **116**:2650–2655.

Farrar, W. L., and Anderson, W. B., 1985, Interleukin-2 stimulates association of protein kinase C with plasma membrane, *Nature* **315**:233–235.

Farrar, J. J., Mizel, S. B., Fuller-Farrar, J., Farrar, W. L., and Hilfiker, M. L., 1980, Macrophage-independent activation of helper T cells: I. Production of interleukin 2, *J. Immunol.* **125**:793–798.

Farrar, W. L., Thomas, T. P., and Anderson, W. B., 1985, Altered cytosol/membrane enzyme redistribution on interleukin-3 activation of protein kinase C, *Nature* **315**:235–237.

Fava, R. A., and Cohen, S., 1984, Isolation of a calcium-dependent 35-kilodalton substrate for the epidermal growth factor receptor/kinase from A-431 cells, *J. Biol. Chem.* **259**:2636–2645.

Fearn, J. C., and King, A. C., 1985, EGF receptor affinity is regulated by intracellular calcium and protein kinase C, *Cell* **40**:991–1000.

Ferrari, S., Marchiori, F., Borin, G., and Pinna, L. A., 1985, Distinct structural requirements of Ca^{2+}/phospholipid-dependent protein kinase (protein kinase C) and cAMP-dependent protein kinase as evidenced by synthetic peptide substrates, *FEBS Lett.* **184**:72–77.

Feuerstein, N., and Cooper, H. L., 1983, Rapid protein phosphorylation induced by phorbol ester in HL-60 cells, *J. Biol. Chem.* **258**:10786–10793.

Feuerstein, N., and Cooper, H. L., 1984, Rapid phosphorylation–dephosphorylation of specific proteins induced by phorbol ester in HL-60 cells, *J. Biol. Chem.* **259**:2782–2788.

Feuerstein, N., Monos, D. S., and Cooper, H. L., 1985, Phorbol ester effect in platelets, lymphocytes, and leukemic cells (HL-60) is associated with enhanced phosphorylation of class I HLA antigens, *Biochem. Biophys. Res. Commun.* **126**:206–213.

Fey, E. G., and Penman, S., 1984, Tumor promoters induce a specific morphological signature in the nuclear matrix–intermediate filament scaffold of Madin–Darby canine kidney (MDCK) cell colonies, *Proc. Natl. Acad. Sci. USA* **81**:4409–4413.

Fibach, E., Marks, P. A., and Rifkind, R. A., 1980, Tumor promoters enhance myeloid and erythroid colony formation by normal mouse hemopoietic cells, *Proc. Natl. Acad. Sci. USA* **77**:4152–4155.

Fisher, S. K., Hottman, S. R., Heacock, A. M., Ernst, S. A., and Agranoff, B. W., 1983, Muscarinic stimulation of phospholipid turnover in dissociated avian salt gland cells, *FEBS Lett.* **155**:43–46.

Fitzgerald, D. J., Knowles, S. E., Ballard, F. J., and Murray, A. W., 1983, Rapid and reversible inhibition of junctional communication by tumor promoters in a mouse cell line, *Cancer Res.* **43**:3614–3618.

Friedman, B., Frackleton, A. R., Jr., Ross, A. H., Connors, J. M., Fujiki, J. M., Sugimura, T., and Rosner, M. R., 1984, Tumor promoters block tyrosine specific phosphorylation of the epidermal growth factor receptor, *Proc. Natl. Acad. Sci. USA* **81**:3034–3038.

Fry, M. J., Gebhardt, A., Parker, P. J., and Foulkes, J. G., 1985, Phosphatidylinositol turnover and transformation of cells by Abelson murine leukaemia virus, *EMBO J.* **12**:3173–3178.

Fujiki, H., Tanaka, Y., Miyake, R., Kikkawa, U., Nishizuka, Y., and Sugimura, T., 1984, Activation of calcium-activated, phospholipid-dependent protein kinase (protein kinase C) by new classes of tumor promoters: Teleocidin and debromoaplysiatoxin, *Biochem. Biophys. Res. Commun.* **120**:339–343.

Fujita, I., Irita, K., Takeshige, K., and Minakami, S., 1984, Diacylglycerol, 1-oleoyl-2-acetyl-glycerol, stimulates superoxide-generation from human neutrophils, *Biochem. Biophys. Res. Commun.* **120**:318–324.

Gainer, H. S. C., and Murray, A. W., 1985, Diacylglycerol inhibits gap junctional communication in cultured epidermal cells: Evidence for a role of protein kinase C, *Biochem. Biophys. Res. Commun.* **126**:1109–1113.

Gallis, B., Lewis, A., Wignall, J., Alpert A., Mochizuki, D. Y., Cosmal, D., Hopp, T., and Urdal, D., 1986, Phosphorylation of the human interleukin-2 receptor and a synthetic peptide identical to its C-terminal, cytoplasmic domain, *J. Biol. Chem.* **261**:5075–5080.

Garcia-Sainz, J. A., Mendlovic, F., and Martinez-Olmedo, M. A., 1985, Effects of phorbol esters on α₁-adrenergic-mediated and glucagon-mediated actions in isolated rat hepatocytes, *Biochem. J.* **228:**277–280.

Garrison, J. C., Johnsen, D. E., and Companile, C. P., 1984, Evidence for the role of phosphorylase kinase, protein kinase C and other Ca^{2+}-sensitive protein kinases in the response of hepatocytes to angiotensin II and vasopressin, *J. Biol. Chem.* **259:**3283–3292.

Garte, S. J., and Belman, S., 1980, Tumour promoter uncouples β-adrenergic receptor from adenylate cyclase in mouse epidermis, *Nature* **284:**171–173.

Geiger, B., 1979, A 130K protein from chicken gizzard: Its localization at the termini of microfilament bundles in cultured chicken cells, *Cell* **18:**193–205.

Geiger, B., Tokuyasu, K. T., Dutton, A. H., and Singer, S. J., 1980, Vinculin, an intracellular protein localized at specialized sites where microfilament bundles terminate at cell membranes, *Proc. Natl. Acad. Sci. USA* **77:**4127–4131.

Gelfand, E. W., Cheung, R. K., Mills, G. B., and Grinstein, S., 1985, Mitogens trigger a calcium-independent signal for proliferation in phorbol ester–treated lymphocytes, *Nature* **315:**419–420.

Gennaro, R., Florio, C., and Romeo, D., 1985, Activation of protein kinase C in neutrophil cytoplasts: Localization of protein substrates and possible relationship with stimulus–response coupling, *FEBS Lett.* **180:**185–190.

Gentry, L. E., Chaffin, K. E., Shoyab, M., and Purchio, A. F., 1986, Novel serine phosphorylation of pp60^{c-src} in intact cells after tumor promoter treatment, *Mol. Cell. Biol.* **6:**735–738.

Gerke, V., and Weber, K., 1984, Identity of p36K phosphorylated upon Rous sarcoma virus transformation with a protein purified from brush borders; calcium-dependent binding to non-erythroid spectrin and F-actin, *EMBO J.* **3:**227–233.

Gerke, V., and Weber, K., 1985, The regulatory chain in the p36-kd substrate complex of viral tyrosine-specific protein kinases is related in sequence to the S-100 protein of glial cells, *EMBO J.* **4:**2917–2920.

Ghosh-Dastidar, P., and Fox, C. F., 1984, cAMP-dependent protein kinase stimulates epidermal growth factor–dependent phosphorylation of epidermal growth factor receptors, *J. Biol. Chem.* **259:**3864–3869.

Gibbs, E. M., Allard, W. J., and Lienhard, G. E., 1986, The glucose transporter in 3T3-L1 adipocytes is phosphorylated in response to phorbol ester but not in response to insulin, *J. Biol. Chem.* **261:**16597–16603.

Gilmore, T., and Martin, G. S., 1983, Phorbol ester and diacylglycerol induce protein phosphorylation at tyrosine, *Nature* **306:**487–490.

Gilmour, S. K., Avdalovic, N., Madara, T., and O'Brien, T. G., 1985, Induction of ornithine decarboxylase by 12-O-tetradecanoylphorbol 13-acetate in hamster fibroblasts: Relationship between levels of enzyme activity, immunoreactive protein, and RNA during the induction process, *J. Biol. Chem.* **260:**16439–16444.

Girard, P. R., Mazzei, G. J., Wood, J. G., and Kuo, J. F., 1985, Polyclonal antibodies to phospholipid/Ca^{2+}-dependent protein kinase and immunocytochemical localization of the enzyme in rat brain, *Proc. Natl. Acad. Sci. USA* **82:**3030–3034.

Girard, P. R., Mazzei, G. J., and Kuo, J. G., 1986, Immunological quantitation of phospholipid/Ca^{2+}-dependent protein kinase and its fragments, *J. Biol. Chem.* **261:**370–375.

Glenney, J. R., Jr., 1985, Phosphorylation of p36 *in vitro* with pp60^{src}: Regulation by Ca^{2+} and phospholipid, *FEBS Lett.* **192:**79–82.

Glenney, J., 1986a, Phospholipid-dependent Ca^{2+} binding by the 36-kDa tyrosine kinase substrate (calpactin) and its 33-kDa core, *J. Biol. Chem.* **261:**7247–7252.

Glenney, J., 1986b, Two related but distinct forms of the M_r 36,000 tyrosine kinase substrate (calpactin) that interact with phospholipid and actin in a Ca^{2+}-dependent manner, *Proc. Natl. Acad. Sci. USA* **83:**4258–4262.

Glenney, J. R., Jr., and Tack, B. F., 1985, Amino-terminal sequence of p36 and associated p10: Identification of the site of tyrosine phosphorylation and homology with S-100, *Proc. Natl. Acad. Sci. USA* **82:**7884–7888.

Goldberg, A. R., Delclos, K. B., and Blumberg, P. M., 1980, Phorbol ester action is independent of viral and cellular *src* kinase levels, *Science* **208:**191–193.

Golden, A., Nemeth, S. P., and Brugge, J. S., 1986, Blood platelets express high levels of the pp60^{c-src}-specific tyrosine kinase activity, *Proc. Natl. Acad. Sci. USA* **83:**852–856.

Goldstein, I. M., Hoffstein, S. T., and Weissmann, G., 1975, Mechanisms of lysosomal enzyme release from human polymorphonuclear leukocytes: Effects of phorbol myristate acetate, *J. Cell Biol.* **66:**647–652.

Gorelick, F. S., Cohn, J. A., Freedman, S. D., Delahunt, N. G., Gershoni, J. M., and Jamieson, J. D., 1983, Calmodulin-stimulated protein kinase activity from rat pancreas, *J. Cell Biol.* **97:**1294–1298.

Gottesman, M. M., and Sobel, M. E., 1980, Tumor promoters and Kirsten sarcoma virus increase synthesis of a secreted glycoprotein by regulating levels of translatable mRNA, *Cell* **19:**449–455.

Gottesman, M. M., and Yuspa, S. H., 1981, Tumor promoters induce the synthesis of a secreted glycoprotein in mouse skin and cultured primary mouse epidermal cells, *Carcinogenesis 2:*971–976.

Gould, K. L., Cooper, J. A., and Hunter, T., 1984, The 46,000-dalton tyrosine protein kinase substrate is widespread, whereas the 36,000-dalton substrate is only expressed at high levels in certain rodent tissues, *J. Cell Biol.* **98:**487–497.

Gould, K. L., Woodgett, J. R., Cooper, J. A., Buss, J. E., Shalloway, D., and Hunter, T., 1985, Protein kinase C phosphorylates pp60^{src} at a novel site, *Cell* **42:**849–857.

Gould, K. L., Woodgett, J. R., Isacke, C. M., and Hunter, T., 1986, The protein–tyrosine kinase substrate, p36, is also a substrate for protein kinase C *in vitro* and *in vivo, Mol. Cell. Biol.* **6:**2738–2744.

Greenberg, M. E., and Ziff, E. B., 1984, Stimulation of 3T3 cells induces transcription of the c-*fos* proto-oncogene, *Nature* **311:**433–438.

Greenberg, M. E., Brackenbury, R., and Edelman, G. M., 1984, Changes in the distribution of the 34-kilodalton tyrosine kinase substrate during differentiation and maturation of chicken cells, *J. Cell Biol.* **98:**473–486.

Grinstein, S., Cohen, S., Goetz, J. D., Rothstein, A., and Gelfand, E. W., 1985, Characterization of the activation of Na^+/H^+ exchange in lymphocytes by phorbol esters: Change in cytoplasmic pH dependence of the antiport, *Proc. Natl. Acad. Sci. USA* **82:**1429–1433.

Grubbs, R. D., and Maguire, M. E., 1986, Regulation of magnesium but not calcium transport by phorbol ester, *J. Biol. Chem.* **261:**12550–12554.

Grunberger, G., and Gordon, P., 1982, Affinity alteration of insulin receptor induced by a phorbol ester, *Am. J. Physiol.* **243:**E319–E324.

Gschwendt, M., Horn, F., Kittstein, W., and Marks, F., 1983, Inhibition of the calcium- and phospholipid-dependent protein kinase activity from mouse brain cytosol by quercetin, *Biochem. Biophys. Res. Commun.* **117:**444–447.

Guigni, T. D., James, L. C., and Haigler, H. T., 1985, Epidermal growth factor stimulates tyrosine phosphorylation of specific proteins in permeabilized human fibroblasts, *J. Biol. Chem.* **260:**15081–15090.

Guy, G. R., Gordon, J., Michell, R. H., and Brown, G., 1985, Synergism between diacylglycerols and calcium ionophore in the induction of human B cell proliferation mimics the inositol lipid polyphosphate breakdown signals induced by crosslinking surface immunoglobulin, *Biochem. Biophys. Res. Commun.* **131:**484–491.

Haarr, L., Kleppe, K., and Lillehaug, J. R., 1986, Changes in polypeptide synthesis and glycosylation in mouse embryonic fibroblast C3H/10T1/2 C18 cells caused by the tumor promoter 12-O-tetradecanoylphorbol 13-acetate, *Biochim. Biophys. Acta* **889:**334–345.

Habenicht, A. J., Glomset, J. A., King, W. C., Nist, C., Mitchell, C., and Ross, R., 1981, Early changes in phosphatidylinositol and arachidonic acid metabolism in quiescent Swiss 3T3 cells stimulated to divide by platelet-derived growth factor, *J. Biol. Chem.* **256**:12329–12335.

Hagino-Yamagishi, K., Ikawa, S., Kawai, S., Hihara, H., Yamamoto, T., and Toyoshima, K., 1984, Characterization of two strains of avian sarcoma virus isolated from avian lymphatic leukosis virus-induced sarcomas, *Virology* **137**:266–275.

Halegoua, S., and Patrick, J., 1980, Nerve growth factor mediates phosphorylation of specific proteins, *Cell* **22**:571–581.

Hallam, T. J., Sanchez, A., and Rink, T. J., 1984, Stimulus–response coupling in human platelets, *Biochem. J.* **218**:819–827.

Hannun, Y. A., Loomis, C. R., and Bell, R. M., 1985, Activation of protein kinase C by Triton X-100 mixed micelles containing diacylglycerol and phosphatidylserine, *J. Biol. Chem.* **260**:10039–10043.

Hannun, Y. A., Loomis, C. R., Merrill, A. H., Jr., and Bell, R. M., 1986, Sphingosine inhibition of protein kinase C activity and of phorbol dibutyrate binding *in vitro* and in human platelets, *J. Biol. Chem.* **261**:12604–12609.

Hardie, D. G., Carling, D., Ferrari, S., Guy, P. S., and Aitken, A., 1986, Characterization of the phosphorylation of rat mammary ATP-citrate lyase and acetyl-CoA carboxylase by Ca^{2+} and calmodulin-dependent multiprotein kinase and Ca^{2+} and phospholipid–dependent protein kinase, *Eur. J. Biochem.* **157**:553–561.

Haring, H., Kirsch, D., Obermaier, B., Ermel, B., and Machicao, F., 1986, Tumor-promoting phorbol esters increase the K_m of the ATP-binding site of the insulin receptor kinase from rat adipocytes, *J. Biol. Chem.* **261**:3869–3875.

Harrison, J. R., Lynch, K. R., and Sando, J. J., 1987, Phorbol esters induce interleukin 2 mRNA in sensitive but not in resistant EL4 cells, *J. Biol. Chem.* **262**:234–238.

Hasegawa-Sasaki, H., and Sasaki, T., 1983, Phytohemagglutinin induces rapid degradation of phosphatidylinositol 4,5-bisphosphate and transient accumulation of phosphatidic acid and diacylglycerol in a human T lymphoblastoid cell line, CCRF-CEM, *Biochim. Biophys. Acta* **754**:305–314.

Haselbacher, G. K., Humbel, R. E., and Thomas, G., 1979, Insulin-like growth factor: Insulin or serum increase phosphorylation of ribosomal protein S6 during transition of stationary chick embryo fibroblasts into early G_i phase of the cell cycle, *FEBS Lett.* **100**:185–190.

Haslam, R. J., and Lynham, J. A., 1977, Relationship between phosphorylation of blood platelet proteins and secretion of platelet granule constitutents. I. Effects of different aggregating agents, *Biochem. Biophys. Res. Commun.* **77**:714–722.

Hawrylowicz, C. M., and Klaus, G. G. B., 1984, Effects of tumour promoter phorbol myristate acetate on mouse lymphocytes: Selective inhibition of B cell activation by mitogens and antigens, *Immunology* **51**:327–332.

Hayashi, K., Fujiki, H., and Sugimura, T., 1983, Effects of tumor promoters on the frequency of metallothionein 1 gene amplification in cells exposed to cadmium, *Cancer Res.* **43**:5433–5436.

Haycock, J. W., Bennett, W. F., George, R. J., and Waymire, J. C., 1982a, Multiple site phosphorylation of tyrosine hydroxylase: Differential regulation *in situ* by 8-bromo-cAMP and acetylcholine, *J. Biol. Chem.* **257**:13699–13703.

Haycock, J. W., Meligeni, J. A., Bennett, W. F., and Waymire, J. C., 1982b, Phosphorylation and activation of tyrosine hydroxylase mediate the acetylcholine-induced increase of catecholamine biosynthesis in adrenal chromaffin cells, *J. Biol. Chem.* **257**:12641–12648.

Helfman, D. M., Appelbaum, B. D., Vogler, W. R., and Kuo, J. F., 1983a, Phospholipid-sensitive Ca^{2+}-dependent protein kinase and its substrates in human neutrophils, *Biochem. Biophys. Res. Commun.* **111**:847–853.

Helfman, D. M., Barnes, K. C., Kinkade, Jr., J. M., Vogler, W. R., Shoji, M., and Kuo, J. F.,

1983b, Phospholipid-sensitive Ca^{2+}-dependent protein phosphorylation system in various types of leukemic cells from human patients and in human leukemic cell lines HL60 and K562, and its inhibition by alkyl-lysophospholipid, *Cancer Res.* **43**:2955–2961.

Hennings, H., and Yuspa, S. H., 1985, Two-stage tumor promotion in mouse skin: An alternative interpretation, *J. Natl. Cancer Inst.* **74**:735–740.

Herman, B., and Pledger, W. J., 1985, Platelet-derived growth factor–induced alterations in vinculin and actin distribution in BALB/c-3T3 cells, *J. Cell Biol.* **100**:1031–1040.

Herschman, H. R., 1985, A 12-O-tetradecanoylphorbol-13-acetate (TPA)-nonproliferative variant of 3T3 cells is resistant to TPA-enhanced gene amplification, *Mol. Cell Biol.* **5**:1130–1135.

Hersh, C. L., Yeh, R. K., Callaway, J. E., Garcia, J. A., Jr., and Gilmore-Hebert, M., 1986, Induction of collagenase production in U937 cells by phorbol ester and partial purification of the induced enzyme, *Biochemistry* **25**:4750–4757.

Hesketh, T. R., Moore, J. P., Morris, J. D. H., Taylor, M. V., Rogers, J., Smith, G. A., and Metcalfe, J. C., 1985, A common sequence of calcium and pH signals in the mitogenic stimulation of eukaryotic cells, *Nature* **313**:481–484.

Hidaka, H., Inagaki, M., Kawamoto, S., and Sasaki, Y., 1984, Isoquinolinesulfonamides, novel and potent inhibitors of cyclic nucleotide dependent protein kinase and protein kinase C, *Biochemistry* **23**:5036–5041.

Hincke, M. T., and Tolnai, S., 1986, Phosphorylation of bovine cardiac calcium-activated neutral protease by protein kinase C, *Biochem. Biophys. Res. Commun.* **137**:559–565.

Hirata, F., Matsuda, K., Notsu, Y., Hattori, T., and Del Carmine, R., 1984, Phosphorylation at a tyrosine residue of lipomodulin in mitogen-stimulated murine thymocytes, *Proc. Natl. Acad. Sci. USA* **81**:4717–4721.

Hirata, M., Sasaguri, T., Hamachi, T., Hashimoto, T., Kukita, M., and Koga, T., 1985, Irreversible inhibition of Ca^{2+} release in saponin-treated macrophages by the photoaffinity derivative of inositol-1,4,5-trisphosphate, *Nature* **317**:723–725.

Hirota, K., Hirota, T., Aguilera, G., and Catt, K. J., 1985, Hormone-induced redistribution of calcium-activated phospholipid-dependent protein kinase in pituitary gonadotrophs, *J. Biol. Chem.* **260**:3243–3246.

Hishikawa, R., Fukase, M., Yamatani, T., Kadowaki, S., and Fujita, T., 1985, Phorbol ester stimulates calcitonin secretion synergistically with A23187, and additively with dibutyryl cyclic AMP in a rat C-cell line, *Biochem. Biophys. Res. Commun.* **132**:424–429.

Hiwasa, T., Fujimura, S., and Sakiyama, S., 1982, Tumor promoters increase the synthesis of a 32,000-dalton protein in BALB/c 3T3 cells, *Proc. Natl. Acad. Sci. USA* **79**:1800–1804.

Hofer, H. W., Schlatter, S., and Graefe, M., 1985, Phosphorylation of phosphofructokinase by protein kinase C changes the allosteric properties of the enzyme, *Biochem. Biophys. Res. Commun.* **129**:892–897.

Hokin, L. E., 1985, Receptors and phosphoinositide-generated second messengers, *Annu. Rev. Biochem.* **54**:205–236.

Hokin, L. E., and Hokin, M., 1953, Enzymatic secretion and the incorporation of P^{32} into phospholipides of pancreas slices, *J. Biol. Chem.* **203**:967–977.

Hollingsworth, E. B., Ukena, D., and Daly, J. W., 1986, The protein kinase C activator phorbol-12-myristate-13-acetate enhances cyclic AMP accumulation in pheochromocytoma cells, *FEBS Lett.* **196**:131–134.

Homma, Y., Henning-Chubb, C. B., and Huberman, E., 1986, Translocation of protein kinase C in human leukemia cells susceptible or resistant to differentiation induced by phorbol 12-myristate 13-acetate, *Proc. Natl. Acad. Sci. USA* **83**:7316–7319.

Horn, F., Gschwendt, M., and Marks, F., 1985, Partial purification and characterization of the calcium-dependent and phospholipid-dependent protein kinase C from chick oviduct, *Eur. J. Biochem.* **148**:533–538.

Hornbeck, P., and Paul, W. E., 1986, Anti-immunoglobulin and phorbol ester induce phosphorylation of proteins associated with the plasma membrane and cytoskeleton in murine B lymphocytes, *J. Biol. Chem.* **261**:14817–14824.

Horne, W. C., Leto, T. L., and Marchesi, V. T., 1985, Differential phosphorylation of multiple sites in protein 4.1 and protein 4.9 by phorbol ester–activated and cyclic AMP–dependent protein kinases, *J. Biol. Chem.* **260**:9073–9076.

Horowitz, A. D., Fujiki, H., Weinstein, B., Jeffrey, A., Okin, E., Moore, R. E., and Sugimura, T., 1983, Comparative effects of aplysiatoxin, debromoaplysiatoxin, and teleocidin on receptor binding and phospholipid metabolism, *Cancer Res.* **43**:1529–1535.

Horwitz, A., Duggan, K., Buck, C., Beckerle, M. C., and Burridge, K., 1986, Interaction of plasma membrane fibronectin receptor with talin—A transmembrane linkage, *Nature* **320**:531–533.

Hoshijima, M., Kikuchi, A., Tanimoto, T., Kaibuchi, K., and Takai, Y., 1986, Formation of a phorbol ester–binding fragment from protein kinase C by proteolytic digestion, *Cancer Res.* **46**:3000–3004.

House, C., Wettenhall, R. E. H., and Kemp, B. E., 1987, The influence of basic residues on the substrate specificity of protein kinase C, *J. Biol. Chem.* **262**:772–777.

Housey, G. M., O'Brian, C. A., Johnson, M. D., Kirschmeier, P., and Weinstein, I. B., 1987, Isolation of cDNA clones encoding protein kinase C: Evidence for a protein kinase C-related gene family, *Proc. Natl. Sci. Acad. USA* **84**:1065–1069.

Hsiao, W.-L. W., Gattoni-Celli, S., and Weinstein, I. B., 1984, Oncogene-induced transformation of C3H 10T1/2 cells is enhanced by tumor promoters, *Science* **226**:552–555.

Hsiao, W.-L. W., Wu, T., and Weinstein, I. B., 1986, Oncogene-induced transformation of a rat embryo fibroblast cell line is enhanced by tumor promoters, *Mol. Cell. Biol.* **6**:1943–1950.

Huang, C., Hill, Jr., J. M., Bormann, B., Mackin, W. M., and Becker, E. L., 1983, Endogenous substrates for cyclic AMP–dependent and calcium-dependent protein phosphorylation in rabbit peritoneal neutrophils, *Biochim. Biophys. Acta* **760**:126–135.

Huang, K. P., and Huang, F. L., 1986, Immunochemical characterization of rat brain protein kinase C, *J. Biol. Chem.* **261**:14781–14787.

Huang, K. P., Chan, K.-F. J., Singh, T. J., Nakabayashi, H., and Huang, F. L., 1986a, Autophosphorylation of rat brain Ca^{2+}-activated and phospholipid-dependent protein kinase, *J. Biol. Chem.* **261**:12134–12140.

Huang, K. P., Nakabayashi, H., and Huang, F. L., 1986b, Isozymic forms of rat brain Ca^{2+}-activated and phospholipid-dependent protein kinase, *Proc. Natl. Acad. Sci. USA* **83**:8535–8539.

Huang, K.-S., Wallner, B. P., Mattaliano, R. J., Tizard, R., Burne, C., Frey, A., Hession, C., McGray, P., Sinclair, L. K., Chow, E. P., Browning, J. L., Ramachandran, K. L., Tang, J., Smart, J. E., and Pepinsky, R. B., 1986c, Two human 35 kd inhibitors of phospholipase A_2 are related to substrates of $pp60^{v-src}$ and of the epidermal growth factor receptor/kinase, *Cell* **46**:191–199.

Huberman, E., and Callaham, M. F., 1979, Induction of terminal differentiation in human promyelocytic leukemia cells by tumor-promoting agents, *Proc. Natl. Acad. Sci. USA* **76**:1293–1297.

Huberman, E., Heckman, C., and Langenbach, R., 1979, Stimulation of differentiated functions in human melanoma cells by tumor-promoting agents and dimethylsulfoxide, *Cancer Res.* **39**:2618–2624.

Huberman, E., Weeks, C., Herrmann, A., Callaham, M. F., and Slaga, T. J., 1981, Alterations in polyamine levels induced by phorbol diesters and other agents that promote differentiation in human promyelocytic leukemia cells, *Proc. Natl. Acad. Sci. USA* **78**:1062–1066.

Huberman, E., Braslawsky, G. R., Callaham, M., and Fugiki, H., 1982, Induction of differentiation of human promyelocytic leukemia (HL-60) cells by teleocidin and phorbol-12-myristate-13-acetate, *Carcinogenesis* **3**:111–114.

Huganir, R. L., Miles, K., and Greengard, P., 1984, Phosphorylation of the nicotinic acetylcholine receptor by an endogenous tyrosine-specific protein kinase, *Proc. Natl. Acad. Sci. USA* **81**:6968–6972.

Humble, E., Heldin, P., Forsberg, P., and Engstrom, L., 1984, Phosphorylation of human fibrinogen *in vitro* by calcium-activated phospholipid-dependent protein kinase from pig spleen, *J. Biochem.* **95**:1435–1443.

Hunter, T., 1984a, Oncogenes and proto-oncogenes: How do they differ? *J. Natl. Cancer Inst.* **73**:773–786.

Hunter, T., 1984b, The epidermal growth factor receptor gene and its product, *Nature* **311**:414–416.

Hunter, T., and Cooper, J. A., 1985, Protein-tyrosine kinases, *Annu. Rev. Biochem.* **54**:897–930.

Hunter, T., Ling, N., and Cooper, J. A., 1984, Protein kinase C phosphorylation of the EGF receptor at a threonine residue close to the cytoplasmic face of the plasma membrane, *Nature* **311**:480–483.

Iba, H., Takeya, T., Cross, F. R., Hanafusa, T., and Hanafusa, H., 1984, Rous sarcoma virus variants that carry the cellular *src* gene instead of the viral *src* gene cannot transform chicken embryo fibroblasts, *Proc. Natl. Acad. Sci. USA* **81**:4424–4428.

Iba, H., Cross, F. R., Garber, E. A., and Hanafusa, H., 1985, Low level of cellular protein phosphorylation by nontransforming overproduced p60^{src}, *Mol. Cell. Biol.* **5**:1058–1066.

Ieyasu, H., Takai, Y., Kaibuchi, K., Sawamur, M., and Nishizuka, Y., 1982, A role of calcium-activated, phospholipid-dependent protein kinase in platelet-activating factor-induced serotonin release from rabbit platelets, *Biochem. Biophys. Res. Commun.* **108**:1701–1708.

Ikebe, M., Inagaki, M., Kanamaru, K., and Hidaka, H., 1985, Phosphorylation of smooth muscle myosin light chain kinase by Ca^{2+}-activated phospholipid-dependent protein kinase, *J. Biol. Chem.* **260**:4547–4550.

Imaoka, T., Lynham, J. A., and Haslam, R. J., 1983, Purification and characterization of the 47,000-dalton protein phosphorylated during degranulation of human platelets, *J. Biol. Chem.* **258**:11404–11414.

Imazu, M., Strickland, W. G., Chrisman, T. D., and Exton, J. H., 1984, Phosphorylation and inactivation of liver glycogen synthase by liver protein kinases, *J. Biol. Chem.* **259**:1813–1821.

Imbra, R. J., and Karin, M., 1987 Metallothionein gene expression is regulated by serum factors and activators of protein kinase C, *Mol. Cell. Biol.* **7**:1358–1363.

Inagaki, M., Kawamoto, S., and Hidaka, H., 1984, Serotonin secretion from human platelets may be modified by Ca^{2+}-activated, phospholipid-dependent myosin phosphorylation, *J. Biol. Chem.* **259**:14321–14323.

Inoue, M., Kishimoto, A., Takai, Y., and Nishizuka, Y., 1977, Studies on a cyclic nucleotide–independent protein kinase and its proenzyme in mammalian tissues, *J. Biol. Chem.* **252**:7610–7616.

Isacke, C. M., Trowbridge, I. S., and Hunter, T., 1986a, Modulation of p36 tyrosine phosphorylation in human cells, *Mol. Cell. Biol.* **6**:2745–2751.

Isacke, C. M., Meisenhelder, J., Brown, K. D., Gould, K. L., Gould, S. J., and Hunter, T., 1986b, Early phosphorylation events following the treatment of Swiss 3T3 cells with bombesin and GRP, *EMBO J.* **5**:2889–2898.

Ishii, D. N., 1978, Effect of tumor promoters on the response of cultured embryonic chick ganglia to nerve growth factor, *Cancer Res.* **38**:3886–3893.

Ishii, D. N., Fibach, E., Yamasaki, H., Weinstein, I. B., 1978, Tumor promoters inhibit morphological differentiation in cultured mouse neuroblastoma cells, *Science* **200**:556–559.

Issandou, M., Bayard, F., and Darbon, J.-M., 1986, Activation of phorbol esters of protein kinase C in MCF-7 human breast cancer cells, *FEBS Lett.* **200**:337–342.

Ito, S., Richert, N., and Pastan, I., 1982, Phospholipids stimulate phosphorylation of vinculin by the tyrosine-specific protein kinase of Rous sarcoma virus, *Proc. Natl. Acad. Sci. USA* **79**:4628–4631.

Ito, S., Werth, D. K., Richert, N. D., and Pastan, I., 1983, Vinculin phosphorylation by the src kinase: Interaction of vinculin with phospholipid vesicles, *J. Biol. Chem.*. **258**:14626–14631.

Iwasa, Y., and Hosey, M. M., 1984, Phosphorylation of cardiac sarcolemma proteins by the calcium-activated phospholipid-dependent protein kinase, *J. Biol. Chem.* **259**:534–540.

Iwasa, Y., Takai, Y., Kikkawa, U., and Nishizuka, Y., 1980, Phosphorylation of calf thymus H1 histone by calcium-activated, phospholipid-dependent protein kinase, *Biochem. Biophys. Res. Commun.* **96:**180–187.

Iwashita, S., and Fox, C. F., 1984, Epidermal growth factor and potent phorbol tumor promoters induce epidermal growth factor receptor phosphorylation in a similar but distinctively different manner in epidermoid carcinoma A431 cells, *J. Biol. Chem.* **259:**2559–2567.

Iwashita, S., Kitamura, N., and Yoshida, M., 1983, Molecular events leading to fusiform morphological transformation by partial *src* deletion mutant of Rous sarcoma virus, *Virology* **125:**419–431.

Jacobs, S., and Cuatrecasas, P., 1986, Phosphorylation of receptors for insulin and insulin-like growth factor. I. Effects of hormones and phorbol esters, *J. Biol. Chem.* **261:**934–939.

Jacobs, S., Sahyoun, N., Saltiel, A. R., and Cuatrecasas, P., 1983a, Phorbol esters stimulate the phosphorylation of receptors for insulin and somatomedin C, *Proc. Natl. Acad. Sci. USA* **80:**6211–6213.

Jacobs, S., Kull, F. C., Jr., Earp, H. S., Svoboda, M. E., Van Wyke, J. J., and Cuatrecasas, P., 1983b, Somatomedin-C stimulates the phosphorylation of the β-subunit of its own receptor, *J. Biol. Chem.* **258:**9581–9584.

Jaken, S., and Kiley, S. C., 1987, Purification and characterization of three types of protein kinase C from rabbit brain cytosol, *Proc. Natl. Acad. Sci. USA* **84:**4418–4422.

Jaken, S., Tashjian, A. H., and Blumberg, P. M., 1981, Characterization of phorbol ester receptors and their down-modulation in GH₄C₁ rat pituitary cells, *Cancer Res.* **41:**2175–2181.

Jakobs, K. H., Bauer, S., and Watanabe, Y., 1985, Modulation of adenylate cyclase of human platelets by phorbol ester, *Eur. J. Biochem.* **151:**425–430.

Jeffrey, A. M., and Liskamp, R. M. J., 1986, Computer-assisted molecular modeling of tumor promoters: Rationale for the activity of phorbol esters, teleocidin B, and aplysiatoxin, *Proc. Natl. Acad. Sci. USA* **83:**241–245.

Jeng, A. Y., Lichti, U., Strickland, J. E., and Blumberg, P. M., 1985, Similar effects of phospholipase C and phorbol ester tumor promoters on primary mouse epidermal cells, *Cancer Res.* **45:**5714–5721.

Jeng, A. Y., Sharkey, N. A., and Blumberg, P. M., 1986, Purification of stable protein kinase C from mouse brain cytosol by specific ligand elution using fast protein liquid chromatography, *Cancer Res.* **46:**1966–1971.

Jetten, A. M., Ganong, B. R., Vandenbark, G. R., Shirley, J. E., and Bell, R. M., 1985, Role of protein kinase C in diacylglycerol-mediated induction of ornithine decarboxylase and reduction of epidermal growth factor binding, *Proc. Natl. Acad. Sci. USA* **82:**1941–1945.

Joh, T. H., Park, D. H., and Reis, D. J., 1978, Direct phosphorylation of brain tyrosine hydroxylase by cyclic AMP–dependent protein kinase: Mechanism of enzyme activation, *Proc. Natl. Acad. Sci. USA* **75:**4744–4748.

Johnson, J. A., Goka, T. J., and Clark, R. B., 1986, Phorbol ester–induced augmentation and inhibition of epinephrine-stimulated adenylate cyclase in S49 lymphoma cells, *J. Cyclic Nucl. Prot. Phos. Res.* **11:**199–216.

Johnson, P. J., Coussens, P. M., Danko, A. V., and Shalloway, D., 1985, Overexpressed pp60^{c-src} can induce focus formation without complete transformation of NIH 3T3 cells, *Mol. Cell. Biol.* **5:**1073–1083.

Johnsson, N., Van, P. N., Soling, H.-D., and Weber, K., 1986, Functionally distinct serine phosphorylation sites of p36, the cellular substrate of retroviral protein kinase; differential inhibition of reassociation with p11, *EMBO J.* **5:**3455–3460.

Kaibuchi, K., Takai, Y., and Nishizuka, Y., 1981, Cooperative roles of various membrane phospholipids in the activation of calcium-activated, phospholipid-dependent protein kinase, *J. Biol. Chem.* **256:**7146–7149.

Kaibuchi, K., Takai, Y., Sawamura, M., Hoshijima, M., Fujikura, T., and Nishizuka, Y., 1983, Synergistic functions of protein phosphorylation and calcium mobilization in platelet activation, *J. Biol. Chem.* **258:**6701–6704.

Kaibuchi, K., Takai, Y., and Nishizuka, Y., 1985, Protein kinase C and calcium ion in mitogenic response of macrophage-depleted human peripheral lymphocytes, *J. Biol. Chem.* **260:**1366–1369.

Kaibuchi, K., Tsuda, T., Kikuchi, A., Tanimoto T., Yamashita, T., and Takai, Y., 1986, Possible involvement of protein kinase C and calcium ion in growth factor–induced expression of c-*myc* oncogene in Swiss 3T3 fibroblasts, *J. Biol. Chem.* **261:**1187–1192.

Kajikawa, N., Kaibuchi, K., Matsubara, T., Kikkawa, U., Takai, Y., and Nishizuka, Y., 1983, A possible role of protein kinase C in signal-induced lysosomal enzyme release, *Biochem. Biophys. Res. Commun.* **116:**743–750.

Kamps, M. P., Buss, J. E., and Sefton, B. M., 1985, Mutation of NH_2-terminal glycine of $p60^{src}$ prevents both myristoylation and morphological transformation, *Proc. Natl. Acad. Sci. USA* **82:**4625–4628.

Kanoh, H., and Ono, T., 1986, Phosphorylation of pig brain diacylglycerol kinase by endogenous protein kinase, *FEBS Lett.* **201:**97–100.

Kasuga, M., Van Obberghen, E., Nissley, S. P., and Rechler, M. M., 1981, Demonstration of two subtypes of insulin-like growth factor by affinity cross-linking, *J. Biol. Chem.* **256:**5305–5308.

Kasuga, M., Zick, Y., Blithe, D. L., Crettaz, M., and Kahn, C. R., 1982, Insulin stimulates tyrosine phosphorylation of the insulin receptor in a cell-free system, *Nature* **298:**667–669.

Katada, T., Gilman, A. G., Watanabe, Y., Bauer, S., and Jakobs, K. H., 1985, Protein kinase C phosphorylates the inhibitory guanine-nucleotide-binding regulatory component and apparently suppresses its function in hormonal inhibition of adenylate cyclase, *Eur. J. Biochem.* **151:**431–437.

Katakami, Y., Kaibuchi, K., Sawamura, M., Takai, Y., and Nishizuka, Y., 1984, Synergistic action of protein kinase C and calcium for histamine release from rat peritoneal mast cells, *Biochem. Biophys. Res. Commun.* **121:**573–378.

Katoh, N., Wrenn, R. W., Wise, B. C., Shoji, M., and Kuo, J. F., 1981, Substrate proteins for calmodulin-sensitive and phospholipid-sensitive Ca^{2+}-dependent protein kinases in heart, and inhibition of their phosphorylation by palmitoylcarnitine, *Proc. Natl. Acad. Sci. USA* **78:**4813–4817.

Katoh, N., Wise, B. C., and Kuo, J. F., 1983, Phosphorylation of cardiac troponin inhibitory subunit (troponin I) and tropomyosin-binding subunit (troponin T) by cardiac phospholipid-sensitive Ca^{2+}-dependent protein kinase, *Biochem. J.* **209:**189–195.

Kawahara, Y., Takai, Y., Minakuchi, R., Sano, K., and Nishizuka, Y., 1980, Phospholipid turnover as a possible transmembrane signal for protein phosphorylation during human platelet activation by thrombin, *Biochem. Biophys. Res. Commun.* **97:**309–317.

Kawamoto, S., and Hidaka, H., 1984, Ca^{2+}-activated, phospholipid-dependent protein kinase catalyzes the phosphorylation of actin-binding proteins, *Biochem. Biophys. Res. Commun.* **118:**736–742.

Kelleher, D. J., and Johnson, G. L., 1986, Phosphorylation of rhodopsin by protein kinase C *in vitro, J. Biol. Chem.* **261:**4749–4757.

Kelleher, D. J., Pessin, J. E., Ruoho, A. E., and Johnson, G. L., 1984, Phorbol ester induces desensitization of adenylate cyclase and phosphorylation of the β-adrenergic receptor in turkey erythrocytes, *Proc. Natl. Acad. Sci. USA* **81:**4316–4320.

Kellie, S., Holme, T. C., and Bissell, M. J., 1985, Interaction of tumour promoters with epithelial cells in culture: An immunofluorescence study, *Exp. Cell Res.* **160:**259–274.

Kelly, K., Cochran, B. H., Stiles, C. D., and Leder, P., 1983, Cell-specific regulation of the c-*myc* gene by lymphocyte mitogens and platelet-derived growth factor, *Cell* **35:**603–610.

Kemp, B. E., Benjamini, E., and Krebs, E. G., 1976, Synthetic hexapeptide substrates and inhibitors of 3':5'-cyclic AMP–dependent protein kinase, *Proc. Natl. Acad. Sci. USA* **73:**1038–1042.

Kemp, B. E., Pearson, R. B., and House, C., 1983, Role of basic residues in the phosphorylation of synthetic peptides by myosin light chain kinase, *Proc. Natl. Acad. Sci. USA* **80:**7471–7475.

Kennedy, I. M., and Leader, D. P., 1981, Increased phosphorylation of ribosomal protein S6 in hamster fibroblasts transformed by polyoma virus and simian virus 40, *Biochem. J.* **198:**235–237.

Khanna, N. C., Tokuda, M., and Waisman, D. M., 1986a, Phosphorylation of lipocortins *in vitro* by protein kinase C, *Biochem. Biophys. Res. Commun.* **141:**5547–554.

Khanna, N. C., Tokuda, M., Chong, S. M., and Waisman, D. M., 1986b, Phosphorylation of p36 *in vitro* by protein kinase C, *Biochem. Biophys. Res. Commun.* **137:**397–403.

Kikkawa, U., Takai, Y., Minakuchi, R., Inohara, S., and Nishizuka, Y., 1982, Calcium-activated, phospholipid-dependent protein kinase from rat brain, *J. Biol. Chem.* **257:**13341–13348.

Kikkawa, U., Minakuchi, R., Takai, Y., and Nishizuka, Y., 1983a, Calcium-activated, phospholipid-dependent protein kinase (protein kinase C) from rat brain, *Methods Enzymol.* **99:**288–298.

Kikkawa, U., Takai, Y., Tanaka, Y., Miyake, R., and Nishizuka, Y., 1983b, Protein kinase C as a possible receptor protein of tumor-promoting phorbol esters, *J. Biol. Chem.* **258:**11442–11445.

Kikkawa, U., Go, M., Koumoto, J., and Nishizuka, Y., 1986, Rapid purification of protein kinase C by high performance liquid chromatography, *Biochem. Biophys. Res. Commun.* **135:**636–643.

Kikuchi, A., Kozawa, O., Hamamori, Y., Kaibuchi, K., and Takai, Y., 1986, Inhibition of chemotactic peptide–induced phosphoinositide hydrolysis by phorbol esters through the activation of protein kinase C in differentiated human leukemia (HL-60) cells, *Cancer Res.* **46:**3401–3406.

Kimura, K., Sakurada, K., and Katoh, N., 1985, Inhibition by gossypol of phospholipid-sensitive Ca^{2+}-dependent protein kinase from pig testis, *Biochim. Biophys. Acta* **839:**276–280.

King, C. S., and Cooper, J. A., 1986, Effects of protein kinase C activation after epidermal growth factor binding on epidermal growth factor receptor phosphorylation, *J. Biol. Chem.* **261:**10073–10078.

Kinoshita, Y., Fukase, M., Yamatani, T., Hishikawa, R., and Fujita, T., 1986, Phorbol esters stimulate phosphate accumulation synergistically with A23187 in cultured renal tubular cells, *Biochem. Biophys. Res. Commun.* **136:**177–182.

Kishimoto, A., Takai, Y., and Nishizuka, Y., 1977, Activation of glycogen phosphorylase kinase by a calcium-activated, cyclic nucleotide–independent protein kinase system, *J. Biol. Chem.* **21:**7449–7452.

Kishimoto, A., Mori, T., Takai, Y., and Nishizuka, Y., 1978, Comparison of calcium-activated, cyclic nucleotide–independent protein kinase and adenosine 3':5'-monophosphate-dependent protein kinase as regards the ability to stimulate glycogen breakdown *in vitro*, *J. Biochem.* **84:**47–53.

Kishimoto, A., Takai, Y, Mori, T., Kikkawa, U., and Nishizuka, Y., 1980, Activation of calcium and phospholipid-dependent protein kinase by diacylglycerol, its possible relation to phosphatidylinositol turnover, *J. Biol. Chem.* **255:**2273–2276.

Kishimoto, A., Kajikawa, N., Shiota, M., and Nishizuka, Y., 1983, Proteolytic activation of calcium-activated phospholipid-dependent protein kinase by calcium-dependent neutral protease, *J. Biol. Chem.* **258:**1156–1164.

Kishimoto, A., Nishiyama, K., Nakanishi, H., Uratsuji, Y., Nomura, H., Takeyama, Y., and Nishizuka, Y., 1985, Studies on the phosphorylation of myelin basic protein by protein kinase C and adenosine 3':5'-monophosphate-dependent protein kinase, *J. Biol. Chem.* **260:**12492–12499.

Kiss, Z., and Luo, Y., 1986, Phorbol ester and 1,2-diolein are not fully equivalent activators of protein kinase C in respect to phosphorylation of membrane proteins *in vitro*, *FEBS Lett.* **198:**203–207.

Kiss, Z., and Steinberg, R. A., 1985, Interactions between cyclic AMP– and phorbol ester–dependent phosphorylation systems in S49 mouse lymphoma cells, *J. Cell. Physiol.* **125:**200–206.

Kitagawa, K., Nishino, H., and Iwashima, A., 1986, Effect of protein kinase C activation and Ca^{2+} mobilization on hexose transport in Swiss 3T3 cells, *Biochim. Biophys. Acta* **887:**100–104.

Kitano, Y., Okada, N., and Adachi, J., 1986, TPA-induced alteration of actin organization in cultured human keratinocytes, *Exp. Cell Res.* **167:**369–375.

Klatzmann, D., Champagne, E., Chamaret, S., Gruest, J., Guetard, D., Hercend, T., Gluckman, J.-C., and Montagnier, L., 1984, T-lymphocyte T4 molecule behaves as the receptor for human retrovirus LAV, *Nature* **312:**767–768.

Klausner, R. D., Harford, J., and van Renswoude, J., 1984, Rapid internalization of the transferrin receptor in K562 cells is triggered by ligand binding or treatment with a phorbol ester, *Proc. Natl. Acad. Sci. USA* **81:**3005–3009.

Kmiecik, T. E., and Shalloway, D., 1987, Activation and suppression of pp60^{c-src} transforming ability by mutation of its primary sites of tyrosine phosphorylation, *Cell* **49:**65–73.

Knight, D. E., and Baker, P. F., 1983, The phorbol ester TPA increases the affinity of exocytosis for calcium in "leaky" adrenal medullary cells, *FEBS Lett.* **160:**98–100.

Knopf, J. L., Lee, M.-H., Sultzman, L. A., Kriz, R. W., Loomis, C. R., Hewick, R. M., and Bell, R. M., 1986, Cloning and expression of multiple protein kinase C cDNAs, *Cell* **46:**491–502.

Koeffler, H. P., Bar-Eli, M., and Territo, M. C., 1981, Phorbol ester effect on differentiation of human myeloid leukemia cell lines blocked at different stages of maturation, *Cancer Res.* **41:**919–926.

Kohno, M., 1985, Diverse mitogenic agents induce rapid phosphorylation of a common set of cellular proteins at tyrosine in quiescent mammalian cells, *J. Biol. Chem.* **260:**1771–1779.

Kojima, I., Lippes, H., Kojima, K., and Rasmussen, H., 1983, Aldosterone secretion: Effect of phorbol ester and A23187, *Biochem. Biophys. Res. Commun.* **116:**555–562.

Kojima, I., Kojima, K., Kreutter, D., and Rasmussen, H., 1984, The temporal integration of the aldosterone secretory response to angiotensin occurs via two intracellular pathways, *J. Biol. Chem.* **259:**14448–14457.

Kojima, I., Shibata, H., and Ogata, E., 1986, Phorbol ester inhibits angiotensin-induced activation of phospholipase C in adrenal glomerulosa cells, *Biochem. J.* **237:**253–258.

Konig, B., DiNitto, P. A., and Blumberg, P. M., 1985a, Phospholipid and Ca^{++} dependency of phorbol ester receptors, *J. Cell. Biochem.* **27:**255–265.

Konig, B., DiNitto, P. A., and Blumberg, P. M., 1985b, Stoichiometric binding of diacylglycerol to the phorbol ester receptor, *J. Cell. Biochem.* **29:**37–44.

Korchak, H. M., Vienne, K., Rutherford, L. E., Wilkenfeld, C., Finkelstein, M. C., and Weissmann, G., 1984, Stimulus response coupling in the human neutrophil. II. Temporal analysis of changes in cytosolic calcium and calcium efflux, *J. Biol. Chem.* **259:**4076–4082.

Kraft, A. S., and Anderson, W. B., 1983, Phorbol esters increase the amount of Ca^{2+}, phospholipid-dependent protein kinase associated with plasma membrane, *Nature* **301:**621–623.

Kraft, A. S., Anderson, W. B., Cooper, H. L., and Sando, J. J., 1982, Decrease in cytosolic calcium/phospholipid-dependent protein kinase activity following phorbol ester treatment of EL4 thymoma cells, *J. Biol. Chem.* **257:**13193–13196.

Kraft, A. S., Smith, J. B., and Berkow, R. L., 1986, Bryostatin, an activator of the calcium phospholipid–dependent protein kinase, blocks phorbol ester–induced differentiation of human promyelocytic leukemia cells HL-60, *Proc. Natl. Acad. Sci. USA* **83:**1334–1338.

Krakauer, T., Mizel, D., and Oppenheim, J. J., 1982, Independent and synergistic thymocyte proliferative activities of PMA and IL 1, *J. Immunol.* **129:**939–941.

Kramer, C. M., and Sando, J. J., 1986, Substrates for protein kinase C in cytosol of EL4 mouse thymoma cells, *Cancer Res.* **46:**3040–3045.

Krebs, E. G., and Beavo, J. A., 1979, Phosphorylation–dephosphorylation of enzymes, *Annu. Rev. Biochem.* **48:**923–959.

Kreutter, D., Caldwell, A. B., and Morin, M. J., 1985, Dissociation of protein kinase C activation from phorbol ester–induced maturation of HL-60 leukemia cells, *J. Biol. Chem.* **260:**5979–5984.

Kristensen, T., Saris, C. J. M., Hunter, T., Hicks, L. J., Noonan, D. J., Glenney, J. R., Jr., and Tack. B. F., 1986, Primary structure of bovine calpactin I heavy chain (p36), a major cellular substrate for retroviral protein–tyrosine kinases: Homology with the human phospholipase A$_2$ inhibitor lipocortin, *Biochemistry* **25:**4497–4503.

Krueger, J. G., Garber, E. A., and Goldberg, A. R., 1983, Subcellular localization of pp60^{src} in RSV-transformed cells, *Curr. Top. Microbiol. Immunol.* **107:**51–124.

Kruijer, W., Cooper, J. A., Hunter, T., and Verma, I. M., 1984, Platelet-derived growth factor induces rapid but transient expression of the c-*fos* gene and protein, *Nature* 312:711–716.

Kuo, J. F., Andersson, R. G. G., Wise, B. C., Mackerlova, L., Salomonsson, I., Brackett, N. L., Katoh, N., Shoji, M., and Wrenn, R. W., 1980, Calcium-dependent protein kinase: Widespread occurrence in various tissues and phyla of the animal kingdom and comparison of effects of phospholipid, calmodulin, and trifluoperazine, *Proc. Natl. Acad. Sci. USA* 77:7039–7043.

Kuo, J. F., Raynor, R. L., Mazzei, J. G., Schatzman, R. C., Turner, R. S., and Wrenn, R. W., 1983, Cobra polypeptide cytotoxin I and marine worm polypeptide cytotoxin A-IV are potent and selective inhibitors of phospholipid-sensitive Ca^{2+}-dependent protein kinase, *FEBS Lett.* 153:183–186.

Kuret, J., Woodgett, J. R., and Cohen, P., 1985, Multisite phosphorylation of glycogen synthase from rabbit skeletal muscle: Identification of the sites phosphorylated by casein kinase-I, *Eur. J. Biochem.* 151:39–48.

Labarca, R., Janowski, J., Patel, J., and Paul, S. M., 1984, Phorbol esters inhibit agonist-induced [^{3}H]inositol-1-phosphate accumulation in rat hippocampal slices, *Biochem. Biophys. Res. Commun.* 123:703–709.

Lagast, H., Pozzan, T., Waldvogel, F. A., and Lew, P. D., 1984, Phorbol myristate acetate stimulates ATP-dependent calcium transport by the plasma membrane of neutrophils, *J. Clin. Invest.* 73:878–883.

Lai, Y., Nairn, A. C., and Greengard, P., 1986, Autophosphorylation reversibly regulates the Ca^{2+}/calmodulin-dependence of Ca^{2+}/calmodulin-dependent protein kinase II, *Proc. Natl. Acad. Sci. USA* 83:4253–4257.

Lam, H.-Y. P., 1984, Tamoxifen is a calmodulin antagonist in the activation of cAMP phosphodiesterase, *Biochem. Biophys. Res. Commun.* 118:27–32.

Lampe, P. D., Bazzi, M. D., Nelsestuen, G. L., and Johnson, R. G., 1986, Phosphorylation of lens intrinsic membrane proteins by protein kinase C, *Eur. J. Biochem.* 156:351–357.

Lapetina, E. G., 1982, Regulation of arachidonic acid production: Role of phospholipases C and A_2, *Trends Pharmacol. Sci.* 3:115–118.

Lapetina, E. G., Reep, B., Ganong, B. R., and Bell, R. M., 1985, Exogenous *sn*-1,2-diacylglycerols containing saturated fatty acids function as bioregulators of protein kinase C in human platelets, *J. Biol. Chem.* 260:1358–1361.

Lastick, S. M., and McConkey, E. H., 1980, Control of ribosomal protein phosphorylation in HeLa cells, *Biochem. Biophys. Res. Commun.* 95:917–923.

Laszlo, A., Radke, K., Chin, S., and Bissell, M. J., 1981, Tumor promoters alter gene expression and protein phosphorylation in avian cells in culture, *Proc. Natl. Acad. Sci. USA* 78:6241–6245.

Lau, A., Rayson, T. C., and Humphreys, T., 1986, Tumor promoters and diacylglycerol activate the Na^+/H^+ antiporter of sea urchin eggs, *Exp. Cell Res.* 166:23–30.

Laudano, A. P., and Buchanan, J. M., 1986, Phosphorylation of tyrosine in the carboxyl-terminal tryptic peptide of pp60^{c-src}, *Proc. Natl. Acad. Sci. USA* 83:892–896.

Lawen, A., and Martini, O. H. W., 1985, A chick embryo fibroblast protein kinase recognizing ribosomal protein S6, *FEBS Lett.* 185:272–276.

Laychock, S. G., 1983, Identification and metabolism of polyphosphoinositides in isolated islets of Langerhans, *Biochem. J.* 216:101–106.

Leach, K. L., James, M. L., and Blumberg, P. M., 1983, Characterization of a specific phorbol ester aporeceptor in mouse brain cytosol, *Proc. Natl. Acad. Sci. USA* 80:4208–4212.

Le Bon, T. R., Jakobs, S., Cuatrecasas, P., Kathuria, S., and Fujita-Yamaguchi, Y., 1986, Purification of insulin-like growth factor I receptor from human placental membranes, *J. Biol. Chem.* 261:7685–7689.

Lee, L., and Weinstein, I. B., 1978, Tumor promoting phorbol esters inhibit binding of epidermal growth factor to receptors, *Science* 202:313–315.

Lee, L., and Weinstein, I. B., 1979, Membrane effects of tumor promoters: Stimulation of sugar uptake in mammalian cell cultures, *J. Cell Physiol.* **99**:451–460.

Lee, M.-Y., and Bell, R. M., 1986, The lipid binding, regulatory domain of protein kinase C, *J. Biol. Chem.* **261**:14867–14870.

Leeb-Lundberg, L. M. F., Cotecchia, S., Lomasney, J. W., DeBernardis, J. F., Lefkowitz, R. J., and Caron, M. G., 1985, Phorbol esters promote α_1-adrenergic receptor phosphorylation and receptor uncoupling from inositol phospholipid metabolism, *Proc. Natl. Acad. Sci. USA* **82**:5651–5655.

Leonard, W. J., Depper, J. M., Crabtree, J. R., Rudikoff, S., Pumphrey, J., Robb, R. J., Kronke, M., Svetlik, P. B., Peffer, N. J., Waldmann, T. A., and Greene, W. C., 1984, Molecular cloning and expression of cDNAs for the human interleukin-2 receptor, *Nature* **311**:626–631.

Leonard, W. J., Depper, J. M., Kronke, M., Robb, R. J., Waldmann, T. A., and Greene, W. C., 1985, The human receptor for T-cell growth factor, *J. Biol. Chem.* **260**:1872–1880.

Le Peuch, C. J., Ballester, R., and Rosen, O., 1983, Purified rat brain calcium- and phospholipid-dependent protein kinase phosphorylates ribosomal protein S6, *Proc. Natl. Acad. Sci. USA* **80**:6858–6862.

Letendre, C. H., MacDonnell, P. C., and Guroff, G., 1977, The biosynthesis of phosphorylated tyrosine hydroxylase by organ cultures of rat adrenal medulla and superior cervical ganglia, *Biochem. Biophys. Res. Commun.* **74**:891–897.

Leung, N. L., Vickers, J. D., Kinlough-Rathbone, R. L., Reimers, H.-J., and Mustard, J. F., 1983, ADP-induced changes in [^{32}P]phosphate labelling of phosphatidylinositol 4,5-bisphosphate in washed rabbit platelets made refactory by prior ADP stimulation, *Biochem. Biophys. Res. Commun.* **113**:483–490.

Lichti, U., Patterson, E., Hennings, H., and Yuspa, S. H., 1981, The tumor promoter 12-O-tetradecanoylphorbol-13-acetate induces ornithine decarboxylase in proliferating basal cells but not in differentiating cells from mouse epidermis, *J. Cell. Physiol.* **107**:261–270.

Liles, W. C., Hunter, D. D., Meier, K. E., and Nathanson, N. M., 1986, Activation of protein kinase C induces rapid internalization and subsequent degradation of muscarinic acetylcholine receptors in neuroblastoma cells, *J. Biol. Chem.* **261**:5307–5313.

Lim, M. S., Sutherland, C., and Walsh, M. P., 1985, Phosphorylation of bovine cardiac C–protein by protein kinase C, *Biochem. Biophys. Res. Commun.* **132**:1187–1195.

Lin, C. R., Chen, W. S., Lazar, C. S., Carpenter, C. D., Gill, G. N., Evans, R. M., and Rosenfeld, M. G., 1986, Protein kinase C phosphorylation at Thr 654 of the unoccupied EGF receptor and EGF binding regulate functional receptor loss by independent mechanisms, *Cell* **44**:839–848.

Ling, E., and Sapirstein, V., 1984, Phorbol ester stimulates the phosphorylation of rabbit erythrocyte band 4.1, *Biochem. Biophys. Res. Commun.* **120**:291–298.

Ling, E., Gardner, K., and Bennett, V., 1986, Protein kinase C phosphorylates a recently identified membrane skeleton–associated calmodulin-binding protein in human erythrocytes, *J. Biol. Chem.* **261**:13875–13878.

Litchfield, D. W., and Ball, E. H., 1986, Phosphorylation of the cytoskeletal protein talin by protein kinase C, *Biochem. Biophys. Res. Commun.* **134**:1276–1283.

Litosch, I., Lee, H. S., and Fain, J. N., 1984, Phosphoinositide breakdown in blowfly salivary glands, *Am. J. Physiol.* **246**:C141–C147.

Litosch, I., Wallis, C., and Fain, J. N., 1985, 5-Hydroxytrypamine stimulates inositol phosphate production in a cell-free system from blowfly salivary glands, *J. Biol. Chem.* **260**:5464–5471.

Lotem, J., and Sachs, L., 1979, Regulation of normal differentiation in mouse and human myeloid leukemic cells by phorbol esters and the mechanism of tumor promotion, *Proc. Natl. Acad. Sci. USA* **76**:5158–5162.

Lowe, M. E., Pacifici, M., and Holtzer, H., 1978, Effects of phorbol-12-myristate-13 acetate on the phenotypic program of cultured chondroblasts and fibroblasts, *Cancer Res.* **38**:2350–2356.

Lynch, C. J., Charest, R., Bocckino, S. B., Exton, J. H., and Blackmore, P. F., 1985, Inhibition of hepatic α_1-adrenergic effects and binding by phorbol myristate acetate, *J. Biol. Chem.* **260:**2844–2851.

Lyons, R. M., and Atherton, R. M., 1979, Chracterization of a platelet protein phosphorylated during the thrombin-induced release reaction, *Biochemistry* **18:**544–552.

Lyons, R. M., Stanford, N., and Majerus, P. W., 1975, Thrombin-induced protein phosphorylation in human platelets, *J. Clin. Invest.* **56:**924–936.

Macara, I. G., Marinetti, G. V., and Balduzzi, P. C., 1984, Transforming protein on avian sarcoma virus UR2 is associated with phosphatidylinositol kinase activity: Possible role in tumorigenesis, *Proc. Natl. Acad. Sci. USA* **81:**2728–2732.

Macfarlane, D. E., 1986, Phorbol diester–induced phosphorylation of nuclear matrix proteins in HL60 promyelocytes: Possible role in differentiation studied by cationic detergent gel electrophoresis systems, *J. Biol. Chem.* **261:**6947–6953.

MacIntyre, D. E., McNicol, A., and Drummond, A. H., 1985, Tumour-promoting phorbol esters inhibit agonist-induced phosphatidate formation and Ca^{2+} flux in human platelets, *FEBS Lett.* **180:**160–164.

MacPhee, C. H., Drummond, A. H., Otto, A. M., and De Asua, L. J., 1984, Prostaglandin F2α stimulates phosphatidylinositol turnover and increases the cellular content of 1,2-diacylglycerol in confluent resting Swiss 3T3 cells, *J. Cell. Physiol.* **119:**35–40.

Madison, D. V., Malenka, R. C., and Nicoll, R. A., 1986, Phorbol esters block a voltage-sensitive chloride current in hippocampal pyramidal cells, *Nature* **321:**695–697.

Makowske, M., Birnbaum, M. J., Ballester, R. M., and Rosen, O. M., 1986, A cDNA encoding protein kinase C identifies two species of mRNA in brain and GH$_3$ cells, *J. Biol. Chem.* **261:**13389–13392.

Malenka, R. C., Madison, D. V., Andrade, R., and Nicoll, R. A., 1986, Phorbol esters mimic some cholinergic actions in hippocampal pyramidal neurons, *J. Neurosci.* **6:**475.

Malkinson, A. M., Beer, D. S., Sadler, A. J., and Coffman, D. S., 1985, Decrease in the protein kinase C–catalyzed phosphorylation of an endogenous lung protein (M_r 36,000) following treatment of mice with the tumor-modulatory agent butylated hydroxytoluene, *Cancer Res.* **45:**5751–5756.

Maller, J. L., Foulkes, J. G., Erikson, E., and Baltimore, D., 1985, Phosphorylation of ribosomal protein S6 on serine after microinjection of the Abelson murine leukemia virus tyrosine-specific protein kinase into Xenopus oocytes, *Proc. Natl. Acad. Sci. USA* **82:**272–276.

Mallorga, P., Tallman, J. F., Hennebury, R. C., Hirata, F., Strittmatter, W. T., and Axelrod, J., 1980, Mepacrine blocks β-adrenergic agonist-induced desensitization in astrocytoma cells, *Proc. Natl. Acad. Sci. USA* **77:**1341–1345.

Martin-Perez, J., Siegmann, M., and Thomas, G., 1984, EGF, PGF$_2\alpha$, and insulin induce the phosphorylation of identical S6 peptides in Swiss mouse 3T3 cells: Effect of cAMP on early sites of phosphorylation, *Cell* **36:**287–294.

Martz, A., Mookerjee, B. K., and Jung, C. Y., 1986, Insulin and phorbol esters affect the maximum velocity rather than the half-saturation constant of 3-O-methylglucose transport in rat adipocytes, *J. Biol. Chem.* **261:**13606–13609.

Massague, J., and Czech, M. P., 1982, The subunit structure of two distinct receptors for insulin-like growth factors I and II and their relationship to the insulin receptor, *J. Biol. Chem.* **257:**5038–5045.

Masters, S. B., Harden, T. K., and Brown, J. H., 1984, Relationships between phosphoinositide and calcium responses to muscarinic agonists in astrocytoma cells, *Mol. Pharmacol.* **26:**149–155.

Mastro, A. M., and Mueller, G. C., 1974, Synergistic action of phorbol esters in mitogen-activated bovine lymphocytes, *Exp. Cell Res.* **88:**40–46.

Mastro, A. M., and Smith, M. C., 1983, Calcium-dependent activation of lymphocytes by ionophore, A23187, and a phorbol ester tumor promoter, *J. Cell. Physiol.* **116:**51–56.

Matsui, T., Nakao, Y., Koizumi, T., Katakami, Y., and Fujita, T., 1986, Inhibition of phorbol ester–induced phenotypic differentiation of HL-60 cells by 1-(5-isoquinolinesulfonyl)-2-methylpiperazine, a protein kinase inhibitor, *Cancer Res.* **46**:583–587.

Matthews, J. T., and Benjamin, T. L., 1986, 12-O-tetradecanoylphorbol-13-acetate stimulates phosphorylation of the 58,000-M_r form of polyoma virus middle T antigen *in vivo:* Implications for a possible role of protein kinase C in middle T function, *J. Virol.* **58**:239–246.

May, W. S., Jacobs, S., and Cuatrecasas, P., 1984, Association of phorbol ester induced hyperphosphorylation and reversible regulation of transferrin membrane receptors in HL60 cells, *Proc. Natl. Acad. Sci. USA* **81**:2016–2020.

May, W. S., Sahyoun, N., Wolf, M., and Cuatrecasas, P., 1985a, Role of intracellular calcium mobilization in the regulation of protein kinase C–mediated membrane processes, *Nature* **317**:549–551.

May, W. S., Sahyoun, N., Jakobs, S., Wolf, M., and Cuatrecasas, P., 1985b, Mechanism of phorbol diester–induced regulation of surface transferrin receptor involves the action of activated protein kinase C and an intact cytoskeleton, *J. Biol. Chem.* **260**:9419–9426.

May, W. S., Lapetina, E. G., and Cuatrecasas, P., 1986, Intracellular activation of protein kinase C and regulation of the surface transferrin receptor by diacylglycerol is a spontaneously reversible process that is associated with rapid formation of phosphatidic acid, *Proc. Natl. Acad. Sci. USA* **83**:1281–1284.

Mazzei, G. J., and Kuo, J. F., 1984, Phosphorylation of skeletal-muscle troponin I and troponin T by phospholipid-sensitive Ca^{2+}-dependent protein kinase and its inhibition by troponin C and tropomyosin, *Biochem. J.* **218**:361–369.

Mazzei, G. J., Katoh, N., and Kuo, J. F., 1982, Polymyxin B is a more selective inhibitor for phospholipid-sensitive Ca^{2+}-dependent protein kinase than for calmodulin-sensitive Ca^{2+}-dependent protein kinase, *Biochem. Biophys. Res. Commun.* **109**:1129–1133.

Mazzei, G. J., Schatzman, R. C., Turner, R. S., Vogler, W. R., and Kuo, J. F., 1984, Phospholipid-sensitive Ca^{2+}-dependent protein kinase inhibition by R-24571, a calmodulin antagonist, *Biochem. Pharmacol.* **33**:125–130.

McCaffrey, P. G., Friedman, B. A., and Rosner, M. R., 1984, Diacylglycerol modulates binding and phosphorylation of the epidermal growth factor, *J. Biol. Chem.* **259**:12502–12507.

McDonald, J. R., and Walsh, M. P., 1985, Ca^{2+}-binding proteins from bovine brain including a potent inhibitor of protein kinase C, *Biochem. J.* **232**:559–567.

McDonald, J. R., and Walsh, M. P., 1986, Regulation of protein kinase C activity by natural inhibitors, *Biochem. Soc. Trans.* **14**:585–586.

McGuinness, T. L., Lai, Y., and Greengard, P., 1985, Ca^{2+}/calmodulin-dependent protein kinase II. Isozymic forms from rat forebrain and cerebellum, *J. Biol. Chem.* **260**:1696–1704.

McMillan, M., Chewnow, B., and Roth, B. L., 1986, Phorbol esters inhibit alpha₁-adrenergic receptor–stimulated phosphoinositide hydrolysis and contraction in rat aorta: Evidence for a link between vascular contraction and phosphoinositide turnover, *Biochem. Biophys. Res. Commun.* **134**:970–974.

McNeil, P. L., McKenna, M. P., and Taylor, D. L., 1985, A transient rise in cytosolic calcium follows stimulation of quiescent cells with growth factors and is inhibitable with phorbol myristate acetate, *J. Cell Biol.* **101**:372–379.

McTigue, M., Cremins, J., and Halegoua, S., 1985, Nerve growth factor and other agents mediate phosphorylation and activation of tyrosine hydroxylase: A convergence of multiple kinase activities, *J. Biol. Chem.* **260**:9047–9056.

Meggio, F., Marchiori, F., Borin, G., Chessa, G., and Pinna, L. A., 1984, Synthetic peptides including acidic clusters as substrates and inhibitors of rat liver casein kinase TS (type-2), *J. Biol. Chem.* **259**:14576–14579.

Meligeni, J. A., Haycock, J. W., Bennett, W. F., and Waymire, J. C., 1982, Phosphorylation and activation of tyrosine hydroxylase mediate the cAMP-induced increase in catecholamine biosynthesis in adrenal chromaffin cells, *J. Biol. Chem.* **257**:12632–12640.

Melloni, E., Pontremoli, S., Michetti, M., Sacco, O., Sparatore, B., Salamino, F., and Horecker, B. L., 1985, Binding of protein kinase C to neutrophil membranes in the presence of Ca^{2+} and its activation by a Ca^{2+} requiring proteinase, *Proc. Natl. Acad. Sci. USA* **82**:6435–6439.

Melloni, E., Pontemoli, S., Michetti, M., Sacco, O., Sparatore, B., and Horecker, B. L., 1986, The involvement of calpain in the activation of protein kinase C in neutrophils stimulated by phorbol myristic acid, *J. Biol. Chem.* **261**:4101–4105.

Mendoza, S. A., Lopez-Rivas, A., Sinnett-Smith, J. W., and Rozengurt, E., 1986a, Phorbol esters and diacylglycerol inhibit vasopressin-induced increases in cytoplasmic-free Ca^{2+} and $^{45}Ca^{2+}$ efflux in Swiss 3T3 cells, *Exp. Cell Res.* **164**:536–545.

Mendoza, S. A., Schneider, J. A., Lopez-Rivas, A., Sinnett-Smith, J. W., and Rozengurt, E., 1986b, Early events elicited by bombesin and structurally related peptides in quiescent Swiss 3T3 cells. II. Changes in Na^+ and Ca^{2+} fluxes, Na^+/K^+ pump activity, and intracellular pH, *J. Cell Biol.* **102**:2223–2233.

Merrill, A. H., Jr., Sereni, A. M., Stevens, V. L., Hannun, Y. A., Bell, R. M., and Kinkade, J. M., Jr., 1986, Inhibition of phorbol ester-dependent differentiation of human promyelocytic leukemic (HL-60) cells by sphinganine and other long-chain bases, *J. Biol. Chem.* **261**:12610–12615.

Merritt, J. E., and Rubin, R. P., 1985, Pancreatic amylase secretion and cytoplasmic free calcium: Effects of ionomycin, phorbol dibutyrate, and diacylglycerols alone and in combination, *Biochem. J.* **230**:151–159.

Miao, R. M., Fieldsteel, A. H., and Fodge, D. W., 1978, Opposing effects of tumour promoters on erythroid differentiation, *Nature* **274**:271–272.

Michell, R. H., 1975, Inositol phospholipids and cell surface receptor function, *Biochim. Biophys. Acta* **415**:81–147.

Michener, M. L., Dawson, W. B., and Creutz, C. E., 1986, Phosphorylation of a chromaffin granule–binding protein in stimulated chromaffin cells, *J. Biol. Chem.* **261**:6548–6555.

Miller, S. G., and Kennedy, M. B., 1985, Distinct forebrain and cerebellar isozymes of type II Ca^{2+}/calmodulin-dependent protein kinase associate differently with the postsynaptic density fraction, *J. Biol. Chem.* **260**:9039–9046.

Miller, S. G., and Kennedy, M. B., 1986, Regulation of brain type II Ca^{2+}/calmodulin-dependent protein kinase by autophosphorylation: A Ca^{2+}-triggered molecular switch, *Cell* **44**:861–870.

Minakuchi, R., Takai, Y., Yu, B., and Nishizuka, Y., 1981, Widespread occurrence of calcium-activated phospholipid-dependent protein kinase in mammalian tissues, *J. Biochem.* **89**:1651–1654.

Mire, A. R., Wichremasinghe, R. G., and Hoffbrand, A. V., 1986, Phytohemagglutinin treatment of T lymphocytes stimulates rapid increases in activity of both particulate and cytosolic protein kinase C, *Biochem. Biophys. Res. Commun.* **137**:128–134.

Misbahuddin, M., Isosaki, M., Houchi, H., and Oka, M., 1985, Muscarinic receptor–mediated increase in cytoplasmic free Ca^{2+} in isolated bovine adrenal medullary cells: Effect of TMB-8 and phorbol ester TPA, *FEBS Lett.* **190**:25–28.

Mitchell, R. L., Zokas, L., Schreiber, R. D., and Verma, I. M., 1985, Rapid induction of the expression of proto-oncogene fos during human monocytic differentiation, *Cell* **40**:209–217.

Mitsuse, S., and Oishi, M., 1985, Induction of a novel nuclear protein (p54) by phorbol esters in mouse erythroleukemia (Friend) cells, *Cancer Res.* **45**:3836–3842.

Miyake, R., Tanaka, Y., Tsuda, T., Kaibuchi, K., Kikkawa, U., and Nishizuka, Y., 1984, Activation of protein kinase C by non-phorbol tumor promoter, mezerein, *Biochem. Biophys. Res. Commun.* **121**:649–656.

Mizel, S. B., Rosenstreich, D., and Oppenheim, J. J., 1978, Phorbol myristic acetate stimulates LAF production by the macrophage cell line P388D$_1$, *Cell Immunol.* **40**:230–235.

Mizuguchi, J., Beaven, M. A., Li, J. H., and Paul, W. E., 1986, Phorbol myristate acetate inhibits anti-IgM-mediated signaling in resting B cells, *Proc. Natl. Acad. Sci. USA* **83:**4474–4478.

Mizuta, K., Hashimoto, E., and Yamamura, H., 1985, Proteolytic activation of protein kinase C by membrane-bound protease in rat liver plasma membrane, *Biochem. Biophys. Res. Commun.* **131:**1262–1268.

Mochly-Rosen, D., and Koshland, D. E., Jr., 1987, Domain structure and phosphorylation of protein kinase C, *J. Biol. Chem.* **262:**2291–2297.

Monroe, J. G., and Kass, M., 1985, Molecular events in B cell activation. I. Signals required to stimulate G_0 to G_1 transition of resting B lymphocytes, *J. Immunol.* **135:**1674–1682.

Montesano, R., and Orci, L., 1985, Tumor-promoting phorbol esters induce angiogenesis *in vitro, Cell* **42:**469–477.

Moolenaar, W. H., Tertoolen, L. G. J., and de Laat, S. W., 1984a, Growth factors immediately raise cytoplasmic free Ca^{2+} in human fibroblasts, *J. Biol. Chem.* **259:**8066–8069.

Moolenaar, W. H., Tertoolen, L. G. J., and de Laat, S. W., 1984b, Phorbol ester and diacylglycerol mimic growth factors in raising cytoplasmic pH, *Nature* **312:**371–374.

Moore, J. P., Todd, J. A., Hesketh, T. R., and Metcalfe, J. C., 1986, c-fos and c-myc gene activation, ionic signals, and DNA synthesis in thymocytes, *J. Biol. Chem.* **261:**8158–8162.

Mori, T., Takai, Y., Minakuchi, R., Yu, B., and Kishizuka, Y., 1980, Inhibitory action of chlor-promazine, dibucaine, and other phospholipid-interacting drugs on calcium-activated, phos-pholipid-dependent protein kinase, *J. Biol. Chem.* **255:**8378–8380.

Moroney, J., Smith, A., Tomei, L. D., and Wenner, C. E., 1978, Stimulation of $^{86}Rb^+$ and $^{32}P_i$ movements in 3T3 cells by prostaglandins and phorbol esters, *J. Cell. Physiol.* **95:**287–294.

Morris, J. D. H., Metcalfe, J. C., Smith, G. A., Hesketh, T. R., and Taylor, M. V., 1984, Some mitogens cause rapid increases in free calcium in fibroblasts, *FEBS Lett.* **169:**189–193.

Mottola, C., and Romeo, D., 1982, Calcium movement and membrane potential changes in the early phase of neutrophil activation by phorbol myristate acetate: A study with ion-selective electrodes, *J. Cell Biol.* **93:**129–134.

Movsesian, M. A., Nishikawa, M., Adelstein, R. S., 1984, Phosphorylation of phospholamban by calcium-activated, phospholipid-dependent protein kinase, *J. Biol. Chem.* **259:**8029–8032.

Mufson, R. A., Simsiman, R. C., and Boutwell, R. K., 1977, The effect of the phorbol ester tumor promoters on the basal and catecholamine-stimulated levels of cyclic adenosine 3′,5′-mono-phosphate in mouse skin and epidermis *in vivo, Cancer Res.* **37:**665–669.

Mufson, R. A., Fisher, P. B., and Weinstein, I. B., 1979a, Effect of phorbol ester tumor promoters on the expression of melanogenesis in B-16 melanoma cells, *Cancer Res.* **39:**3915–3919.

Mufson, R. A., Kulkarni, P., Eakins, K. E., and Weinstein, I. B., 1979b, Effects of phorbol ester tumor promoters in platelet aggregation and platelet production of cyclooxygenase products, *Cancer Res.* **39:**3602–3606.

Muller, R., Muller, D., and Guilbert, L., 1984, Differential expression of c-fos in hemato-poietic cells: Correlation with differentiation of monomyelocytic cells *in vitro, EMBO J.* **3:**1887–1890.

Murakami, K., Chan, S. Y., and Routtenberg, A., 1986, Protein kinase C activation by *cis*-fatty acid in the absence of Ca^{2+} and phospholipids, *J. Biol. Chem.* **261:**15424–15429.

Murao, S., Gemmell, M. A., Callaham, M. F., Anderson, N. L., and Huberman, E., 1983, Control of macrophage cell differentiation in human promyelocytic HL-60 leukemia cells by 1,25-dihydroxyvitamin D_3 and phorbol-12-myristate-13-acetate, *Cancer Res.* **43:**4989–4996.

Murdoch, G. H., Waterman, M., Evans, R. M., and Rosenfeld, M. G., 1985, Molecular mechanisms of phorbol ester, thyrotropin-releasing hormone, and growth factor stimulation of prolactin gene transcription, *J. Biol. Chem.* **260:**11852–11858.

Murray, A. W., and Fitzgerald, D. J., 1979, Tumor promoters inhibit metabolic cooperation in cocultures of epidermal and 3T3 cells, *Biochem. Biophys. Res. Commun.* **91:**395–401.

Naccache, P. H., Molski, T. F. P., Borgeat, P., White, J. R., and Sha'afi, R. I., 1985, Phorbol esters inhibit fMet-Leu-Phe- and leukotriene B_4-stimulated calcium mobilization and enzyme secretion in rabbit neutrophils, *J. Biol. Chem.* **260**:2125–2131.

Nagasawa, K., and Mak, T., 1980, Phorbol esters induce differentiation in human malignant T lymphoblasts, *Proc. Natl. Acad. Sci. USA* **77**:2964–2968.

Nagatsu, T., Levitt, M., and Udenfriend, S., 1964, Tyrosine hydroxylase: The initial step in norepinephrine biosynthesis, *J. Biol. Chem.* **239**:2910–2917.

Nagle, D. S., Jaken, S., Castagna, M., and Blumberg, P. M., 1981, Variation with embryonic development and regional localization of specific [^{3}H]phorbol 12,13-dibutyrate binding to brain, *Cancer Res.* **41**:89–93.

Nairn, A. C., Hemmings, H. C., Jr., and Greengard, P., 1985, Protein kinases in the brain, *Annu. Rev. Biochem.* **54**:931–976.

Naka, M., Nishikawa, M., Adelstein, R. S., and Hidaka, H., 1983, Phorbol ester–induced activation of human platelets is associated with protein kinase C phosphorylation of myosin light chains, *Nature* **306**:490–492.

Nakanishi, H., Nomura, H., Kikkawa, U., Kishimoto, A., and Nishizuka, Y., 1985, Rat brain and liver soluble phospholipase C: Resolution of two forms with different requirements for Ca^{2+}, *Biochem. Biophys. Res. Commun.* **132**:582–590.

Naor, Z., and Eli, Y., 1985, Synergistic stimulation of luteinizing hormone (LH) release by protein kinase C activators and Ca^{2+}-ionophore, *Biochem. Biophys. Res. Commun.* **130**:848–853.

Naor, Z., Zer, J., Zakut, H., and Hermon, J., 1985, Characterization of pituitary calcium-activated, phospholipid-dependent protein kinase: Redistribution by gonadotropin-releasing hormone, *Proc. Natl. Acad. Sci. USA* **82**:8203–8207.

Negro-Vilar, A., and Lapetina, E. G., 1985, 1,2-didecanoylglycerol and phorbol 12,13-dibutyrate enhance anterior pituitary hormone secretion *in vitro, Endocrinology* **117**:1559–1564.

Nel, A. E., Wooten, M. W., Landreth, G. E., Goldschmidt-Clermont, P. J., Stevenson, H. C., Miller, P. J., and Galbraith, R. M., 1986, Translocation of phospholipid/Ca^{2+}-dependent protein kinase in B-lymphocytes activated by phorbol ester or cross-linking of membrane immunoglobulin, *Biochem. J.* **233**:145–149.

Nestler, E. J., Walaas, S. I., and Greengard, P., 1984, Neuronal phosphoproteins: Physiological and clinical implications, *Science* **225**:1357–1364.

Nettelblad, F. A., Forsberg, P., Humble, E., and Egstrom, L., 1986, Aspects on the phosphorylation of muscle phosphofructokinase by protein kinase C—Inhibition by phosphofructokinase stabilisers, *Biochem. Biophys. Res. Commun.* **136**:445–453.

Niedel, J. E., Kuhn, L. J., and Vandenbark, G. R., 1983, Phorbol diester receptor copurifies with protein kinase C, *Proc. Natl. Acad. Sci. uSA* **80**:36–40.

Nielsen, P. J., Thomas, G., and Maller, J. L., 1982, Increased phosphorylation of ribosomal protein S6 during meiotic maturation of Xenopus oocytes, *Proc. Natl. Acad. Sci. USA* **79**:2937–2941.

Nigg, E. A., Sefton, B. M., Singer, S. J., and Vogt, P. K., 1986, Cytoskeletal organization, vinculin–phosphorylation, and fibronectin expression in transformed fibroblasts with different cell morphologies, *Virology* **151**:50–65.

Nikaido, T., Shimizu, A., Ishida, N., Sabe, H., Teshigawara, K., Maeda, M., Uchiyama, T., Todoi, J., and Honjo, T., 1984, Molecular cloning of cDNA encoding human interleukin-2 receptor, *Nature* **311**:631–635.

Nishihara, J., McPhail, L. C., and O'Flaherty, J. T., 1986, Stimulus-dependent mobilization of protein kinase C, *Biochem. Biophys. Res. Commun.* **134**:587–594.

Nishikawa, M., Hidaka, H., and Adelstein, R. S., 1983, Phosphorylation of smooth muscle heavy meromyosin by calcium-activated, phospholipid-dependent protein kinase, *J. Biol. Chem.* **258**:14069–14072.

Nishikawa, M., Sellers, J. R., Adelstein, R. S., and Hidaka, H., 1984, Protein kinase C modulates *in vitro* phosphorylation of the smooth muscle heavy meromyosin by myosin light chain kinase, *J. Biol. Chem.* **259**:8808–8814.

Nishikawa, M., Shirakawa, S., and Adelstein, R. S., 1985, Phosphorylation of smooth muscle myosin light chain kinase by protein kinase C, *J. Biol. Chem.* **260**:8978–8983.

Nishizuka, Y., 1983, Phospholipid degradation and signal translation for protein phosphorylation, *Trends Biochem. Sci.* **8**:13–16.

Nishizuka, Y., 1984, The role of protein kinase C in cell surface signal transduction and tumour promotion, *Nature* **308**:693–698.

Nishizuka, Y., 1986, Studies and perspectives of protein kinase C, *Science* **233**:305–312.

Noda, S., Horn, F., Linder, D., and Schoner, W., 1986, Purified pyruvate kinases type M_2 from unfertilized hen's egg are substrates of protein kinase C, *Eur. J. Biochem.* **155**:643–651.

Nose, P. S., Griffith, L. C., and Schulman, H., 1985, Ca^{2+}-dependent phosphorylation of tyrosine hydroxylase in PC12 cells, *J. Cell Biol.* **101**:1182–1190.

Novak-Hofer, I., and Thomas, G., 1984, An activated S6 kinase in extracts from serum- and epidermal growth factor-stimulated Swiss 3T3 cells, *J. Biol. Chem.* **259**:5995–6000.

Novak-Hofer, I., and Thomas, G., 1985, Epidermal growth factor–mediated activation of an S6 kinase in Swiss mouse 3T3 cells, *J. Biol. Chem.* **260**:10314–10319.

O'Brian, C. A., Lawrence, D. S., Kaiser, E. T., and Weinstein, I. B., 1984, Protein kinase C phosphorylates the synthetic peptide ARG-ARG-LYS-ALA-SER-GLY-PRO-PRO-VAL in the presence of phospholipid plus either Ca^{2+} or a phorbol ester tumor promoter, *Biochem. Biophys. Res. Commun.* **124**:296–302.

O'Brian, C. A., Liskamp, R. M., Solomon, D. H., and Weinstein, I. B., 1985, Inhibition of protein kinase C by tamoxifen, *Cancer Res.* **45**:2462–2465.

O'Brien, T. G., 1982, Hexose transport in undifferentiated and differentiated BALB/c 3T3 preadipose cells: Effects of 12-O-tetradecanoylphorbol-13-acetate and insulin, *J. Cell. Physiol.* **110**:63–71.

O'Brien, T. G., Lewis, M. A., and Diamond, L., 1979a, Ornithine decarboxylase activity and DNA synthesis after treatment of cells in culture with 12-O-tetradecanoylphorbol-13-acetate, *Cancer Res.* **39**:4477–4480.

O'Brien, T. G., Saladik, D., and Diamond, L., 1979b, The tumor promoter 12-O-tetradecanoyl-phorbol-13-acetate stimulates lactate production in BALB/c 3T3 preadipose cells, *Biochem. Biophys. Res. Commun.* **88**:103–110.

Oettgen, H. C., Pettey, C. L., Maloy, W. L., and Terhorst, C., 1986, A T3-like protein complex associated with the antigen receptor on murine T cells, *Nature* **320**:272–275.

O'Flaherty, J. T., Redman, J. F., and Jacobson, D. P., 1986, Protein kinase C regulates leukotriene B_4 receptors in human neutrophils, *FEBS Lett.* **206**:279–282.

Ohno, S., Kawasaki, H., Imajoh, S., Suzuki, K., Inagaki, M., Yokokura, H., Sakoh, T., and Hidaka, H., 1987, Tissue-specific expression of three distinct types of rabbit protein kinase C, *Nature* **325**:161–166.

Ohtsuka, T., Okamura, N., and Ishibashi, S., 1986, Involvement of protein kinase C in the phosphorylation of 46 kDa proteins which are phosphorylated in parallel with activation of NADPH oxidase in intact guinea-pig polymorphonuclear leukocytes, *Biochim. Biophys. Acta* **888**:332–337.

Ojakian, G. K., 1981, Tumor promoter–induced changes in the permeability of epithelial cell tight junctions, *Cell* **23**:95–103.

Ono, Y., Kurokawa, T., Fujii, T., Kawahara, K., Igarashi, K., Kikkawa, U., Ogita, K., and Nishizuka, Y., 1986a, Two types of complementary DNAs of rat brain protein kinase C: Heterogeneity determined by alternative splicing, *FEBS Lett.* **206**:347–352.

Ono, Y., Kurokawa, T., Kawahara, K., Nishimura, O., Marumoto, R., Igarashi, K., Sugino, Y., Kikkawa, U., Ogita, K., and Nishizuka, Y., 1986b, Cloning of rat brain protein kinase C complementary DNA, *FEBS Lett.* **203**:111–115.

Orellana, S. A., Solski, P. A., and Brown, J. H., 1985, Phorbol ester inhibits phosphoinositide hydrolysis and calcium mobilization in cultured astrocytoma cells, *J. Biol. Chem.* **260**:5236–5239.

Osborne, C. K., Hamilton, B., Nover, M., and Ziegler, J., 1981, Antagonism between epidermal growth factor and phorbol ester tumor promoters in human breast cancer cells, *J. Clin. Invest.* **67**:943–951.

Osborne, R., and Tashjian, Jr., A. H., 1981, Tumor-promoting phorbol esters affect production of prolactin and growth hormone by rat pituitary cells, *Endocrinology* **108**:1164–1170.

Otani, S., Kuramoto, A., and Morisawa, S., 1985, Induction of ornithine decarboxylase in guinea-pig lymphocytes: Synergistic effect of diacylglycerol and calcium, *Eur. J. Biochem.* **147**:27–31.

Owen, N. E., 1985, Effect of TPA on ion fluxes and DNA synthesis in vascular smooth muscle cells, *J. Cell Biol.* **101**:454–459.

Padel, U., and Soling, H.-D., 1985, Phosphorylation of the ribosomal protein S6 during agonist-induced exocytosis in exocrine glands is catalyzed by calcium-phospholipid-dependent protein kinase (protein kinase C), *Eur. J. Biochem.* **151**:1–10.

Pahlman, S., Ruusala, A.-I., Abrahamsson, L., Odelstad, L., and Nilsson, K., 1983, Kinetics and concentration effects of TPA-induced differentiation of cultured human neuroblastoma cells, *Cell Differ.* **12**:165–170.

Palfrey, H. C., and Waseem, A., 1985, Protein kinase C in the human erythrocyte, *J. Biol. Chem.* **260**:16021–16029.

Pandol, S. J., and Schoeffield, M. S., 1986, 1,2-diacylglycerol, protein kinase C, and pancreatic enzyme secretion, *J. Biol. Chem.* **261**:4438–4444.

Papanikolaou, P., Humble, E., and Engstrom, L., 1982, Phosphorylation of human fibrinogen *in vitro* with calcium-activated, phospholipid-dependent protein kinase and [^{32}P]ATP, *FEBS Lett.* **143**:199–204.

Papayannopoulou, T., Nakamoto, B., Yokochi, T., Chait, A., and Kannagi, R., 1983, Human erythroleukemia cell line (HEL) undergoes a drastic macrophage-like shift with TPA, *Blood* **62**:832–845.

Papini, E., Grzeskowiak, M., Bellavite, P., and Rossi, F., 1985, Protein kinase C phosphorylates a component of NADPH oxidase of neutrophils, *FEBS Lett.* **190**:204–208.

Park, S., and Rasmussen, H., 1985, Activation of tracheal smooth muscle contraction: Synergism between Ca^{2+} and activators of protein kinase C, *Proc. Natl. Acad. Sci. USA* **82**:8835–8839.

Parker, P. J., Stabel S., and Waterfield, M. D., 1984a, Purification to homogeneity of protein kinase C from bovine brain—Identity with the phorbol ester receptor, *EMBO J.* **3**:953–959.

Parker, R. C., Varmus, H. E., and Bishop, J. M., 1984b, Expression of v-*src* and chicken c-*src* in rat cells demonstrates qualitative differences between pp60^{v-src} and pp60^{c-src}, *Cell* **37**:131–139.

Parker, P. J., Katan, M., Waterfield, M. D., and Leader, D. P., 1985, The phosphorylation of eukaryotic ribosomal protein S6 by protein kinase C, *Eur. J. Biochem.* **148**:579–586.

Parker, P. J., Coussens, L., Totty, N., Rhee, L., Young, S., Chen, E., Stabel, S., Waterfield, M. D., and Ullrich, A., 1986, The complete primary structure of protein kinase C—The major phorbol ester receptor, *Science* **233**:853–859.

Parsons, S. J., and Creutz, C. E., 1986, pp60^{c-src} activity detected in the chromaffin granule membrane, *Biochem. Biophys. Res. Commun.* **134**:736–742.

Patschinsky, T., Hunter, T., Esch, F. S., Cooper, J. A., and Sefton, B. M., 1982, Analysis of the sequence of amino acids surrounding sites of tyrosine phosphorylation, *Proc. Natl. Acad. Sci. USA* **79**:973–977.

Patschinsky, T., Hunter, T., and Sefton, B. M., 1986, Phosphorylation of the transforming protein of Rous sarcoma virus: Direct demonstration of phosphorylation of serine 17 and identification of an additional site of tyrosine phosphorylation in pp60^{v-src} of Prague Rous sarcoma virus, *J. Virol.* **59**:73–81.

Patskan, G. J., and Baxter, C. S., 1985, Specific stimulation of histone H2B and H4 phosphorylation in mouse lymphocytes by 12-O-tetradecanoylphorbol 13-acetate, *J. Biol. Chem.* **260**:12899–12903.

Payette, R., Biehl, J., Toyama, Y., Holtzer, S., and Holtzer, H., 1980, Effects of 12-O-tetradecanoylphorbol-13-acetate on the differentiation of avian melanocytes, *Cancer Res.* **40**:2465–2474.

Pearson, R. B., Woodgett, J. R., Cohen, P., and Kemp, B. E., 1985, Substrate specificity of a multifunctional calmodulin-dependent protein kinase, *J. Biol. Chem.* **260:**14471–14476.

Pearson, R. B., Misconi, L. Y., and Kemp, B. E., 1986, Smooth muscle myosin light chain kinase requires residues on the COOH-terminal side of the phosphorylation site, *J. Biol. Chem.* **261:**25–27.

Pegoraro, L., Abrahm, J., Cooper, R. A., Levis, A., Lange, B., Meo, P., and Rovera, G., 1980, Differentiation of human leukemias in response to 12-O-tetradecanoylphorbol-13-acetate *in vitro, Blood* **55:**859–862.

Penfield, A., and Dale, M. M., 1984, Synergism between A23187 and 1-oleoyl-2-acetyl-glycerol in superoxide production by human neutrophils, *Biochem. Biophys. Res. Commun.* **125:**332–336.

Pepinsky, R. B., and Sinclair, L. K., 1986, Epidermal growth factor–dependent phosphorylation of lipocortin, *Nature* **321:**81–84.

Perisic, O., and Traugh, J. A., 1983, Protease-activated kinase II mediates multiple phosphorylation of ribosomal protein S6 in reticulocytes, *J. Biol. Chem.* **258:**13998–14002.

Petruzzelli, L. M., Ganguly, S., Smith, C. J., Cobb, M. H., Rubin, C. S., and Rosen, O. M., 1982, Insulin activates a tyrosine-specific protein kinase in extracts of 3T3-L1 adipocytes and human placenta, *Proc. Natl. Acad. Sci. USA* **79:**6792–6796.

Pierschbacher, M. D., and Ruoslahti, E., 1984, Cell attachment activity of fibronectin can be duplicated by small synthetic fragments of the molecule, *Nature* **309:**30–33.

Pilch, P. F., and Czech, M. P., 1979, Interaction of cross-linking agents with the insulin effector systems of isolated fat cells, *J. Biol. Chem.* **254:**3375–3381.

Piwnica-Worms, H., Saunders, K. B., Roberts, T. M., Smith, A. E., and Cheng, S. H., 1987, Tyrosine phosphorylation regulates the biochemical and biological properties of pp60^{c-src}, *Cell* **49:**75–82.

Pocotte, S. L., and Holz, R. W., 1986, Effects of phorbol ester on tyrosine hydrolxylase phosphorylation and activation in cultured bovine adrenal chromaffin cells, *J. Biol. Chem.* **261:**1873–1877.

Pocotte, S. L., Frye, R. A., Senter, R. A., TerBush, D. R., Lee, S. A., and Holz, R. W., 1985, Effects of phorbol ester on catecholamine secretion and protein phosphorylation in adrenal medullary cell cultures, *Proc. Natl. Acad. Sci. USA* **82:**930–934.

Poll, C., and Westwick, J., 1986, Phorbol esters modulate thrombin-operated calcium mobilisation and dense granule release in human platelets, *Biochim. Biophys. Acta* **886:**434–440.

Pollock, W. K., and Rink, T. J., 1986, Thrombin and ionomycin can raise platelet cytosolic Ca^{2+} to micromolar levels by discharge of internal Ca^{2+} stores: Studies using fura-2, *Biochem. Biophys. Res. Commun.* **139:**308–314.

Pollock, W. K., Sage, S. O., and Rink, T. J., 1987, Stimulation of Ca^{2+} efflux from fura-2-loaded platelets activated by thrombin or phorbol myristate acetate, *FEBS Lett.* **210:**132–136.

Pontremoli, S., Melloni, E., Michetti, M., Sacco, O., Salamino, F., Sparatore, B., and Horecker, B. L., 1986a, Biochemical responses in activated human neutrophils mediated by protein kinase C and a Ca^{2+}-requiring proteinase, *J. Biol. Chem.* **261:**8309–8313.

Pontremoli, S., Melloni, E., Michetti, M., Sparatore, B., Salamino, F., Damiani, G., and Horecker, B. L., 1986b, Cytolytic effects of neutrophils: Role for a membrane-bound neutral proteinase, *Proc. Natl. Acad. Sci. uSA* **83:**1685–1689.

Pozzan, T., Lew, D. P., Wollheim, C. B., and Tsien, R. Y., 1983, Is cytosolic ionized calcium regulating neutrophil activation? *Science* **221:**1413–1415.

Pozzan, T., Gatti, G., Dozio, N., Vicentini, L. M., and Meldolesi, J., 1984, Ca^{2+}-dependent and -independent release of neurotransmitters from PC12 cells: A role for protein kinase C activation? *J. Cell Biol.* **99:**628–638.

Preiss, J., Loomis, C. R., Bishop, W. R., Stein, R., Niedel, J. E., and Bell, R. M., 1986, Quantitative measurement of *sn*-1,2-diacylglycerols present in platelets, hepatocytes, and *ras*- and *sis*-transformed normal rat kidney cells, *J. Biol. Chem.* **261:**8597–8600.

Presti, C. F., Scott, B. T., and Jones, L. R., 1985, Identification of an endogenous protein kinase

C activity and its intrinsic 15-dilodalton substrate in purified canine cardiac sarcolemmal vesicles, *J. Biol. Chem.* **260**:13879–13889.

Purchio, A. F., Shoyab, M., and Gentry, L. E., 1985, Site-specific increased phosphorylation of pp60^{v-src} after treatment of RSV-transformed cells with a tumor promoter, *Science* **229**:1393–1395.

Purchio, A. F., Gentry, L., and Shoyab, M., 1986, Phosphorylation of pp60^{v-src} by the TPA receptor kinase (protein kinase C), *Virology* **150**:524–529.

Pytela, R., Pierschbacher, M. D., and Ruoslahti, E., 1985, A 125/115-kDa cell surface receptor specific for vitronectin interacts with the arginine-glycine-aspartic acid adhesion sequence derived from fibronectin, *Proc. Natl. Acad. Sci. USA* **82**:5766–5770.

Qi, D.-F., Schatzman, R. C., Mazzei, G. J., Turner, R. S., Raynor, R. L., Leao, S., and Kuo, J. F., 1983, Polyamines inhibit phospholipid-sensitive and calmodulin-sensitive Ca^{2+}-dependent protein kinases, *Biochem. J.* **213**:281–288.

Quigley, J. P., 1979, Phorbol ester–induced morphological changes in transformed chick fibroblasts: Evidence for direct catalytic involvement of plasminogen activator, *Cell* **17**:131–141.

Quintanilla, M., Brown, K., Ramsden, M., and Balmain, A., 1986, Carcinogen-specific mutation and amplification of Ha-*ras* during mouse skin carcinogenesis, *Nature* **322**:78–80.

Rabin, M. S., Doherty, P. J., and Gottesman, M. M., 1986, The tumor promoter phorbol 12-myristate 13-acetate induces a program of altered gene expression similar to that induced by platelet-derived growth factor and transforming oncogenes, *Proc. Natl. Acad. Sci. USA* **83**:357–360.

Rackoff, W. R., Rubin, R. A., and Earp, H. S., 1984, Phosphorylation of the hepatic EGF receptor with cAMP-dependent protein kinase, *Mol. Cell. Endocrinol.* **34**:113–119.

Radke, K., Gilmore, T., and Martin, G. S., 1980, Transformation by Rous sarcoma virus: A cellular substrate for transformation-specific protein phosphorylation contains phosphotyrosine, *Cell* **21**:821–828.

Ralston, R., and Bishop, J. M., 1985, The product of the protooncogene c-*src* is modified during the cellular response to platelet-derived growth factor, *Proc. Natl. Acad. Sci. USA* **82**:7845–7849.

Ramachandran, C., Yau, P., Bradbury, E. M., Shyamala, G., Yasuda, H., and Walsh, D. A., 1984, Phosphorylation of high-mobility-group proteins by the calcium-phospholipid-dependent protein kinase and the cyclic AMP–dependent protein kinase, *J. Biol. Chem.* **259**:13495–13503.

Rance, A. J., Thonnes, M., and Issinger, O.-G., 1985, Ribosomal protein S6 phosphorylation and morphological changes in response to the tumour promoter 12-O-tetradecanoylphorbol 13-acetate in primary human tumour cells, established and transformed cell lines, *Biochim. Biophys. Acta* **847**:128–131.

Rane, S. G., and Dunlap, K., 1986, Kinase C activator 1,2-oleoylacetylglycerol attenuates voltage-dependent calcium current in sensory neurons, *Proc. Natl. Acad. Sci. USA* **83**:184–188.

Rebecchi, M. J., and Gershengorn, M. C., 1983, Thyroliberin stimulates rapid hydrolysis of phosphatidylinositol 4,5-bisphosphate by a diesterase in rat mammotropic pituitary cells, *Biochem. J.* **216**:287–294.

Rebois, R. V., and Patel, J., 1985, Phorbol ester causes desensitization of gonadotropin-responsive adenylate cyclase in a murine Leydig tumor cell line, *J. Biol. Chem.* **260**:8026–8031.

Reed, J. C., Nowell, P. D., and Hoover, R. G., 1985, Regulation of c-*myc* mRNA levels in normal human lymphocytes by modulators of cell proliferation, *Proc. Natl. Acad. Sci. USA* **82**:4221–4224.

Reiners, J. J., Jr., and Slaga, T. J., 1983, Effects of tumor promoters on the rate and commitment to terminal differentiation of subpopulations of murine keratinocytes, *Cell* **32**:247–255.

Reinherz, E. L., Meuer, S., Fitzgerald, K. A., Hussey, R. E., Levine, H., and Schlossman, S. F., 1982, Antigen recognition by human T lymphocytes is linked to surface expression of the T3 molecular complex, *Cell* **30**:735–743.

Rhodes, D., Prpic, V., Exton, J. H., and Blackmore, P. F., 1983, Stimulation of phosphatidylinositol 4,5-bisphosphate hydrolysis in hepatocytes by vasopressin, *J. Biol. Chem.* **258**:2770–2773.

Rickard, J. E., and Sheterline, P., 1985, Evidence that phorbol ester interferes with stimulated Ca^{2+} redistribution by activating Ca^{2+} efflux in neutrophil leucocytes, *Biochem. J.* **231**:623–628.

Rider, M. H., and Hue, L., 1986, Phosphorylation of purified bovine heart and rat liver 6-phosphofructo-2-kinase by protein kinase C and comparison of the fructose-2,6-bisphosphatase activity of the two enzymes, *Biochem. J.* **240:**57–61.

Rifkin, D. B., Crowe, R. M., and Pollack, R., 1979, Tumor promoters induce changes in the chick embryo fibroblast cytoskeleton, *Cell* **18:**361–368.

Rink, T. J., Smith, S. W., and Tsien, R. Y., 1982, Cytoplasmic free Ca^{2+} in human platelets: Ca^{2+} thresholds and Ca-independent activation for shape-change and secretion, *FEBS Lett.* **148:**21–26.

Rink, T. J., Sanchez, A., and Hallam, T. J., 1983, Diacylglycerol and phorbol ester stimulate secretion without raising cytoplasmic free calcium in human platelets, *Nature* **305:**317–319.

Rittenhouse, S. E., and Sasson, J. P., 1985, Mass changes in myoinositol trisphosphate in human platelets stimulated by thrombin, *J. Biol. Chem.* **260:**8657–8660.

Rittenhouse-Simmons, S., 1981, Differential activation of platelet phospholipases by thrombin and ionophore A23187, *J. Biol. Chem.* **256:**4153–4155.

Rittenhouse, S. E., and Sasson, J. P., 1985, Mass changes in myoinositol trisphosphate in human platelets stimulated by thrombin, *J. Biol. Chem.* **260:**8657–8660.

Roach, P. J., 1981, Glycogen synthase and glycogen synthase kinases, *Curr. Top. Cell. Reg.* **20:**45–105.

Roach, P. J., and Goldman, M., 1983, Modification of glycogen synthase activity in isolated rat hepatocytes by tumor-promoting phorbol esters: Evidence for differential regulation of glycogen synthase and phosphorylase, *Proc. Natl. Acad. Sci. USA* **80:**7170–7172.

Robb, R. J., and Rusk, C. M., 1986, High and low affinity receptors for interleukin 2: Implications of pronase, phorbol ester, and cell membrane studies upon the basis for differential ligand affinities, *J. Immunol.* **137:**142–149.

Robinson, J. M., Badwey, J. A., Karnovsky, M. L., and Karnovsky, M. J., 1984, Superoxide release by neutrophils: Synergistic effects of a phorbol ester and a calcium ionophore, *Biochem. Biophys. Res. Commun.* **122:**734–739.

Rodriguez-Pena, A., and Rozengurt, E., 1985, Serum, like phorbol esters, rapidly activates protein kinase C in intact quiescent fibroblasts, *EMBO J.* **4:**71–76.

Rodriguez-Pena, A., and Rozengurt, E., 1986, Phosphorylation of an acidic mol. wt. 80 000 cellular protein in a cell-free system and intact Swiss 3T3 cells: A specific marker of protein kinase C activity, *EMBO J.* **5:**77–83.

Rohrschneider, L., and Rosok, M. J., 1983, Transformation parameters and pp60*src* localization in cells infected with partial transformation mutants of Rous sarcoma virus, *Mol. Cell. Biol.* **3:**731–746.

Ronning, S. A., and Martin, T. F. J., 1986, Characterization of phorbol ester- and diacylglycerol-stimulated secretion in permeable GH_3 pituitary cells: Interaction with Ca^{2+}, *J. Biol. Chem.* **17:**7840–7845.

Rosenstreich, D. L., and Mizel, S. B., 1979, Signal requirements for T lymphocyte activation. I. Replacement of macrophage function with phorbol myristic acetate, *J. Immunol.* **123:**1749–1754.

Roskoski, R., Jr., 1986, Regulation of tyrosine hydroxylase by cyclic GMP-dependent phosphorylation, *Trans. Am. Soc. Neurochem.* **17:**263.

Roskoski, R., Jr., and Roskoski, L. M., 1987, Activation of tyrosine hydroxylase in PC 12 cells by the cyclic GMP and cyclic AMP second messenger systems, *J. Neurochem.* **48:**236–242.

Rosoff, P. M., and Cantley, L. C., 1985, Lipopolysaccharide and phorbol esters induce differentiation but have opposite effects on phosphatidylinositol turnover and Ca^{2+} mobilization in 70Z/3 pre-B lymphocytes, *J. Biol. Chem.* **260:**9209–9215.

Rosoff, P. M., Stein, L. F., and Cantley, L. C., 1984, Phorbol esters induce differentiation in a pre-B-lymphocyte cell line by enhancing Na^+/H^+ exchange, *J. Biol. Chem.* **259:**7056–7060.

Rosok, M. J., and Rohrschneider, L. R., 1983, Increased phosphorylation of vinculin on tyrosine does not occur during the release of stress fibers before mitosis in normal cells, *Mol. Cell. Biol.* **3:**475–479.

Rouis, M., Thomopoulos, P., Haziot, A., and Broquet, C., 1986, Phosphorylation of class I HLA antigen in U-937 monocyte-like cells, *Exp. Cell Res.* **164:**556–561.

Rovera, G., O'Brien, T. G., and Diamond, L., 1977, Tumor promoters inhibit spontaneous differentiation of Friend erythroleukemia cells in culture, *Proc. Natl. Acad. Sci. USA* **74:**2894–2898.

Rovera, G., O'Brien, T. G., and Diamond, L., 1979a, Induction of differentiation in human promyelocytic leukemia cells by tumor promoters, *Science* **204:**868–870.

Rovera, G., Santoli, D., and Damsky, C., 1979b, Human promyelocytic leukemia cells in culture differentiate into macrophage-like cells when treated with a phorbol diester, *Proc. Natl. Acad. Sci. USA* **76:**2779–2783.

Rovera, G., Ferrero, D., Pagliardi, G. L., Vartikar, J., Pessano, S., Bottero, L., Abraham, S., and Lebman, D., 1982, Induction of differentiation of human myeloid leukemias by phorbol diesters: Phenotypic changes and mode of action, *Ann. NY Acad. Sci.* **379:**211–220.

Rozengurt, E., Rodriguez-Pena, M., and Smith, K., 1983, Phorbol ester, phospholipase C, and growth factors rapidly stimulate the phosphorylation of a M_r 80,000 protein in intact quiescent 3T3 cells, *Proc. Natl. Acad. Sci. USA* **80:**7244–7248.

Rozengurt, E., Rodriguez-Pena, A., Coombs, M., and Sinnett-Smith, J., 1984, Diacylglycerol stimulates DNA synthesis and cell division in mouse 3T3 cells: Role of Ca^{2+}-sensitive phospholipid-dependent protein kinase, *Proc. Natl. Acad. Sci. USA* **81:**5748–5752.

Rubin, J. B., Shia, M. A., and Pilch, P. F., 1983, Stimulation of tyrosine-specific phosphorylation *in vitro* by insulin-like growth factor. I. *Nature* **305:**438–440.

Russell, J. H., McCulley, D. E., and Taylor, A. S., 1986, Antagonistic effects of phorbol esters on lymphocyte activation: Evidence that protein kinase C provides an early signal associated with lytic function, *J. Biol. Chem.* **261:**12643–12648.

Ryffel, B., Henning, C. B., and Huberman, E., 1982, Differentiation of human T-lymphoid leukemia cells into cells that have a suppressor phenotype is induced by phorbol 12-myristate 13-acetate, *Proc. Natl. Acad. Sci. USA* **79:**7336–7340.

Sagi-Eisenberg, R., Lieman, H., and Pecht, I., 1985, Protein kinase C regulation of the receptor-coupled calcium signal in histamine-secreting rat basophilic leukaemia cells, *Nature* **313:**59–60.

Sahai, A., Feuerstein, N., Cooper, H. L., and Salomon, D. S., 1986, Effect of epidermal growth factor and 12-O-tetradecanoylphorbol-13-acetate on the phosphorylation of soluble acidic proteins in A431 epidermoid carcinoma cells, *Cancer Res.* **46:**4143–4150.

Sahyoun, N., Wolf, M., Besterman, J., Hsieh, T.-S., Sander, M., LeVine, H., III, Chang, K.-J., and Cuatrecasas, P., 1986, Protein kinase C phosphorylates topoisomerase II. Topoisomerase activation and its possible role in phorbol ester–induced differentiation of HL-60 cells, *Proc. Natl. Acad. Sci. USA* **83:**1603–1607.

Sakamoto, C., Matozaki, T., Nagao, M., and Baba, S., 1985, Combined effect of phorbol ester and A23187 or dibutryl cyclic AMP on pepsinogen secretion from isolated gastric glands, *Biochem. Biophys. Res. Commun.* **131:**314–319.

Sando, J. J., and Young, M., 1983, Identification of high-affinity phorbol ester receptor in cytosol of EL4 thymoma cells: Requirement for calcium, magnesium, and phospholipids, *Proc. Natl. Acad. Sci. USA* **80:**2642–2646.

Sano, K., Takai, Y., Yamanishi, J., and Nishizuka, Y., 1983, A role of calcium-activated phospholipid-dependent protein kinase in human platelet activation, *J. Biol. Chem.* **258:**2010–2013.

Sano, K., Voelker, D. R., and Mason, R. J., 1985a, Involvement of protein kinase C in pulmonary surfactant secretion from alveolar type II cells, *J. Biol. Chem.* **260:**12725–12729.

Sano, K., Nakamura, H., Matsuo, T., Kawahara, Y., Fukuzaki, H., Kaibuchi, K., and Takai, Y., 1985b, Comparison of the modes of action of Ca^{2+} ionophore A23187 and thrombin in protein kinase C activation in human platelets, *FEBS Lett.* **192:**4–8.

Saris, C. J. M., Tack, B. F., Kristensen, T., Glenney, J. R., Jr., and Hunter, T., 1986, The cDNA sequence for the protein–tyrosine kinase substrate p36 (calpactin 1 heavy chain) reveals a multidomain protein with internal repeats, *Cell* **46:**201–212.

Sato, C., Nishizawa, K., Nakayama, T., and Kobayashi, T., 1985, Effect upon mitogenic stimulation of calcium-dependent phosphorylation of cytoskeleton-associated 350,000- and 80,000-mol-wt polypeptides in quiescent 3Y1 cells, *J. Cell Biol.* **100:**748–753.

Sawyer, S. T., and Cohen, S., 1981, Enhancement of calcium uptake and phosphatidylinositol turnover by epidermal growth factor in A-431 cells, *Biochemistry* **20:**6280–6286.

Sawyer, S. T., and Cohen, S., 1985, Epidermal growth factor stimulates the phosphorylation of the calcium-dependent 35,000-dalton substrate in intact A-431 cells, *J. Biol. Chem.* **260:**8233–8236.

Schatzman, R. C., Wise, B. C., and Kuo, J. F., 1981, Phospholipid-sensitive calcium-dependent protein kinase: Inhibition by anti-psychotic drugs, *Biochem. Biophys. Res. Commun.* **98:**669–676.

Schatzman, R. C., Grifo, J. A., Merrick, W. C., and Kuo, J. F., 1983a, Phospholipid-sensitive Ca^{2+}-dependent protein kinase phosphorylates the β-subunit of eukaryotic initiation factor 2 (EIF-2), *FEBS Lett.* **159:**167–170.

Schatzman, R. C., Raynor, R. L., Fritz, R. B., and Kuo, J. F., 1983b, Purification to homogeneity, characterization, and monoclonal antibodies of phospholipid-sensitive Ca^{2+}-dependent protein kinase from spleen, *Biochem. J.* **209:**435–443.

Scheurich, P., Unglaub, R., Maxeiner, B., Thoma, B., Zugmaier, G., and Pfizenmaier, K., 1986, Rapid modulation of tumor necrosis factor membrane receptors by activators of protein kinase C, *Biochem. Biophys. Res. Commun.* **141:**855–860.

Schimmel, R. J., Dzierzanowski, D., Elliott, M. E., and Honeyman, T. W., 1986, Stimulation of phosphoinositide metabolism in hamster brown adipocytes exposed to α_1-adrenergic agents and its inhibition with phorbol esters, *Biochem. J.* **236:**757–764.

Schleimer, R. P., Gillespie, E., Daiuta, R., and Lichtenstein, L. M., 1982, Release of histamine from human leukocytes stimulated with the tumor-promoting phorbol diesters. II. Interaction with other stimuli, *J. Immunol.* **128:**136–140.

Schliwa, M., Nakamura, T., Porter, K. R., and Euteneuer, U., 1984, A tumor promoter induces rapid and coordinated reorganization of actin and vinculin in cultured cells, *J. Cell Biol.* **99:**1045–1059.

Schneider, C., Zanetti, M., and Romeo, D., 1981, Surface-reactive stimuli selectively increase protein phosphorylation in human neutrophils, *FEBS Lett.* **127:**4–8.

Schwantke, N., and Le Peuch, C. J., 1984, A protein kinase C inhibitory activity is present in rat brain homogenate, *FEBS Lett.* **177:**36–40.

Schwartz, D. E., Tizard, R., and Gilbert, W., 1983, Nucleotide sequence of Rous sarcoma virus, *Cell* **32:**853–869.

Scott, J. D., Fischer, E. H., Takio, K., Demaille, J. G., and Krebs, E. G., 1985, Amino acid sequence of the heat-stable inhibitor of the cAMP-dependent protein kinase from rabbit skeletal muscle, *Proc. Natl. Acad. Sci. USA* **82:**5732–5736.

Sefton, B. M., Hunter, T., Ball, E. H., and Singer, S. J., 1981, Vinculin: A cytoskeletal target of the transforming protein of Rous sarcoma virus, *Cell* **24:**165–174.

Seyfred, M. A., and Wells, W. W., 1984, Subcellular site and mechanism of vasopressin-stimulated hydrolysis of phosphoinositides in rat hepatocytes, *J. Biol. Chem.* **259:**7666–7672.

Sha'afi, R. I., White, J. R., Molski, T. F. P., Shefcyk, J., Volpi, M., Naccache, P. H., and Feinstein, M. B., 1983, Phorbol 12-myristate 13-acetate activates rabbit neutrophils without an apparent rise in the level of intracellular free calcium, *Biochem. Biophys. Res. Commun.* **114:**638–645.

Shackelford, D. A., and Trowbridge, I. S., 1984, Induction of expression and phosphorylation of the human interleukin 2 receptor by a phorbol ester, *J. Biol. Chem.* **259:**11706–11712.

Shackelford, D. A., and Trowbridge, I. S., 1986, Identification of lymphocyte integral membrane proteins as substrates for protein kinase C: Phosphorylation of the interleukin-2 receptor, class 1 HLA antigens, and T200 glycoprotein, *J. Biol. Chem.* **261:**8334–8341.

Shalloway, D., Coussens, P. M., and Yaciuk, P., 1984, Overexpression of the c-*src* protein does not induce transformation of NIH 3T3 cells, *Proc. Natl. Acad. Sci. USA* **81:**7071–7075.

Shenolikar, S., Karbon, E. W., and Enna, S. J., 1986, Phorbol esters down-regulate protein kinase C in rat brain cerebral cortical slices, *Biochem. Biophys. Res. Commun.* **139:**251–258.

Sherr, C. J., Rettenmier, C. W., Sacca, R., Roussel, M. F., Look, A. T., and Stanley, E. R., 1985, The c-*fms* proto-oncogene product is related to the receptor for the mononuclear phagocyte growth factor, CSF-1, *Cell* **41:**665–676.

Shimizu, Y., Fujiki, H., Sugimura, T., and Shimizu, N., 1986, Mouse 3T3-L1 cell variants unable to respond to mitogenic stimulation of dihydroteleocidin B: Genetic evidence for the synergism of tumor promoters with growth factors, *Cancer Res.* **46:**4027–4031.

Shinohara, O., Knecht, M., and Catt, K. J., 1985a, Differential actions of phorbol ester and diacylglycerol on inhibition of granulosa cell maturation, *Biochem. Biophys. Res. Commun.* **133:**468–474.

Shinohara, O., Knecht, M., and Catt, K. J., 1985b, Inhibition of gonadotropin-induced granulosa cell differentiation by activation of protein kinase C, *Proc. Natl. Acad. Sci. USA* **82:**8518–8522.

Shoji, M., Vogler, W. R., and Kuo, J. F., 1985, Inhibition of phospholipid/Ca^{2+}-dependent protein kinase and phosphorylation of leukemic cell proteins by CP-46,665-1, a novel antineoplastic lipoidal amine, *Biochem. Biophys. Res. Commun.* **127:**590–595.

Shoji, M., Girard, P. R., Mazzei, G. J., Vogler, W. R., and Kuo, J. F., 1986, Immunocytochemical evidence for phorbol ester–induced protein kinase C translocation in HL 60 cells, *Biochem. Biophys. Res. Commun.* **135:**1144–1149.

Shoyab, M., 1985, Inhibition of protein kinase activity of phorboid and ingenoid receptor by di(adenosine-5')oligophosphate, *Arch. Biochem. Biophys.* **236:**441–444.

Shoyab, M., and Todaro, G. J., 1980, Vitamin K₃ and related quinones, like tumor-promoting phorbol esters, alter the affinity of epidermal growth factor for its membrane receptors, *Nature* **288:**451–455.

Shoyab, M., and Todaro, G. J., 1980, Vitamin K₃ and related quinones, like tumor-promoting phorbol esters, alter the affinity of epidermal growth factor for its membrane receptors, *Nature* **288:**451–455.

Shoyab, M., DeLarco, J. E., and Todaro, G. J., 1979, Biologically active phorbol esters specifically alter affinity of epidermal growth factor membrane receptors, *Nature* **279:**387–391.

Shoyab, M., Warren, T. C., and Todaro, G. J., 1981, Tissue and species distribution and developmental variation of specific receptors for biologically active phorbol and ingenol esters, *Carcinogenesis* **2:**1273–1276.

Shriver, K., and Rohrschneider, L., 1981, Organization of pp60^{src} and selected cytoskeletal proteins within adhesion plaques and junctions of Rous sarcoma virus–transformed rat cells, *J. Cell Biol.* **89:**525–535.

Shukla, S. D., and Hanahan, D. J., 1983, An early transient decrease in phosphatidylinositol 4,5-bisphosphate upon stimulation of rabbit platelets with acetylglycerylether phosphorylcholine (platelet activating factor), *Arch. Biochem. Biophys.* **227:**626–629.

Sibley, D. R., Nambi, P., Peters, J. R., and Lefkowitz, R. J., 1984, Phorbol diesters promote β-adrenergic receptor phosphorylation and adenylate cyclase desensitization in duck erythrocytes, *Biochem. Biophys. Res. Commun.* **121:**973–979.

Sieber-Blum, M., and Sieber, F., 1981, Tumor-promoting phorbol ester promote melanogenesis and prevent expression of the adrenergic phenotype in quail neural crest cells, *Differentiation* **20:**117–123.

Siebert, P. D., and Fukuda, M., 1986, Human glycophorin A and B are encoded by separate, single copy genes coordinately regulated by a tumor-promoting phorbol ester, *J. Biol. Chem.* **261:**12433–12436.

Siffert, W., and Scheid, P., 1986, A phorbol ester and 1-oleoyl-2-acetylglycerol induce Na^+/H^+ exchange in human platelets, *Biochem. Biophys. Res. Commun.* **141**:13–19.

Sinnett-Smith, J. W., and Rozengurt, E., 1985, Diacylglycerol treatment rapidly decreases the affinity of the epidermal growth factor receptors of Swiss 3T3 cells, *J. Cell. Physiol.* **124**:81–86.

Sisskin, E. E., and Barrett, J. C., 1981, Inhibition of terminal differentiation of hamster epidermal cells in culture by the phorbol ester 12-O-tetradecanoylphorbol-13-acetate, *Cancer Res.* **41**:593–603.

Slaga, T. J., 1984, Mechanisms of tumor promotion, *Tumor Promotion and Skin Carcinogenesis,* Vol. 2, CRC Press, Boca Raton, Florida.

Smart, J. E., Oppermann, H., Czernilofsky, A. P., Purchio, A. F., Erikson, R. L., and Bishop, J. M., 1981, Characterization of sites for tyrosine phosphorylation in the transforming protein of Rous sarcoma virus ($pp60^{c-src}$) and its normal cellular homologue ($pp60^{c-src}$), *Proc. Natl. Acad. Sci. USA* **78**:6013–6017.

Smigel, M. D., Ferguson, K. M., and Gilman, A. G., 1985, Control of adenylate cyclase activity by G proteins, *Adv. Cyclic. Nucleotide Res.* **19**:103–112.

Smith, J. B., Smoth, L., Brown, E. R., Barnes, D., Sabir, M. A., Davis, J. S., and Farese, R. V., 1984, Angiotensin II rapidly increases phosphatidate–phosphoinositide synthesis and phosphoinositide hydrolysis and mobilizes intracellular calcium in cultured arterial muscle cells, *Proc. Natl. Acad. Sci. USA* **81**:7812–7816.

Smith, R. J., and Iden, S. S., 1979, Phorbol myristate acetate–induced release of granule enzymes from human neutrophils: Inhibition by the calcium antagonist, 8-(N,N-diethylamino)-octyl 3,4,5-trimethoxybenzoate hydrochloride, *Biochem. Biophys. Res. Commun.* **91**:263–271.

Sobel, M. E., Dion, L. D., Vuust, J., and Colburn, N. H., 1983, Tumor-promoting phorbol esters inhibit procollagen synthesis at a pretranslational level in JB-6 mouse epidermal cells, *Mol. Cell. Biol.* **3**:1527–1532.

Solanki, V., and Slaga, T. J., Callaham, M., and Huberman, E., 1981, Down regulation of specific binding of [20-³H]phorbol 12,13-dibutyrate and phorbol ester–induced differentiation of human promyelocytic leukemia cells, *Proc. Natl. Acad. Sci. USA* **78**:1722–1725.

Solomon, D. H., O. Brian, C. A., and Weinstein, I. B., 1985, N-α-tosyl-L-lysine chloromethyl ketone and N-α-tosyl-L-phenylalanine chloromethyl ketone inhibit protein kinase C, *FEBS Lett.* **190**:342–344.

Sonnenberg, A., Sulbecco, R., and Okada, S., 1983, Tumor promoter binding to rat mammary cell cultures: Role of receptors in the inhibition of dome formation, *Cancer Res.* **43**:1059–1065.

Sorge, L. K., Levy, B. T., and Maness, P. F., 1984, $pp60^{c-src}$ is developmentally regulated in the neural retina, *Cell* **36**:249–257.

Spat, A., Bradford, P. G., McKinney, J. S., Rubin, R. P., and Putney, J. W., Jr., 1986, A saturable receptor for ³²P-inositol-1,4,5-trisphosphate in hepatocytes and neutrophils, *Nature* **319**:514–516.

Spivack, J. G., Erikson, R. L., and Maller, J. L., 1984, Microinjection of $pp60^{v-src}$ into Xenopus oocytes increases phosphorylation of ribosomal protein S6 and accelerates the rate of progesterone-induced meiotic maturation, *Mol. Cell. Biol.* **4**:1631–1634.

Stefanovic, D., Erikson, E., Pike, L., and Maller, J. L., 1986, Activation of a ribosomal protein S6 protein kinase in Xenopus oocytes by insulin and insulin-receptor kinase, *EMBO J.* **5**:157–160.

Stern, D. F., Heffernan, P. A., and Weinberg, R. A., 1986, p185, a product of the *neu* proto-oncogene, is a receptorlike protein associated with tyrosine kinase activity, *Mol. Cell. Biol.* **6**:1729–1740.

Stevenson, M. A., Calderwood, S. K., and Hahn, G. M., 1986, Rapid increases in inositol trisphosphate and intracellular Ca^{2+} after heat shock, *Biochem. Biophys. Res. Commun.* **137**:826–833.

Stoppelli, M. P., Verde, P., Grimaldi, G., Locatelli, E. K., and Blase, F., 1986, Increase in urokinase plasminogen activator mRNA synthesis in human carcinoma cells is a primary effect of the potent tumor promoter, phorbol myristate acetate, *J. Cell Biol.* **102**:1235–1241.

Stralfors, P., Hiraga, A., and Cohen, P., 1985, The protein phosphatases involved in cellular regulation: Purification and characterisation of the glycogen-bound form of protein phosphatase-1 from rabbit skeletal muscle, *Eur. J. Biochem.* **149**:295–303.

Streb, H., Irvine, R. F., Berridge, M. J., and Schulz, I., 1983, Release of Ca^{2+} from a non-mitochondrial intracellular store in pancreatic acinar cells by inositol-1,4,5-trisphosphate, *Nature* **306**:67–69.

Stuart, R. K., and Hamilton, J. A., 1980, Tumor-promoting phorbol esters stimulate hematopoietic colony formation *in vitro, Science* **208**:402–404.

Sturani, E., Vicentini, L. M., Zippel, R., Toschi, L., Pandiella-Alonso, A., Comoglio, P. M., and Meldolesi, J., 1986, PDGF-induced receptor phosphorylation and phosphoinositide hydrolysis are unaffected by protein kinase C activation in mouse swiss 3T3 and human skin fibroblasts, *Biochem. Biophys. Res. Commun.* **137**:343–350.

Su, H.-D., Mazzei, G. J., Vogler, W. R., and Kuo, J. G., 1985, Effect of tamoxifen, a nonsteroidal antiestrogen, on phospholipid/Ca^{2+}-dependent protein kinase and phosphorylation of its endogenous substrate proteins from the rat brain and ovary, *Biochem. Pharmacol.* **34**:3645–3653.

Su, H.-D., Kemp, B. E., Turner, R. S., and Kuo, J. F., 1986a, Synthetic myelin basic protein peptide analogs are specific inhibitors of phospholipid/calcium-dependent protein kinase (protein kinase C), *Biochem. Biophys. Res. Commun.* **134**:78–84.

Su, H.-D., Shoji, M., Mazzei, J. G., Vogler, W. R., and Kuo, J. F., 1986b, Effects of selenium compounds on phospholipid/Ca^{2+}-dependent protein kinase (protein kinase C) system from human leukemic cells, *Cancer Res.* **46**:3684–3687.

Sugden, D., Vanecek, J., Klein, D. C., Thomas, T. P., and Anderson, W. B., 1985, Activation of protein kinase C potentiates isoprenaline-induced cyclic AMP accumulation in rat pinealocytes, *Nature* **314**:359–361.

Summers, S. T., and Cronin, M. J., 1986, Phorbol esters enhance basal and stimulated adenylate cyclase activity in a pituitary cell line, *Biochem. Biophys. Res. Commun.* **135**:276–281.

Summers, T. A., and Creutz, C. E., 1985, Phosphorylation of a chromaffin granule-binding protein by protein kinase C, *J. Biol. Chem.* **260**:2437–2443.

Suzuki, Y., and Lehrer, R. I., 1980, NAD(P)H oxidase activity in human neutrophils stimulated by phorbol myristate acetate, *J. Clin. Invest.* **66**:1409–1418.

Swann, K., and Whitaker, M., 1985, Stimulation of the Na/H exchanger of sea urchin eggs by phorbol ester, *Nature* **314**:274–277.

Tabarini, D., Heinrich, J., and Rosen, O. M., 1985, Activation of S6 kinase activity in 3T3-L1 cells by insulin and phorbol ester, *Proc. Natl. Acad. Sci. USA* **82**:4369–4373.

Taffet, S. M., Greenfield, A. R. L., and Haddox, M. K., 1983, Retinal inhibits TPA activated, calcium-dependent, phospholipid-dependent protein kinase ("C kinase"), *Biochem. Biophys. Res. Commun.* **114**:1194–1199.

Taguchi, M., Thomas, T. P., Anderson, W. B., and Farrar, W. L., 1986, Direct phosphorylation of the IL-2 receptor Tac antigen epitope by protein kinase C, *Biochem. Biophys. Res. Commun.* **135**:239–247.

Takai, Y., Kishimoto, A., Inoue, M., and Nishizuka, Y., 1977, Studies on a cyclic nucleotide–independent protein kinase and its proenzyme in mammalian tissues. 1. Purification and characterization of an active enzyme from bovine cerebellum, *J. Biol. Chem.* **252**:7603–7609.

Takai, Y., Kishimoto, A., Kikkawa, U., Mori, T., and Nishizuka, Y., 1979a, Unsaturated diacylglycerol as a possible messenger for the activation of calcium-activated, phospholipid-dependent protein kinase system, *Biochem. Biophys. Res. Commun.* **91**:1218–1224.

Takai, Y., Kishimoto, A., Iwasa, M., Kawahara, Y., Mori, T., and Nishizuka, Y., 1979b, Calcium-dependent activation of a multifunctional protein kinase by membrane phospholipids, *J. Biol. Chem.* **254**:3692–3695.

Takai, Y., Kaibuchi, K., Matsubara, T., and Nishizuka, Y., 1981, Inhibitory action of guanosine 3',5'-monophosphate on thrombin-induced phosphatidylinositol turnover and protein phosphorylation in human platelets, *Biochem. Biophys. Res. Commun.* **101**:61–67.

Takai, Y., Kaibuchi, K., Tsuda, T., and Hoshijima, M., 1985, Role of protein kinase C in transmembrane signalling, *J. Cell. Biochem.* **29**:143–155.

Takayama, S., White, M. F., Lauris, V., and Kahn, C. R., 1984, Phorbol esters modulate insulin
receptor phosphorylation and insulin action in cultured hepatoma cells, *Proc. Natl. Acad. Sci.
USA* **81**:7797–7801.

Takeda, A., Hashimoto, E., Yamamura, H., and Shimazu, T., 1986, Phosphorylation of liver gap
junction protein by protein kinase C, *FEBS Lett.* **210**:169–172.

Takeya, T., and Hanafusa, H., 1982, DNA sequence of the viral and cellular src gene of chickens.
2. Comparison of the src genes of two strains of avian sarcoma virus and of the cellular homolog.
J. Virol. **44**:12–18.

Takeya, T., and Hanafusa, H., 1983, Structure and sequence of the cellular gene homologous
to the RSV src gene and the mechanism for generating the transforming virus, *Cell*
32:881–890.

Takeya, T., Feldman, R. A., and Hanafusa, H., 1982, DNA sequence of the viral and cellular *src*
gene of chickens. 1. Complete nucleotide sequence of an EcoR1 fragment of recovered avian
sarcoma virus which codes for gp37 and pp60^{src}, *J. Virol.* **44**:1–11.

Takuma, T., and Ichida, T., 1986, Phorbol ester stimulates amylase secretion from rat parotid cells,
FEBS Lett. **199**:53–56.

Tamaoki, T., Nomoto, H., Takahashi, I., Kato, Y., Morimoto, M., and Tomita, F., 1986, Staurosporine, a potent inhibitor of phospholipid/Ca^{2+} dependent protein kinase, *Biochem. Biophys.
Res. Commun.* **135**:397–402.

Tamura, T., Friis, R. R., and Bauer, H., 1984, pp60^{c-src} is a substrate for phosphorylation when
cells are stimulated to enter cycle, *FEBS Lett.* **177**:151–156.

Tanaka, C., Taniyama, K., and Kusunoki, M., 1984, A phorbol ester and A23187 act synergistically
to release acetylcholine from the guinea pig ileum, *FEBS Lett.* **175**:165–169.

Tanaka, T., Ohmura, T., and Hidaka, H., 1982, Hydrophobic interactions of the Ca^{2+}–calmodulin
complex with calmodulin antagonists, *Mol. Pharmacol.* **22**:403–407.

Tapley, P. M., and Murray, A. W., 1984a, Modulation of Ca^{2+}-activated, phospholipid-dependent
protein kinase in platelets treated with a tumor-promoting phorbol ester, *Biochem. Biophys.
Res. Common.* **122**:158–164.

Tapley, P. M., and Murray, A. W., 1984b, Platelet Ca^{2+}-activated, phospholipid-dependent protein
kinase: Evidence for proteolytic activation of the enzyme in cells treated with phospholipase
C, *Biochem. Biophys. Res. Commun.* **118**:835–841.

Tapley, P. M., and Murray, A. W., 1985, Evidence that treatment of platelets with phorbol ester
causes proteolytic activation of Ca^{2+}-activated, phospholipid-dependent protein kinase, *Eur.
J. Biochem.* **151**:419–423.

Thomas, G., Siegmann, M., and Gordon, J., 1979, Multiple phosphorylation of ribosomal protein
S6 during transition of quiescent 3T3 cells into early G$_1$, and cellular compartmentalization of
the phosphate donor, *Proc. Natl. Acad. Sci. USA* **76**:3952–3956.

Thomas, G., Martin-Perez, J., Siegman, M., and Otto, A. M., 1982, The effect of serum, EGF,
PGF$_2\alpha$, and insulin on S6 phosphorylation and the initiation of protein and DNA synthesis,
Cell **30**:235–242.

Thomopoulos, P., Testa, U., Gourdin, M. F., Hervy, C., Titeux, M., and Vainchenker, W., 1982,
Inhibition of insulin receptor binding by phorbol esters, *Eur. J. Biochem.* **129**:389–393.

Tohmatsu, T., Hattori, H., Nagao, S., Ohki, K., and Nozawa, Y., 1986, Reversal by protein kinase
C inhibitor of suppressive actions of phorbol-12-myristate-13-acetate on polyphosphoinositide
metabolism and cytosolic Ca^{2+} mobilization in thrombin-stimulated human platelets, *Biochem.
Biophys. Res.Commun.* **134**:868–875.

Totterman, T. H., Nilsson, K., and Sundstrom, C., 1980, Phorbol ester–induced differentiation of
chronic lymphocytic leukaemia cells, *Nature* **288**:176–178.

Tran, P. L., Castagna, M., Sala, M., Vassent, G., Horowitz, A. D., Schachter, D., and Weinstein,

I. B., 1983, Differential effect of tumor promoters on phorbol-ester-receptor binding and membrane fluorescence anisotropy in C3H 10T1/2 cells, *Eur. J. Biochem.* **130**:155–160.

Trevillyan, J. M., Kulkarni, R. K., and Byus, C. V., 1984, Tumor-promoting phorbol esters stimulate the phosphorylation of ribosomal protein S6 in quiescent Reuber H35 hepatoma cells, *J. Biol. Chem.* **259**:897–902.

Trevillyan, J. M., Perisic, O., Traugh, J. A., and Byus, C. V., 1985, Early steps of lymphocyte activation bypassed by synergy between calcium ionophores and phorbol ester, *Nature* **313**:318–320.

Tsien, R. Y., Pozzan, T., and Rink, T. J., 1982, T-cell mitogens cause early changes in cytoplasmic free Ca^{2+} and membrane potential in lymphocytes, *Nature* **295**:68–71.

Tsuda, T., Kaibuchi, K., Kawahara, Y., Fukuzaki, H., and Takai, Y., 1985, Induction of protein kinase C activation and Ca^{2+} mobilization by fibroblast growth factor in Swiss 3T3 cells, *FEBS Lett.* **191**:205–210.

Tsuyama, S., Bramblett, G. T., Huang, K.-P., and Flavin, M., 1986, Calcium/phospholipid-dependent kinase recognizes sites in microtubule-associated protein 2 which are phosphorylated in living brain and are not accessible to other kinases, *J. Biol. Chem.* **261**:4110–4116.

Turgeon, J. L., and Cooper, R. H., 1986, Protein kinase C and an endogenous substrate associated with adenohypophyseal secretory granules, *Biochem. J.* **237**:53–61.

Turgeon, J. L., Ashcroft, S. J. H., Waring, D. W., Milewski, M. A., and Walsh, D. A., 1984, Characteristics of the adenohypophyseal Ca^{2+}-phospholipid-dependent protein kinase, *Mol. Cell. Endocrinol.* **34**:107–122.

Turner, P. R., Sheetz, M. P., and Jaffe, L. A., 1984a, Fertilization increases the polyphosphoinositide content of sea urchin eggs, *Nature* **310**:414–419.

Turner, R. S., Chou, C.-H. J., Kibler, R. F., and Kuo, J. F., 1982, Basic protein in brain myelin is phosphorylated by endogenous phospholipid-sensitive Ca^{2+}-dependent protein kinase, *J. Neurochem.* **39**:1397–1404.

Turner, R. S., Chou, C.-H. J., Mazzei, G. J., Dembure, P., and Kuo, J. F., 1984b, Phospholipid-sensitive Ca^{2+}-dependent protein kinase preferentially phosphorylates serine-115 of bovine myelin basic protein, *J. Neurochem.* **43**:1257–1264.

Turner, R. S., Taynor, R. L., Mazzei, G. J., Girard, P. R., and Kuo, J. G., 1984c, Developmental studies of phospholipid-sensitive Ca^{2+}-dependent protein kinase and its substrates and of phosphoprotein phosphatases in rat brain, *Proc. Natl. Acad. Sci. USA* **81**:3143–3147.

Turner, R. S., Kemp, B. E., Su, H.-D., and Kuo, J. F., 1985, Substrate specificity of phospholipid/Ca^{2+}-dependent protein kinase as probed with synthetic peptide fragments of bovine myelin basic protein, *J. Biol. Chem.* **260**:11503–11507.

Uchida, T., and Filburn, C. R., 1984, Affinity chromatography of protein kinase C–phorbol ester receptor on polyacrylamide-immobilized phosphatidylserine, *J. Biol. Chem.* **259**:12311–12314.

Ullrich, A., Coussens, L., Hayflick, J. S., Dull, T. J., Gray, A., Tam, A. W., Lee, J., Yarden, Y., Liberman, T. A., Schlessinger, J., Downward, J., Mayes, E. L. V., Waterfield, D., Whittle, M., and Seeburg, P. H., 1984, Human epidermal growth factor receptor cDNA sequence and aberrant expression of the amplified gene in A431 epidermoid carcinoma cells, *Nature* **309**:418–425.

Umekawa, H., and Hidaka, H., 1985, Phosphorylation of caldesmon by protein kinase C, *Biochem. Biophys. Res. Commun.* **132**:56–62.

Uratsuji, Y., Nakanishi, H., Takeyama, Y., Kishimoto, A., and Nishizuka, Y., 1985, Activation of cellular protein kinase C and mode of inhibitory action of phospholipid-interacting compounds, *Biochem. Biophys. Res. Commun.* **130**:654–661.

Ushiro, H., and Cohen, S., 1980, Identification of phosphotyrosine as a product of epidermal growth factor–activated protein kinase in A-431 cell membranes, *J. Biol. Chem.* **255**:8363–8365.

Vandenburg, C. A., and Montal, M., 1984, Light-regulated biochemical events in invertebrate photoreceptors. 2. Light-regulated phosphorylation of rhodopsin and phosphoinositides in squid photoreceptor membranes. *Biochemistry* **23**:2347–2352.

Van Obberghen, E., Rossi, B., Kowalski, A., Gazzano, H., and Ponzio, G., 1983, Receptor-mediated phosphorylation of the hepatic insulin receptor: Evidence that the M_r 95.000 receptor subunit is its own kinase, *Proc. Natl. Acad. Sci. USA* **80:**945–949.

Vara, F., and Rozengurt, E., 1985, Stimulation of Na^+/H^+ antiport activity by epidermal growth factor and insulin occurs without activation of protein kinase C, *Biochem. Biophys. Res. Commun.* **130:**646–653.

Vara, F., Schneider, J. A., and Rozengurt, E., 1985, Ionic responses rapidly elicited by activation of protein kinase C in quiescent Swiss 3T3 cells, *Proc. Natl. Acad. Sci. USA* **82:**2384–2388.

Varshavsky, A., 1981, Phorbol ester dramatically increases incidence of methotrexate-resistant mouse cells: Possible mechanisms and relevance to tumor promotion, *Cell* **25:**561–572.

Vartanian, T., Szuchet, S., Dawson, G., and Campagnoni, A. T., 1986, Oligodendrocyte adhesion activates protein kinase C–mediated phosphorylation of myelin basic protein, *Science* **234:**1395–1398.

Veldhuis, J. D., and Demers, L. M., 1986, An inhibitory role for the protein kinase C pathway in ovarian steroidogenesis: Studies with cultured swine granulosa cells, *Biochem. J.* **239:**505–511.

Verma, A. K., Erickson, D., and Dolnick, B. J., 1986, Increased mouse epidermal ornithine decarboxylase activity by the tumour promoter 12-O-tetradecanoylphorbol 13-acetate involves increased amounts of both enzyme protein and messenger RNA, *Biochem. J.* **237:**297–300.

Vicentini, L. M., and Villereal, M. L., 1984, Serum, bradykinin, and vasopressin stimulate release of inositol phosphates from human fibroblasts, *Biochem. Biophys. Res. Commun.* **123:**663–670.

Vicentini, L. M., and Villereal, M. L., 1985, Activation of Na^+/H^+ exchange in cultured fibroblasts: Synergism and antagonism between phorbol ester, Ca^{2+} ionophore, and growth factors, *Proc. Natl. Acad. Sci. USA* **82:**8053–8056.

Vicentini, L. M., Di Virgilio, F., Ambrosini, A., Pozzan, T., and Meldolesi, J., 1985, Tumor promoter phorbol 12-myristate, 13-acetate inhibits phosphoinositide hydrolysis and cytosolic Ca^{2+} rise induced by the activation of muscarinic receptors in PC12 cells, *Biochem. Biophys. Res. Commun.* **127:**310–317.

Vigne, P., Frelin, C., and Lazdunski, M., 1985, The Na^+/H^+ antiport is activated by serum and phorbol esters in proliferating myoblasts but not in differentiated myotubes: Properties of the activation process, *J. Biol. Chem.* **260:**8088–8013.

Vilgrain, I., Cochet, C., and Chambaz, E. M., 1984a, Hormonal regulation of a calcium-activated, phospholipid-dependent protein kinase in bovine adrenal cortex, *J. Biol. Chem.* **259:**3403–3406.

Vilgrain, I., Defaye, G., and Chambaz, E. M., 1984b, Adrenocortical cytochrome P-450 responsible for cholesterol side chain cleavage (P-450$_{scc}$) is phosphorylated by the calcium-activated, phospholipid-sensitive protein kinase (protein kinase C), *Biochem. Biophys. Res. Commun.* **125:**554–561.

Volpi, M., Yassin, R., Naccache, P. H., and Sha'afi, R. I., 1983, Chemotactic factor causes rapid decreases in phosphatidyl-inositol 4,5-bisphosphate and phosphatidylinositol 4-monophosphate in rabbit neutrophils, *Biochem. Biophys. Res. Commun.* **112:**957–964.

Vorbrodt, A., Meo, P., and Rovera, G., 1979, Regulation of acid phosphatase activity in human promyelocytic leukemic cells induced to differentiate in culture, *J. Cell Biol.* **83:**300–307.

Vulliet, P. R., Langan, T. A., and Weiner, N., 1980, Tyrosine hydroxylase: A substrate of cyclic AMP-dependent protein kinase, *Proc. Natl. Acad. Sci. USA* **77:**92–96.

Vulliet, P. R., Woodgett, J. R., and Cohen, P., 1984, Phosphorylation of tyrosine hydroxylase by calmodulin-dependent multiprotein kinase, *J. Biol. Chem.* **259:**13680–13683.

Vulliet, P. R., Woodgett, J. R., Ferrari, S., and Hardie, D. G., 1985, Characterization of the sites phosphorylated on tyrosine hydroxylase by Ca^{2+} and phospholipid-dependent protein kinase, calmodulin-dependent multiprotein kinase, and cyclic AMP–dependent protein kinase, *FEBS Lett.* **182:**335–339.

Wakade, A. R., Malhotra, R. K., and Wakade, T. D., 1986, Phorbol ester facilitates ^{45}Ca accumulation and catecholamine secretion by nicotine and excess K⁺ but not by muscarine in rat adrenal medulla, *Nature* **321**:698–700.

Wallner, B. P., Mattaliano, R. J., Hession, C., Cate, R. L., Tizard, R., Sinclair, L. K., Foeller, C., Chow, E. P., Browning, J. L., Ramachandran, K. L., and Pepinsky, R. B., 1986, Cloning and expression of human lipocortin, a phospholipase A₂ inhibitor with potential anti-inflammatory activity, *Nature* **320**:77–81.

Walsh, M. P., Valentine, K. A., Ngai, P. K., Carruthers, C. A., and Hollenberg, M. D., 1984, Ca²⁺-dependent hydrophobic-interaction chromatography, *Biochem. J.* **224**:117–127.

Wang, J. L., McClain, D. A., and Edelman, G. M., 1975, Modulation of lymphocyte mitogenesis, *Proc. Natl. Acad. Sci. USA* **72**:1917–1921.

Wang, Y., Camici, M., Lee, F.-T., Ahmad, Z., De Paoli-Roach, A. A., and Roach, P. J., 1986, Multiple phosphorylation sites of rat liver glycogen synthase, *Biochim. Biophys. Acta* **888**:225–236.

Ware, J. A., Johnson, P. C., Smith, M., and Salzman, E. W., 1985, Aequorin detects increased cytoplasmic calcium in platelets stimulated with phorbol ester or diacylglycerol, *Biochem. Biophys. Res. Commun.* **133**:98–104.

Warren, G., Davoust, J., and Cockcroft, A., 1984, Recycling of transferrin receptors in A431 cells is inhibited during mitosis, *EMBO J.* **3**:2217–2225.

Watanabe, Y., Horn, F., Bauer, S., and Jakobs, K. H., 1985, Protein kinase C interferes with Nᵢ-mediated inhibition of human platelet adenylate cyclase, *FEBS Lett.* **192**:23–27.

Watson, S. P., and Lapetina, E. G., 1985, 1,2-diacylglycerol and phorbol ester inhibit agonist-induced formation of inositol phosphates in human platelets: Possible implications for negative feedback regulation of inositol phospholipid hydrolysis, *Proc. Natl. Acad. Sci. USA* **82**:2623–2626.

Welch, W. J., 1985, Phorbol ester, calcium ionophores, or serum added to quiescent rat embryo fibroblast cells all result in the elevated phosphorylation of two 28,000-dalton mammalian stress proteins, *J. Biol. Chem.* **260**:3058–3062.

Wender, P. A., Koehler, K. F., Sharkey, N. A., Dell'Aquila, M. L., and Blumberg, P. M., 1986, Analysis of the phorbol ester pharmacophore on protein kinase C as a guide to the rationale design of new classes of analogs, *Proc. Natl. Acad. Sci. USA* **83**:4214–4218.

Werth, D. K., and Pastan, I., 1984, Vinculin phosphorylation in response to calcium and phorbol esters in intact cells, *J. Biol. Chem.* **259**:5264–5270.

Werth, D. K., Niedel, J. E., and Pastan, I., 1983, Vinculin, a cytoskeletal substrate of protein kinase C, *J. Biol. Chem.* **258**:11423–11426.

Wettenhall, R. E. H., and Cohen, P., 1982, Isolation and characterisation of cyclic AMP–dependent phosphorylation sites from rat liver ribosomal protein S6, *FEBS Lett.* **140**:263–269.

White, J. G., Rao, G. H. R., and Estensen, R. D., 1974, Investigation of the release reaction in platelets exposed to phorbol myristate acetate, *Am. J. Pathol.* **75**:301–314.

White, J. R., Huang, C.-K., Hill, Jr., J. M., Naccache, P. H., Becker, E. L., and Sha'afi, R. I., 1984, Effect of phorbol 12-myristate 13-acetate and its analogue 4β-phorbol 12,13-didecanoate on protein phosphorylation and lysosomal enzyme release in rabbit neutrophils, *J. Biol. Chem.* **259**:8605–8611.

White, J. R., Pluznik, D. H., Ishizaka, K., and Ishizaka, T., 1985, Antigen-induced increase in protein kinase C activity in plasma membrane of mast cells, *Proc. Natl. Acad. Sci. USA* **82**:8193–8197.

Whiteley, B., Cassel, D., Zhuang, Y.-X., and Glaser, L., 1984, Tumor promoter phorbol 12-myristate 13-acetate inhibits mitogen-stimulated Na⁺/H⁺ exchange in human epidermoid carcinoma A431 cells, *J. Cell Biol.* **99**:1162–1166.

Whiteley, B., Deuel, T., and Glaser, L., 1985, Modulation of the activity of the platelet-derived growth factor by phorbol myristate acetate, *Biochem. Biophys. Res. Commun.* **129**:854–861.

Whitin, J. C., Chapman, C. E., Simons, E. R., Chovaniec, M. E., and Cohen, H. J., 1980, Correlation between membrane potential changes and superoxide production in human granulocytes stimulated by phorbol myristate acetate: Evidence for defective activation in chronic granulomatous disease, *J. Biol. Chem.* **255:**1874–1878.

Wiener, E., Dubyak, G., and Scarpa, A., 1986, Na$^+$/H$^+$ exchange in Ehrlich ascites tumor cells: Regulation by extracellular ATP and 12-O-tetradecanoylphorbol 13-acetate, *J. Biol. Chem.* **261:**4529–4534.

Wightman, P. C., and Raetz, C. R. H., 1984, The activation of protein kinase C by biologically active lipid moieties of lipopolysaccharide, J. Biol. Chem. **259:**10048–10052.

Wigler, M., and Weinstein, I. B., 1976, Tumour promotor induces plasminogen activator, *Nature* **259:**232–233.

Wilson, D. B., Bross, T. E., Hofmann, S. L., and Majerus, P. W., 1984, Hydrolysis of polyphosphoinositides by purified sheep seminal vesicle phospholipase C enzymes, *J. Biol. Chem.* **259:**11718–11724.

Wilson, E., Olcott, M. C., Bell, R. M., Merrill, A. H., Jr., and Lambeth, J. D., 1986, Inhibition of the oxidative burst in human neutrophils by sphingoid long-chain bases, *J. Biol. Chem.* **261:**12616–12623.

Wise, B. C., Raynor, R. L., and Kuo, J. F., 1982a, Phospholipid-sensitive Ca^{2+}-dependent protein kinase from heart, *J. Biol. Chem.* **257:**8481–8488.

Wise, B. C., Glass, D. B., Chou, C.-H., J., Raynor, R. L., Katoh, N., Schatzman, R. C., Turner, R. S., Kobler, R. F., and Kuo, J. F., 1982b, Phospholipid-sensitive Ca^{2+}-dependent protein kinase from heart, *J. Biol. Chem.* **257:**8489–8495.

Witte, O. N., Dasgupta, A., and Baltimore, D., 1980, Abelson murine leukaemia virus protein is phosphorylated *in vitro* to form phosphotyrosine, *Nature* **283:**826–831.

Witters, L. A., Vater, C. A., Lienhard, G. E., 1985, Phosphorylation of the glucose transporter *in vitro* and *in vivo* by protein kinase C, *Nature* **315:**777–778.

Wolf, M., and Sahyoun, N., 1986, Protein kinase C and phosphatidylserine bind to M$_r$ 110,000/ 115,000 polypeptides enriched in cytoskeletal and postsynaptic density preparations, *J. Biol. Chem.* **261:**13327–13332.

Wolf, M., Sahyoun, L., LeVine, H., III., and Cuatrecasas, P., 1984, Protein kinase C: Rapid enzyme purification and substrate-dependence of the diacylglycerol effect, *Biochem. Biophys. Res. Commun.* **122:**1268–1275.

Wolf, M., Cuatrecasas, P., and Sahyoun, N., 1985a, Interaction of protein kinase C with membranes is regulated by Ca^{2+}, phorbol esters, and ATP, *J. Biol. Chem.* **260:**15718–15722.

Wolf, M., LeVine, H., III., May, W. S., Cuatrecasas, P., and Sahyoun, N., 1985b, A model for intracellular translocation of protein kinase C involving synergism between Ca^{2+} and phorbol esters, *Nature* **317:**546–549.

Woodgett, J. R., and Hunter, T., 1987a, Isolation and characterization of two distinct forms of protein kinase C, *J. Biol. Chem.* **262:**4836–4843.

Woodgett, J. R., and Hunter, T., 1987b, Immunological evidence for two physiological forms of protein kinase C, *Mol. Cell. Biol.* **7:**85–96.

Woodgett, J. R., Gould, K. L., and Hunter, T., 1986, Substrate specificity of protein kinase C, *Eur. J. Biochem.* **161:**177–184.

Wooten, M. W., Nel, A. E., Goldschmidt-Clermont, P. J., Galbraith, R. M., and Wrenn, R. W., 1985, Identification of a major endogenous substrate for phospholipid/Ca^{2+}-dependent kinase in pancreatic acini as Gc (vitamin D-binding protein), *FEBS Lett.* **191:**97–101.

Wooten, M. W., Vandenplas, M., and Nel, A. E., 1987, Rapid purification of protein kinase-C from rat brain: A novel method employing protamine agarose affinity column chromatography, *Eur. J. Biochem.* **164:**461–467.

Wright, C. D., and Hoffman, M. D., 1986, The protein kinase C inhibitors H-7 and H-9 fail to inhibit human neutrophil activation, *Biochem. Biophys. Res. Commun.* **135:**749–755.

Wright, D. G., Bralove, D. A., and Gallin, J. I., 1977, The differential mobilization of human neutrophil granules: Effects of phorbol myristate acetate and ionophore A23187, *Am. J. Path.* **87:**273–282.

Wu, W. C.-S., Walaas, S. I., Nairn, A. C., and Greengard, P., 1982, Calcium/phospholipid regulates phosphorylation of a M_r "87k" substrate protein in brain synaptosomes, *Proc. Natl. Acad. Sci. USA* **79:**5249–5253.

Xu, Y.-H., Ishii, S., Clark, A. J. L., Sullivan, M., Wilson, R. K., Ma, D. P., Roe, B. A., Merlino, G. T., and Pastan, I., 1984, Human epidermal growth factor receptor cDNA is homologous to a variety of RNAs overproduced in A431 carcinoma cells, *Nature* **309:**806–810.

Yamamoto, S., Gotoh, H., Aizu, E., and Kato, R., 1985, Failiure of 1-oleoyl-2-acetylglycerol to mimic the cell-differentiation action of 12-O-tetradecanoyl 13-acetate in HL-60 cells, *J. Biol. Chem.* **260:**14230–14234.

Yamanishi, J., Takai, Y., Kaibuchi, K., Sano, K., Castagna, M., and Nishizuka, Y., 1983, Synergistic functions of phorbol ester and calcium in serotonin release from human platelets, *Biochem. Biophys. Res. Commun.* **112:**778–786.

Yamasaki, A., Kurokawa, T., Dan'ura, T., Higashi, K., and Ishibashi, S., 1986, Induction of desensitization by phorbol ester to β-adrenergic agonist stimulation in adenylate cyclase system of rat reticulocytes, *Biochem. Biophys. Res. Commun.* **138:**125–130.

Yamasaki, H., Fibach, E., Nudel, U., Weinstein, I. B., Rifkind, R. A., and Marks, P. A., 1977, Tumor promoters inhibit spontaneous and induced differentiation of murine erythroleukemia cells in culture, *Proc. Natl. Acad. Sci. USA* **74:**3451–3455.

Yamasaki, H., Devron, C., and Martel, N., 1982, Specific binding of phorbol esters to Friend erythroleukemia cells—General properties, down regulation, and relationship to cell differentiation, *Carcinogenesis,* :905–910.

Yamashita, A., Kruokawa, T., Dan'ura, T., Higashi, K., and Ishibashi, S., 1986, Induction of desensitization by phorbol ester to β-adrenergic agonist stimulation in adenylate cyclase system of rat reticulocytes, *Biochem. Biophys. Res. Commun.* **138:**125–130.

Yamauchi, T., and Fujisawa, H., 1979, *In vitro* phosphorylation of bovine adrenal tyrosine hydroxylase by adenosine 3':5'-monophosphate-dependent protein kinase, *J. Biol. Chem.* **254:**503–507.

Yancey, S. B., Edens, J. E., Trosko, J. E., Chang, C.-C., and Revel, J.-P., 1982, Decreased incidence of gap junctions between Chinese hamster V-79 cells upon exposure to the tumor promoter 12-O-tetradecanoyl phorbol-13-acetate, *Exp. Cell Res.* **139:**329–340.

Yano, K., Nakashima, S., and Nozawa, Y., 1983, Coupling of polyphosphoinositide breakdown with calcium efflux in formyl-methionyl-leucyl-phenylalanine-stimulated rabbit neutrophils, *FEBS Lett.* **161:**296–300.

Yano, K., Higashida, H., Inoue, R., and Nozawa, Y., 1984, Bradykinin-induced rapid breakdown of phosphatidylinositol 4,5-bisphosphate in neuroblastoma × glioma hybrid NG108-15 cells, *J. Biol. Chem.* **259:**10201–10207.

Yarden, Y., Escobedo, J. A., Kuang, W.-J., Yang-Feng, T. L., Daniel, T. O., Tremble, P. M., Chen, E. Y., Ando, M. E., Harkins, R. N., Francke, U., Fried, V. A., Ullrich, A., and Williams, L. T., 1986, Structure of the receptor for platelet-derived growth factor helps define a family of closely related growth factor receptors, *Nature* **323:**226–232.

Yonemoto, W., Jarvis-Morar, M., Brugge, J. S., Bolen, J. B., and Israel, M. A., 1985, Tyrosine phosphorylation within the amino-terminal domain of $pp60^{c-src}$ molecules associated with polyoma virus middle-sized tumor antigen, *Proc. Natl. Acad. Sci. USA* **82:**4568–4572.

Yoshida, K., Dubyak, G., and Nachmias, V. T., 1986, Rapid effects of phorbol ester on platelet shape change, cytoskeleton, and calcium transient, *FEBS Lett.* **206:**273–278.

Yotti, L. P., Chang, C. C., and Trosko, J. E., 1979, Elimination of metabolic cooperation in Chinese hamster cells by a tumor promoter, *Science* **206:**1089–1091.

Young, S., Parker, P. J., Ullrich, A., and Stabel, S., 1987, Down-regulation of protein kinase C is due to an increased rate of degradation, *Biochem. J.* **244:**775–779.

Yuan, S., and Sen, A. K., 1986, Characterization of the membrane-bound protein kinase C and its substrate proteins in canine cardiac sarcolemma, *Biochim. Biophys. Acta* **886:**152–161.

Yun, K., and Scott, R., 1983, Biological mechanisms of phorbol myristate acetate–induced inhibition of proadipocyte differentiation, *Cancer Res.* **43:**88–96.

Yuspa, S. H., Ben, T., Hennings, H., and Lichti, U., 1982, Divergent responses in epidermal basal cells exposed to the tumor promoter 12-O-tetradecanoylphorbol-13-acetate, *Cancer Res.* **42:**2344–2349.

Yuspa, S. H., Kilkenny, A. E., Stanley, J., and Lichti, U., 1985, Keratinocytes blocked in phorbol ester–responsive early stage of terminal differentiation by sarcoma viruses, *Nature* **314:**459–462.

Zachary, I., Sinnett-Smith, J. W., and Rozengurt, E., 1986, Early events elicited by bombesin and structurally related peptides in quiescent Swiss 3T3 cells. 1. Activation of protein kinase C and inhibition of epidermal growth factor binding, *J. Cell Biol.* **102:**2211–2222.

Zanders, E., Lamb, J., Feldman, M., Green, N., and Beverley, P., 1983, Tolerance of T cell clones is associated with membrane antigen changes, *Nature* **303:**625–627.

Zavoico, G. B., Halenda, S. P., Sha'afi, R. I., and Feinstein, M. B., 1985, Phorbol myristate acetate inhibits thrombin-stimulated Ca^{2+} mobilization and phosphatidylinositol 4,5-bisphosphate hydrolysis in human platelets, *Proc. Natl. Acad. Sci. USA* **82:**3859–3862.

Zawalich, W., Brown, C., and Rasmussen, H., 1983, Insulin secretion: Combined effects of phorbol ester and A23187, *Biochem. Biophys. Res. Commun.* **117:**448–455.

Zetterqvist, O., Ragnarsson, U., Humble, E., Berglund, L., and Engstrom, L., 1976, The minimum substrate of cyclic AMP–stimulated protein kinase, as studied by synthetic peptides representing the phosphorylatable site of pyruvate kinase (type L) of rat liver, *Biochem. Biophys. Res. Commun.* **70:**696–703.

Zick, Y., Sagi-Eisenberg, R., Pines, M., Gierschik, P., and Spiegel, A. M., 1986, Multisite phosphorylation of the α-subunit of transducin by the insulin receptor kinase and protein kinase C, *Proc. Natl. Acad. Sci. USA* **83:**9294–9297.

Zucker, M. B., Troll, W., and Belman, S., 1974, The tumor-promoter phorbol ester (12-O-tetradecanoyl-phorbol-13-acetate), a potent aggregating agent for blood platelets, *J. Cell Biol.* **60:**325–336.

Zurgil, N., and Zisapel, N., 1985, Phorbol ester and calcium act synergistically to enhance neurotransmitter release by brain neurons in culture, *FEBS Lett.* **185:**257–261.

Zwiller, J., Revel, M.-O., and Malviya, A. N., 1985, Protein kinase C catalyzes phosphorylation of guanylate cyclase *in vitro*, *J. Biol. Chem.* **260:**1350–1353.

7

MODULATION OF THE EXTRACELLULAR MATRIX BY TUMOR CELL–FIBROBLAST INTERACTIONS

Chitra Biswas and Bryan P. Toole

1. INTRODUCTION

Tumor cell invasion and metastasis involve numerous cellular phenomena, e.g., changes in cytoskeletal and cell-surface characteristics, interactions between tumor cells and cells of the vascular and immune systems, and penetration of extracellular matrices. In relation to the last of these phenomena, two major classes of extracellular matrices are traversed by malignant tumor cells during invasion of host tissues, viz., interstitial matrices and basement membranes. It is apparent that successful penetration of extracellular matrices by tumor cells would depend on at least two types of alterations in the structure of these matrices: (1) destruction of preexisting extracellular matrix barriers, and (2) reconstruction of a suitable extracellular environment for cell movement and proliferation. Both of these requirements are, in turn, certain to be multifaceted. However, important in the former would be production of degradative enzymes, such as type I collagenase (Gross *et al.*, 1981; Biswas, 1982a; Woolley, 1982, 1984), type IV collagenase (Liotta *et al.*, 1983; Liotta, 1986), heparanase (Kramer *et al.*, 1982; Nakajima *et al.*, 1983), cathepsin B (Poole *et al.*, 1978; Graf *et al.*, 1981; Sloane *et al.*, 1986), and plasminogen activator (Ossowski and Reich, 1983). Important in the latter is likely to be production of those extracellular macromolecules characteristic of the pericellular milieu of migratory and proliferating cells in embryonic tissues, e.g., hyaluronate (Toole, 1981; Toole *et al.*, 1984) and

---
Chitra Biswas and Bryan P. Toole ● Department of Anatomy and Cellular Biology, Tufts University Schools of Medicine, Dental Medicine and Veterinary Medicine, Boston, Massachusetts 02111.
Work from these laboratories was funded, in part, by grants from the National Institutes of Health, CA 38817, and CA 41701 (to C. B.), and DE 05838 (to B. P. T.).

fibronectin (Hynes, 1981; Thiery, 1984). It is evident that tumor cells themselves have the capacity to produce many of these matrix-degrading enzymes and extracellular macromolecules. However, several recent studies have shown that interactions between tumor cells and normal fibroblasts may be a major factor in modulation of extracellular matrices during tumor cell invasion in that they lead to the production by fibroblasts of elevated levels of degradative enzymes, such as collagenase, and of matrix components, such as hyaluronate and proteoglycans. These studies are discussed in this chapter. Interaction of tumor cells with cells of the vascular and immune systems may also play a role in modulating the composition of the extracellular matrix (Folkman, 1986; Biswas, 1982a; Woolley, 1984; Pauli *et al.*, 1983), but this is not considered here.

2. EFFECT OF TUMOR CELL–FIBROBLAST INTERACTIONS ON PRODUCTION OF MATRIX-DEGRADING ENZYMES

The major structural constituents of extracellular matrices are the collagens, elastin, proteoglycans, and multifunctional glycoproteins such as fibronectin and laminin. These molecules interact extensively to form a variety of dense networks of fibrillar and nonfibrillar structures that constitute barriers to invasion by tumor cells. These barriers can be modified to become more conducive to cellular immigration by either synthetic or degradative means. In relation to degradation, the networks can be modified significantly by degradation of any one of their major components. Degradation of the proteoglycans, glycoproteins, elastin, and many of the collagens can be effected by a variety of neutral proteases, whereas type I collagen, the major fibrillar component of most interstitial matrices, is cleaved only by collagenase.

Most animal collagenases described to data degrade types I, II, and III collagen, giving rise to two reaction products, three-quarters and one-quarter the length of the intact molecule. In this chapter we will call this group of enzymes "type I collagenase." Other collagens are usually resistant to type I collagenases but are susceptible to various other neutral proteases (Miller *et al.*, 1976; Sage *et al.*, 1979; Mainardi *et al.*, 1980a,b, 1981; Liotta *et al.*, 1981a). Liotta and co-workers (1979, 1981b) have isolated an enzyme characteristic of metastatic tumor cells that degrades type IV collagen but not other collagen substrates and that we term here "type IV collagenase." The properties of these collagenases have been reviewed elsewhere extensively (Gross *et al.*, 1980; Biswas, 1982a; Woolley, 1984; Liotta *et al.*, 1983; Liotta, 1986).

Since most of the work implicating tumor cell–fibroblast interactions in modulating fibroblast production of matrix-degrading enzymes has involved studies of type I collagenase, Section 2.1 deals mainly with that enzyme.

2.1. Type I Collagenase

Studies from several laboratories indicate that increased type I collagenase activity is associated with many types of tumors (Dresden *et al.*, 1972; Hashimoto *et al.*, 1973; Abramson *et al.*, 1975; McCroskery *et al.*, 1975; Bauer *et al.*, 1977; Dabbous *et al.*, 1977; Biswas *et al.*, 1978). These results have been confirmed, in several cases, by morphological studies demonstrating interstitial fiber dissolution (Gross *et al.*, 1981) and immunolocalization of collagenase in these tumors (Bauer *et al.*, 1977; Woolley *et al.*, 1980; Woolley, 1982). The latter approach has shown, however, that the enzyme is often distributed in small foci or is even absent, despite the ability of the tumor tissue to produce collagenase when excised and cultured; this has led Woolley (1984) to conclude that collagenase production *in vivo* may be "transient or intermittent." Nevertheless, it is clearly apparent from the extensive literature on this subject that both the endogenous levels of collagenase and the potential to produce collagenase are very often elevated in tumors. Also, when detected by immunohistochemical methods, collagenase is frequently seen to be concentrated at the outer edges or invasion zone of the tumors (Bauer *et al.*, 1977; Woolley, 1984).

Bauer and co-workers (1979) have reported that fibroblasts derived from human basal cell carcinoma produce collagenase in amounts that are considerably greater than those produced by normal skin fibroblasts but that these levels of production do not persist on repeated passage of the fibroblasts. They subsequently demonstrated that extracts of this tumor stimulate normal fibroblasts to produce elevated levels of collagenase (Goslen *et al.*, 1985). It has also been shown that collagenase levels are stimulated in mixed cultures of explants of rabbit V_2 carcinoma and normal subcutaneous tissue (Baici *et al.*, 1984). Using species-specific antibodies to mouse and rabbit collagenases, one of us has shown that collagenase present in the rabbit V_2 carcinoma implanted in the nude mouse is at least in part produced by cells of the nude mouse host (Biswas *et al.*, 1982). Although it is clear that some tumor cells themselves are capable of production of type I collagenase (Paranjpe *et al.*, 1980; O'Grady *et al.*, 1981), the above studies suggest quite strongly that in many cases, a significant factor in the regulation of collagenase production in tumors may be stimulation of fibroblasts by factors derived from the tumor cells. This has led to investigation of the role of tumor cell–fibroblast interactions in the production of collagenase by the fibroblasts in more defined systems *in vitro*.

In initial studies (Biswas, 1982b, 1984), cocultures of various types of tumor cells with normal fibroblasts were used to determine whether interactions between the two types of cells would lead to increased production of collagenase. Dramatic increases in collagenase activity were observed in the media of cocultures of five different types of human carcinoma or melanoma cells with normal human skin fibroblasts (Biswas, 1984), as well as with cocultures of

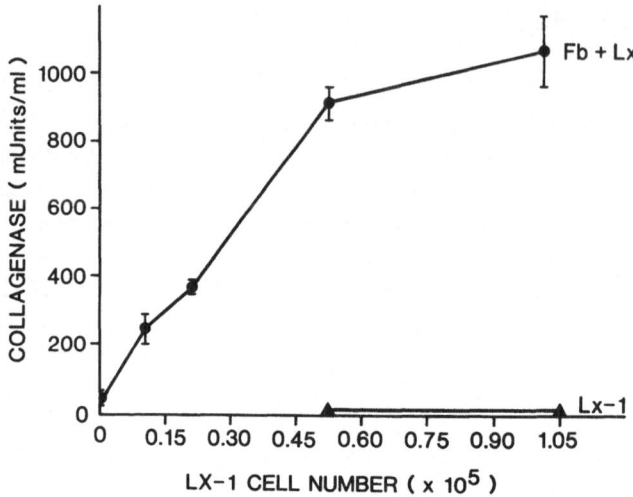

FIGURE 1. Stimulation of collagenase activity in cocultures of human LX-1 lung carcinoma cells and human skin fibroblasts. Increasing numbers of LX-1 cells were cultured in the presence (Fb + LX) or absence (LX-1) of a constant number (1×10^5) of fibroblasts. (Reproduced from Biswas, 1984, with permission of Alan Liss, Inc.)

murine B16 melanoma or A10 adenocarcinoma cells with normal rabbit synovial, muscle, or skin fibroblasts (Biswas, 1982b) (e.g., see Figure 1). Conditioned media obtained from the tumor cells were found to stimulate fibroblast collagenase production, but the reverse combination was ineffective (Figure 2), indicating that the tumor cells were the source of the stimulatory factor and the fibroblasts were the responding cells. The response in the fibroblasts was due to stimulation of new synthesis of collagenase, since cycloheximide and α-amanitin blocked the response.

Attempts to characterize the collagenase–stimulatory factor from B16 melanoma cells led to the observation that the presence of the factor in conditioned medium depends on the culture conditions (Biswas, 1985a). Although stimulation of collagenase was consistently observed in cocultures of the B16 cells with fibroblasts, the B16 cells alone did not always condition their medium with the stimulatory factor. In those cases where the B16 cells did not condition their medium, contact between the two cell types seemed to be necessary to elicit the stimulation of fibroblast collagenase production, unless the B16 cells were first cultured on fibroblast matrix (Figure 3). Thus, fibroblast matrix interacts with the tumor cells to cause shedding, secretion, or increased synthesis of the collagenase–stimulatory factor. This effect of matrix could not be duplicated by plating the cells on substrata of purified matrix components, such as fibronectin or collagen, nor was it blocked by treatment of the matrix with chondroitinase

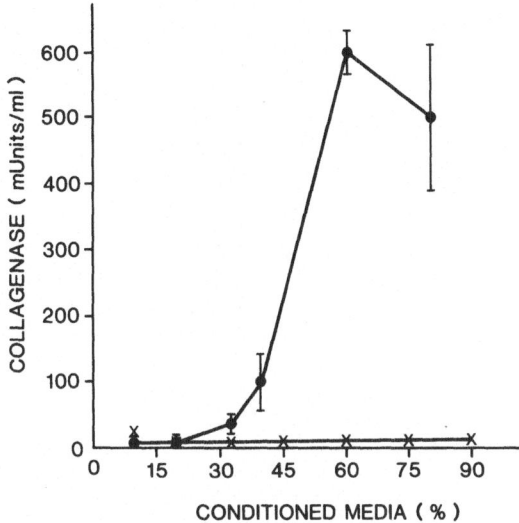

FIGURE 2. Stimulation of human fibroblast collagenase activity by conditioned medium from cultures of human LX-1 lung carcinoma cells. (●) Fibroblasts plus LX-1 conditioned medium; (×) LX-1 cells plus fibroblast conditioned medium. (Data from Biswas, 1984.)

ABC or collagenase (Biswas, 1985a) (Figure 3). More recent preliminary experiments have shown, however, that the matrix effect is removed, at least in part, by treatment with heparanase. This finding suggests the possibility that heparan sulfate proteoglycan in the fibroblast-derived matrix interacts with the surface of the tumor cells to cause the synthesis and/or release of the collagenase–stimulatory factor. To test this, we have performed experiments to determine whether there are heparan sulfate-binding sites on the surface of the B16 melanoma cells. Binding of ^{35}S-labeled heparan sulfate and ^{3}H-labeled heparin to B16 cells and to membranes derived from these cells was obtained. Scatchard analyses of these data indicate that the binding has a K_d of approximately 1 μM (Figure 4A). These analyses and measurements of relative degrees of competition for binding of heparan sulfate by various glycosaminoglycans (Figure 4B) suggest that the binding sites have similar properties to those described previously for hepatocytes (Kjellen *et al.*, 1980).

The data discussed above, especially the apparent requirement in some cases for cell contact between the tumor cells and fibroblasts, suggested the possibility that the collagenase–stimulatory factor was bound to the plasma membrane and that the conditioning of medium observed in some cultures was due to shedding of the factor from the plasma membrane either by vesiculation or by limited proteolytic cleavage. Consequently, membranes were isolated from mechanically harvested or trypsin-treated B16 cells and fibroblasts and tested for stimulatory

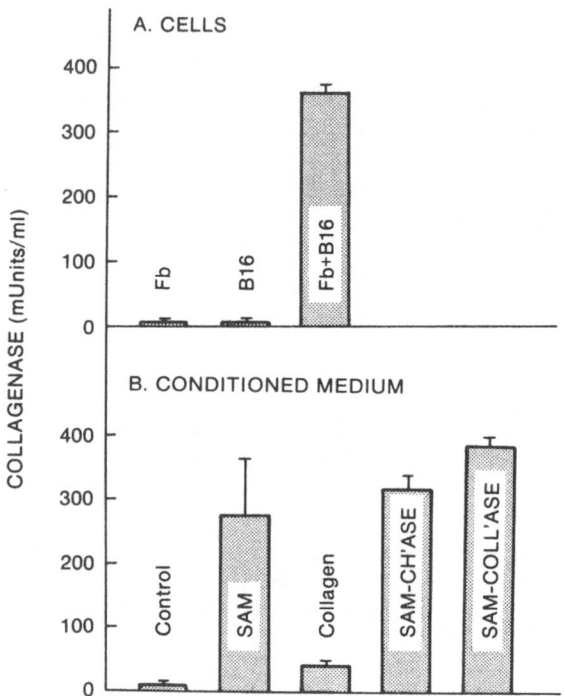

FIGURE 3. Effect of fibroblast matrix on secretion of collagenase stimulatory factor by murine B16 melanoma cells. (A) Collagenase activity in medium from cultures of fibroblasts alone (Fb) or B16 melanoma cells alone (B16), or from cocultures of fibroblasts plus B16 cells (Fb + B16). (B) Collagenase activity in medium from cultures of fibroblasts treated with conditioned medium from B16 cells plated on plastic (control), substrate-attached matrix from fibroblast cultures (SAM), collagen, chondroitinase-treated SAM (SAM–CH'ASE), and collagenase-treated SAM (SAM–COLL'ASE). (Data from Biswas, 1985a.)

activity (Biswas, 1985b; Biswas and Nugent, 1987). Addition of the B16 membrane preparation to cultures of fibroblasts stimulated collagenase production (Figure 5). This result, coupled with the data discussed above, suggests strongly that most of the factor is attached to the plasma membrane with its active site projecting externally. If it was attached to internal membranes, it would have been protected from trypsin during the treatment of the intact cells.

The membrane preparations described above were extracted with octylglucoside, and the dialyzed extracts also found to contain stimulatory activity (Figure 5). This activity is trypsin sensitive, heat labile, and nondialyzable, indicating that it is a protein. This protein can be reconstituted quantitatively into lipid vesicles (Figure 5), suggesting that it contains a hydrophobic domain that presumably intercalates it into the plasma membrane.

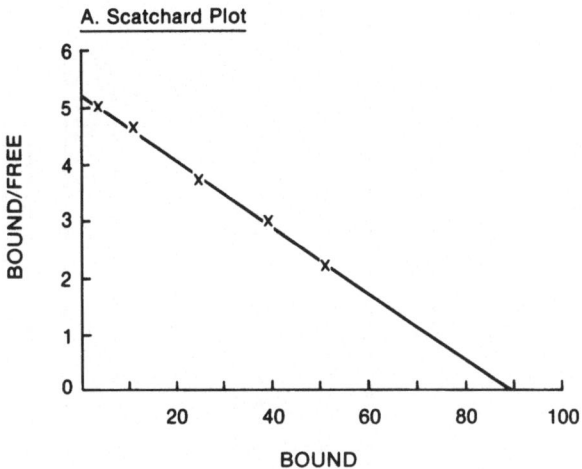

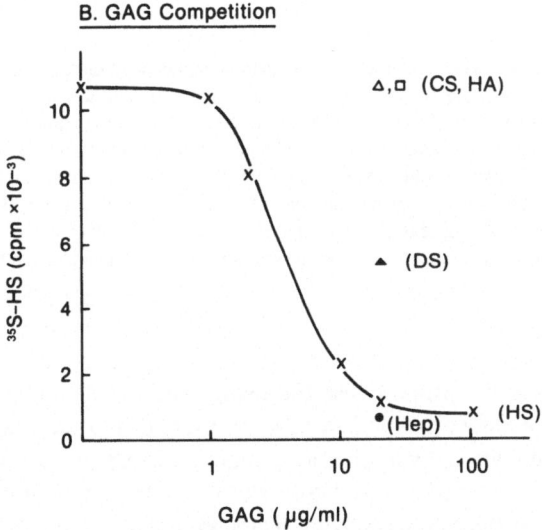

FIGURE 4. Heparan sulfate binding to B16 melanoma cells. (A) Scatchard plot of the binding of ^{35}S-labeled heparan (sulfate (K_d approx. 1 μM). (B) Competition for binding of ^{35}S-labeled heparan sulfate by heparan sulfate (HS) and heparin (Hep); dermatan sulfate (DS) was partially effective at 20 μg/ml, whereas chondroitin sulfate (CS) and hyaluronate (HA) gave no competition.

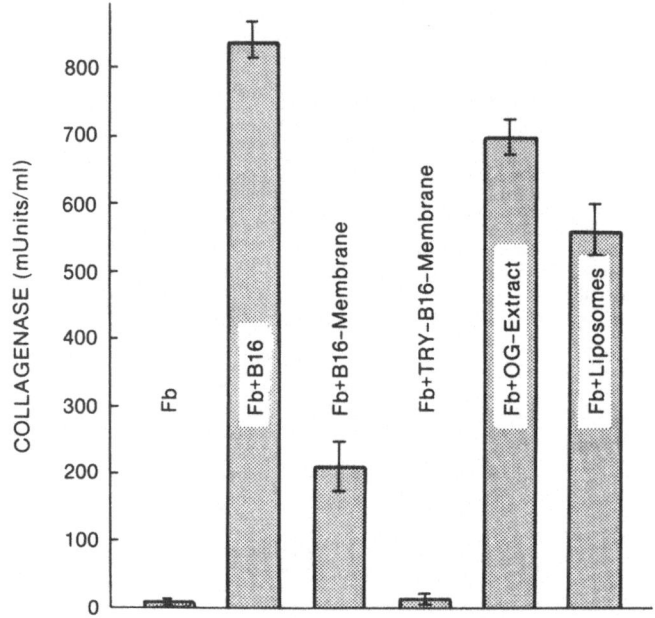

FIGURE 5. Membrane association of the collagenase–stimulatory factor of B16 melanoma cells. In each case, the collagenase activity obtained in medium from cultures of 1×10^5 fibroblasts is given. (Fb) Fibroblasts alone; (Fb + B-16) cocultures of fibroblasts plus 1×10^5 B16 cells; (Fb + B16–Membrane) fibroblast plus membrane prepared from the 1×10^5 B16 cells; (Fb + TRY–B16–Membrane) fibroblasts plus membranes prepared from 1×10^5 cells subsequent to pretreatment of the cells with trypsin to remove cell surface–associated factor; (Fb + OG–Extract) fibroblasts plus octylglucoside extract of membrane from 1×10^5 B16 cells; (Fb + Liposomes) fibroblasts plus liposomes reconstituted from the octylglucoside extract of membranes from 1×10^5 B16 cells.

In summary, we propose the following working hypothesis: (1) that the collagenase–stimulatory factor is a protein inserted into the plasma membrane of the tumor cell via a hydrophobic domain; (2) that an external hydrophilic domain contains the active stimulatory site that interacts with receptors on the target fibroblast surface, thus causing increased collagenase production: (3) that the active external domain is cleaved or shed from the cell surface under certain conditions, such as exposure to fibroblast matrix, thus appearing in soluble or vesicular form in conditioned medium. This latter process may be stimulated by an interaction between heparan sulfate proteoglycan in the fibroblast matrix with heparan sulfate–binding sites in the plasma membrane of the tumor cells. Thus, it is envisioned that *in vivo*, tumor cells at the invasion zone would interact with heparan sulfate proteoglycan of the surrounding extracellular matrix, causing increased production or release of the collagenase–stimulatory factor. The factor would, in turn, interact with adjacent fibroblasts, causing them to alter their matrix by collagenolysis. An analogous influence of tumor cell–matrix interaction has been invoked for the control of type IV collagenase production. In this case, interaction of human melanoma and fibrosarcoma cells with laminin, a component of the basement membrane, led to an increased level of type IV collagenase (Turpeenniemi-Hujanen *et al.*, 1986). Here, however, the response was production of the enzyme itself by the tumor cell, rather than production of a stimulatory factor that acts on host fibroblasts.

2.2. Other Matrix-Degrading Enzymes

Movement of tumor cells through normal tissue involves penetration of both basement membranes and interstitial stroma. In addition to type I collagenase, increased levels of other degradative enzymes have been implicated in these invasive events, e.g., type IV collagenase (Barsky *et al.*, 1983; Turpeenniemi-Hujanen *et al.*, 1985), heparanase (Nakajima *et al.*, 1983), plasminogen activator (Ossowski and Reich, 1983), and cathepsin B (Recklies *et al.*, 1980; Graf *et al.*, 1981; Sloane *et al.*, 1986). It appears that tumor cells themselves produce the enzymes required for the degradation of the characteristic basement membrane components, type IV collagen (Liotta *et al.*, 1979, 1981b) and heparan sulfate proteoglycan (Kramer *et al.*, 1982; Nakajima *et al.*, 1983). Increased production of type IV collagenase by tumor cells has been found to correlate very closely with the invasive potential of these cells and may be linked genetically to the metastatic photype (Liotta, 1986). However, there is some evidence that the elevated levels of cathepsin B found at the invasion zone of some tumors (Poole *et al.*, 1978; Graf *et al.*, 1981) are due to tumor cell–host interactions. In this case, conflicting reports have appeared wherein, for murine B16 melanomas, the enzyme was produced by tumor cells themselves (Sloane *et al.*, 1982, 1986), but for the rabbit V2 carcinoma, the enzyme was present in fibroblasts, leukocytes, and adjacent extracellular matrix associated with the tumors but not in the tumor cells themselves (Graf *et al.*, 1981).

3. EFFECT OF TUMOR CELL–FIBROBLAST INTERACTIONS ON PRODUCTION OF EXTRACELLULAR MACROMOLECULES

A second means by which extracellular matrices may become more conducive to tumor cell invasion is via altered synthesis of matrix components. It is quite clear that tumor cells produce different types and proportions of matrix macromolecules to their normal counterparts, but it has also been observed that the composition of extracellular matrices produced by tumor cells alone in culture often does not correspond to the composition of the matrices of the tumors from which the cells were derived. Thus, tumor cell–host cell interactions may be important in regulating synthesis of the extracellular matrix macromolecules of tumors. Since embryonic tissues and tumors often share common properties and since hyaluronate is a ubiquitous component of embryonic mesenchymal tissues, we have focused on the effect of tumor cell–fibroblast interactions on hyaluronate production.

3.1. Hyaluronate

Hyaluronate is a major component of the extracellular matrix surrounding migratory and proliferating mesenchymal cells during development of many embryonic tissues and organs, as well as during regeneration and healing of adult tissues (Toole, 1981; Toole and Underhill, 1983; Toole et al., 1984). It is also a major component of many types of tumors. For example, in the human, high concentrations are present in kidney (Hopwood and Dorfman, 1978), mammary (Takeuchi et al., 1976), hepatic (Kojima et al., 1975), lung (Horai et al., 1981; Kawai et al., 1985), and parotid gland (Takeuchi et al., 1981) tumors.

Several studies have indicated that hyaluronate can influence the behavior of cells, especially in regard to modulation of cell aggregation and cell movement. For both of these types of cell behavior, hyaluronate can have stimulatory or inhibitory influences, depending on the cell type being studied, the concentrations and molecular weight of hyaluronate employed, and the mode of interaction of hyaluronate with the cell. For example, hyaluronate inhibits the migration of endothelial cells (Feinberg and Beebe, 1983) but promotes that of several other cell types (Turley and Roth, 1979; Bernanke and Markwald, 1979; Abatangelo et al., 1983; Turley et al., 1985; Doillon and Silver, 1986). Hyaluronate blocks aggregation of several types of cells at high concentrations, whereas it can mediate aggregation at low concentrations (Fraser and Clarris, 1970; Pessac and Defendi, 1972; Wasteson et al., 1973; Underhill and Dorfman, 1978; McBride and Bard, 1979; Love et al., 1979; Forrester and Lackie, 1981; Wright et al., 1981). In studies of certain virally transformed cells, it has been shown that aggregation is due to cross bridging by hyaluronate of hyaluronate receptors present in the plasma membrane of the cells. However, these receptors become saturated at high concentrations of hyaluronate, thus inhibiting cross bridging (Underhill and Toole, 1981; Underhill, 1982; Toole and Underhill, 1983).

It is thought that hyaluronate promotes mesenchymal cell movement in three ways. First, high concentrations of hyaluronate form a meshwork that exerts considerable osmotic pressure (Meyer, 1983; Meyer et al., 1983). This "swelling pressure" may lead to formation of hydrated pathways separating barriers to cell migration, such as collagen fibers and cell layers (Toole, 1981; Toole *et al.,* 1984). Examples of this are seen in the accumulation of hyaluronate and the concomitant tissue hydration and separation of tissue structures at the time of migration of embryonic corneal mesenchyme (Toole and Trelstad, 1971) and neural crest cells (Pratt *et al.,* 1975). Second, hyaluronate is essential to the structure of hydrated pericellular coats (see Figure 6) which, in several cases, have been shown to prevent close associations of cells (Fraser and Clarris, 1970; McBride and Bard, 1979; Underhill and Toole, 1982). Prevention of tight intercellular adhesion is necessary for continued cell movement or proliferation.

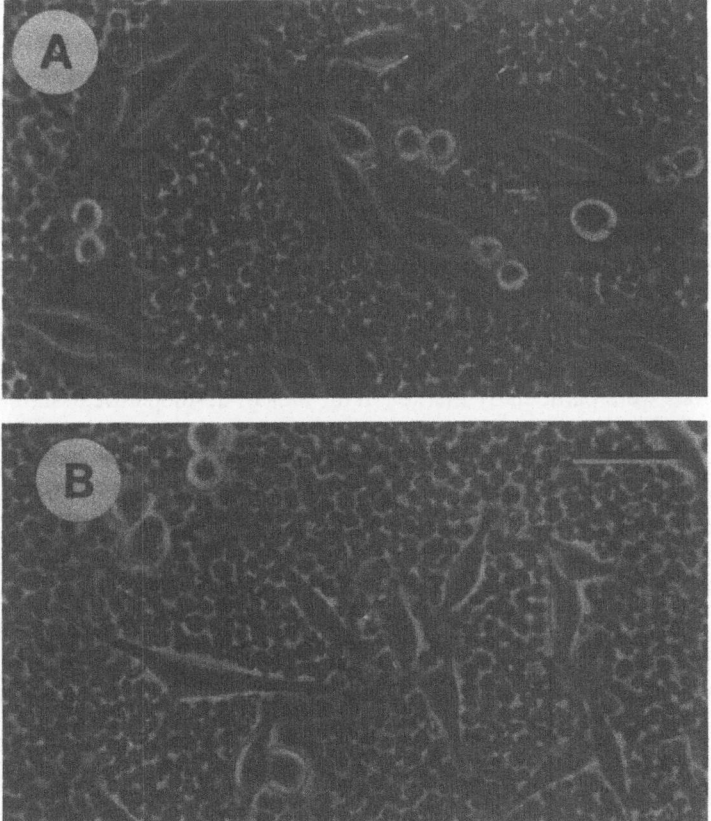

FIGURE 6. Hyaluronate-dependent coats surrounding rat fibrosarcoma cells. (A) The coats have been visualized by exclusion of particles (fixed red blood cells) (Fraser and Clarris, 1970; McBride and Bard, 1979; Underhill and Toole, 1982). (B) Removal of pericellular coat by treatment with Streptomyces hyaluronidase.

For example, embryonic myoblasts exhibit a prominent hyaluronate-dependent coat, whereas this coat is lost on fusion of the cells to form myotubes (Orkin *et al.*, 1985). Constant exposure of myoblasts to a hyaluronate-conjugated cell culture substratum prevents their fusion and permits the myoblasts to continue to proliferate (Kujawa and Tepperman, 1983; Kujawa *et al.*, 1986). Third, hyaluronate is associated with reduced adhesion to substrata (Culp *et al.*, 1979; Barnhart *et al.*, 1979; Fisher and Solursh, 1979; Mikuni-Takagaki and Toole, 1980; Abatangelo *et al.*, 1982; Turley *et al.*, 1985). Notably, Turley *et al.* (1985) have shown that addition of a hyaluronate-binding protein and hyaluronate together to fibroblasts increases their motility, reduces cell spreading, and increases cell underlapping. The hyaluronate-binding protein was shown to be localized at the leading lamellae and retraction processes of motile cells, leading to the conclusion that these phenomena were due to reduced adhesion resulting from hyaluronate–substratum interactions (Turley and Torrance, 1984; Turley *et al.*, 1985). Hyaluronate receptors are present in the plasma membrane of virally transformed cells (Underhill and Toole, 1979, 1981; Underhill *et al.*, 1983, 1985; Goldberg *et al.*, 1984), but their role in cell motility has not yet been established.

Because of the probable involvement of hyaluronate in cell migration and its high concentration in many tumors, we compared the levels of hyaluronate in the rabbit V2 carcinoma under circumstances where it exhibited invasive and noninvasive phenotypes (Toole *et al.*, 1979). We showed that high concentrations were present in this highly invasive tumor when it was grown subcutaneously or intramuscularly in its natural host, the rabbit. The elevated hyaluronate concentrations were particularly marked in the area of desmoplastic tissue surrounding the tumor mass (Figure 7). Similar results have been obtained recently by Iozzo and Muller-Glauser (1985) using the V2 carcinoma implanted in the rabbit mesentery. When the same tumor is implanted subcutaneously or intramuscularly in the nude mouse, it grows rapidly but is not invasive or metastatic (Biswas *et al.*, 1982). Tumors grown in nude mice were found to contain much lower amounts of hyaluronate (Figure 7). The concentrations of hyaluronate in the capsules of these tumors were similar to normal connective tissue of the host mice and much lower than that in the tissue surrounding the invasive tumors in the rabbit (Toole *et al.*, 1979). This suggested the possibility that pericellular hyaluronate facilitates the invasion of tumor cells through adult tissues. Consequently, we have begun to investigate the regulation of hyaluronate metabolism in a fashion similar to that outlined above for type I collagenase.

It has been shown that some types of tumor and transformed cells actively synthesize hyaluronate in cell culture (Hopwood and Dorfman, 1977; Takeuchi *et al.*, 1981; Mikuni-Takagaki and Toole, 1981; Ullrich and Hawkes, 1983). It has also been shown that certain variants of mouse mammary carcinoma cells that are highly metastatic *in vivo* produce more hyaluronate in culture than less

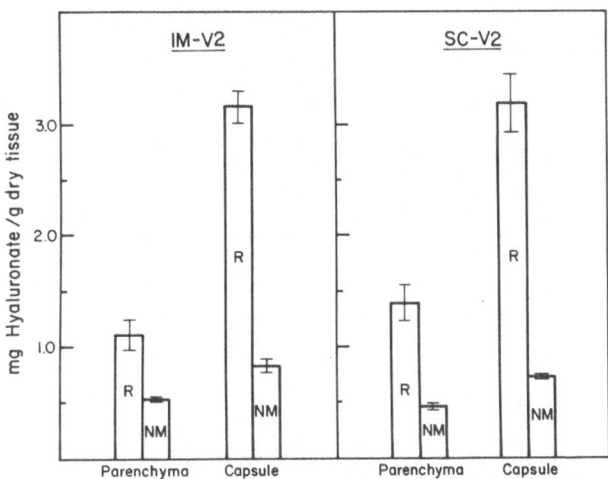

FIGURE 7. Elevated levels of hyaluronate in invasive rabbit V2 carcinomas. Concentrations of hyaluronate were compared in tumors grown intramuscularly (IM-V2) or subcutaneously (SC-V2) in the rabbit (R) and nude mouse (NM). The central (parenchyma) and peripheral (capsule) regions were analyzed separately. The tumors grown in the rabbit were invasive, whereas those in the nude mouse were noninvasive. (Reproduced from Toole *et al.*, 1979.)

metastic variants (Angello *et al.*, 1982; Kimata *et al.*, 1983). However, transformed cell lines, such as simian virus–transformed 3T3 cells and polyoma-transformed BHK cells produce considerably less hyaluronate in culture than their nontransformed counterparts (Hamerman *et al.*, 1965; Underhill and Toole, 1982). Also, several types of tumor cells have been shown to produce only small amounts of hyaluronate in culture, even when the the the tumors from which they were derived contain high levels of this glycosaminoglycan (Satoh *et al.*, 1974; Bhavanandan, 1981; Knudson *et al.*, 1984a,b). Thus, we examined the synthesis of hyaluronate in cocultures of tumor cells and fibroblasts. No increase in hyaluronate production was observed in murine cocultures, although extracts of the original murine tumors greatly stimulated the amount of hyaluronate produced by murine fibroblasts (Knudson *et al.*, 1984a). In cocultures of human tumor cells and fibroblasts, however, marked stimulation was observed (Figure 8). This stimulation was obtained with all three human tumor types tested, viz. LX-1 lung carcinoma, DAN pancreatic carcinoma, and TRIG melanoma and appeared to require contact between the tumor cells and the fibroblasts (Knudson *et al.*, 1984b). In further studies (Knudson and Toole, 1985) it has been shown that membranes derived from LX-1 carcinoma cells and detergent extracts thereof mimic the stimulatory effect of the intact LX-1 cells, thus inferring that the stimulatory factor is localized in the tumor cell plasma membrane in similar

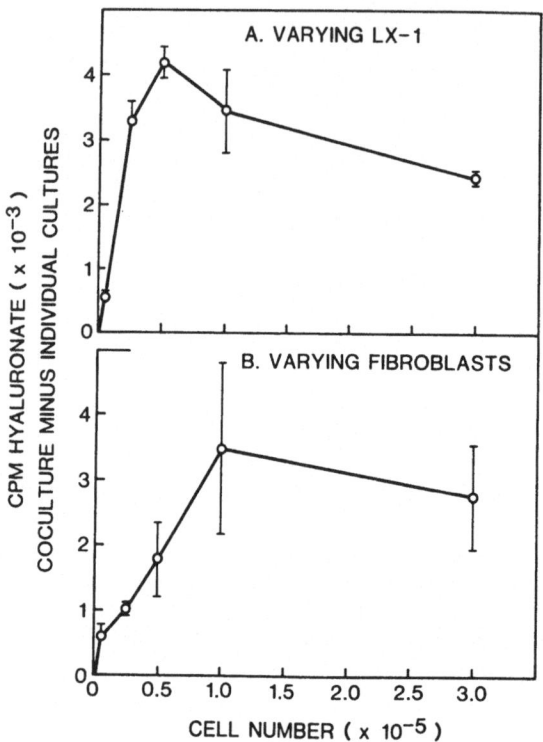

FIGURE 8. Stimulation of hyaluronate synthesis in cocultures of human LX-1 lung carcinoma cells and human fibroblasts. (A) Increasing numbers of LX-1 cells were cocultured with 1×10^5 fibroblasts. (B) Increasing numbers of fibroblasts were cocultured with 1×10^5 LX-1 cells. Results are expressed as the increase in amount of incorporation of ^{3}H-labeled acetate into hyaluronate in the cocultures, as compared with the sum of incorporation in separate cultures of LX-1 cells and fibroblasts. (Reproduced from Knudson *et al.*, 1984b.)

fashion to the collagenase–stimulatory protein described above (Section 2.1). Attempts to duplicate these results with conditioned medium obtained from cultures of the LX-1 tumor cells have not been successful (Knudson *et al.*, 1984b), but Merrilees and Finlay (1985) have obtained marked stimulation of production of hyaluronate by fibroblasts on treatment with conditioned medium from several other human tumor cell lines.

3.2. Other Extracellular Macromolecules

Many invasive neoplasms exhibit accumulation of connective tissue at their borders, a process termed desmoplasia (Vasiliev, 1958; Liotta *et al.*, 1983; Iozzo, 1984). Thus, the accumulation of hyaluronate due to tumor cell–fibroblast interactions described above could be a reflection of the desmoplastic response *in vivo*. In this regard, it was observed that the stimulation of hyaluronate *in vivo* and in culture, as described above, was accompanied by increases in chondroitin sulfate at a lower level than hyaluronate, in some cases (Toole *et al.*, 1979; Knudson *et al.*, 1984a,b; Merrilees and Finlay, 1985) and at a higher level, in other cases (Iozzo and Muller-Glauser, 1985; Knudson *et al.*, 1984a). Iozzo and co-workers (Iozzo, 1984, 1985a,b; Iozzo *et al.*, 1982; Iozzo and Wight, 1982) have studied in detail the increases in chondroitin sulfate proteoglycan production associated with human colon carcinoma *in vivo* and the stimulation of production of the proteoglycan by fibroblasts in response to interaction with colon carcinoma cells. This proteoglycan is localized mainly in the connective tissue immediately surrounding the tumor rather than within the tumor mass itself (Iozzo *et al.*, 1982), and, recently, it has been demonstrated that colon carcinoma cells release protein factors that stimulate colon fibroblasts to produce elevated levels of the chondroitin sulfate proteoglycan (Iozzo, 1985b). It is not yet known whether these factors are derived from the plasma membrane, as seems to be the case for the collagenase—and hyaluronate—stimulatory factors described above.

Several recent studies have also documented increases in deposition of collagen in tumor-associated connective tissue (Barsky *et al.*, 1982; Iozzo and Muller-Glauser, 1985). Of particular interest is the finding that extracts of rat mammary adenocarcinomas contain factors that stimulate collagen production by normal fibroblasts or epithelial cells (Bano *et al.*, 1983).

4. CELL–CELL INTERACTIONS AND EXTRACELLULAR MATRIX MODULATION IN EMBRYONIC DEVELOPMENT, TISSUE REGENERATION, AND TUMOR INVASION

Remodeling of the extracellular matrix is a prominent feature of embryogenesis and regeneration, as well as tumorigenesis, and we would propose that the phenomena that we have described above are related to events that occur during morphogenesis and regeneration. As in tumor invasion, both degradative and synthetic processes are important in matrix remodeling in embryonic or regenerative systems. For example, synthesis and removal of hyaluronate (Toole, 1981; Toole and Underhill, 1983; Toole *et al.*, 1984), fibronectin (Hynes, 1981; Thiery, 1984), and collagens (Toole and Linsenmayer, 1977; Hay *et al.*, 1979; Hay, 1981; Reddi, 1984) are commonly associated with tissue morphogenesis and differentiation. Of particular interest are epithelial–mesenchymal interactions, which are crucial to developmental processes and which have been shown to lead to increases in both matrix degradation and synthesis. It has been shown, for example, that corneal and skin epithelia produce factors that stimulate fibroblasts to produce elevated levels of type I collagenase (Johnson-Muller and Gross, 1978; Johnson-Wint, 1980; Johnson-Wint and Bauer, 1985) and that various epithelial and endothelial cells stimulate hyaluronate production in cells of mesenchymal origin (Merrilees and Scott, 1980, 1981; Merrilees et al., 1983). These are also of special interest since the stimulation of fibroblast collagenase and hyaluronate production by tumor cells described above has only been observed so far with tumor cells of epithelial origin and not with sarcomas (Biswas, 1982b, 1984; Knudson et al., 1984b; Iozzo, 1985a,b; Merrilees and Finlay, 1985).

A particularly instructive example from embryonic development is the formation of kidney tubules from metanephric mesenchyme in response to contact with ureteric epithelium (Ekblom, 1984). The earliest morphological event observed to take place during induction of tubule formation is condensation of mesenchymal cells around the ureteric bud. The earliest molecular event so far described is the loss of types I and III collagen and fibronectin from the matrix of the condensed mesenchyme (Ekblom, 1981; Ekblom et al., 1981). The mesenchyme goes on to differentiate into epithelial precursors of the kidney tubules with concomitant production of basement membrane components (Ekblom, 1981, 1984). Loss of hyaluronate and increase in hyaluronidase activity also accompany the differentiation of the tubules (Reeves et al., 1980; Belsky and Toole, 1983). Thus, it seems likely that interaction of the ureteric epithelium with the mesenchyme may give rise to production of a collagenase (and possibly other degradative enzymes) in the mesenchyme in an analogous fashion to the tumor cell–fibroblast interaction described above. Possibly, also, the initial ingrowth of ureteric epithelium into and contact with the mesenchymal matrix might lead to production of a stimulatory factor in a similar manner to the matrix effect on B16 melanoma cells (Figure 3). Similar interactions are likely to be involved in the spatial control of extracellular matrix deposition and degradation during branching of embryonic epithelia, such as in the developing salivary gland (Bernfield and Banerjee, 1972; Cohn et al., 1977; Bernfield et al., 1984).

Regeneration, as well as development, characteristically involves matrix remodeling analogous to that described above for tumorigenesis. Early events in newt limb regeneration include the onset of increased collagenase activity (Grillo et al., 1968; Mailman and Dresden, 1979) and hyaluronate synthesis (Toole and Gross, 1971; Smith et al., 1975; Mescher and Munaim, 1986). These both occur at the crucial stage of blastemal cell "dedifferentiation," proliferation, and migration. As in tumorigenesis, regeneration appears to require removal of preexisting structural barriers that may interfere with cell proliferation and migration, as well as reconstruction of an embryoniclike extracellular milieu prior to redifferentiation. An important structure in the control of regeneration is the wound epithelium that grows over the end of the amputated stump (Tassava and Mescher, 1975). Possibly, interaction of this epithelium with the underlying mesenchyme may be involved in regulation of matrix remodeling. In this regard, evidence has been obtained suggesting that mammalian would epithelium–mesenchyme interaction stimulates collagenase production (Grillo and Gross, 1967).

In summary, it is apparent that tumor cell invasion is accompanied by degradative and synthetic alterations in the extracellular matrix. These changes give rise to removal of structural barriers to tumor cell movement and reconstruction of an embryoniclike matrix more conducive to movement. It is proposed that these changes are, in some instances, regulated by tumor cell–fibroblast interactions analogous to cell interactions that occur in embryonic development and tissue regeneration.

REFERENCES

Abatangelo, G., Cortivo, R., Martelli, M., and Vecchia, P., 1982, Cell detachment mediated by hyaluronic acid, *Exp. Cell Res.* **137**:73–78.

Abatangelo, G., Martelli, M., and Vecchia, P., 1983, Healing of hyaluronic acid-enriched wounds: Histological observations, *J. Surg. Res.* **35**:410–416.

Abramson, M., Schilling, R. W., Huang, C. C., Salome, R. G., 1975, Collagenase activity in epidermoid carcinoma of the oral cavity and larynx. *Ann. Otol. Rhinol. Laryngol.* **84**:158–163.

Angello, J. C., Danielson, D. G., Anderson, L. W., and Hosick, H. L., 1982, Glycosaminoglycan synthesis by subpopulations of epithelial cells from a mammary adenocarcinoma, *Cancer Res.* **42**:2207–2210.

Baici, A., Gyger-Marazzi, M., and Strauli, P., 1984, Extracellular cysteine proteinase and collagenase activities as a consequence of tumor–host interaction in the rabbit V₂-carcinoma, *Invasion Metastasis* **4**:13–27.

Bano, M., Zwiebel, J. A., Salomon, D. S., and Kidwell, W. R., 1983, Detection and partial characterization of collagen synthesis stimulating activities in rat mammary adenocarcinomas, *J. Biol. Chem.* **258**:2729–2735.

Barnhart, B. J., Cox, S. H., and Kraemer, P. M., 1979, Detachment variants of Chinese hamster cells. Hyaluronic acid as a modulator of cell detachment. *Exp. Cell Res.* **119**:327–332.

Barsky, S. H., Rao, C. N., Grotendorst, G. R., and Liotta, L. A., 1982, Increased content of type V collagen in desmoplasia of human breast carcinoma, *Am. J. Pathol.* **108**:276–283.

Barsky, S. H., Siegal, G. P., Janotta, F., and Liotta, L. A., 1983, Loss of basement membrane components by invasive tumors but not by their benign counterparts, *Lab. Invest.* **49**:140–147.

Bauer, E. A., Gordon, J. M., Reddick, M. E., and Eisen, A. Z., 1977, Quantitation and immunocytochemical localization of human skin collagenase in basal cell carcinoma, *J. Invest. Dermatol.* **29**:363–367.

Bauer, E. A., Uitto, J., Walters, R. C., and Eisen, A. Z., 1979, Enhanced collagenase production by fibroblasts derived from human basal cell carcinomas, *Cancer Res.* **39**:4594–4599.

Belsky, E., and Toole, B. P., 1983, Hyaluronate and hyaluronidase in the developing chick embryo kidney, *Cell Differ.* **12**:61–66.

Bernanke, D. H., and Markwald, R. R., 1979, Effects of hyaluronic acid on cardiac cushion tissue cells in collagen matrix cultures, *Tex. Rep. Biol. Med.* **39**:271–285.

Bernfield, M. R., and Banerjee, S., 1972, Acid mucopolysaccharide (glycosaminoglycan) at the epithelial–mesenchymal interface of mouse embryo salivary glands, *J. Cell Biol* **52**:664–673.

Bernfield, M. R., Banerjee, S. P., Koda, J. E., and Rapraeger, A. C., 1984, Remodeling of the basement membrane as a mechanism of morphogenetic tissue interaction, in: *The Role of*

Extracellular Matrix in Development (R. L. Trelstad, ed.), Alan R. Liss, New York, pp. 545–572.

Bhavanandan, V. P., 1981, Glycosaminoglycans of cultured human fetal uveal melanocytes and comparison with those produced by cultured human melonoma cells, *Biochemistry* **20:** 5595–5602.

Biswas, C., 1982a, Host–tumor cell interactions and collagenase activity, in: *Tumor Invasion and Metastasis* (L. A. Liotta and I. R. Hart, eds.), Martinus Nijhoff, The Hague, p. 405–425.

Biswas, C., 1982b, Tumor cell stimulation of collagenase production by fibroblasts, *Biochem. Biophys. Res. Commun.* **109:**1026–1034.

Biswas, C., 1984, Collagenase stimulation in cocultures of human fibroblasts and human tumor cells, *Cancer Lett.* **24:**201–207.

Biswas, C., 1985a, Matrix influence on the tumor cell stimulation of fibroblast collagenase production, *J. Cell Biochem.* **28:**39–45.

Biswas, C., 1985b, Tumor cell stimulation of fibroblast collagenase production: Membrane association of the tumor cell stimulator, *Fed. Proc. 44:*1337.

Biswas, C., and Nugent, M. A., 1987, Membrane association of the collagenase stimulatory factor from B-16 melanoma cells *J. Cell. Biochem.* (in press).

Biswas, C., Moran, W. P., Bloch, K. J., and Gross, J., 1978, Collagenolytic activity of rabbit V_2-carcinoma growing at multiple sites, *Biochem. Biophys. Res. Commun.* **80:**33–38.

Biswas, C., Block, K. J., and Gross, J., 1982, Collagenolytic activity of rabbit V_2-carcinoma implanted in the nude mouse, *J. Natl. Cancer Inst.* **69:**1229–1336.

Cohn, R. H., Banerjee, S. D., and Bernfield, M. R., 1977, Basal lamina of embryonic salivary epithelia. Nature of glycosaminoglycan and organization of extracellular materials, *J. Cell Biol.* **73:**464–478.

Culp, L. A., Murray, B. A., and Rollins, B. J., 1979, Fibronectin and proteoglycans as determinants of cell–substratum adhesion, *J. Supramol. Struct.* **11:**401–427.

Dabbous, M. K., Roberts, A. N., and Brinkley, B., 1977, Collagenase and neutral protease activities in cultures of rabbit VX-2 carcinoma, *Cancer Res.* **37:**3537–3544.

Doillon, C. J., and Silver, F. H., 1986, Collagen-based wound dressing: Effects of hyaluronic acid and fibronectin on would healing, *Biomaterials* **7:**3–8.

Dresden, M. H., Heilman, S. A., and Schmidt, J. D., 1972, Collagenolytic enzymes in human neoplasms, *Cancer Res* **32:**993–996.

Ekblom, P., 1981, Formation of basement membranes in the embryonic kidney: An immunohistological study, *J. Cell Biol.* **91:**1–10.

Ekblom, P., 1984, Basement membrane proteins and growth factors in kidney differentiation, in: *The Role of Extracellular Matrix in Development* (R. L. Trelstad, ed.), Alan R. Liss, New York, pp. 173–206.

Ekblom, P., Lehtonen, E., Saxen, L. and Timpl, R., 1981, Shift in collagen type as an early response to induction of the metanephric mesenchyme, *J. Cell Biol.* **89:**276–283.

Feinberg, R. N., Beebe, D. C., 1983, Hyaluronate in vasculogenesis, *Science* **220:**1177–1179.

Fisher, M., and Solursh, M., 1979, The influence of the substratum on mesenchyme spreading *in vitro*, *Exp. Cell Res.* **123:**1–14.

Folkman, J., 1986, How is blood vessel growth regulated in normal and neoplastic tissue?, *Cancer Res.* **46:**467–473.

Forrester, J. V., and Lackie, J. M., 1981, Effect of hyaluronic acid on neutrophil adhesion, *J. Cell Sci.* **50:**329–344.

Fraser, J. R., and Clarris, B. J., 1970, On the reactions of human synovial cells exposed to homologous leucocytes *in vitro*, *Clin. Exp. Immunol.* **6:**211–225.

Goldberg, R. L., Seidman, J. D., Chi-Rosso, G., and Toole, B. P., 1984, Endogenous hyaluronate–cell surface interactions in 3T3 and simian virus–transformed 3T3 cells, *J. Biol. Chem.* **259:**9440–9446.

Goslen, J. B., Eisen, A. Z., and Bauer, E. A., 1985, Stimulation of skin fibroblast collagenase production by a cytokine derived from basal cell carcinomas, *J. Invest. Dermatol.* **85**:161–164.

Graf, M., Baici, A., and Strauli, P., 1981, Histochemical localization of cathepsin B at the invasion front of the rabbit V_2 carcinoma, *Lab. Invest.* **45**:587–596.

Grillo, H. C., and Gross, J., 1967, Collagenolytic activity during mammalian wound repair, *Dev. Biol.* **15**:300–317.

Grillo, H. C., Lapiere, C. M., Dresden, M. H., and Gross, J., 1968, Collagenolytic activity in regenerating forelimbs of the adult newt. *Dev. Biol.* **17**:571–583.

Gross, J., Highberger, J. H., Johnson-Wint, B., and Biswas, C., 1980, Mode of action and regulation of tissue collagenases, in: *Collagenase in Normal and Pathological Connective Tissues* (D. E. Woolley and J. M. Evanson, eds.), John Wiley and Sons, New York, pp. 11–35.

Gross, J., Azizkhan, R. G., Biswas, C., Bruns, R. R., Hsieh, D-S.T. and Folkman, J., 1981, Inhibition of tumor growth, vascularization, and collagenolysis in the rabbit cornea by medroxyprogesterone, *Proc. Natl. Acad. Sci. USA* **78**:1176–1180.

Hamerman, D., Todaro, G. J., and Green, H., 1965, The production of hyaluronate by spontaneously established cell lines and virally transformed lines of fibroblastic origin, *Biochim. Biophys. Acta* **101**:343–351.

Hashimoto, K., Yamanishi, Y., Maeyens, E., Dabbous, M. K., and Kansaki, T., 1973, Collagenolytic activities of squamous cell carcinoma of the skin, *Cancer Res.* **33**:2790–2801.

Hay, E. D., 1981, Collagen and embryonic development, in: *Cell Biology of Extracellular Matrix* (E. D. Hay, ed.), Plenum Press, New York, pp. 379–409.

Hay, E. D., Linsenmayer, T. F., Trelstad, R. L., and von der Mark, K., 1979, Origin and distribution of collagens in the developing avian cornea, *Curr. Top. Eye Res.* **1**:1–31.

Hopwood, J. D., and Dorfman, A., 1977, Glycosaminoglycan synthesis by cultured human skin fibroblasts after transformation with simian virus 40, *J. Biol. Chem.* **252**:4777–4785.

Hopwood, J. J. and Dorfman, A., 1978, Glycosaminoglycan synthesis by Wilm's tumor, *Pediatr. Res.* **12**:52–56.

Horai, T., Nakamura, N., Tateishi, R. and Hattori, S., 1981, Glycosaminoglycans in human lung cancer, *Cancer* **48**:2016–2021.

Hynes, R. O., 1981, Fibronectin and its relation to cellular structure and behavior, in: *Cell Biology of Extracellular Matrix* (E. D. Hay, ed.) Plenum Press, New York, pp. 295–334.

Iozzo, R. V., 1984, Proteoglycans and neoplastic–mesenchymal cell interactions, *Hum. Pathol.* **15**:1–10.

Iozzo, R. V., 1985a, Proteoglycans: Structure, function, and role in neoplasia, *Lab. Invest.* **53**:373–396.

Iozzo, R. V., 1985b, Neoplastic modulation of extracellular matrix: Colon carcinoma cells release polypeptides that alter proteoglycan metabolism in colon fibroblasts, *J. Biol. Chem.* **260**:7464–7473.

Iozzo, R. V., and Muller-Glauser, W., 1985, Neoplastic modulation of extracellular matrix: Proteoglycan changes in the rabbit mesentery induced by V_2 carcinoma cells, *Cancer Res.* **45**:5677–5687.

Iozzo, R. V., and Wight, T. N., 1982, Isolation and characterization of proteoglycans synthesized by human colon and colon carcinoma, *J. Biol. Chem.* **257**:11135–11144.

Iozzo, R. V., Bolender, R. P., and Wight, T. N., 1982, Proteoglycan changes in the intercellular matrix of human colon carcinoma. An integrated biochemical and stereologic analysis, *Lab. Invest.* **47**:124–138.

Johnson-Muller, B., and Gross, J., 1978, Regulation of corneal collagenase production: Epithelial stromal cell interactions. *Proc. Natl. Acad. Sci. USA* **75**:4417–4421.

Johnson-Wint, B., 1980, Regulation of stromal cell collagenase production in adult rabbit cornea: *In vitro* stimulation and inhibition by epithelial cell products, *Proc. Natl. Acad. Sci. USA* **77**:5331–5335.

Johnson-Wint, B., and Bauer, E. A., 1985, Stimulation of collagenase synthesis by a 20,000-dalton epithelial cytokine: Evidence for pretranslational regulation, *J. Biol. Chem.* **260**:2080–2085.

Kawai, T., Suzuki, M., Shinmei, M., Maenaka, Y., and Kageyama, D., 1985, Glycosaminoglycans of malignant diffuse mesothelioma, *Cancer* **56**:576–574.

Kimata, K., Honma, Y., Okayama, M., Oguri, K., Hozumi, M. and Suzuki, S., 1983, Increased synthesis of hyaluronic acid by mouse mammary carcinoma cell variants with high metastatic potential, *Cancer Res.* **43**:1347–1352.

Kjellen, L., Oldberg, A., and Hook, M., 1980, Cell surface heparan sulfate: Mechanisms of proteoglycan–cell association, *J. Biol. Chem.* **255**:10407–10413.

Knudson, W. and Toole, B. P., 1985, Properties of a membrane-bound tumor cell factor which stimulates hyaluronate synthesis in fibroblasts, *J. Cell Biol.* **101**:339a.

Knudson, W., Biswas, C., and Toole, B. P., 1984a, Stimulation of glycosaminoglycan production in murine tumors, *J. Cell Biochem.* **25**:183–196.

Knudson, W., Biswas, C., and Toole, B. P., 1984b, Interactions between human tumor cells and fibroblasts stimulate hyaluronate synthesis, *Proc. Natl. Acad. Sci. USA* **81**:6767–6771.

Kojima, J., Nakamura, N., Kanatani, M. and Ohmori, K., 1975, The glycosaminoglycans in human hepatic cancer, *Cancer Res.* **35**:542–547.

Kramer, R. H., Vogel, K. G., and Nicholson, G. L., 1982, Solubilization and degradation of subendothelial matrix glycoproteins and proteoglycans by metastatic tumor cells, *J. Biol. Chem.* **257**:2678–2686.

Kujawa, M. J., and Tepperman, K., 1983, Culturing chick muscle cells on glycosaminoglycan substrates: Attachment and differentiation. *Dev. Biol.* **99**:277–286.

Kujawa, M. J., Pechak, D. J., Fizman, M. Y., and Caplan, A. I., 1986, Hyaluronic acid bonded to cell culture surfaces inhibits the program of myogenesis, *Dev. Biol.* **113**:10–16.

Liotta, L. A., 1986, Tumor invasion and metastasis—Role of the extracellular matrix, *Cancer Res.* **46**:1–7.

Liotta, L. A., Abe, S., Robey, P. G., and Martin, G. R., 1979, Preferential digestion of basement membrane collagen by an enzyme derived from a metastatic murine tumor, *Proc. Natl. Acad. Sci. USA* **76**:2268–2272.

Liotta, L. A., Lanzer, W. L., and Garbisa, S., 1981a, Identification of a type V collagenolytic enzyme, *Biochem. Biophys. Res. Commun.* **98**:184–190.

Liotta, L. A., Tryggvason, K., Garbisa, S., Robey, P. G., and Abe, S., 1981b, Partial purification and characterization of a neutral protease which cleaves type IV collagen, *Biochemistry* **20**:100–104.

Liotta, L. A., Rao, C. N., and Barsky, S. H., 1983, Tumor invasion and the extracellular matrix, *Lab. Invest.* **49**:636–649.

Love, S. H., Shannon, B. T., Myrvik, Q. N., and Lynn, W. S., 1979, Characterization of macrophage agglutinating factor as a hyaluronic acid–protein complex, *J. Reticuloendothel. Soc.* **25**:269–282.

Mailman, M. L., and Dresden, M. H., 1979, Denervation effects on newt limb regeneration: Collagen and collagenase, *Dev. Biol.* **71**:60–70.

Mainardi, C. L., Dixit, S. N., and Kang, A. H., 1980a, Degradation of type IV (basement membrane) collagen by a proteinase isolated from human polymorphonuclear leukocyte granules, *J. Biol. Chem.* **255**:5435–5441.

Mainardi, C. L., Seyer, J. M., and Kang, A. H., 1980b, Type-specific collagenolysis: A type V collagen-degrading enzyme from macrophages, *Biochem. Biophys. Res. Commun.* **97**:1108–1115.

Mainardi, C. L., Hasty, D. L., Seyer, J. M., and Kang, A. H., 1981, Specific cleavage of human type III collagen by human polymorphonuclear leukocyte elastase, *J. Biol. Chem.* **255**:12006–12010.

McBride, W. H., and Bard, J. B., 1979, Hyaluronidase-sensitive halos around adherent cells. Their role in blocking lymphocyte-mediated cytolysis, *J. Exp. Med.* **149**:507–515.

McCroskery, P. A., Richards, J. F., and Harris, E. D., Jr., 1975, Purification and characterization of a collagenase extracted from rabbit tumours, *Biochem. J.* **152**:131–142.

Merrilees, M. J., and Finlay, G. L., 1985, Human tumor cells in culture stimulate glycosaminoglycan synthesis by human skin fibroblasts, *Lab. Invest.* **53**:30–36.

Merrilees, M. J., and Scott, L., 1980, Interaction of epithelial cells and fibroblasts in culture: Effect on glycosaminoglycan levels, *Dev. Biol.* **76**:396–409.

Merrilees, M. J., and Scott, L., 1981, Interaction of aortic endothelial cells and smooth muscle cells in culture: Effect on glycosaminoglycan levels, *Artherosclerosis* **39**:147–161.

Merrilees, M. J., Sodek, J., and Aubin, J. E., 1983, Effect of cells of epithelial rests of Malassez and endothelial cells on synthesis of glycosaminoglycans by periodontal ligament fibroblasts *in vitro*, *Dev. Biol.* **97**:146–153.

Mescher, A. L., and Munaim, S. I., 1986, Changes in the extracellular matrix and glycosaminoglycan synthesis during the initiation of regeneration in adult newt forelimbs, *Anat. Rec.* **214**:424–431.

Meyer, F. A., 1983, Macromolecular basis of globular protein exclusion and of swelling pressure in loose connective tissue (umbilical cord), *Biochim. Biophys. Acta* **755**:388–399.

Meyer, F. A., Laver-Rudich, Z., and Tanenbaum, R., 1983, Evidence for a mechanical coupling of glycoprotein microfibrils with collagen fibrils in Wharton's jelly, *Biochim. Biophys. Acta* **755**:376–387.

Mikuni-Takagaki, Y., and Toole, B. P., 1980, Cell–substratum attachment and cell surface hyaluronate of Rous sarcoma virus–transformed chondrocytes, *J. Cell. Biol.* **85**:481–488.

Mikuni-Takagaki, Y., and Toole, B. P., 1981, Hyaluronate–protein complex of Rous sarcoma virus–transformed chick embryo fibroblasts, *J. Biol. Chem.* **256**:8463–8469.

Miller, E. J., Finch, J. E., Chung, E., Butler, W. T. and Robertson, P. B., 1976, Specific cleavage of the native type III collagen molecule with trypsin, *Arch. Biochem. Biophys.* **173**:631–637.

Nakajima, M., Irimura, T., DiFerrante, D., DiFerrante, N., and Nicholson, G., 1983, Heparan sulfate degradation: Relation to tumor invasive and metastatic properties of B16 melanoma sublines, *Science* **220**:611–613.

O'Grady, R. L., Upfold, L. I., and Stephens, R. W., 1981, Rat mammary carcinoma cells secrete active collagenase and activate latent enzyme in the stroma via plasminogen activator, *Int. J. Cancer* **28**:509–515.

Orkin, R. W., Knudson, W., and Toole, B. P., 1985, Loss of hyaluronate-dependent coat during myoblast fusion, *Dev. Biol.* **107**:527–530.

Ossowski, L., and Reich, E., 1983, Antibodies to plasminogen activator inhibit tumor metastasis, *Cell* **35**:611–619.

Paranjpe, M., Engel, L., Young, N., and Liotta, L. A., 1980, Activation of breast carcinoma collagenase through plasminogen activator, *Life Sci.* **26**:1223–1231.

Pauli, B. U., Schwartz, D. E., Thomas, E., and Kuettner, K., 1983, Tumor invasion and host extracellular matrix, *Cancer Metastasis Rev.* **2**:129–152.

Pessac, B., and Defendi, V., 1972, Cell aggregation: Role of acid mucopolysaccharides, *Science* **175**:898–900.

Poole, A. R., Tiltman, K. J., Recklies, A. D., Stoker, T. A. M., 1978, Differences in secretion of the proteinase cathepsin B at the edges of human breast carcinomas and fibroadenomas, *Nature* **273**:545–547.

Pratt, R. M., Larsen, M. A., and Johnston, M. C., 1975, Migration of cranial neural crest cells in a cell-free hyaluronate-rich matrix, *Dev. Biol.* **44**:298–305.

Recklies, A. D., Tiltman, K. J., Stoker, A. M., and Poole, A. R., 1980, Secretion of proteinases from malignant and nonmalignant human breast tissue, *Cancer Res.* **40**:550–556.

Reddi, A. H., 1984, Extracellular matrix and development, in: *Extracellular Matrix Biochemistry* (K. A. Piez and A. H. Reddi, eds.), Elsevier, New York, pp. 375–404.

Reeves, W. H., Kanwar, Y. S., and Farquhar, M. G., 1980, Assembly of the glomerular filtration surface. Differentiation of anionic sites in glomerular capillaries of newborn rat kidney, *J. Cell. Biol.* **85**:735–753.

Sage, H., Woodbury, R. G., and Bornstein, P., 1979, Structural studies on human type IV collagen, *J. Biol. Chem.* **254**:9893–9900.

Satoh, C., Banks, J., Horst, P., Kreider, J. W., and Davidson, E. A., 1974, Polysaccharide production by cultured B-16 mouse melanoma cells, *Biochemistry* **13**:1233–1241.

Sloane, B. F., Honn, K. V., Sadler, J. G., Turner, W. A., Kimpson, J. J., and Taylor, J. D., 1982, Cathepsin B activity in B16 melanoma cells: A possible marker for metastatic potential, *Cancer Res.* **42**:980–986.

Sloane, B. F., Rozhin, J., Johnson, K., Taylor, H., Crissman, J. D. and Honn, K. V., 1986, Cathepsin B: Association with plasma membrane in metastatic tumors, *Proc. Natl. Acad. Sci. USA* **83**:2483–2487.

Smith, G. N., Toole, B. P., and Gross, J., 1975, Hyaluronidase activity and glycosaminoglycan synthesis in the amputated newt limb. Comparison of denervated, non-regenerating limbs with regenerates, *Dev. Biol.* **42**:221–232.

Takeuchi, J., Sobue, M., Sato, E., Shamato, M., Miura, K., and Nakagaki, 1976, Variation in glycosaminoglycan components of breast tumors, *Cancer Res.* **36**:2133–2139.

Takeuchi, J., Sobue, M., Sato, E., Yoshida, M., Uchibori, N., and Miura, K., 1981, A high level of glycosaminoglycan synthesis of squamous cell carcinoma of the parotid gland, *Cancer* **47**:2030–2035.

Tassava, R. A., and Mescher, A. L., 1975, The role of injury, nerves, and the wound epidermis during the initiation of amphibian limb regeneration, *Differentiation* **4**:23–24.

Thiery, J. P., 1984, Mechanisms of cell migration in the vertebrate embryo, *Cell. Differ.* **15**:1–15.

Toole, B. P., 1981, Glycosaminoglycans in morphogenesis, in: *Cell Biology of The Extracellular Matrix* (E. D. Hay, ed.), Plenum Press, New York, pp. 259–294.

Toole, B. P., and Gross, J., 1971, The extracellular matrix of the regenerating newt limb: Synthesis and removal of hyaluronate prior to differentiation, *Dev. Biol.* **25**:57–77.

Toole, B. P., and Linsenmayer, T. F., 1977, Newer knowledge of skeletogenesis: Macromolecular transitions in the extracellular matrix, *Clin. Orthop. Rel. Res.* **129**:258–278.

Toole, B. P., and Trelstad, R. L., 1971, Hyaluronate production and removal during corneal development in the chick, *Dev. Biol.* **26**:28–35.

Toole, B. P., and Underhill, C. B., 1983, Regulation of morphogenesis by the pericellular matrix, in: *Cell Interactions and Development: Molecular Mechanisms* (K. Yamada, ed.), John Wiley and Sons, New York, pp. 203–220.

Toole, B. P., Biswas, C., and Gross, J., 1979, Hyaluronate and invasiveness of the rabbit V_2 carcinoma, *Proc. Natl. Acad. Sci. USA* **76**:6299–6303.

Toole, B. P., Goldberg, R. L., Chi-Rosso, G., Underhill, C. B., and Orkin, R. W., 1984, Hyaluronate-cell interactions, in: *The Role of Extracellular Matrix in Development* (R. L. Trelstad, ed.), Alan R. Liss, Inc., New York, pp. 43–66.

Turley, E. A., and Roth, S., 1979, Spontaneous glycosylation of glycosaminoglycan substrates by adherent fibroblasts, *Cell* **17**:109–115.

Turley, E. A., and Torrance, J., 1984, Localization of hyaluronate and hyaluronate-binding protein on motile and non-motile fibroblasts, *Exp. Cell. Res.* **161**:17–28.

Turley, E. A., Bowman, P., and Kytryk, M. A., 1985, Effects of hyaluronate and hyaluronate-binding proteins on cell motile and contact behavior, *J. Cell Sci.* **78**:133–145.

Turpeenniemi-Hujanen, T., Thorgeirsson, U. P., Hart, I. R., Grant, S. S., and Liotta, L. A., 1985, Expression of collagenase IV (basement membrane collagenase) activity in murine tumor cell hybrids that differ in metastatic potential, *J. Natl. Cancer Inst.* **75**:99–103.

Turpeenniemi-Hujanen, T., Thorgeirsson, U. P., Rao, C. N., and Liotta, L. A., 1986, Laminin increases the release of type IV collagenase from malignant cells, *J. Biol. Chem.* **261**:1883–1889.

Ullrich, S. J., and Hawkes, S. P., 1983, The effect of the tumor promoter phorbol myristate acetate on hyaluronic acid synthesis by chicken embryo fibroblasts, *Exp. Cell Res.* **148**:377–386.

Underhill, C. B., 1982, Interaction of hyaluronate with the surface of simian virus 40–transformed 3T3 cells: Aggregation and binding studies, *J. Cell. Sci.* **56**:177–189.
Underhill, C. B., and Dorfman, A., 1978, The role of hyaluronic acid in intercellular adhesion of cultured mouse cells, *Exp. Cell. Res.* **117**:155–164.
Underhill, C. B., and Toole, B. P., 1979, Binding of hyaluronate to the surface of cultured cells, *J. Cell. Biol.* **82**:475–484.
Underhill, C. B., and Toole, B. P., 1981, Receptors for hyaluronate on the surface of parent and virus-transformed cell lines. Binding and aggregation studies, *Exp. Cell. Res.* **131**:419–423.
Underhill, C. B., and Toole, B. P., 1982, Transformation-dependent loss of the hyaluronate-containing coats of cultured cells, *J. Cell. Physiol.* **110**:123–128.
Underhill, C. B., Chi-Rosso, G., and Toole, B. P., 1983, Effects of detergent solubilization on the hyaluronate-binding protein from membranes of simian virus 40–transformed 3T3 cells, *J. Biol. Chem.* **258**:8086–8091.
Underhill, C. B., Thurn, A. L., and Lacey, B. E., 1985, Characterization and identification of the hyaluronate-binding site from membranes of SV-3T3 cells, *J. Biol. Chem.* **260**:8128–8133.
Vasiliev, J. M., 1958, The role of connective tissue proliferation in invasive growth of normal and malignant tissues: A review, *Br. J. Cancer* **12**:524–536.
Wasteson, A., Estermark, B., Lindahl, U., and Ponten, J., 1973, Aggregation of feline lymphoma cells by hyaluronic acid, *Int. J. Cancer* **12**:169–178.
Woolley, D. E., 1982, Collagenase immunolocalization studies of human tumors, in: *Tumor Invasion and Metastasis* (L. A. Liotta and I. R. Hart, eds.), Martinus Nijhoff Publishers, The Hague, pp. 391–404.
Woolley, D. E., 1984, Collagenolytic mechanisms in tumor cell invasion, *Cancer Metastasis Rev.* **3**:361–372.
Woolley, D. E., Tetlow, L. C., and Evanson, J. M., 1980, Collagenase immunolocalization studies of rheumatoid and malignant tissue, in: *Collagenase in Normal and Pathological Connective Tissue* (D. E. Wooley and J. M. Evanson, eds.), John Wiley and Sons, New York, pp. 105–126.
Wright, T. C., Underhill, C. B., Toole, B. P., and Karnovsky, M. J., 1981, Divalent cation-independent aggregation of rat-1 fibroblasts infected with a temperature-sensitive mutant of Rous sarcoma virus, *Cancer Res.* **41**:5107–5113.

EARLY CYTOPLASMIC SIGNALS AND CYTOSKELETAL RESPONSES INITIATED BY GROWTH FACTORS IN CULTURED CELLS

Paul L. McNeil and D. Lansing Taylor

1. INTRODUCTION

Growth factors activate quiescent cells in a stimulus–response coupling process that is initiated by binding the growth factor to its receptor on the cell surface (see Carpenter, 1984). The ultimate result of such activation is DNA synthesis and cell division (mitogenesis). However, such activated cells exhibit a host of earlier responses, some of which begin seconds poststimulation. These "early" cellular responses to growth factors include endocytosis, changes in the cytoskeleton, cell motility, phospholipid turnover, Na^+/H^+ exchanger activation, fluxes of a variety of ions, calcium transients, protein synthesis, changes in cyclic nucleotide levels, phosphorylation of specific proteins, and gene transcription. The connection, if any, of these early cellular responses with the later DNA synthetic response is obscure at present.

The goal of this chapter is twofold. First, we will review the literature on measurements of the generation of putative cytoplasmic signals as a consequence of growth factor stimulation. A variety of such signals are hypothesized as the temporal and spatial links between initial stimulation by growth factors of the quiescent cell at its surface and subsequent cytoplasmic and nuclear responses. The available evidence for linking such putative signals with mitogenesis will

Paul L. McNeil and D. Lansing Taylor • Department of Biological Sciences and Center for Fluorescence Research in Biomedical Sciences, Carnegie-Mellon University, Pittsburgh, Pennsylvania 15213. *Present address for P.L.M.:* Department of Anatomy and Cellular Biology, Harvard Medical School, Boston, Massachusetts 02115.

be discussed since this has been a major historical reason for attempting such measurements. Second, we will review studies on one of the earliest observable responses to growth factors, which is a change in cell shape and motility. We shall attempt to link, where possible, data on second messengers with observed cytoskeletal responses to growth factors.

We will not cover many other interesting aspects of growth factor biology, and refer the reader to various excellent reviews by Carpenter and Cohen, 1984; Scher *et al.*, 1979; Gospodarowicz and Moran, 1976; Heldin and Westermark, 1984; Hunter, 1985; Ross and Vogel, 1978; Carpenter and Cohen, 1979; Stiles, 1983; Heldin *et al.*, 1985; and Ross *et al.*, 1986. The possible role of the cytoskeleton as a definite component in the signaling pathway leading to DNA synthesis, although certainly related to the area reviewed here, will not be discussed either but has been reviewed recently (Maness, 1981; Marceau and Swierenga, 1985; Thyberg, 1984).

Studies of the effects of growth factors on cultured cells have made use of different cell types, growth factors, methods for inducing quiescence, and even different culturing environments. Hence, the difficulty, as will be evident in the first part of this chapter, is in making absolute generalizations concerning the signals used by cells responding to growth factors. Moreover, analysis of the cytoskeletal responses of cells to growth factors is particularly difficult since there are few published investigations at present. Therefore, one purpose of the latter part of this chapter is to explore how growth factor binding to receptor and generation of secondary messengers might be coupled to cellular responses involving the cytoskeleton, the regulation of which has been particularly well characterized *in vitro*.

Two major signal pathways are invoked presently as general explanations of how cells respond to exogenous polypeptide ligands. These are the cyclic adenosine monophosphate (cAMP) and the Ca^{2+}-inositol triphosphate–diacylglycerol pathways. A common and possibly fundamental consequence of both pathways is the activation of protein kinases and, hence, phosphorylation of numerous cellular proteins. We provide a highly simplified summary of each pathway in Figure 1.

Not surprisingly, perhaps, each of these pathways has been invoked as a cytoplasmic basis for activation of quiescent, cultured animal cells by growth factors. For ease of presentation, we have treated various components of these signal pathways separately in the following sections. We recognize, however, that each pathway must be viewed as an orderly, integrated mechanism, highly dependent for its functioning on each of its various parts. Moreover, it must be kept in mind that these two pathways can interact with one another, providing complex positive and negative feedback regulation of each other's activity (reviewed by Cohen, 1982; Rasmussen *et al.*, 1984; Means *et al.*, 1982). Whether such reciprocal interactions between pathways occur in activated fibroblasts is not known. Finally, we wish to stress an important point at the outset: It remains to be established whether either of these signal pathways (or any one of their known components) is directly responsible for cell activation by growth factors.

2. IONS AS SIGNALS

2.1. Ca^{2+}

The calcium ion has been called a "universal regulator" (Campbell, 1983). Though this obviously may overstate the case, strong support for Ca^{2+} as a regulator or cytoplasmic second messenger takes three major forms. First, whereas free intracellular Ca^{2+} ion concentration $[Ca^{2+}]_i$ is about 10^{-7} M in resting cells, considerable reserves of Ca^{2+} are stored within certain organelles (i.e., endoplasmic reticulum and mitochondria), and the Ca^{2+} concentration outside the cell ($[Ca^{2+}]_o$) is about 1.8 mM, 10^4-fold greater than $[Ca^{2+}]_i$. Hence, gradients of Ca^{2+} concentration are established across the plasma and organelle membranes, priming the cell for rapid and large changes in $[Ca^{2+}]_i$ through dissipation of these gradients. Second, activities of numerous cytosolic proteins, primarily calmodulin, are regulated by $[Ca^{2+}]_i$, providing the internal effector mechanisms for coupling rises in $[Ca^{2+}]_i$ to cellular responses. Third, a rise in $[Ca^{2+}]_i$ is a measured intracellular event leading to many and varied cellular responses to external stimuli. In the present context, fertilization of sea urchin eggs (Whitaker and Steinhardt, 1982) and mitogenic stimulation of lymphocytes (Hesketh et al., 1984) are highly relevant examples because activation of these quiescent cells is followed immediately by a rise in $[Ca^{2+}]_i$, which appears to be an obligatory component of the normal signal pathway leading to DNA synthesis.

2.1.1. $[Ca^{2+}]_o$ and Cell Growth

Extracellular Ca^{2+} is essential for continued growth of all cultured animal cells (reviewed in Whitfield et al., 1976; Whitfield, 1982). $[Ca^{2+}]_o$ is normally 1.8 mM in culture medium, and its reduction to 0.1–0.01 mM severely inhibits growth of most cells in culture (see next paragraph). Such reductions in $[Ca^{2+}]_o$ do not appear to affect cells in the DNA synthetic or mitotic phases of the cell cycle (reviewed in Pardee et al., 1978), but appear to arrest cells in early

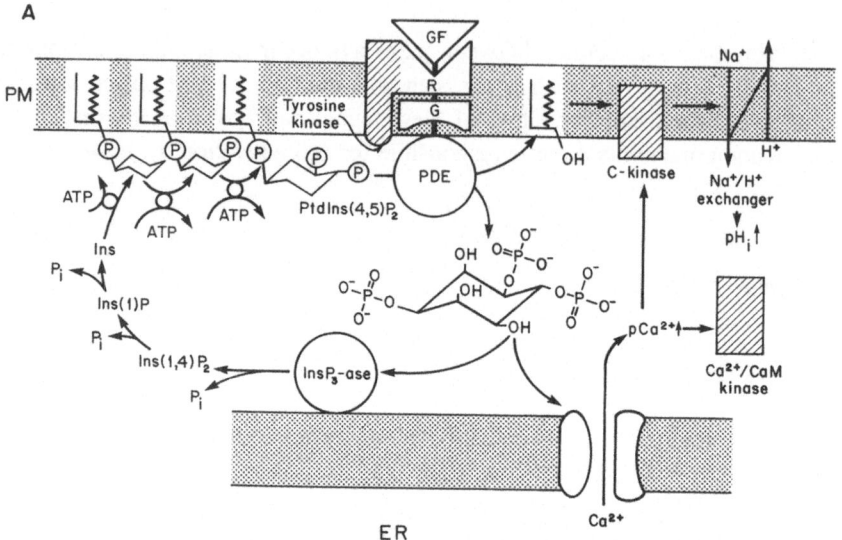

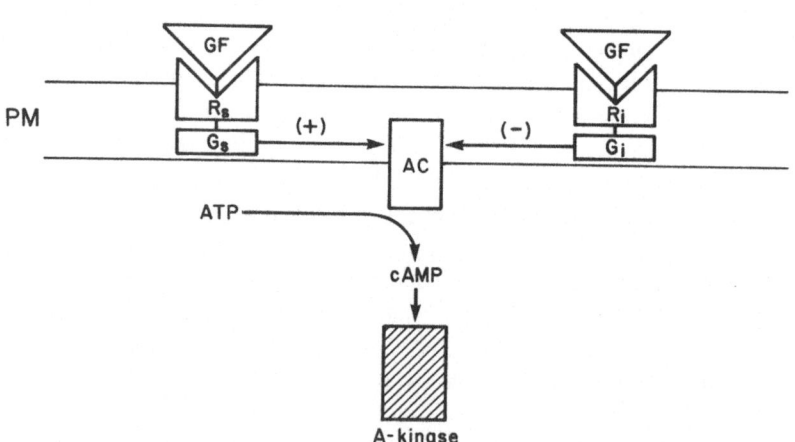

FIGURE 1. Generalized, hypothetical depictions of two possible signal pathways for the activation of quiescent cells by growth factors. (A) The Ca^{2+}–inositol lipid pathway. Growth factor (GF) binding to receptor (R) leads to activation via a GTP binding protein (G) of polyphosphoinositide phosphodiesterase (PDE). PDE then hydrolyzes phosphatidylinositol 4,5-biphosphate [PtdIns(4,5)P2], producing inositol triphosphate (IP$_3$) and diaclyglycerol (DAG), which promote, respectively, a rise in cytosolic Ca^{2+} and H^+ under some conditions. Three protein kinases that might be activated by this pathway are indicated by diagonal hatching. Other abbreviations: (ER) endoplasmic reticulum; (PM) plasma membrane. (B) The cAMP pathway. GF binding to a stimulatory receptor (R$_S$) activates via a GTP binding protein (G$_S$) adenylate cyclase (AC), causing hydrolysis of ATP to cAMP and, thereby, activation of A kinase. Inhibitor receptors (R$_i$) coupled to AC via another G protein (G$_i$) are known to inhibit AC in other stimulus–response systems utilizing cAMP as a second messenger.

(Hazelton *et al.*, 1979) and/or late (Boynton *et al.*, 1976) prereplicative phase (G_1).

These observations alone do not distinguish Ca^{2+} from numerous other ions (such as Mg^{2+}; see Kamine and Rubin, 1976) that are essential nutrients for growth; but additional observations warrant special attention to the role of Ca^{2+} ions in growth control. First, most transformed cells require considerably less (1/10) $[Ca^{2+}]_o$ than "normal" cells do of the same line (Nicholson *et al.*, 1984; Balk *et al.*, 1973). Furthermore, in some cell lines, at least, the requirement for exogenous Ca^{2+} decreases with repeated passage (Boynton *et al.*, 1974). Third, addition of supranormal levels (10–50 mM) of Ca^{2+} (Dulbecco and Elkington, 1975; Boynton and Whitfield, 1976), or of precipitates of Ca^{2+} and pyrophosphate (Bowen-Pope and Rubin, 1983), to culture medium induces growth. Last, in subconfluent cultures, growth factors such as platelet derived growth factor (PDGF) and epidermal growth factor (EGF) lower the growth requirement for extracellular Ca^{2+} (McKeehan *et al.*,1980, 1982; Betsholtz and Westermark, 1984). It would appear that a Ca^{2+}-sensitive component of G_1 exists, that this component is lost or is of diminished sensitivity in transformed cells or cells passaged repeatedly, and that the Ca^{2+}-sensitive component of G_1 is related to the serum or growth-factor component of G_1, such that depletion of growth factors can, in some cases, be overcome by an excessive supply of Ca^{2+}, and vice versa. Any satisfactorily complete hypothesis to explain growth of normal and transformed cells must, therefore, explain the apparently unique role of $[Ca^{2+}]_o$ identified by the experiments described above.

2.1.2. $[Ca^{2+}]_i$ and Growth Factor Activation

The resting $[Ca^{2+}]_i$ of quiescent fibroblasts appears to be slightly higher than that of growing fibroblasts (McNeil *et al.*, 1985). It is now clear that serum and some growth factors raise the $[Ca^{2+}]_i$ of quiescent fibroblasts immediately and transiently, implying an active role for Ca^{2+} in the signaling pathway initiated by growth factors. Disagreement exists, however, over exactly which growth factors raise $[Ca^{2+}]_i$, and the magnitude and duration of the rise in $[Ca^{2+}]_i$ induced. As measured with the fluorescent indicator quin 2, serum stimulated a rapid rise from 140 to 300 nM in the $[Ca^{2+}]_i$ of quiescent human fibroblasts, which declined to 200 nM over a period of 3 min, after which time values for $[Ca^{2+}]_i$ were not reported (Moolenaar *et al.*, 1984b). As measured with aequorin, serum stimulated an immediate rise from 89 nM to 5.3 μM in the $[Ca^{2+}]_i$ of quiescent 3T3 fibroblasts, $[Ca^{2+}]_i$ declined to below prestimulatory levels within 3 min thereafter, and remained constant at this level(≤ 50 nM) for the next 12 hr of continuous measurement (McNeil, *et al.*, 1985). Quin 2 (Moolenaar *et al.*, 1984b) and aequorin (McNeil *et al.*, 1985) both measured a rise in $[Ca^{2+}]_i$ upon

PDGF stimulation, but again the rise over resting levels recorded by aequorin was considerably greater ($\sim$ 15-fold) than that measured by quin 2 ($\sim$ 2.1-fold); and the PDGF rise measured with quin 2 lasted for at least 15 min, whereas that measured with aequorin lasted only 3.5 min. These differences can be explained by the greater sensitivity of aequorin to changes in $[Ca^{2+}]$ in the range 0.1–10 μM (Johnson et al., 1985) and by calcium buffering by quin 2 (Tsien et al., 1982). EGF produced a small (1.3–2.3-fold) rise in $[Ca^{2+}]_i$ in two quin 2 studies (Moolenaar et al., 1984b; Morris et al., 1984) but not in the aequorin study. Fibroblast growth factor (FGF) caused a rise (12-fold) in the $[Ca^{2+}]_i$ of 3T3 fibroblasts (McNeil et al., 1985). The source of Ca^{2+} mobilized by serum and PDGF (but not EGF) appears to be primarily internal, perhaps the endoplasmic reticulum.

Not all growth factors cause rises in $[Ca^{2+}]_i$: The comitogens, insulin and phorbol 12-myristate-13-acetate (PMA), are without affect on $[Ca^{2+}]_i$ of quiescent cells (Morris et al., 1984; McNeil et al., 1985). Thus, the mitogenic action of insulin and PMA would appear to be by Ca^{2+}-independent signaling pathways or by the bypassing of a Ca^{2+}-dependent step.

In lymphocytes, a rise in $[Ca^{2+}]_i$ appears to be sufficient for DNA synthesis, since such synthesis occurs after treatment with Ca^{2+} ionophores (Hesketh et al., 1982, 1984). In fibroblasts, on the other hand, there is at least one published report (Dulbecco and Elkington, 1975) that Ca^{2+} ionophores do not initiate mitogenesis. It is therefore an open question whether Ca^{2+} may be sufficient for initiating DNA synthesis in fibroblasts. However, as stated above, PDGF causes an increase in the $[Ca^{2+}]$ of quiescent fibroblasts, and recently it has been found (Tsuda et al., 1985) that Ca^{2+} ionophore mimics another action of PDGF—the induction of c-*myc* gene expression (Kelly et al., 1983). Thus, Ca^{2+} may be a sufficient signal for inducing expression of the c-*myc* gene, whose protein product is located in the nucleus and binds DNA (for review, see Heldin and Westermark, 1984).

There is some evidence that the rise in $[Ca^{2+}]_i$ evoked by growth factors is not necessary for inducing DNA synthesis under conditions in which protein kinase C is activated simultaneously. Thus, if quiescent fibroblasts are first treated with PMA, which by itself causes no change in $[Ca^{2+}]_i$ and subsequently stimulated with serum (McNeil et al., 1985) or EGF (Hesketh et al., 1985), the expected rises in $[Ca^{2+}]_i$ are observed to be reduced dramatically or abolished. Yet, in the case of serum or PDGF at least, such PMA "pretreatment" does not inhibit growth factor stimulation of DNA synthesis (McNeil et al., 1985). Moreover, release of inositol triphosphate, which is normally induced by the mitogen, thrombin, in NIL fibroblasts and which is believed to cause the mobilization of cytosolic Ca^{2+} (see Section 3), can be blocked by neomycin without inhibition of DNA synthesis (Carney et al., 1985).

In sum, it would appear that if calcium is involved as a cytoplasmic signal for mitogenic activation of fibroblasts by some growth factors, its action is early, transient, and possibly not obligatory in the face of protein kinase C activation by PMA or diacyglycerol (DAG) (Neidel *et al.*, 1983; Leach *et al.*, 1983). This last condition may be a normal consequence of growth factor stimulation of cells (see Section 3). It is, however, far too early to rule out Ca^{2+} as an important component of the normal pathway for growth factor activation. PMA treatments may, for example, cause protein kinase C activation far in excess of that normally present in cells stimulated with growth factors. Further tests of calcium's role in mitogenesis, as well as the other cellular responses, are definitely in order. These might include novel means for the artificial, transient elevation of the $[Ca^{2+}]_i$ of quiescent cells, as well as selective inhibition of growth factor–induced rises in $[Ca^{2+}]_i$, followed by measurements of DNA synthesis consequent on these treatments to test, respectively, the sufficiency and necessity of rises in $[Ca^{2+}]_i$ for DNA synthesis and other cellular responses. The use of high-quality calcium indicators must be emphasized (McNeil *et al.*, 1985).

2.2. H^+

It is now apparent that intracellular pH (pH_i) is not invariant with time (for review, see Busa and Nuccitelli, 1984). Rather, in a variety of cells, pH_i changes by 0.1–1.6 units as a function of cell developmental and metabolic state. Moreover, the activities of a variety of proteins are greatly affected by pH. It is of particular interest that Ca^{2+} binding by calmodulin is affected strongly by pH, since many cellular stimuli, including growth factors under some conditions, change both $[Ca^{2+}]_i$ and pH_i (see Busa and Nuccitelli, 1984). It may be that these two cations, H^+ and Ca^{2+}, are interrelated and obligatory signals for cell activation and, indeed, this would seem to be the case in the sea urchin egg (Whitaker and Steinhardt, 1982).

2.2.1. pH_o and Cell Growth

The pH outside cells (pH_o) is normally maintained in laboratory cultures at 7.2–7.4, the optimum range for growth. In fact, at pH_o 6.9, the cell density reached at quiescence is two- to four-fold less than that reached at pH_o 7.4, but cell density can be increased readily by this factor, simply by raising pH_o to 7.4 (Ceccarini and Eagle, 1971a,b). Provocatively, quiescent, serum-starved cultures of Swiss 3T3 cells ($pH_o = 7.4$) can be stimulated to proliferate merely by addition of alkaline culture medium (Rubin 1971; Zetterberg and Engstrom,

1981). In such continuous exposures, pH_o 8.1 is maximally effective, causing 50% of the cells to proliferate (Zetterberg and Engstrom, 1981). Higher levels of continuous alkalinity are toxic to cells. Alternatively, by raising pH_o briefly (for 2–20 min) to between 8.5 and 10, cell death is minimized, and a mitogenic stimulus, fully 80% of that caused by fresh serum, is observed (Zetterberg and Engstrom, 1981). However, Burroni and Ceccarini (1984) could not reproduce the results of Zetterberg and Engstrom (1981). Unfortunately, this is an example of an inconclusive comparison based on the use of different cell lines. Burroni and Ceccarini (1984) did not study Swiss 3T3 cells. There are several possible mechanisms by which raising pH_o could initiate mitogenesis in the absence of added growth factors, the most intriguing of which, suggested by recent work (see section 2.2.2), is that alkaline pH_o elevates intracellular pH_i. Important in this context, therefore, are observations that the ability of cultured cells to regulate pH_i declines above $pH_o > 8$ (reviewed in Gillies, 1981).

Recently, mutant fibroblasts have been selected for possession of inactive membrane Na^+/H^+ exchangers (Pouyssegur et al., 1984). Growth of these mutants, incapable of normal pH_i regulation in an HCO_3-free medium, was optimal in that medium at alkaline pH_o (8.0–8.3) and completely inhibited at $pH_o < 7.2$. Moreover, when stimulated with growth factors, these mutants failed to elevate pH_i and did not synthesize DNA unless $pH_o > 7.2$ (Pouyssegur et al., 1985). Thus, a pH_i above a critical threshold (7.2) appears to be a prerequisite for initiation of DNA synthesis by quiescent fibroblasts.

2.2.2. pH_i and Growth Factor Activation

Na^+/H^+ exchange at the plasma membrane is a major means by which cells regulate pH_i (Roos and Boron, 1981). Hence, the finding that serum and various growth factors caused $^{22}Na^+$ influx was the earliest indication that mitogenic activation of cultured cells might alter pH_i (for reviews of earlier studies, see Rozengurt, 1981a, 1982a). Direct measurements have since confirmed that serum and most growth factors (including PDGF, EGF, vassopressin, thrombin, and insulin) cause a rise in the pH_i of quiescent cells under ionic conditions where HCO_3^- is absent (Burns and Rozengurt, 1983; Moolenaar et al., 1983; Hesketh et al., 1985; Moolenaar et al., 1984a; Rothenberg et al., 1983; Cassel et al., 1985; Whitely et al., 1984; Taylor et al., 1986). This rise in pH_i normally becomes measurable within 1–2 min after stimulation with serum or growth factors, peaks 5–15 min later at 0.1–0.3 units above prestimulatory levels, and remains stable at this elevated level for at least 15 min thereafter. Since amiloride inhibits such mitogen-induced rises in pH_i, it is inferred that a Na^+/H^+ exchanger is indeed responsible for alkalinization (Moolenaar, et al., 1983).

Simple kinetic analysis indicates that growth factors may activate alkalinization by increasing the Na^+/H^+ exchanger's sensitivity to pH_i. This increased sensitivity is proposed to result from a conformational change in the allosteric site of the exchanger itself (Moolenaar *et al.*, 1983). Importantly, PMA can cause cytoplasmic alkalinization (Moolenaar *et al.*, 1984a; Besterman and Cuatrecasas, 1984; Whitely *et al.*, 1984) without causing a change in Ca^{2+} flux or $[Ca^{2+}]_i$ (see Section 2.1.2). Therefore, activation of protein kinase C, which is believed to be an early consequence of growth factor addition to quiescent cells, and is induced experimentally by the PMA treatment of cells (Nishizuka, 1984), is suggested as a Ca^{2+}-independent trigger for mitogen-induced rises in pH_i. However, a Ca^{2+}-dependent pathway for activation of the Na^+/H^+ antiport is also suggested both for fibroblasts (Owen and Villereal, 1982) and T lymphocytes (Rosoff and Cantley, 1985). Recently, in fact, PMA alone was shown to have no effect on Na^+/H^+ exchange in a human foreskin fibroblast strain (HSWP cells) but did act synergistically with the calcium ionophore A23187 in causing Na^+ influx in these cells (Vicentini and Villereal, 1985). This suggests that activation of protein kinase C may not be sufficient for activation of Na^+/H^+ exchange in all fibroblast types and reemphasizes the possibility of a Ca^{2+}-dependent mechanism for activation of the Na^+/H^+ exchanger. Activation of Na^+/H^+ exchange in A431 cells by hyperosmotic conditions is suggested to be independent of protein kinase C (Whitely and Glaser, 1986).

Is elevation of pH_i sufficient for initiation of DNA synthesis? It would appear not, since PMA by itself, though capable of elevating pH_i by 0.2 units (Moolenaar *et al.*, 1984a; Whitely *et al.*, 1984), is not capable of initiating DNA synthesis. PMA is a comitogen that is strongly mitogenic when presented to fibroblasts with EGF (also not mitogenic for fibroblasts by itself), or with several other molecules (Rozengurt and Dicker, 1978; and Dicker and Rozengurt, 1980). It would seem, therefore, that initiation of DNA synthesis requires some signal (or signals) additional to elevated pH_i.

Ca^{2+} might be such a signal. First, EGF and PDGF, and perhaps other growth factors, raise $[Ca^{2+}]_i$ in many types of quiescent cells, in addition to elevating pH_i (Hesketh *et al.*, 1985). Second, when quiescent fibroblasts are activated mitogenically by alkaline treatments (see Section 2.2.1), there is a rapid translocation of cellular $^{45}Ca^{2+}$ from one more to one less exchangeable internal compartment. And when such $^{45}Ca^{2+}$ exchange is inhibited by depletion of Ca^{2+} from the more exchangeable compartment, alkaline treatment fails to cause DNA synthesis (Engstrom *et al.*, 1982). Third, activation of protein kinase C by PMA, together with elevation of $[Ca^{2+}]_i$ by ionophores, activates synergistically such cellular responses as neutrophil degranulation (White *et al.*, 1984; and Dale and Penfield, 1984) and platelet aggregation (Rink and Hallam, 1984) in the absence of stimulatory ligand. However, the possibility of such synergism between protein kinase C (pH_i) and Ca^{2+} in initiating DNA synthesis in quiescent fibroblasts remains to be demonstrated directly.

Is activation of Na^+/H^+ exchange, with or without alkalinization of pH_i, necessary for initiation of DNA synthesis? Again, it would appear not, since activation of Na^+/H^+ exchange can be inhibited by amiloride in bicarbonate-buffered medium without concomitant inhibition of the DNA synthesis initiated by growth factors (L'Allemain et al., 1984a,b). It is noteworthy, however, that in a carbonate-free medium, such as that employed predominantly in measurements of growth factor–induced rises in pH_i, amiloride can inhibit DNA synthesis, implying that under such artificial conditions Na^+/H^+ exchange may be necessary for initiation of DNA synthesis (L'Allemain et al., 1984a,b). One report at least has shown that Na^+/H^+ exchange initiated by growth factors in bicarbonate-buffered medium does not lead to an increase in pH_i (Cassel et al., 1985). Obviously, it will be important to determine whether, Cl^-/HCO_3^- exchange is of importance in regulating pH_i, as in some invertebrate cells, since fibroblasts are standardly gown and some experiments with growth factors are performed in bicarbonate-buffered culture medium. There is, in fact, evidence that fibroblasts possess a Na^+-dependent, Cl^-/HCO_3^- system capable of regulating pH_i (Rothenberg et al., 1983; L'Allemain et al., 1985). Finally, it has been pointed out that enhanced Na^+/H^+ exchange could play a "housekeeping," rather than a regulatory, role; that is, it might provide the means by which the cell copes with the increased metabolism and, hence, a predicted tendency toward acidosis that follows quiescent cell activation by growth factors (Boron, 1984). A pH-regulated anion antiporter exists in Vero and in L cells (Olsnes et al., 1986), which is suggested to decrease the cytoplasmic pH when the cytoplasm becomes too alkaline. Therefore, it may not be alkalinization of cytoplasm that is important in the activated cell but the ability to maintain a constant buffering of the cytoplasm. Whether pH_i plays an active or purely a passive role in the activation of quiescent cells remains to be elucidated.

3. INOSITOL LIPID METABOLITES AS SIGNALS

Many stimulatory ligands acting on various cell types cause a polyphosphoinositide phosphodiesterase to hydrolyze the membrane inositol lipid, phosphatidylinositol 4,5-bisphosphate [PtdIns (4,5)P2] (reviewed in Michell, 1975; Berridge and Irvine, 1984; Hokin, 1985; Majerus et al., 1984). Hydrolysis of PtdIns(4,5)P2 yields two products, (DAG) and inositol 1,4,5-trisphosphate or (IP$_3$), which are believed to be the two signals responsible for initiating a bifurcating pathway (Berridge and Irvine, 1984). Thus, DAG activates protein kinase C and is believed to activate the Na^+/H^+ exchanger (but see Section 2.2.2), whereas IP$_3$ causes release of Ca^{2+} from a nonmitochondrial organelle (probably the endoplasmic reticulum (Prentki et al., 1984) and thus raises $[Ca^{2+}]_i$.

It has been proposed that this bifurcating pathway initiated by inositol lipid hydrolysis and resulting in rises in both pH_i and $[Ca^{2+}]_o$ could regulate cell growth (Berridge *et al.*, 1984a). This is not a new hypothesis in the sense that degranulation, DNA synthesis, and mitotic division by the fertilized sea urchin egg have been proposed previously to be under the "dual control" of pH_i and $[Ca^{2+}]_i$ (Whitaker and Steinhardt, 1982).

Enhanced metabolism of phosphatidylinositols is the measured consequence of growth factor stimulation of a variety of cells in tissue culture (Ristow *et al.*, 1973; Hoffman *et al.*, 1974; Diringer and Friis, 1977; Sawyer and Cohen, 1981; Habenicht *et al.*, 1981; Berridge *et al.*, 1984b; Carney *et al.*, 1985). Importantly, such enhanced metabolism of inositol lipids is stimulated in Swiss 3T3 cells by PDGF and has been shown to produce DAG (Habenicht *et al.*, 1981; Whitely and Glaser, 1986). Moreover, further work with 3T3 cells has shown that PDGF and, to a leser extent, EGF cause cytoplasmic accumulation of IP_3 (Berridge *et al.*, 1984a,b). These are essential observations, indicating that some growth factors stimulate hydrolysis of PtdIns(4,5)P2 to produce the two key signal transducers DAG and IP_3 and therefore suggesting a mechanism for the observed elevation of $[Ca^{2+}]_i$ and activation of the Na^+/H^+ exchanger by growth factors (see Sections 2.1 and 2.2.1).

If IP_3 and/or DAG are in fact involved in regulating growth, conditions that enhance their concentration in cells should be mitogenic. This seems to be the case for IP_3. Thus Li^+, which inhibits 1-phosphomonoesterase, generally amplifies formation of IP_3 in ligand-stimulated cells (Berridge *et al.*, 1982), and this also is the case in cells stimulated with PDGF (Berridge *et al.*, 1984a). In fact, Li^+ is mitogenic for Balb/c–3T3 cells when added with various growth factors but not by itself (Ryback and Stockdale, 1981), supporting a role for IP_3 in growth. This conclusion rests on the degree of specificity of Li^+ as an inhibitor of 1-phosphomonoesterase, which remains to be established. There is similar evidence that DAG may act as a mitogenic signal. An exogenously added DAG (1-oleoyl-2-acetyglycerol), which freely incorporates into cell plasma membranes, acts synergistically with insulin and other growth factors in promoting DNA synthesis by quiescent cells (Rozengurt *et al.*, 1984).

If DAG and/or IP_3 are necessary for mitogenesis, inhibition of their formation should inhibit mitogenesis. Neomycin binds to PtdIns(4,5)P2 and inhibits its breakdown, and it also inhibits breakdown of PtdInsP at higher concentrations (Lang *et al.*, 1977; Lodhi *et al.*, 1979; and Downes *et al.*, 1981). It can thus be used to inhibit both DAG and IP_3 formation, depending on the concentration employed. When used at lower concentrations, neomycin inhibited IP_3 accumulation demonstrably in hamster fibroblasts stimulated with thrombin but only slightly affected the DNA snythesis induced by this mitogen (Carney *et al.*, 1985). In contrast, when used at higher concentrations to also inhibit DAG formation, DNA synthesis induced by thrombin was inhibited (Carney *et al.*,

1985). Thus, DAG formation from phosphatidylinositols, but not IP_3 formation from PtdIns(4,5)P2, may be required for mitogenesis induced by thrombin. As for Li^+, the specificity of neomycin remains to be established.

Although Ca^{2+} ionophores do not seem to induce growth in fibroblasts (see Section 2.1.1), recent evidence suggests that calcium mobilizing agents acting specifically through phospholipase C may provide signals sufficient for fibroblast growth (Moolenaar et al., 1986). Thus, phosphatidic acid causes expression of the c-fos and c-myc proto-oncogenes, and it also stimulates DNA synthesis. Moreover, phosphatidic acid raises $[Ca^{2+}]_i$, probably by causing release of Ca^{2+} from intracellular stores, and it also raises pH_i (see Section 2.2.2). This combined action suggests the as yet untested hypothesis that phosphatidic acid triggers hydrolysis of phosphoinositides in those cells in which it has stimulated growth. Again, therefore, IP_3 and DAG mobilization, a rise in $[Ca^{2+}]_i$ and pH_i, and cell growth are all linked as a common sequence of events. Whether such linkage is direct, or merely fortuitous, remains to be shown.

4. cAMP AS A SIGNAL

Many stimulatory ligands (prominently epinephrine) activate adenylate cyclase (A kinase) and, hence, produce increased concentrations of cAMP in the cytoplasm of responsive cells. cAMP, in turn, binds to and activates A kinase. Phosphorylation of specific cellular proteins (enzymes) by activated A kinase then elicits, directly in some cases (e.g., lipase phosphorylation in fat cells), the observable cellular response to exogenous ligand (reviewed in Sutherland, 1972; Cohen, 1982).

Most, though not all, of the evidence from earlier studies suggested that cAMP inhibited initiation of DNA synthesis, both in lymphocytes and in fibroblasts (earlier studies are reviewed in Pastan and Johnson, 1974; Pastan et al., 1975; Willingham, 1976). Serum addition to quiescent fibroblast cultures generally led to a rapid fall in measured cAMP levels; analogues of cAMP and other treatments that raised cAMP levels in cells inhibited growth, both in logarithmic cultures and in quiescent cultures stimulated with serum; and arrest of growth by elevating cAMP often was detected at G_1, prior to DNA synthesis. A fall in cAMP levels generally was believed to be a necessary prelude for the initiation of DNA synthesis in most cell types. Rebhun (1977), reviewing the role of cAMP in cell division, found examples of both stimulatory and inhibitory effects.

A recent carefully controlled study seems to confirm that cAMP is an inhibitory signal for lymphocyte mitogenesis (Kaever and Resch, 1985). Thus, whereas optimal mitogenic dosages of concanavalin A produced small increases in cellular cAMP, "supraoptimal" doses (which elicited three-fold less DNA synthesis than optimal doses did) led to a ten- to 20-fold elevation of cAMP. Moreover, mutant CHO strains, selected for the ability to grow in conditions (cholera toxin or 8-bromo-cAMP added to culture medium) that normally raise cAMP levels, were found to be defective in A-kinase activity (Gottesman et al., 1980; Singh et al.,1985). One interpretation is that the usual inhibition of growth caused by cAMP is dependent on phosphorylations mediated by A kinase and that mutants but not normal cells elude the inhibitory effects of cAMP because they have defective A-kinase activity.

Recent work on Swiss 3T3 fibroblasts, however, has contradicted the proposition that an early decline in cAMP levels and, hence, A-kinase activity is necessary for DNA synthesis. Thus, in contrast to earlier reports, agents that elevate cAMP, including cholera toxin (Pruss and Herschman, 1979; Rozengurt, 1981b; Rozengurt et al., 1981), adenosine agonists (Rozengurt, 1982a), or cAMP analogues (Paris and Rozengurt, 1982) were found to enhance DNA synthesis by fibroblasts when added with other growth factors but not by themselves. Similarly, cholera toxin and other cAMP-elevating agents are mitogenic for epidermal cells (Green, 1978) and lymphocytes (Whitfield et al., 1970), as well as other cells. One possible explanation for the apparently contradictory nature of the literature discussed thus far is that cAMP may have opposing, concentration-dependent regulatory effects on mitogenesis. Thus, cholera toxin alone caused a two to three-fold increase in fibroblast cAMP, and potentiated the DNA synthesis stimulated by PDGF, apparently by potentiating competence formation (Wharton et al., 1982). In contrast, cholera toxin and isobutylmethylxanthine together caused a ten- to fifteen-fold increase in cAMP, which inhibited DNA synthesis apparently by inhibiting cell progression through G_1 (Leof et al., 1982).

Increased levels of cAMP have been measured in quiescent fibroblasts as a consequence of stimulation with one growth factor only—PDGF—and this rise was significant (greater than two-fold) only when cells were incubated in phosphodiesterase inhibitors during PDGF presentation (Rozengurt et al., 1983). EGF is reported not to alter cAMP levels of quiescent fibroblasts, even in the presence of phosphodiesterase inhibitor (Pruss and Herschman, 1979). Obviously, further work and/or improved methods will be required to resolve the fundamental question of whether cAMP is growth stimulatory or inhibitory.

5. PROTEIN PHOSPHORYLATION AS A SIGNAL

Protein phosphorylation has been described as "the major general mechanism by which intracellular events respond to external stimuli" (Cohen, 1982). Indeed, both of the signal pathways discussed herein—the cAMP and Ca^{2+}–

DAG pathways—promote protein phosphorylations. Protein kinase C is activated by DAG (and Ca^{2+}), Ca^{2+}/calmodulin kinase by Ca^{2+}, and A kinase by cAMP. These protein kinases all have specificity for serine and threonine residues, but tyrosine-specific protein kinases are now well known and are also believed to be important cellular regulators (reviewed in Sefton and Hunter, 1984). Since the activity of such tyrosine kinases is often intrinsic to surface receptors, ligand–receptor interaction can lead directly to protein phosphorylation on tyrosine, without the requirement for intermediary messengers. Hormonal control of glycogen metabolism is the paradigm of cellular regulation by protein phosphorylations. Glycogen phosphorylase, which breaks down glycogen, is activated in stimulated muscle by phosphorylase kinase–mediated phosphorylation of one of its serine residues (for review, see Cohen, 1982). Unfortunately, the molecular basis of many cellular responses is not understood as well as that of glycogen metabolism. This makes less rewarding interpretations of the role of phosphorylations in many cellular responses, including, certainly, cell growth.

Growth factors stimulate the rapid (< 5 min) phosphorylation of numerous proteins in quiescent fibroblasts. Enhanced phosphorylation on tyrosine is detected readily and has been examined most extensively (for general review, see Sefton and Hunter, 1984). For example, PDGF increased relative phosphotyrosine levels by 2.6-fold in quiescent, NR-6 3T3 cells, whereas relative levels of phosphoserine and phosphotyrosine were changed only slightly (Cooper *et al.*, 1982). Minor but potentially regulatory levels of enhanced phosphorylations on serine and threonine can go undetected in such measurements since they must be measured against high background of 90% (relative level) phosphoserine and 10% phosphothreonine. By comparison, the relative level of phosphotyrosine in unstimulated quiescent fibroblasts is 0.054% (Cooper *et al.*, 1982). Nevertheless, enhanced serine and/or threonine phosphorylations of specific proteins have been detected in quiescent cells stimulated with PDGF (Nishimura and Duel, 1981; Cooper *et al.*, 1984) and EGF (Hunter and Cooper, 1981). EGF, serum, trypsin, and phorbol esters all stimulate phosphorylation of proteins on tyrosine in quiescent fibroblasts (Gilmore and Martin, 1983; Cooper *et al.*, 1984; Hunter and Cooper, 1981; Nakamura *et al.*, 1983).

Provocatively, phosphorylation on tyrosine of a 42,000-dalton protein substrate is a common denominator of stimulation by all of the growth factors mentioned above (Nakamura *et al.*, 1983; Cooper *et al.*, 1984). This protein is therefore suggested to be a possible regulator of cell growth. Clearly, identification of the functional activity of the 42,000-dalton protein and how its activity is affected by phosphorylation will be of great interest.

The identity of the protein kinase(s) responsible for phosphorylations in the mitogenically activated cell is unclear, though some tentative suggestions have been made. Receptor for PDGF appears to be associated with a tyrosine kinase activity, since purified membranes containing this receptor phosphorylate protein upon addition of PDGF (Ek *et al.*, 1982; Nishimura *et al.*, 1982). Moreover, receptor for EGF clearly possesses an EGF-activatable, intrinsic tyrosine kinase activity (Cohen *et al.*, 1980; Ushiro and Cohen, 1980). Possibly, therefore, it is receptor-associated tyrosine kinase activity that is stimulated by EGF and PDGF. Phosphorylation on tyrosine stimulated by PMA cannot be explained as readily since PMA is not known to activate a tyrosine kinase directly. However, PMA does activate protein kinase C and so, conceivably, could activate a tyrosine kinase indirectly by causing serine and/or threonine phosphorylations of that kinase. Since protein kinase C causes phosphorylation on serine of the EGF receptor and thereby inhibits its intrinsic tyrosine kinase activity (Cochet *et al.*, 1984), it would seem unlikely that PMA-activated protein kinase C phosphorylation of the receptor for EGF accounts for PMA-stimulated phosphorylations on tyrosine. PMA, PDGF, and FGF do, however, cause phosphorylations on serine and threonine of an 80,000-dalton protein, and protein kinase C (but not cAMP-activated A kinase) clearly is implicated in such phosphorylations (Blackshear *et al.*, 1985). The mode of action of trypsin in causing tyrosine phosphorylation remains entirely speculative. It has been suggested that tyrosine kinases additional to those associated with PDGF and EGF receptors may be involved in growth factor–stimulated tyrosine phosphorylations (Cooper *et al.*, 1984).

6. MECHANISMS OF DESENSITIZATION TO GROWTH FACTOR STIMULATION

Desensitization mechanisms are often initiated by the exogenous stimulus itself, causing the cell's response in the continuing presence of that stimulus to wane over time. Because growth factors are usually administered as a continuous dose, activation of quiescent cells might be expected to elicit desensitization mechanisms. Indeed, such desensitization mechanisms might bring about an orderly progression of signals and regulators in cytoplasm, a logical prerequisite for the orderly progression of the quiescent cell back "into" and "through" the cell cycle.

Just as stimulation of signal pathways can explain how a cell's response is turned on, so inhibition of the normal signal pathway can explain how the response is turned off. As far as is known, desensitization results most commonly from a disruption at the proximal end of signal pathways, that is, by alterations in ligand–receptor binding. This mechanism of desensitization is sometimes termed receptor down-regulation, especially when a loss of receptors from the cell surface is involved. It is perhaps best characterized in the β-adrenergic receptor–coupled adenylate cyclase system (for review, see Sibley and

Lefkowitz, 1985), but also appears to operate in cells stimulated with growth factors.

It is now clear that one consequence of exposure of cells to EGF at 37°C is a decrease in cellular binding capacity for this ligand (Krupp *et al.*, 1982; Carpenter and Cohen, 1976). This decrease in binding capacity is due to loss of receptors, which apparently results from their rapid removal from the cell surface by receptor-mediated endocytosis, delivery into lysosomes, and degradation. Thus, 2.5 min after cells are exposed to a conjugate of EGF and ferritin, 32% of this conjugate (and hence, presumably, EGF receptor) is observed in electron micrographs of cells within intracellular vesicles or endosomes (Haigler *et al.*, 1979). And by 30 min after presentation of the conjugate, 84% is present within multivesicular bodies (lysosomes). Biochemical studies provide direct evidence for degradation of 80% of cellular EGF receptors (to the point of unrecognizability by antibodies) within 2 hr of EGF stimulation of cells (Stoscheck and Carpenter, 1984; Bequinot *et al.*, 1985), and photoaffinity-labeled, putative EGF receptor has been detected in the lysosomal fraction from broken cells (Das and Fox, 1978).

PDGF also may induce down-regulation of its receptor. Binding of ^{125}I-PDGF to a variety of responsive cells is saturable and of high affinity and results in PDGF internalization and degradation at 37°C (Heldin *et al.*, 1981; Bowen-Pope and Ross, 1982; Huang *et al.*, 1982; Rosenfeld *et al.*, 1984; Nilsson *et al.*, 1983). Internalization (endocytosis) and delivery to lysosomes of PDGF is clearly revealed in electron micrographs of cells exposed to PDGF adsorbed to colloidal gold (Rosenfeld *et al.*, 1984). Therefore, if the receptor for PDGF follows its ligand along this route, it too would become internalized and degraded. Proof of such down-regulation of the PDGF receptor will await its isolation and the production of antibodies directed against it, so that the receptors' fate can then be traced directly by standard biochemical and morphological means.

What is the signal for initiating such growth factor–induced loss of receptors? Because phorbol esters induce loss of EGF receptor (Lee and Weinstein, 1978; Salomon, 1981) or a decrease in their affinity (Shoyab *et al.*, 1979) and because the EGF receptor is a target for protein kinase C (Iwashita and Fox, 1984; Cochet *et al.*, 1984), it is suggested that activation of protein kinase C and, hence, phosphorylation of EGF receptor at threonine 654 (Hunter *et al.*, 1984) drives EGF-induced receptor internalization (Beguinot *et al.*, 1985). However, EGF, but not phorbol esters, induce internalization of a mutated EGF receptor in which alanine replaces threonine at residue 654 (Lin *et al.*, 1986). Thus, two independent mechanisms appear to drive phorbol ester- and EGF-induced receptor internalization. The intracytoplasmic domain of the EGF receptor remains as the suspect regulatory site of internalization. PDGF similarly (to EGF) appears to modulate EGF receptors by a mechanism independent of that used by phorbol esters (Olashaw *et al.*, 1986).

Another, perhaps related, effect of protein kinase C activation is on Ca^{2+} and H^+ mobilization that is normally induced by growth factors. PMA treatment of quiescent fibroblasts abolishes or diminishes greatly the rise in $[Ca^{2+}]_i$ normally evoked by serum (McNeil *et al.*, 1985) or EGF (Hesketh *et al.*, 1985). In fact, PMA inhibits agonist-induced rises in $[Ca^{2+}]_i$ in a variety of systems (Naccache *et al.*, 1985; Sagi-Eisenberg, *et al.*, 1985; Drummond, 1985). Provocatively, in platelets (MacIntyre and Drummond, 1985; Watsona and Lapetina, 1985) and astrocytoma cells (Orellana *et al.*, 1985), PMA inhibits inositol lipid metabolism and, most importantly, formation of IP_3 (Orellana *et al.*, 1985; Watson and Lapetina, 1985). As stated in Section 3, IP_3 is believed to be the second messenger that causes Ca^{2+} release from internal stores. Activation of C kinase may provide a negative feedback signal serving to limit the magnitude and duration of growth factor and other agonist-induced increases in $[Ca^{2+}]_i$, and it may do so by inhibiting inositol lipid metabolism. Importantly, PMA treatments can also inhibit EGF- and PDGF-induced rises in pH_i (Whitely *et al.*, 1984, 1985). Therefore, a negative feedback system centering around protein kinase C and perhaps inhibiting inositol lipid metabolism may not only accomplish regulation of the Ca^{2+} but also of the Na^+ and H^+ ion fluxes associated with growth factor activation of quiescent fibroblasts.

There is one further possible role of protein kinase C in desensitization of cells to growth factors. We have mentioned that the EGF receptor itself has a tyrosine kinase activity that catalyzed auto- and other phosphorylations (Cohen *et al.*, 1980). Protein kinase C–mediated phosphorylation of the EGF receptor inhibits the tyrosine kinase activity subsequently stimulated by EGF (Cochet *et al.*, 1984). Insofar as this tyrosine kinase activity of the EGF receptor is involved in regulating growth, which is an open question at present, its inhibition by phosphorylation is yet another potential form of desensitization evoked by growth factors.

7. CELL SHAPE AND MOTILITY

This chapter will now shift to one of the early responses of the quiescent cell to growth factor stimulation: a change in cell shape and motility. Where feasible, we will point out correlations of changes in cell shape and motility with fluctuations in proposed cytoplasmic signals.

That ligand stimulation can lead to changes in cell shape and motility is well documented, as, for example, in phagocytosing macrophages (Silverstein *et al.*, 1979) and chemotaxing leukocytes (Zigmond, 1982). Hormones, too, are known to affect the morphology of target cells (Lawrence *et al.*, 1979; Miller *et al.*, 1976). Chinkers *et al.*, (1979) described the extensive ruffling and extension of filopodia within 5 min of exposing density-arrested A431 cells to

EGF. This early morphological response was transient, lasting only 10–15 min. Furthermore, the timing of the early shape changes was related to the stimulation of pinocytosis (Gey, 1955). In a subsequent study, Chinkers et al. (1981) indicated that cell-adhesion properties, as well as energy-dependent cytoskeletal processes, were involved in cell-shape changes in response to EGF stimulation. These investigators were the first to note that the morphological alterations induced by EGF were similar to those induced by avian sarcoma virus in chick fibroblasts (Ambros et al., 1975; Wang and Goldberg, 1976). Moreover, they also noted the possible correlation between the phosphorylation of several proteins, including myosin, with changes in cell shape and motility. Mellstrom et al. (1983) showed that serum-starved glial cells exhibited both changes in shape and increased motility when stimulated with PDGF. These events were transient, with ruffling activity occurring for up to 20 min after stimulation and returning to prestimulatory levels after about 1 hr. These observations paved the way for investigating the role of cytoskeletal proteins in the changes of cell shape and motility.

Recently, O'Neill et al. (1985) examined the relationship between cell motility and cell proliferation using serum-starved Swiss 3T3 cells stimulated with EGF and vasopressin in the presence of insulin. The mean rate of cell movement of quiescent cells, measured by quantifying cine records, was less than 2 μm/hr. A combination of EGF, vasopressin, and insulin, which has been shown to induce cell proliferation, increased the mean rate of cell movement to 30 μm/hr. It would be valuable to know the combined effects of EGF, vasopressin, and insulin on the ionic parameters discussed earlier.

The most significant finding in the study by O'Neill et al. (1985) was the observation that agents that stimulate a rise in intracellular cAMP level, such as cholera toxin, inhibitors of cyclic nucleotide phosphodiesterase, and insulin (Rozengurt, 1982b; Boynton and Whitfield, 1983), actually depress the mean rate of cell motility. These same agents also promote cell proliferation. Therefore, the cellular responses of cell motility and cell proliferation are not always coupled directly (Gail et al., 1972).

It is an intriguing possibility that one class of growth-promoting factors may stimulate a rise in cAMP that could depress cell motility by mechanisms similar to that observed in heart muscle (Guidotti et al., 1977), smooth muscle (Bolton, 1979), and possibly in some nonmuscle cells (Willingham, 1976). The second class induces a rapid rise in $[Ca^{2+}]_i$ that would stimulate cell motility. Subsequent cell proliferation in these two pathways could occur by either calcium-dependent or -independent mechanisms. These possibilities would require two types of receptors, as described in Figure 1. It will be important in the future to correlate changes in pH_i, $[Ca^{2+}]_i$, cAMP, and cGMP in cells stimulated by different growth factors in order to extend the relationships defined between cell motility and cell proliferation.

One possible interrelationship between the cAMP signal pathway and the Ca^{2+} pathway is an antagonistic effect on cytoskeletal structure. Two observations are consistent with this possibility. cAMP appears to be responsible for "reverse transformation" in Chinese hamster ovary cells. A rounded, "transformed" shape was reversed to the normal flat shape by exposure to cAMP analogues (Hsei and Puck, 1971; Puck, 1977). In addition, elevation of cytoplasmic cAMP with Bt_2 cAMP caused flattening of 3T3 cells and increased the formation of microfilament bundles at the cell periphery (Willingham and Pastan, 1975). In contrast, agents that induce a rise in Ca_i^{2+} tend to cause cell motility and loss of stress fibers (see Section 11).

Unlike the other growth factors, PDGF appears to play a significant role in wound healing by inducing chemotaxis of fibroblasts (Seppä et al., 1982; Grotendorst, 1984). Interestingly, other factors that stimulate cell proliferation, such as EGF, FGF, and complete serum, inhibited the chemotactic activity induced by PDGF. In contrast, fibronectin continued to stimulate chemotaxis in the presence of these peptides. This is another example of the ability to uncouple the response of cell motility and growth control (Grotendorst, 1984). Although NIH 3T3 cells respond to PDGF stimulation by exhibiting reffling and the formation of microspikes within a few minutes, it requires hours to demonstrate chemotaxis using modified Boyden chambers (Grotendorst, 1984). In addition, RNA synthesis and protein synthesis, but not DNA synthesis, are required for extended chemotaxis (Grotendorst et al., 1981, 1982; Seppä et al., 1982). Therefore, whereas events involved in initiating chemotaxis, such as ruffling, probably occur soon after stimulation, directed net cell locomotion up a gradient of PDGF requires hours.

Clearly, it will be important to develop a microscopic chemotaxis chamber for fibroblasts so that the chemotactic and cell proliferative responses of individual cells can be analyzed with the optimal temporal and spatial resolution. The chamber developed by Zigmond and Hirsch (1973) for leukocyte chemotaxis may not be adequate due to the large time difference between the lifetime of the chemotactic gradient in the chamber and the apparent response time of fibroblasts.

8. CYTOSKELETON

Schlessinger and Geiger (1981) examined the role of specific cytoskeletal proteins in the morphological changes exhibited by A-431 cells in response to EGF (Chinkers et al., 1979). Unfortunately, this report did not indicate whether the cells tested were quiescent. However, using immunofluorescence techniques, they demonstrated that actin containing stress fibers became much less organized and that actin and α-actin in staining generally became more diffuse with some punctuate areas upon incubating A431 cells in EGF for 30–45 min. This change in the cytoskeleton was transient, the maximum effect being observed after 2 hr, with a return to prestimulatory organization within 8 hr. The actin-based cytoskeleton appeared to be the target for the EGF-induced shape change, since neither microtubular nor intermediate filament structure changed in response to EGF. The time course for changes in actin structure was much longer than reported subsequently and may reflect the level of quiescence under the conditions employed by Schlessinger and Geiger (1981). The role of calcium in EGF-induced cytoskeletal change is not clear since some cell types have exhibited a rise in $[Ca^{2+}]_i$ in response to EGF (Moolenaar et al., 1984b), whereas others have not (McNeil et al., 1985). However, the correlation of decreased stress fiber content with increased rate of cell motility in cells stimulated with EGF is consistent with the generalization that highly motile cells exhibit less organized actin structures (Taylor and Condeelis, 1979; Herman et al., 1981; Taylor et al., 1980.)

The addition of insulin, which does not alter $[Ca^{2+}]_i$ by itself (see above, this section) had no obvious effect on actin and α-actinin (Schlessinger and Geiger, 1981). Schlessinger and Geiger (1981) also noted the similarity of the disorganization of the actin structures produced by EGF stimulation to the loss of actin filament bundles induced by avian viruses (Ash et al., 1976; Edelman and Yahara, 1976; Pollack et al., 1975). This loss of actin organization has also been observed in cells infected with other viruses that cause cellular transformation (Jackson and Bellett, 1985).

Within 1–2 min of stimulating starved glial cells with PDGF, activity of lamellipodia and filopodia in these cells increased, and the lamellipodia and filopodia migrated toward the cell center (Mellstrom et al., 1983). During this same time period, cellular organelles (particles) were transported toward the cell center. Immunofluorescence revealed the formation of circular arrays of actin in close association with membrane folds formed by the migration of lamellipodia and filopodia toward the cell center. These rings of actin fluorescence could represent the activity of contractile proteins involved in pinocytosis (see Taylor et al., 1980). Ventral stress fibers were not affected dramatically at early times, but unusual patterns of actin fibers were observed 1 hr after stimulation. Mellstrom et al. (1983) suggested that the formation of lamellipodia and filpodia involved both actin polymerization and changes in actin filament cross-linking.

Bockus and Stiles (1984) made the first attempt to define the specific action of individual growth factors on cytoskeletal components. Quiescent Balb/c–3T3 cells (density arrested) exhibited extensive stress fibers, and the usual filament tangles contained microtubules and vimentin. There was a sequential effect of serum stimulation on the reorganization of the cytoskeletal components. Stress fibers dispersed between 5 and 15 min following stimulation with serum, leaving the actin disorganized. This is a faster response than that of A431 cells to EGF (Schlessinger and Geiger, 1981). Vimentin networks condensed around the nucleus between 1.5 and 3 hr following stimulation, and the tubulin staining became diffuse after 3 hr. The organization of the actin-based cytoskeleton, microtubules, and intermediate filaments returned to the prestimulation state between 5 and 6 hr after serum addition. Addition of PDGF alone induced the dispersal of actin stress fibers within 5 min, and cells returned to a prestimulation organization of actin within 15–90 min. EGF, insulin, and EGF plus insulin had no effect on the actin in these Balb/c–3T3 cells. In addition, PDGF, but not the other growth factors, stimulated the phosphorylation of the myosin light chain within the same time period. In contrast, the addition of insulin to quiescent cells caused the microtubules to disperse within 90 min, whereas neither PDGF nor EGF had any effect. Finally, the addition of EGF and insulin following prestimulation with PDGF caused a maximal reorganization of vimentin. The prestimulation organization of vementin returned within 4–6 hr following stimulation. The requirement of multiple factors for the maximal effect suggests that changes in vimentin may be a secondary event. These results taken together also suggest that PDGF stimulates the actin-based cytoskeleton on a fast time scale, whereas insulin affects microtubules on a much longer time scale. The data on vimentin are too complex to interpret clearly at this time.

The apparent differential effects of PDGF, insulin, and EGF on the cytoskeleton may be related to at the initial pathway of secondary messengers that each may stimulate. It is intriguing that all measurements of $[Ca^{2+}]_i$, whether with aequorin or quin 2, indicate that $[Ca^{2+}]_i$ increases rapidly in response to PDGF, although the kinetics and amplitudes vary (McNeil et al., 1985; Moolenaar et al., 1984b). Using the well-characterized calcium indicator aequorin, McNeil et al. (1985) demonstrated that PDGF alone stimulated a peak rise in $[Ca^{2+}]_i$ of Swiss 3T3 cells within 2 min after stimulation. The time course of the rise in $[Ca^{2+}]_i$ in response to PDGF can be correlated closely with the effects of PDGF on the actin-based cytoskeleton (Bockus and Stiles, 1984). The actin-based cytoskeleton, including stress fibers, contains many calcium-regulated molecules, such as gelsolin (Janmey et al., 1985), caldesmon (Kakiuchi et al., 1983; Bretscher and Lynch, 1985), α-actinin (Burridge and Feramisco, 1981), and calmodulin (Manalan and Klee, 1984). The loss of stress fibers could be due to the action of one or a combination of these and possibly other calcium-regulated cytoskeletal proteins. At elevated $[Ca^{2+}]_i$, α-actinin would be predicted to dissociate maximally from actin (Fechheimer et al., 1982), gelsolin would be predicted to restrict the length of actin filaments (Janmey et al., 1985), and calmodulin would be predicted to complex with many target molecules including myosin light chain kinase (MLCK), caldesmon, (Kakiuchi et al., 1983), and the spectrinlike molecule TW240/260 (fodrin) (Glenney et al., 1982). Therefore, loss of stress fibers could be due to solation (decrease in the structure of actin cross-linking) based on all of these mechanisms (Taylor and Condeelis, 1979; Taylor and Fechheimer, 1982; see also Mittal and Bereiter-Hahn, 1985). The fact that the actin-stabilizing drug, phalloidin, can inhibit actin reorganization in response to PDGF (Bockus and Stiles, 1984) suggests that a gelsolin like activity may play a fundamental part in the loss of stress fibers. In addition, the phosphorylation of many nonmuscle myosin light chains is regulated by a cal-cium-dependent mechanism involving calmodulin that can activate the actin-activated Mg-ATPase of the myosin (Hathaway and Adelstein, 1979). Cell mo-tility might be optimized by the formation of less rigidly cross-linked actin networks (Taylor and Fechheimer, 1982) and maximally activated myosin. Such a model is consistent with the Ca^{2+} pathway mentioned in Section 7, where both motility and cell proliferation are activated (O'Neill et al., 1985).

The absence of a direct effect of PDGF on microtubules and intermediate filaments suggests that a calcium transient induced by PDGF is not sufficient to affect these filament types. It is interesting in this regard that Marcum et al., (1978) have suggested that the calcium–calmodulin complex can induce the depolymerization of microtubules. In addition, Keith et al. (1983) presented evidence that calcium–calmodulin complexes can depolymerize microtubules when injected into cells. The relationship between calcium, calmodulin, and microtubule stability requires more investigation.

Insulin in the presence of agents known to increase the intracellular cAMP levels inhibits cell motility while permitting cell proliferation (Rozengurt, 1982b). Consistent with the role of Ca^{2+} in regulating growth factor–induced changes in actin (mentioned in this section), insulin does not have an early effect on the actin-based cytoskeleton in Balb/c–3T3 cells (Bockus and Stiles, 1984) and does not stimulate a rise in $[Ca^{2+}]_i$ (Morris et al., 1984). A class of initially calcium independent pathways to cell proliferation might exist, as suggested by O'Neill et al. (1985). The loss of microtubules about 90 min following stimulation with insulin suggests that this cytoskeletal change involves either a slow-acting process or a sequence of steps over time. Clearly, it is not an early event of growth factor stimulation in these cells, and apparently it does not depend directly on an early calcium transient. It will be valuable to monitor $[Ca^{2+}]_i$ over extended periods of time in cells stimulated with insulin.

PDGF stimulated a rise in intracellular cAMP in serum-starved Swiss 3T3 cells in the presence of phosphodiesterase inhibitors, which reached a maximum 45 min after addition of PDGF (Rozengurt et al., 1983), long after the calcium transient measured in this cell type. The level of cAMP then declined, but even 24 hr later the cAMP level was much higher than prestimulation levels. It is possible that a rapid rise in $[Ca^{2+}]_i$ is followed by a slower rise in cAMP. This possible sequence of signaling is reminiscent of the interrelated roles of Ca^{2+} and cAMP defined in other stimulus–response coupling events (Rasmussen,

1980; Means *et al.*, 1982) and suggests a possible feedback mechanism for coregulation of $[Ca^{2+}]_i$, cAMP, and different cytoskeletal elements.

Herman and Pledger (1985) extended the study of Bockus and Stiles (1984) on the effects of PDGF on the actin-based cytoskeleton of contact-inhibited and quiescent Balb/c–3T3 cells. Unstimulated cells exhibited actin in stress fibers, as well as vinculin in both adhesion plaques and diffusely in the cytoplasm. These confluent cell cultures exhibited no dramatic changes in cell shape upon stimulation with PDGF, which is a result distinct from nonconfluent, serum-starved cells (see above, this section). Vinculin disappeared from adhesion plaques within 2.5 min after stimulation, whereas actin was unaffected at this time point. Actin stress fibers became disrupted within 10 min, whereas no change in the distribution of talin, vimentin, or microtubules was detected within the first 20 min. The loss of vinculin from adhesion plaques was dose dependent and occurred at concentrations that induced competence for cell proliferation but not DNA synthesis. Recent evidence suggests that $[Ca^{2+}]_i$, calmodulin, protein kinase C, and/or proteolysis may be involved in the removal of vinculin from adhesion plaques (Herman *et al.*, 1986). Subsequent addition of EGF, insulin, or other growth factors caused progression through G_o/G_i. Other competence-inducing factors such as FGF, choleragen, and Ca^{2+} precipitates also caused the disappearance of vinculin from adhesion plaques.·

A close correlation exists between PtdIns metabolism induced by ligand–receptor interactions and actin assembly in extracts of a variety of cells (Lassing and Lindberg, 1985). As described in this section, Mellstrom *et al.*, (1983) showed that a very early event in the stimulation of serum-starved glial cells with PDGF is the formation of lamellipodia and microspikes, which are believed to be formed in part by local actin assembly. Activation of these cells with PDGF also increases PtdIns turnover, including the initial phosphorylation of PtdIns. Recently, a possibly direct relationship between the PtdIns cycle and actin assembly has been suggested (Lassing and Lindberg, 1985). The cytosolic protein profilin is a putative regulator of actin assembly in cells (Malm, *et al.*, 1981; Markey *et al.*, 1982). When complexed with profilin, actin is not capable of assembly. Until recently, no physiological condition could release this profilactin complex. Lassing and Lindberg (1985) have demonstrated that PtdIns(4,5)P2 dissociates the profilactin complex into free actin and profilin, which allows actin assembly. The profilactin complex may be membrane associated in quiescent cells and, following stimulating, PtdIns(4,5)P2 might cause the release of actin for local assembly into lamellipodia and microspikes. This is an exciting possibility but clearly needs more investigation.

The paradigm in cell motility of the relationship between cell stimulation and actin assembly at the membrane–cytoplasm interface is the formation of the acrosomal process in *Thione* sperm (Tilney, 1975). This paradigm might be operational during growth factor stimulation of serum-starved cells and deserves further analysis. Recently, Wang (1985) demonstrated actin assembly at such a membrane–cytoplasm interface in the leading edge of cells using fluorescent analog cytochemistry (Taylor et al., 1986).

9. PHORBOL ESTERS AND THE CYTOSKELETON

The phorbol ester tumor promotors have a wide variety of effects on cells, some of which mimic growth factor stimulation and cell transformation (see Rifkin et al., 1979; Schliwa et al., 1984). Many investigators have noted changes in cytoskeletal structures in response to phorbol ester treatments (Quigley, 1979; Boreiko et al., 1980; Rifkin et al., 1979; Ojakian, 1981). Schliwa et al. (1984) noted that in response to PMA, actin was dispersed from stress fibers in BSC-1 cells, PtK-1 cells, and 3T3 fibroblasts, that actin ribbons formed in the cell perperphery within 2 min, and that this actin dispersion was maximal at 40–50 min. Reversal required 7–10 hr. Vinculin was also lost from attachment plaques and associated with the actin ribbons. However, in contrast to the effect of PDGF stimulation (Herman and Pledger, 1985), PMA caused vinculin to be lost from attachment points after an initial loss of actin. Cycloheximide and cell enucleation had no effect on the actin and vinculin dispersal, which conflicted with the results reported by Rifkin et al. (1979). Finally, calmodulin inhibitors and cAMP had no effect on the PMA-induced reorganization of actin and vinculin. Meigs and Wang (1986) used fluorescent analog cytochemistry (Taylor et al., 1986) to follow the distribution of vinculin and α-actinin in living BSC-1 cells during treatment with PMA. They found that α-actinin was lost from adhesion plaques within 20 min, whereas the adhesion plaques remained attached to the substrate, as indicated by reflection interference microscopy. Vinculin did not change during this time period. This result is consistent with those of Schliwa et al. (1984) on the loss of actin from adhesion plaques.

The mode of action of PMA on the cytoskeleton does not appear to involve Ca^{2+}. McNeil et al. (1985) reported that treatment of serum-starved Swiss 3T3 cells with the phorbol ester PMA did not elicit a calcium transient, although changes in cell shape were observed in the same population of cells from which $[Ca^{2+}]_i$ had been measured. Furthermore, subsequent stimulation of PMA-treated cells with serum elicited almost no rise in $[Ca^{2+}]_i$. These results indicated that phorbol esters could induce some cell activities, including selected cytoskeletal changes, in the absence of a rise in $[Ca^{2+}]_i$ and that some feedback mechanism might exist between the DAG and IP_3 branches of one of the signaling pathways. As discussed in Section 5, phorbol esters can activate protein kinase C directly to phosphorylate cellular proteins (Gilmore and Martin, 1983; Kikkawa et al., 1983; Nishizuka, 1983), including vinculin (Werth and Pastan, 1984). Therefore,

PMA-induced changes in the fibroblast cytoskeleton may depend upon protein phosphorylations that are independent of calcium transients.

One indication of the possible interrelationship between the DAG and IP_3 branches of the signaling pathway comes from a study on platelet activation with either thrombin or PMA. Platelets activated with PMA exhibit phosphorylation of the myosin light chain, which appears to be mediated by protein kinase C and not by the Ca^{2+}/calmodulin-dependent enzyme, MLCK (Naka *et al.*, 1983). In addition, the site of phosphorylation is distinct from that mediated by MLCK. Since PMA treatment causes platelet aggregation, it is assumed that the phosphorylated myosin is functional. It will be important to compare the actin-activated myosin-ATPase activity of the different phosphorylated myosins. Another question is whether protein kinase C normally phosphorylates proteins believed to be the sole substrate for other enzymes. In this regard, phorbol esters will continue to be valuable tools in dissecting the signaling pathways.

10. CYTOPLASMIC pH AND THE CYTOSKELETON

Experiments with phorbol esters have indicated that some of the actin-based cytoskeletal changes observed apparently can occur in the absence of a calcium transient. There are three possibly explanations of these results: (1) A cascade of phosphorylation is responsible primarily for regulating the cytoskeleton, and large calcium transients activating calcium-dependent kinases and other calcium-dependent proteins are not necessary; (2) phorbol esters stimulate the phosphorylation of substrates usually regulated by a calcium-mediated enzyme, and this is thus not a physiologically significant phenomenon; or (3) a physiological parameter other than phosphorylation can alter the affinity of calcium to target molecules without involving a large calcium transient.

Cytoplasmic pH might be such an alternative to phosphorylation as a physiological parameter that effects the actin-based cytoskeleton. There is at least one example of a cytoskeletal protein whose activity is affected directly by changes in pH. The *Dictyostelium discoidem* α-actinin shows extremely sensitive binding to actin, depending on either the $[Ca^{2+}]$ or pH (Fechheimer *et al.*, 1982). The pH sensitivity to binding occurs over a physiological range and is completely reversible. The pH sensitivity of additional structural proteins, as well as enzymes, should be determined to permit an evaluation of the general significance of this one observation.

The early studies on cytoplasmic pH were performed primarily in the absence of external HCO_3^-, which appears to be crucial for pH homeostasis (see Section 2.2). Therefore, the published work on the alkalinization of quiescent cells in response to growth factors is now questioned (Cassel *et al.*, 1985). These studies must now be repeated under a variety of external ionic conditions to determine whether a transient rise in pH_i occurs invariably following stimulation. For example, the presence of a Cl^-/HCO_3^- exchanger might minimize the extent and/or rate of alkalinization, but if the Na^+/H^+ and Cl^-/HCO_3^- exchangers function on a different temporal and/or spatial scale than a transient pH_i, change could occur.

A transient change in pH could occur at the membrane–cytoplasm interface, involving only a small fraction of the cell volume. Although the mobility of H^+ is extremely high in solution, it is possible that a highly structured (gelled) cytoplasm could alter the activity coefficient for H^+ being exchanged rapidly into a small functional cell volume. This point is hypothetical at this time, and the basic physical chemical properties of cytoplasm are not well understood at this time (Luby-Phelps *et al.*, 1986) and need more extensive investigation.

It is also critical to consider that calcium-binding proteins such as calmodulin are sensitive to pH. Therefore, in the absence of a calcium transient, but with a change in pH, proteins like calmodulin might have an increased calcium affinity and an increased affinity for their target. An analogous phenomenon is illustrated by the $[Ca^{2+}]_i$ requirement of protein kinase C. A rise in calcium $[Ca^{2+}]_i$ may not be required for activation of protein kinase C, since the binding of DAG to this enzyme raises its calcium affinity greatly (Kaibuchi *et al.*, 1981; Kishimoto *et al.*, 1980). This can be considered a passive or permissive calcium regulatory mechanism. It is feasible that similar mechanisms could couple other calcium-binding proteins into a signaling pathway in the absence of a calcium transient.

11. MODEL OF THE ROLE OF THE Ca^{2+} SIGNALING PATHWAY IN REGULATING THE ACTIN-BASED CYTOSKELETON

It is tempting to suggest a sequence of molecular events that might be involved in the growth factor–stimulated cell motility of quiescent fibroblasts (see O'Neill *et al.*, 1985). Figure 2 depicts the potential components of such a pathway leading to the early response of the actin-based cytoskeleton in cells upon stimulation with growth factors that induce a rapid and transient rise in $[Ca^{2+}]_i$. A good example is the PDGF stimulation of quiescent 3T3 fibroblasts. All of the signaling events depicted in this working model occur within minutes and could initiate the changes in cell shape and motility observed within 5–20 min. The transient nature of the signaling and feedback mechanisms involved in the different pathways are consistent with the transient nature of the actin-based cytoskeletal responses. It is intriguing to contemplate the interrelationship between an initial $[Ca^{2+}]_i$ response (McNeil *et al.*, 1985) and a possible slower cAMP response (Rozengurt *et al.*, 1983), as discussed by Rasmussen (1980).

The temporal sequence of the effects of serum on the structure of actin, micro-tubules, and perhaps even intermediate filaments could be explained by one of these mechanisms. Clearly, the time course of changes in the major presumptive cellular signaling components must be correlated directly with specific cytoskeletal changes in a single well-defined system.

12. PROSPECTUS

Many fundamental questions remain concerning the stimulus–response pathway governing growth factor activation of quiescent cells. These questions span the full time domain from the earliest observable cellular responses to DNA synthesis and cell division.

Foremost among these are two interrelated questions: What is the mechanism for the mitogenic activation of quiescent cells by each of the many growth factors? Do the different growth factors utilize different mechanisms or is there a common signal pathway? The data reviewed here suggest that metabolism of inositol phospholipids possibly is initiated three known growth factors (EGF, PDGF, and thrombin) and that as a direct consequence of such metabolism, transient increases in $[Ca^{2+}]_i$ and possibly pH_i are two early cytoplasmic events common to fibroblasts activated by these and other growth factors. However, this knowledge does not explain mitogenesis or other aspects of activation. If, for example, Ca^{2+} and H^+ are to be accepted as cytoplasmic signals for DNA synthesis, it will be necessary to show directly that these ions are obligatory components of the activation pathway leading to the DNA synthetic response and to identify for each ion a protein target whose activity is regulated by ion binding and can be related directly or indirectly to the nuclear event in question. Similarly, enhanced phosphorylation of cellular proteins is a characteristic common to fibroblasts stimulated with growth factors. Yet, until the function of the protein substrates involved is discovered, the significance of their phosphorylation to DNA synthesis must remain as a most provocative mystery.

Given the rapidity with which inositol lipid metabolism and changes in $[Ca^{2+}]_i$ and pH_i are stimulated, it seems clear that many early cytoplasmic events are initiated by binding of growth factors to cell-surface receptors and that such early events are *not* dependent on receptor-mediated internalization of growth factor and its receptor. We have not, however, touched on an important question relating to "long-term" (hours rather than minutes) effects of growth factors: Is synthesis and accumulation of one or more labile proteins in G_1 the major regulatory event leading to DNA synthesis? This hypothesis is based on considerable evidence that growth factor–initiated progression of fibroblasts through the first 3–4 hr of G_1 is strongly dependent on a high level of protein synthesis

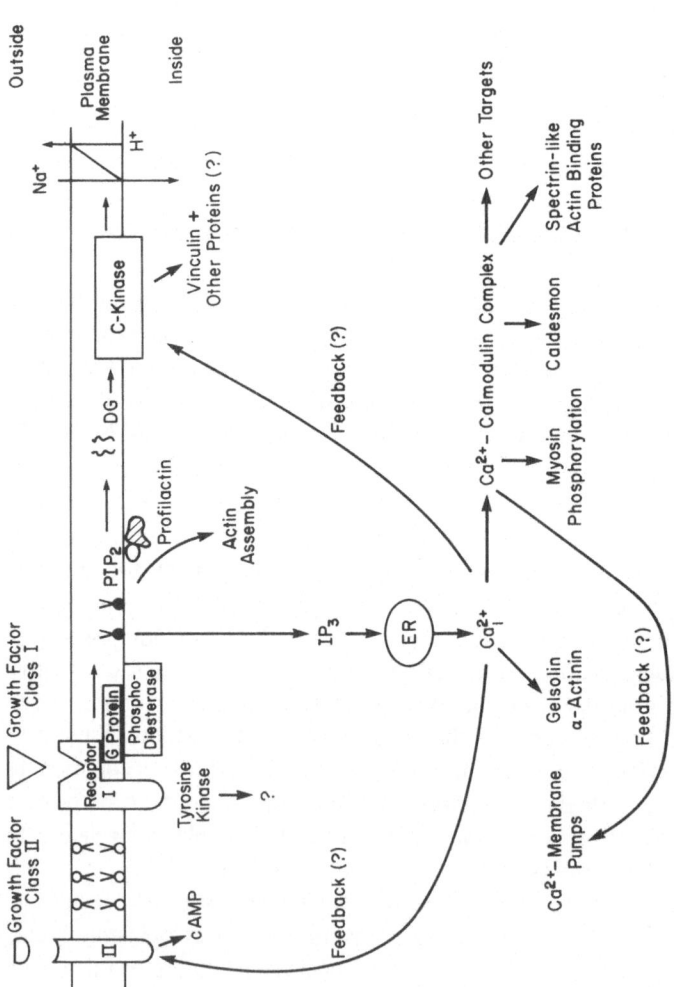

FIGURE 2. Generalized, hypothetical depiction of some of the possible changes in the membrane and cytoplasmic cytoskeletal proteins when a cell is stimulated with a class I growth factor that elevates $[Ca^{2+}]_i$. PtdIns(4,5)P2 may induce actin assembly by dissociation of the profilactin complex. Activation of C kinase by DAG may lead to the phosphorylation of vinculin and other cytoskeletal proteins, in addition to the Na^+/H^+ exchanger. The diffusion of IP3 into the cytoplasm may stimulate the release of calcium from the ER on other intracellular stores. Elevation of $[Ca^{2+}]_i$ should then alter the activity of a large number of cytoskeletal and contractile proteins that would be predicted to affect cell shape and motility.

by the fibroblast (Highfield and Dewey, 1972; Rossow *et al.*, 1979; Croy and Pardee, 1979; Zetterberg and Larson, 1985). The identity of this labile protein (or proteins) is clearly of great interest.

What are the molecular links between the signaling pathways for normal growth control and oncogenesis? There is a vast, rapidly evolving literature not reviewed here on the relationship between normal growth control and cell transformation. The most exciting result in recent years has been the identification of some oncogene products that are chemically similar to growth factor receptors (Hunter, 1985).

It is evident that there are many potential components of the signaling pathway for eliciting the rapid (5–20 min) responses of at least the actin-based cytoskeleton. There are three major issues relating growth control and cell motility that seem to be unresolved: Do growth factors involve single signaling pathways or are there interrelationships between the resumptive Ca^{2+}, cAMP pathways, and any pathways yet to be defined? This is a complex question that has been answered in other stimulus–response processes. For example, it is possible that the earliest cellular responses, such as cell shape and motility, could be triggered by one pathway, whereas the recovery from the stimulation might involve another pathway. Finally, it is likely that the long-term responses, such as DNA synthesis, involve many coordinated temporal and even spatial signaling pathways. A potential tool for investigating this question is the chemotactic and cell proliferative activity of PDGF for fibroblasts under the right conditions. This is a case where cell motility and cell proliferation could occur by using part of the same signaling pathway.

Does the cytoskeleton affect the long-term signaling pathway to cell proliferation? Although this question has been addressed, there does not seem to be a definitive answer, since the cytoskeleton is required for many cell functions and the effects could be indirect or direct. However, it is possible that the cytoskeleton could play a direct role, as originally suggested by Edelman and colleagues (for review, see Edelman, 1976). This would require direct physical coupling of cytoskeletal components with cellular components involved in cell proliferation or an indirect role by altering the physical chemical properties of cytoplasm, which could affect all cellular chemical processes (Luby-Phelps *et al.*, 1986).

Are there spatial, as well as temporal, variations involved in the signaling pathway and cellular responses of stimulated cells? Receptors can change in distribution in the plane of the membrane (see Axelrod, *et al.*, 1976; Jacobson *et al.*, 1983); therefore, cytoskeletal proteins and sources for some cytoplasmic signals (e.g., endoplasmic reticulum) also may not be distributed homogeneously in cytoplasm. We therefore believe that it is crucial to define the spatial, as well as the temporal, dynamics of those processes leading to activation of quiescent cells by growth factors.

The techniques for quantifying the properties of specific molecular constituents in and on single living cells are now available. Quantitative fluorescence microscopy, especially when integrated with modern fluorescence microscopic imaging techniques, should play an important role in defining the molecular details of these and other cellular processes (Taylor and Wang, 1980; Taylor *et al.*, 1986). Fluorescent indicators of ionic signals, such as pH, pCa, and membrane potential and functional fluorescent analogues of selected macromolecules should yield the spatial, as well as the temporal, information required (Taylor *et al.*, 1986). The future of this technology falls into two major categories: (1) the development of new fluorescent probes, over a range of wavelengths specific for cellular parameters, such as selected enzymes (Waggoner, 1986); and (2) the combination of multiple fluorescent indicators separable spectroscopically in the same cells to permit multiple parameter analysis. For example, it should be possible to quantify pH, pCa, membrane potential, one enzyme activity, a lipid, and a structural protein in the same cell. Correlation of spatial with temporal molecular activities may lead to a more detailed understanding of growth factor activation of the quiescent cell.

ACKNOWLEDGMENTS. We would like to thank Dr. Jesse Siskin for reading the manuscript critically. He stimulated a great deal of thought during his sabbatical leave in our laboratory. Helpful comments were also made by A. Pardee and W. Pledger.

REFERENCES

Ambros, V., Chen, L., and Buchanan, J., 1975, Surface ruffles as markers for studies of cell transformation by Rous sarcoma virus, *Proc. Natl. Acad. Sci. USA* **72**:144–148.

Ash, J., Voyt, P., and Singer, S., 1976, Reversion from transformed to normal pheno-type by inhibition of protein synthesis in rat kidney cells infected with a temperature-sensitive mutant of Rous sarcoma virus, *Proc. Natl. Acad. Sci. USA* **73**:3603–3607.

Axelrod, D., Keppel, D., Schlessinger, J., Elson, E., Webb, W., 1976, Mobility measurements by analysis of fluorescence photobleaching recovery kinetics, *Biophysics* **16**:1055–1060.

Balk, S. D., Whitfield, J. F., Youdale, T., and Braun, A. C., 1973, Roles of calcium, serum, plasma, and folic acid in the control of proliferation of normal and Rous sarcoma virus–infected chicken fibroblasts, *Proc. Natl. Acad. Sci. USA* **70**:675–679.

Beguinot, L., Hanover, J. A., Ito, S., Richert, N. D., Willingham, M. C., and Paston, I., 1985, Phorbol esters induce transient internalization without degradation of unoccupied epidermal growth factor receptors, *Proc. Natl. Acad. Sci. USA* **82**:2774–2778.

Berridge, M. J., and Irvine, R. F., 1984, Inositol triphosphate, a novel second messenger in cellular signal transduction, *Nature* **312**:315–321.

Berridge, M. J., Downes, C. P., and Hanley, M. R., 1982, Lithium amplifies agonist-dependent phosphatidylinositol responses in brain and salivary glands, *Biochem. J.* **206**:587–595.

Berridge, M. J., Heslop, J. P., Irvine, R. F. and Brown, K. D., 1984a, Inositol lipids and cell proliferation, *Biochem. Soc. Trans.* **13**:67–71.

Berridge, M. J., Heslop, J. P., Irvine, R. F., and Brown, K. D., 1984b, Inositol triphosphate formation and calcium mobilization in Swiss 3T3 cells in response to platelet-derived growth factor, *Biochem. J.* **222**:195–201.

Besterman, J. M., and Cuatrecasas, P., 1984, Phorbol esters rapidly stimulate amiloride-sensitive Na$^+$/H$^+$ exchange in a human leukemic cell line, *J. Cell Biol.* **99**:340–343.

Betsholtz, C., and Westermark, B., 1984, Growth factor–induced proliferation of human fibroblasts in serum-free culture depends on cell density and extracellular calcium concentration, *J. Cell Physiol.* **118**:203–210.

Blackshear, P. J., Witters, L. A., Girard, P. R., Kuo, J. F., and Quama, S. N., 1985, Growth factor–stimulated protein phosphorylation in 3T3-L1 cells. Evidence for protein kinase C–dependent and –independent pathways, *J. Biol. Chem.* **260**:13304–13315.

Bockus, B., and Stiles, C., 1984, Regulation of cytoskeletal architecture by platelet-derived growth factor, insulin, and epidermal growth factor, *Exp. Cell Res.* **153**:186–197.

Bolton, T. B., 1979, Mechanisms of action of transmitters and other substances on smooth muscle, *Physiol. Rev.* **59**:607–618.

Boreiko, C., Mondal, S., Narajan, K., and Heidelberger, C., 1980, Effect of 12-O-tetradecanoyl-phorbol-13-acetate on the morphology and growth of C3H/10T1/2 mouse embryo cells, *Cancer Res.* **40**:4709–4716.

Boron, W. F., 1984, Cell activation: The "basic" connection, *Nature* **312**:312.

Bottom, T., 1974, Mechanisms of action of transmitters and other substances on smooth muscle, *Physiol. Rev.* **59**:606–718.

Bowen-Pope, D. F., and Ross, R., 1982, Platelet-derived growth factor. II. Specific binding to cultured cells, *J. Cell Biol.* **96**:679–683.

Bowen-Pope, D. F., and Rubin, H., 1983, Growth stimulatory precipitates of Ca^{2+} and pyrophosphate, *J. Cell. Physiol.* **117**:51–61.

Boynton, A. L., and Whitfield, J. F., 1976, The different actions of normal and supranormal calcium concentrations on the proliferation of Balb/c 3T3 mouse cells, *In Vitro* **12**:479–484.

Boynton, A., and Whitfield, J., 1983, The role of cyclic AMP in cell proliferation: A critical assessment of the evidence, *Adv. Cyclic Neucleotide Res.* **15**:193–294.

Boynton, A. L., Whitfield, J. F., Isaacs, R. J., and Morton, H. J., 1974, Control of 3T3 cell proliferation by calcium. *In Vitro* **10**:12–17.

Boynton, A. L., Whitfield, J. F., and Isaacs, R. J., 1976, The different roles of serum and calcium in the control of proliferation of Balb/c 3T3 mouse cells, *In Vitro* **12**:120–123.

Bretscher, A., and Lynch, W., 1985, Identification and localization of immunoreactive forms of caldesmon in smooth and nonmuscle cells: A comparison with the distributions of tropomyosin and α-actinin, *J. Cell Biol.* **100**:1656–1663.

Burns, C. P., and Rozengurt, E., 1983, Serum, platelet-derived growth factor, vassopressin, and phorbol esters increase intracellular pH in Swiss 3T3 cells, *Biochem. Biophys. Res. Commun.* **116**:931–938.

Burridge, K., and Feramisco, J., 1981, Non-muscle α-actinins are calcium sensitive actin-binding proteins, *Nature* **294**:565–567.

Burroni, D., and Ceccarini, C., 1984, The effect of alkaline pH on the cell growth of six different mammalian cells in tissue culture, *Exp. Cell Res.* **150**:505–508.

Busa, W. B., and Nuccitelli, R., 1984, Metabolic regulation via intracellular pH, *Am. J. Physiol.* **246**:R409–R438.

Campbell, A. K., 1983, *Intracellular calcium. Its Universal Role as Regulator,* John Wiley and Sons, New York, p. 556.

Carney, D. H., Scott, D. L., Gordon, E. A., and LaBelle, E. F., 1985, Phosphoinositides in mitogenesis: Neomysin inhibits thrombin-stimulated phosphoinositide turnover and initiation of cell proliferation, *Cell* **42**:479–488.

Carpenter, G., 1984, Properties of the receptor for epidermal growth factor, *Cell* **37**:357–358.

Carpenter, G., and Cohen, S., 1976, ^{125}I-labeled human epidermal growth factor. Binding, internalization, and degradation in human fibroblasts, *J. Cell Biol.* **71**:159–171.

Carpenter, G., and Cohen, S., 1979, Epidermal growth factor, *Annu. Rev. Biochem.* **48**:193–216.

Carpenter, G., and Cohen, S., 1984, Peptide growth factors, *Trends Biochem. Sci.* **9**:169–171.

Cassel, D., Whiteley, B., Zhuang, Y. X., and Glaser, L., 1985, Mitogen-independent activation of Na^+/H^+ exchange in human epidermoid carcinoma A431 cells: Regulation by medium osmolarity, *J. Cell Physiol.* **122**:178–186.

Ceccarini, C., and Eagle, H., 1971a, pH as a determinant of cellular growth and contact inhibition, *Proc. Natl. Acad. Sci. USA* **68**:229–233.

Ceccarini, C. and Eagle, H., 1971b, Induction and reversal of contact inhibition of growth by pH modification, *Nature New Biol.* **233**:271–273.

Chinkers, M., McKenna, J., and Cohen, S., 1979, Rapid induction or morphological changes in human carcinoma cells A-431 by epidermal growth factor, *J. Cell Biol.* **83**:260–265.

Chinkers, M., McKenna, J., and Cohen, S., 1981, Rapid rounding of human epidermoid carcinoma cells A-431 induced by epidermal growth factor, *J. Cell Biol.* **88**:422–429.

Cochet, C., Gill, G. N., Meisenhelder, J., Cooper, J. A., and Hunter, T., 1984, C-kinase phosphorylates the epidermal growth factor receptor and reduces its epidermal growth factor–stimulated tyrosine protein kinase activity, *J. Biol. Chem.* **259**:2553–2558.

Cohen, P., 1982, The role of protein phosphorylation in neural and hormonal control of cellular activity, *Nature* **296**:613–620.

Cohen, S., Carpenter, G.,and King, L., 1980, Epidermal growth factor–receptor–proteinkinase interactions, *J. Biol. Chem.* **255**:4834–4842.

Cooper, J. A., Bowen-Pope, D. F., Raines, E., Ross, R., and Hunter, T., 1982, Similar effects of platelet-derived growth factor and epidermal growth factor on the phosphorylation of tyrosine in cellular proteins, *Cell* **31**:263–273.

Cooper, J. A., Sefton, B. M., and Hunter, T., 1984, Diverse mitogenic agents induce the phosphorylation of two related 42,000 dalton proteins on tyrosine in quiescent cells, *Mol. Cell. Biol.* **4**:30–37.

Croy, R., and Pardee, A. B., 1979, Enhanced synthesis and stabilization of Mr 68,000 protein in transformed Balb/c–3T3 cells: Candidate for restriction point control of cell growth, *Proc. Natl. Acad. Sci. USA* **80**:4699–4703.

Dale, M. M., and Penfield, A., 1984, Synergism between phorbol ester and A23187 in superoxide production by neutrophils, *Fed. Eur. Biochem. Sci.* **175**:170–178.

Das, M., and Fox, C. F., 1978, Molecular mechanism of mitogen action: Processing of receptor induced by epidermal growth factor, *Proc. Natl. Acad. Sci. USA* **75**:2644–2648.

Dicker, P., and Rozengurt, E., 1980, Phorbol esters and vassopressin stimulate DNA synthesis by a common mechanism, *Nature* **287**:607–612.

Diringer, H., and Friis, R. R., 1977, Changes in phosphatidylinositol metabolism correlated to growth state of normal and Rous sarcoma virus–transformed Japanese quail cells, *Cancer Res.* **37**:2978–2984.

Downes, C. P., and Michell, R. H., 1981, The polyphosphoinositide phosphodiesterase of erythorocyte membranes, *Biochem. J.* **198**:133–140.

Drummond, A. H., 1985, Bidirectional control of cytosolic free calcium by thyrotropin-releasing hormone in pituitary cells, *Nature* **315**:752–755.

Dulbecco, R., Elkington, J., 1975, Induction of growth in resting fibroblastic cell cultures by Ca^{2+}, *Proc. Natl. Acad. Sci. USA* **72**:1584–1588.

Edelman, G., 1976, Surface modulation in cell recognition and cell growth, *Science* **192**:218.

Edelman, G., and Yahara, I., 1976, Temperature-sensitive changes in surface modulating assemblies of fibroblasts transformed by mutants of Rous sarcoma virus, *Proc. Natl. Acad. Sci. USA* **73**:2047–2051.

Ek, B., Westermark, B., Wasteson, A., and Heldin, C. H., 1982, Stimulation of tyrosine-specific phosphorylation by platelet derived growth factor, *Nature* **295**:419–420.

Engstrom, W., Zetterberg, A., and Auer, G., 1982, Calcium, phosphate, and cell proliferation, in: *Ions, Cell Proliferation, and Cancer* (A. L. Boynton, W. L. McKeehan, and J. F. Whitfield, eds.), Academic Press, New York, pp. 259–281.

Fechheimer, M., Brier, J., Rockwell, M., Luna, E., and Taylor, D., 1982, A calcium and pH regulated actin binding protein from *D. discoideum, Cell Motility* **2**:287–308.

Gail, M., Scher, C., and Boone, C., 1972, Dissociation of cell motility from cell proliferation in BALB/c–3T3 fibroblasts, *Exp. Cell Res.* **70**:439–443.

Gey, C., 1955, Some aspects of the constitution and behavior of normal and malignant cells maintained in continuous culture, *Harvey Lect.* **50**:154–229.

Gillies, R. J., 1981, Intracellular pH and growth control in eukaryotic cells, in: *The Transformed Cell* (I. Cameron and T. B. Poole, eds.), Academic Press, New York, pp. 347–395.

Gilmore, T., and Martin, G. S., 1983, Phorbol ester and diacylglycerol induce protein phosphorylation at tyrosine, *Nature* **294**:771–773.

Glenney, J., Glenney, P., and Weber, K., 1982, Erythroid spectrin, brain fodrin, and intestinal brush border protein (TW-260/240) are related molecules containing a common calmodulin-binding subunit bound to a variant cell type-specific subunit, *Proc. Natl. Acad. Sci. USA* **79**:4002.

Gospodarowicz, D., and Moran, J. S., 1976, Growth factors in mammalian cell culture, *Annu. Rev. Biochem.* **45**:531–558.

Gottesman, M. M., Singh, T., LeCarn, A., Roth, C., Nicholas, J. C., Cabral, F., and Pastan, I., 1980, Cyclic-AMP-dependent phosphorylation in cultured fibroblasts: A genetic approach, *Cold Spring Harbor Conf. Cell Prolif.* **8**:195–209.

Green, H., 1978, Cyclic AMP in relation to proliferation of the epidermal. A new view, *Cell* **15**:801–811.

Grotendorst, G., 1984, Alteration of the chemotactic response of NIH/3T3 cells to PDGF by growth factors, transformation, and tumor promotors, *Cell* **36**:279–285.

Grotendorst, G., Seppä, H., Kleinman, H., and Martin, G., 1981, Attachment of smooth muscle cells to collagen and their migration toward platelet-derived growth factor, *Proc. Natl. Acad. Sci. USA* **78**:3669–3672.

Grotendorst, G., Chang, T., Seppä, H., Kleinman, J., and Martin, G., 1982, Platelet-derived growth factor is a chemoattractant for vascular smooth muscle cells, *J. Cell Physiol.* **113**:261–266.

Guidotti, A., Hanbauer, I., and Costa, E., 1977, Nuclear translocation of catolytic subunits of cytosol cAMP-dependent protein kinase in the transgraphic induction of medullary tyrosine hydroxylase, *Adv. Cyclic Nucleotide Res.* **9**:185–194.

Habenicht, A. J. R., Glomset, J. A., King, W. C., Nist, C., Mitchell, C. D., and Ross, R., 1981, Early changes in phosphatidylinositol and arachidonic acid metabolism in quiescent Swiss 3T3 cells stimulated to divide by platelet-derived growth factor, *J. Biol. Chem.* **256**:12329–12335.

Haigler, H. T., McKenna, J. A., and Cohen, S., 1979, Rapid stimulation of pinocytosis inhuman carcinoma cells A-431 by epidermal growth factor, *J. Cell Biol.* **83**:82–90.

Hathaway, D. R., and Adelstein, R. S., 1979, Human platelet myosin light chain kinase requires the calcium-binding protein calmodulin for activity, *Proc. Natl. Acad. USA* **76**:1653–1657.

Hazelton, B., Mitchell, B., and Tupper, J., 1979, Calcium, magnesium, and growth control in the WI-38 human fibroblast cell, *J. Cell Biol.* **83**:487–498.

Heldin, C. H., and Westermark, B., 1984, Growth factors: Mechanism of action and relation to oncogenes, *Cell* **37**:19–20.

Heldin, C. H., Westermark, B., and Wasteson, A., 1981, Specific receptors for platelet-derived growth factor on cells derived from connective tissue and glia, *Proc. Natl. Acad. Sci. USA* **78**:3664–3668.

Heldin, C. H., Westeson, A., and Westermark, B., 1985, Platelet-derived growth factor, *Mol. Cell. Endocrinol.* **39**:169–187.

Herman, B., and Pledger, W., 1985, Platelet-derived growth factor–induced alterations in vinculin and actin distribution in BALB/c–33 cells, *J. Cell Biol.* **100**:1031–1040.

Herman, B., Harrington, M., Olashaw, N., and Pledger, W., 1986, Identification of the cellular mechanisms responsible for platelet-derived growth factor induced alterations in cytoplasmic vinculin distribution, *J. Cell. Physiol.* **126**:115–125.

Herman, I., Crisona, N., and Pollard, T., 1981, Relation between cell activity and distribution of cytoplasmic actin and myosin, *J. Cell Biol.* **90**:84–91.

Hesketh, T. R., Smith G. A., and Metcalfe, J. C., 1982, Calcium and lymphocyte activation, in: *Ions, Cell Proliferation, and Cancer* (A. L. Boynton, W. L. McKeehan and J. F. Whitfield, eds.), Academic press, New York, pp. 397–415.

Hesketh, T. R., Smith, G. A., Moore, J. P., Taylor, M. V., and Metcalfe, J. C., 1984, Free cytoplasmic calcium concentration and the mitogenic stimulation of lymphocytes, *J. Biol. Chem.* **258**:4876–4882.

Hesketh, T. R., Moore, J. P., Morris, J. D. H., Taylor, M. V., Rogers, J., Smith, G. A., and Metcalfe, J. C., 1985, A common sequence of calcium of pH signals in the mitogenic stimulation of eukaryatic cells, *Nature* **313**:481–484.

Highfield, D. P., and Dewey, W. C., 1972, Inhibition of DNA synthesis by synchronized Chinese hamster ovary cells treated in G1 or early S phase with cytohexamide or puromycin, *Exp. Cell Res.* **75**:314–320.

Hoffman, R., Ristow, H.-J., Packowsky, H., and Frank, W., 1974, Phospholipid metabolism in embryonic rat fibroblasts following stimulation by a combination of the serum proteins S1 and S2, *Eur. J. Biochem.* **49**:317–324.

Hokin, L. E., 1985, Receptors and phosphoinositide-generated second messengers, *Annu. Rev. Biochem.* **54**:205–235.

Hollenberg, M. D., and Cautrecasas, P., 1973, Epidermal growth factor: Receptors in human fibroblasts and modulation of action by cholera toxin, *Proc. Natl. Acad. Sci. USA.* **70**:2964–2968.

Hsei, A., and Puck, T., 1971, Morphological transformation of Chinese hamster cells by dibutyryl andenosine cyclic 3′:5′:monophosphate and testosterone, *Proc. Natl. Acad. Sci. USA* **68**:358–361.

Huang, J. S., Huang, S. S., Kennedy, B., and Devel, T. F., 1982, Platelet-derived growth factor: Specific binding to target cells, *J. Biol. Chem.* **257**:8130–8136.

Hunter, T., 1985, Oncogenes and growth control, *Trends Biochem. Sci.* **10**:275–280.

Hunter, T., and Cooper, J. A., 1981, Epidermal growth factor induces rapid tyrosine phosphorylation of proteins in A431 human tumor cells, *Cell* **24**:741–752.

Hunter, T., Ling, N., and Cooper, J. A., 1984, Protein kinase C phosphorylationof the EGF receptor at a threonine residue close to the cytoplasmic face of the plasma membrane, *Nature* **311**:480–483.

Iwashita, S., Fox, C. F., 1984, Epidermal growth factor and potent phorbol tumor promoters induce epidermal growth factor receptor phosphorylation in a similar but distinctly different manner in human epidermoid carcinoma A431 cells, *J. Biol. Chem.* **259**:2559–2567.

Jackson, P., and Bellet, A., 1985, Reduced microfilament organization in adenovirus type 5-infected rat embryo cells: A function of early region 1a, J. Virol. 55:644-650.

Jacobsen, K., Elson, E., Koppel, D., Webb, W., 1983, International workshop on the application of fluorescence photobleaching techniques to problems in cell biology, Fed. Proc. 42: 72-79.

Janmey, P., Chaponnier, C., Lind, S., Zaner, K., Stossel, T., and Yin, H., 1985, Interactions of gelsolin and gelsolin-actin complexes with actin. Effects of calcium on actin nucleation, filament severing, and end blocking, Biochemistry 24:3714-3723.

Johnson, P. C., Ware, J. A., Clivedon, P. B., Smith, M., Dvorak, A. M., and Salzman, E. W., 1985, Measurement of ionized calcium in blood platelets with the photoprotein aequorin, J. Biol. Chem. 260:2069-2076.

Kaever, V., and Resch, K., 1985, Are cyclic nucleotides involved in the initiation of mitogenic activation of human lymphocytes?, Biochem. Biophys. Acta 846:216-225.

Kaibuchi, K., Takai, Y., and Nishizuka, Y., 1981, Cooperative roles of various membrane phospholipids in the activation of calcium-activated, phospholipid-dependent protein kinase, J. Biol. Chem. 256:7146-7149.

Kakiuchi, R., Inui, M., Morimoto, K., Kanda, K., Sobue, K., and Kakiuchi, S., 1983, Caldesmon, a calmodulin-binding, F-actin-interacting protein, is present in aorta, uterus, and platelets, FEBS Lett. 154:351-356.

Kamine, J., and Rubin, H., 1976, Magnesium required for serum stimulation of growth in cultures of chick embryo fibroblasts, Nature 263:143-145.

Keith, C., DiPaola, M., Maxfield, F., and Shelanski, M., 1983, Microinjection of Ca^{2+}–calmodulin causes a localized depolymerization of microtubules, J. Cell Biol. 97:1918-1924.

Kelly, K., Cochran, B. H., Stiles, C. D., and Leder, P., 1983, Cell-specific regulation of the c-myc gene by lymphocyte mitogens and platelet-derived growth factor, Cell 35:603-610.

Kikkawa, U., Takai, Y., Tanaka, Y., Miyake, R., and Nishizuka, Y., 1983, Protein kinase C as a possible receptor protein of tumor-promoting phorbol esters, J. Biol. Chem. 258:11442-11445.

Kishimoto, A., Takai, Y., Mori, T., Kikkawa, U., Nishizuka, Y., 1980, Activation of calcium and phospholipid-dependent protein kinase by diacylglycerol, its possible relation to phosphatidylinositol turnover, J. Biol. Chem. 255:2273-2276.

Krupp, M. N., Connolly, D. T., and Lane, M. D., 1982, Synthesis, turnover, and down regulation of epidermal growth factor receptors in human A431 epidermoid carcinoma cells and skin fibroblasts, J. Biol. Chem. 257:11489-11496.

L'Allemain, G., Franchi, A., Cragoe, E. J., Jr., and Pouyssegur, J., 1984a, Blockade of the Na^+/H^+ antiport abolishes growth factor-induced DNA synthesis in fibroblasts. Structure-activity relationships in the amiloride series, J. Biol. Chem. 259:4313-4319.

L'Allemain, G., Paris, S., Pouyssegur, J., 1984b, Growth factor action and intracellular pH regulation in fibroblasts, J. Biol. Chem. 259:5809-5815.

L'Allemain, G., Paris, S., and Pouyssegur, J., 1985, Role of a Na-dependent Cl^-/HCO_3^- exchange in regulation of intracellular pH in fibroblasts, J. Cell Biol. 260:4877-4883.

Lang, V., Pryhitka, C., and Buckley, J. T., 1977, Effect on neomycin and ionophore A23187 and ATP levels and turnover of polyphosphoinositides in human erythrocytes, Can. J. Biochem. 55:1007-1012.

Lassing, I., and Lindberg, V., 1985, Specific interaction between phosphatidylinositol 4,5-biphosphate and profilactin, Nature 314:472-474.

Lawrence, T., Ginzberg, R., Gilula, N., Beers, W., 1979, Hormonally induced cell shape changes in cultured rat ovarian granulosa cells, J. Cell Biol. 80:21-36.

Leach, K., James, M., and Blumberg, P., 1983, Characterization of a specific phorbol ester aporeceptor in mouse brain cytosol, Proc. Natl. Acad. Sci. USA 80:4208-4212.

Lee, L. S., and Weinstein, I. B., 1978, Tumor-promoting phorbol esters inhibit binding of epidermal growth factor to cellular receptors, *Science* **202:**313–315.

Leof, E. B., Wharton, N., O'Keefe, E., and Pledger, W. J., 1982, Elevated intracellular concentrations of cyclic AMP inhibited serum-stimulated, density arrested Balb/c-3T3 cells in mid G1, *J. Cell. Biochem.* **19:**93–103.

Lin, C. R., Chen, W. S., Lazar, C. S., Carpenter, C. D., Gill, G. N., Evans, R. M., and Rosenfeld, M. G., 1986, Protein kinase C phosphorylation at the 654 of the unoccupied EGF receptor and EGF binding regulate functional receptor loss by independent mechanisms, *Cell* **44:**839–848.

Lodhi, S., Weiner, N. D., and Schacht, J., 1979, Interactions of neomycin with monomolecular films of polyphosphoinositides and other lipids, *Biochem. Biophys. Acta* **557:**1–8.

Luby-Phelps, K., Taylor, D., and Lanni, F., 1986, Probing the structure of cytoplasm, *J. Cell Biol.* **102:**2015–2022.

MacIntyre, D. E., and Drummond, A. H., 1985, Tumour-promoting phorbol esters inhibit agonist-induced phosphotidate formation and Ca^{2+} flux in human platelets, *FEBS Lett.* **180:**160–164.

Majerus, P. W., Newfeld, E. J., and Wilson, D. B., 1984, Production of phosphoinositide-derived messengers, *Cell* **37:**701–703.

Malm, B., Persson, T., and Lindberg, U., 1981, Characterization of platelet extracts before and after stimulation with respect to the possible role of profilactin as microfilament precursor, *Cell* **23:**145–153.

Manalan, A., and Klee, C., 1984, Calmodulin, *Adv. Cyclic Nucleotide Protein Phosphorylation Res.* **18:**227–279.

Maness, P., 1981, Actin structure in fibroblasts: Its possible role in transformation and tumoregenesis, *Cell Muscle Motil.* **1:**335–373.

Marceau, N., and Swierenga, S., 1985, Cytoskeletal events during Ca^{2+}- or EGF-induced initiation of DNA synthesis in cultured cells, in: *Cell Muscle Motility*, Vol. 6 (J. Shay, ed.), Plenum Press, New York, pp. 97–140.

Marcum, J., Dedman, J., Brankley, B., and Means, A., 1978, Control of microtubule assembly-disassembly by calcium-dependent regulator protein, *Proc. Natl. Acad. Sci. USA* **75:**3771–3775.

Markey, F., Larsson, H., Weber, K., and Lindberg, U., 1982, Nucleation of actin polymerization from profilactin opposite effects of different nuclei, *Biochem. Biophys. Acta* **704:**43–51.

McKeehan, W. L., and McKeehan, K. A., 1980, Serum factors modify the cellular requirement for Ca^{2+}, K^+, Mg^{2+}, phosphate ions, and 2-oxocarboxylic acids for multiplication of normal human fibroblasts, *Proc. Natl. Acad. Sci. USA* **77:**3417–3421.

McKeehan, W. L., McKeehan, K. A., and Calkins, D., 1982, Epidermal growth factor modifies Ca^{2+}, Mg^{2+}, and 2-oxocarboxylic acid, but not K^+ and phosphate ion requirement for multiplication of human fibroblasts, *Exp. Cell Res.* **140:**25–30.

McNeil, P. L., McKenna, M. D., and Taylor, D. L., 1985, A transient rise in cytosolic calcium follows stimulation of quiescent cells with growth factors and is inhibitable with phorbol myristate acetate, *J. Cell Biol.* **101:**372–379.

Means, A., Tash, J., Chafouleas, J., Legace, L., and Guerriero, V., 1982, Regulation of the cytoskeleton by Ca^{2+}-calmodulin and cAMP, *Ann. N.Y. Acad. Sci.* **383:**69–84.

Meigs, J., and Wang, Y.-L., 1986, Reorganization of alpha-actinin and vinculin induced by a phorbol ester in living cells, *J. Cell Biol.* **102:**1430–1438.

Mellstrom, K., Hoglung, A., Nister, M., Heldin, C., Westermark, B., and Lindberg, U., 1983, The effect of platelet-derived growth factor on morphology and motility of human glial cells, *J. Muscle Res. Cell. Motil.* **4:**589–609.

Michell, R. H., 1975, Inositol phospholipids and cell surface receptor function, *Biochem. Biophys. Acta.* **415:**81–147.

Miller, S., Wolf, A., and Arnaud, C., 1976, Bone cells in culture: Morphologic transformation by hormones, *Science* **192:**1340–1343.

Mittal, A., and Bereiter-Hahn, J., 1985, Ionic control of locomotion and shape of epithelial cells: 1. Role of calcium influx, *Cell Motil.* **5:**123–136.

Moolenaar, W. H., Kruijer, W., Tilly, B. C., Verlaan, I., Bierman, A. J., and de Lat, S. W., 1986, Growth factor-like action of phosphatidic acid, *Nature* **323:**171–173.

Moolenaar, W. H., Tsien, R. Y., van der Saag, P. T., and de Laat, S. W., 1983, Na$^+$/H$^+$ exchange and cytoplasmic pH in the action of growth factors in human fibroblasts, *Nature* **304:** 645–648.

Moolenaar, W. H., Tertoolen, L. G. J., de Laat, S. W., 1984a, Phorbol ester and diacylglycerol mimic growth factors in raising cytoplasmic pH, *Nature* **312:**371–376.

Moolenaar, W. H., Tertoolen, L. G. J., and de Lat, S. W., 1984b, Growth factors immediately raise cytoplasmic free Ca^{2+} in human fibroblasts, *J. Biol. Chem.* **259:**8066–8069.

Morris, J. D. H., Metcalfe, J. C., Smith, G. A., Heskieth, T. R., and Taylor, M. V., 1984, Some mitogens cause rapid increases in free calcium in fibroblasts, *FEBS Lett.* **169:**189–193.

Mroczkoski, R., Mosig, G., and Cohen, S., 1984, ATP-stimulated interaction between epidermal growth factor and super-coiled DNA, *Nature* **309:**270–273.

Naccache, P. H., Molski, T. F. P., Borgeat, P., White, J. R., and Shaafi, R. I., 1985, Phorbol esters inhibit the fMet-Leu-Phe- and leukotrinene B$_4$ stimulated calcium mobilization and enzyme secretion in rabbit neutrophils, *J. Biol. Chem.* **260:**2125–2131.

Naka, M., Nishikawa, M., Adelstein, R., and Hidaka, H., 1983, Phorbol ester–induced activation of human platelets is associated with protein kinase C phosphorylation of myosin light chains, *Nature* **306:**490–492.

Nakamura, K. D., Martinez, R., and Weber, M. J., 1983, Tyrosine phosphorylation of specific proteins after mitogen stimulation of chicken embryo fibroblasts, *Mol. Cell. Biol.* **3:**380–390.

Nicholson, N. B., Chen, S., Blank, G., and Pollack, R., 1984, SV40 transformation of Swiss 3T3 cells can cause a stable reduction in the calcium requirement for growth, *J. Cell Biol.* **99:**2314–2321.

Niedel, J., Kuhn, L., and Vanderbank, G., 1983, Phorbol diester receptor copurifies with protein kinase C, *Proc. Natl. Acad. Sci. USA* **80:**36–40.

Nilsson, J., Thyberg, J., Heldin, C. H., Westermark, B., and Wasteson, A., 1983, Surface binding and internalization of platelet-derived growth factor in human fibroblasts, *J. Biol. Chem.* **80:**5592–5596.

Nishimura, J., and Deuel, T. F., 1981, Stimulation of protein phosphorylation in Swiss mouse 3T3 cells by platelet-derived growth factor, *Biochem. Biophys. Res. Commun.* **103:**355–361.

Nishimura, J., Huang, J. S., and Deuel, T. F., 1982, Platelet-derived growth factor stimulates tyrosine-specific protein kinase activity in Swiss 3T3 cells membranes, *Proc. Natl. Acad. Sci. USA* **79:**4303–4307.

Nishizuka, Y., 1983, Phospholipid turnover and cyclic nucleotides in hormone research, in: *Evolution of Hormone-Receptor Systems*, Alan R. Liss, New York, pp. 425–439.

Nishizuka, Y., 1984, The role of protein kinase C in cell surface signal transduction and tumour production, *Nature* **308:**693–698.

Ojakian, G., 1981, Tumor promotor–induced changes in the permeability of epithelial cell tight junctions, *Cell* **23:**95–103.

Olashaw, N. E., O'Keefe, E. J., and Pledger, W. J., 1986, Platelet-derived growth factor modulates epidermal growth factor receptors by a mechanism distinct from that of phorbol esters, *Proc. Natl. Acad. Sci. USA* **83:**3834–3838.

Olsnes, S., Tonnessen, T., and Sanding, K., 1986, pH-regulated anion antiport in nucleated mammalian cells, *J. Cell Biol.* **102:**967–971.

O'Neill, C., Riddle, P., and Rozengurt, E., 1985, Stimulating the proliferation of quiescent 3T3 fibroblasts by peptide growth factors or by agents which elevate cellular cAMP level has opposite effects on motility, *Exp. Cell Res.* **156**:65–78.

Orellana, S. A., Solski, P. A., and Brown, J. H., 1985, Phorbol ester inhibits phosphoinositide hydrolysis and calcium mobilization in cultured astrocytoma cells, *J. Biol. Chem.* **260**:5236–5239.

Owen, N. E., and Villereal, M. L., 1982, Evidence for a role of calmodulin in serum stimulation of Na^+ influx in human fibroblasts, *Proc. Natl. Acad. Sci. USA* **79**:3537–3541.

Pardee, A. B., Dubrow, R., Hamlin, J. L., and Kletzeen, R. F., 1978, Animal cell cycle, *Annu. Rev. Biochem.* **47**:715–750.

Paris, S., and Rozengurt, E., 1982, Cyclic AMP stimulation of Na-K pump activity in quiescent Swiss 3T3 cells, *J. Cell Physiol.* **112**:273–280.

Pastan, I., and Johnson, G. S., 1974, Cyclic AMP and the transformation of fibroblasts, in: *Advances in Cancer Research* (G. Klein, S. Weinhouse and A. Haddow, eds.), Academic Press, New York, pp. 303–329.

Pastan, I. H., Johnson, G. S., and Anderson, W. B., 1975, Role of cyclic nucleotides in growth control, *Annu. Rev. Biochem.* **44**:491–668.

Pledger, W., Hart, C., Locatell, K., Scher, L., 1981, Platelet-derived growth factor–modulated proteins: Constitutive synthesis by a transformed cell line, *Proc. Natl. Acad. Sci. USA* **78**:4358–4362.

Pollack, R., Osborn, M., and Weber, K., 1975, Patterns of organization of actin and myosin in normal and transformed cultured cells, *Proc. Natl. Acad. Sci. USA* **72**:994–998.

Pouyssegur, J., Sardet, C., Franchi, A., L'Allemain, G., and Paris, S., 1984, A specific mutation abolishing Na^+/H^+ antiport activity in hamster fibroblast precludes growth at neutral and aridic pH, *Proc. Natl. Acad. Sci. USA* **81**:4833–4837.

Pouyssegur, J., Franchi, A., L'Allemain, G., and Paris, S., 1985, Cytoplasmic pH, a key determinant of growth factor–induced DNA synthesis in quiescent fibroblasts, *Fed. Eur. Biochem. Soc.* **190**:115–119.

Prentki, M., Biden, T. J., Janjic, D., Irvine, R. F., Berridge, M. J., and Wollheim, C. B., 1984, Rapid mobilization of Ca^{2+} from rat insulinima microsomes by inositol-1,4,5-triphosphate, *Nature* **309**:562–564.

Pruss, R. M., and Herschman, H. R., 1979, Cholera toxin stimulates division of 3T3 cells, *J. Cell Physiol.* **98**:469–473.

Puck, T., 1977, Cyclic AMP, the microtubule–microfilament system, and cancer, *Proc. Natl. Acad. Sci. USA* **74**:4491–4495.

Quigley, J., 1979, Phorbol ester induced morphological changes in transformed chick fibroblasts: Evidence for direct catalytic involvement of plasminogen activator, *Cell* **17**:131–141.

Rasmussen, H., 1980, Calcium and cAMP in stimulus–response coupling, *Ann. N.Y. Acad. Sci.* **356**:346–353.

Rasmussen, H., Koijima, I., Koijima, K., Zawalich, N., and Apfeldorf, W., 1984, Calcium as intracellular messenger: Sensitivity modulation, C-kinase pathway, and sustained cellular release, *Adv. Cyclic Nucleotide Protein. Phosphorylation Res.* **18**:159–193.

Rebhun, L., 1977, Cyclic nucleotides, calcium, and cell division, *Int. Rev. Cytol.* **49**:1–54.

Rifkin, D., Crowe, R., and Pollack, R., 1979, Tumor promoters induce changes in the chick embryo fibroblast cytoskeleton, *Cell* **18**:361–368.

Rink, T. J., and Hallam, T. J., 1984, What turns platelets on? *Trends Biochem. Sci.* **9**:215–219.

Ristow, H.-J., Frank, W., Frohlich, M., 1973, Stimulation of embryonic rat cells by calf serum. V. Metabolism of inositol and choline phospholipids. *Z. Naturforsch.* **28**:188–194.

Roos, A., and Boron, F., 1981, Intracellular pH, *Physiol. Rev.* **61**:296–433.

Rosenfeld, M. E., Bowen-Pope, D. F., and Ross, R., 1984, Platelet-derived growth factor–morphologic and biochemical studies of binding, internalization, and degradation, *J. Cell Physiol.* **121**:263–274.

Rosoff, P. M., and Cantley, L. C., 1985, Stimulation of the T3–T cell receptor–associated Ca^{2+} influx enhances the activity of the Na^+/H^+ exchanges in a leukemic human T cell line, *J. Biol. Chem.* **260**:14053–14059.

Ross, R., and Vogel, A., 1978, The platelet-derived growth factor, *Cell* **14**:203–210.

Ross, R., Raines, E. W., and Bowen-Pope, D. F., 1986, The biology of platelet-derived growth factor, *Cell* **46**:155–169.

Rossow, P. W., Riddle, V. G. H., and Pardee, A. B., 1979, Synthesis of labile serum dependent protein in early G1 controls animal cell growth, *Proc. Natl. Acad. Sci. USA* **76**:4446–4450.

Rothenberg, P., Glaser, L., Schlesinger, P., and Cassel, P., 1983, Activation of Na^+/H^+ exchange by epidermal growth factor elevates intracellular pH in A431 cells, *J. Biol. Chem* **258**:12644–12653.

Rozengurt, E., 1981a, Stimulation of Na influx, Na-K pump activity, and DNA synthesis in quiescent cultured cells, *Adv. Enzyme Regul.* **19**:61–85.

Rozengurt, E., 1981b, Cyclic AMP: A growth-promoting signal for mouse 3T3 cells, *Adv. Cyclic Nucleotide Res.* **14**:429–442.

Rozengurt, E., 1982a, Monovalent ion fluxes, cyclic nucleotides, and the stimulation of DNA synthesis in quiescent cells, in: *Ions, Cell Proliferation, and Cancer* (A. L. Boynton, W. L. McKeehan, and J. F. Whitfield, eds.), Academic Press, New York, pp. 259–281.

Rozengurt, E., 1982b, Synergistic stimulation of DNA synthesis by cyclic AMP derivatives and growth factors in mouse 3T3 cells, *J. Cell Physiol.* **112**:243–250.

Rozengurt, E., and Dicker, P., 1978, Stimulation of DNA synthesis by tumor promoter and pure mitogenic factors, *Nature* **276**:723–726.

Rozengurt, E., Legg, A., Strang, G., Courtenay-Luck, N., 1981, Cyclic AMP: A mitogenic signal for Swiss 3T3 cells, *Proc. Natl. Acad. Sci. USA* **78**:4392–4396.

Rozengurt, E., Stroobant, P., Waterfield, M. D., Deuel, T. F., and Keehan, M., 1983, Platelet-derived growth factor elicits cyclic AMP accumulation in Swiss 3T3 cells: Role of prostaglandin production, *Cell* **34**:265–272.

Rozengurt, E., Rodriguez-Pena, A., Coombs, M., and Sinnett-Smith, J., 1984, Diacylglycerol stimulated DNA synthesis and cell division in mouse 3T3 cells: Role of Ca^{2+} sensitive phospholipid-dependent protein kinase, *Proc. Natl. Acad. Sci. USA* **81**:5748–5752.

Rubin, H., 1971, pH and population density in the regulation of animal cell multiplication, *J. Cell. Biol.* **51**:686–702.

Ryback, S. M., and Stockdale, F. E., 1981, Growth effects of lithium chloride in Balb/c 3T3 fibroblasts and Madin–Darby canine kidney epithelial cells, *Exp. Cell Res.* **136**:263–270.

Sagi-Eisenberg, R., Lieman, H., and Pecht, S., 1985. Protein kinase C regulation of the receptor-coupled calcium signal in histamine-secreting rat basophilic cells, *Nature* **313**:59–60.

Salomon, P. S., 1981, Inhibition of epidermal growth factor binding to mouse embryonal carcinoma cells by phorbol esters mediated by specific phorbol ester receptors, *J. Biol. Chem.* **256**:7958–7966.

Sawyer, S. T., and Cohen, S., 1981, Enhancement of calcium uptake and phosphatidylinositol turnover by epidermal growth factor in A-431 cells, *Biochemistry* **20**:6280–6286.

Scher, C. D., Shepard, R. C., Antoniades, H. N., and Stiles, C. D., 1979, Platelet-derived growth factor and the regulation of the mammalian fibroblast cell cycle, *Biochem. Biophys. Acta* **560**:217–241.

Schlessinger, J., and Geiger, B., 1981, Epidermal growth factor induces redistribution of actin and actinin in human epidermal carcinoma cells, *Exp. Cell Res.* **134**:273–279.

Schliwa, M., Nakamura, T., Porter, K. R., and Euteneur, U., 1984, A tumor promoter induces rapid and coordinated reorganization of actin and vinculin in cultured cells, *J. Cell Biol.* **99**:1045–1059.

Sefton, B. M., and Hunter, T., 1984, Tyrosine protein kinases, *Adv. Cyclic Nucleotide Protein Phosphorylation Res.* **18**:195–225.

Seppä, H. E. J., Grotendorst, G. R., Seppä, S. I., Schiffman, E., Martin, G. R., 1982, The platelet-derived growth factor is a chemoattractant for fibroblasts, *J. Cell Biol.* **92**:584–588.

Shoyab, M., De Larco, J. E., and Todaro, G. J., 1979, Biologically active phorbol esters specifically alter affinity of epidermal growth factor membrane receptors, *Nature* **279**:387–391.

Sibley, D. R., and Lefkowitz, R. J., 1985, Molecular mechanisms of receptor desensitization using the α-adrenergic receptor-coupled adenylate cyclase system as a model, *Nature* **317**:124–129.

Silverstein, S., Steinman, R., and Cohn, Z., 1979, Endocytosis, *Annu. Rev. Biochem.* **46**:669–722.

Singh, T. J., Hochman, J., Verna, R., Chapman, M., Abraham, I., Pastan, I., and Gottesman, M. J., 1985, Characterization of a cyclic AMP-resistant Chinese hamster ovary cell mutant containing both wild-type and mutant species of type I regular subunit of cyclic AMP-dependent protein kinase, *J. Biol. Chem.* **260**:13927–13933.

Stiles, C. D., 1983, The molecular biology of platelet-derived growth factor, *Cell* **33**:653.

Stoscheck, C. M., and Carpenter, G., 1984, Down regulation of epidermal growth factor receptors: Direct demonstration of receptor degradation in human fibroblasts, *J. Cell Biol.* **98**:1048–1053.

Sutherland, E. W., 1972, Studies on the mechanisms of hormone action, *Science* **177**:401–408.

Taylor, D., and Condeelis, J., 1979, Cytoplasmic structure and contractibility in amoeboid cells, *Int. Rev. Cytol.* **56**:57–144.

Taylor, D. L., and Fecheimer, M., 1982, Cytoplasmic structure and contractility: the solation-contraction coupling hypothesis, *Philos. Trans. R. Soc. Lond. [Biol.]* **299**:185–197.

Taylor, D., and Wang, Y.-L., 1980, Fluorescently labelled molecules as probes of the structure and function of living cells, *Nature* **284**:405–410.

Taylor, D., Wang, Y.-L., and Heiple, J., 1980, The contractile basis of amoeboid movement. VII. The distribution of fluorescently labeled actin in living amebas, *J. Cell Biol.* **86**:590–598

Taylor, D., Amato, P., McNeil, P., Luby-Phelps, K., and Tanasugarn, L., 1986, Spatial and temporal dynamics of specific molecules and ions in living cells, in: *Applications of Fluorescence in the Biomedical Sciences* (D. Taylor, A. Waggoner, R. Murphy, F. Lanni, and R. Birge, eds.), Alan R. Liss, New York, pp. 347–376.

Thyberg, J., 1984, The microtubular cytoskeleton and the initiation of DNA synthesis, *Exp. Cell. Res.* **155**:1–8.

Tilney, L., 1975, Actin filaments in the acrosomal reaction of Limulus sperm, *J. Cell Biol.* **64**:289–310.

Tsien, R. Y., Pozzan, T., and Rink, T. J., 1982, T-cell mitogens cause early changes in cytoplasmic free calcium monitored with a new, intracellularly trapped fluorescent indicator, *J. Cell Biol.* **94**:325–334.

Tsuda, T., Kaibuchi, K., West, B., and Takai, Y., 1985, Involvement of Ca^{2+} in platelet-derived growth factor–induced expression of c-*myc* oncogene in Swiss 3T3 fibroblasts, *Fed. Eur. Biochem. Soc.* **187**:43–46.

Ushiro, H., and Cohen, S., 1980, Identification of phosphotyrosine as a product of epidermal growth factor–activated protein kinase in A-431 cell membranes, *J. Biol. Chem.* **255**:8363–8365.

Vicentini, L. M., and Villereal, M. L., 1985, Activation of Na^+/H^+ exchange in cultured fibroblasts: Synergism and antagonism between phorbol ester, Ca^{2+} ionophore, and growth factors, *Proc. Natl. Acad. Sci. USA* **82**:8053–8056.

Waggoner, A., 1986, Fluorescent probes for analysis of cell structure, function, and health by flow and image cytometry, in: *Applications of Fluorescence in the Biomedical Sciences* (D. Taylor, A. Waggoner, R. Murphy, F. Lanni, and R. Birge, eds.), Alan R. Liss, New York, pp. 3–28.

Wang, E., and Goldberg, A., 1976, Changes in microfilament organization and surface topography upon transformation of chick embryo fibroblasts with Rous sarcoma virus, *Proc. Natl. Acad. Sci. USA* **73:**4065–4069.

Wang, Y.-L., 1985, Exchange of actin subunits at the leading edge of living fibroblasts; possible role of treadmilling, *J. Cell Biol.* **101:**597–602.

Watson, S. P., and Lapetina, E. G., 1985, 1,2-diacylglycerol and phorbol ester inhibit agonist-induced formation of inositol phosphates in human platelets: Possible implications for negative feedback regulation of inositol phospholipid hydrolysis, *Proc. Natl. Acad. Sci. USA* **82:**2623–2626.

Werth, D., and Pastan, I., 1984, Vinculin phosphorylation in response to calcium and phorbol esters in intact cells, *J. Biol. Chem.* **259:**5264–5270.

Wharton, W., Leof, E. B., Olashaw, N., Earp, H. S., and Pledger, W. J., 1982, Increases in cyclic AMP potentiate competence formation in Balb/C–3T3 cells, *J. Cell. Physiol.* **111:**201–206.

Whitaker, M. J., and Steinhardt, R. A., 1982, Ionic regulation of egg activation, *Q. Rev. Biophys.* **15:**593–666.

White, J. R., Huang, L. K., Hill, J., Naccache, P. H., Becker, E. L., and Shaafi, R. I., 1984, Effect of phorbol 12-myristate 13-acetate and its analogue 4,-phorbol 12, 13-didecanoate on protein phosphorylation and lysosomal enzyme release in rabbit neutrophils, *J. Biol. Chem.* **259:**8605–8611.

Whitely, B., Cassel, D., Zuang, U., and Glaser, L., 1984, Tumor promoter phorbol 12-myristate acetate inhibits mitogen-stimulated Na^+/H^+ exchange in human epidermal carcinoma A431 cells, *J. Cell Biol.* **99:**1162–1166.

Whitely, B., Duel, T., and Glaser, L., 1985, Modulation of the activity of the platelet-derived growth factor receptor by phorbol myristate acetate, *Biochem. Biophys. Res. Commun.* **129:**854–861.

Whitfield, J. F., 1982, The roles of calcium and magnesium in cell proliferation: An overview, in: *Ions, Cell Proliferation, and Cancer* (A. L. Boynton, W. L. McKeehan, and J. F., Whitfield, eds.), Academic Press, New York, pp. 283–294.

Whitfield, J. F., MacManus, J. P., and Gillan, D. J., 1970, The possible mediation by cyclic AMP of the stimulation of thymocyte proliferation by vassopressin and the inhibition of this mitogenic action by throcalcitonin, *J. Cell Physiol.* **76:**65–76.

Whitfield, J. F., MacManus, J. P., Rixon, R. H., Boynton, A. L., Youdale, T., and Swierenga, S. H. H., 1976, The positive control of cell proliferation by the interplay of calcium ions and cyclic nucleotides—A review, *In Vitro* **12:**1–18.

Willingham, M., 1976, Cyclic AMP and cell behavior in cultured cells, *Int. Rev. Cytol.* **44:**319–363.

Willingham, M., and Pastan, I., 1975, Cyclic AMP and cell morphology in cultured fibroblasts, *J. Cell Biol.* **67:**146–159.

Zetterberg, A., and Engstrom, W., 1981, Mitogenic effect of alkaline pH on quiescent, serum-starved cells, *Proc. Natl. Acad. Sci. USA* **78:**4334–4338.

Zetterberg, A. W., and Larsson, O., 1985, Kinetic analysis of regulatory events in G1 leading to proliferation or quiescence of Swiss 3T3 cells, *Proc. Natl. Acad. Sci. USA* **82:**5365–5369.

Zigmond, S., 1982, Polymorphonuclear leukocyte response to chemotactic gradients, in: *Cell Behavior* (R. Bellaris, A. Curtis, and G. Dunn, eds.), Cambridge University Press, London, p. 183.

Zigmond, S., and Hirsch, J., 1973, Leukocyte locomotion and chemotaxis: New methods for evaluation and demonstration of cell-derived chemotactic factor, *J. Exp. Med.* **137:**387–400.

INDEX